Methods in Molecular Biology™

Series Editor
John M. Walker
School of Life Sciences
University of Hertfordshire
Hatfield, Hertfordshire, AL10 9AB, UK

For further volumes:
http://www.springer.com/series/7651

Bacterial Molecular Networks

Methods and Protocols

Edited by

Jacques van Helden and Ariane Toussaint

Laboratoire de Bioinformatique des Génomes et des Réseaux (BiGRe), Université Libre de Bruxelles, Bruxelles, Belgium

Denis Thieffry

Institut de Biologie de l'Ecole Normale Supérieure (IBENS), UMR ENS, CNRS 8197, INSERM 1024, Paris, France

Humana Press

Editors
Jacques van Helden
Laboratoire de Bioinformatique des Génomes
et des Réseaux (BiGRe)
Université Libre de Bruxelles
Bruxelles, Belgium
jvhelden@ulb.ac.be

Ariane Toussaint
Laboratoire de Bioinformatique des Génomes
et des Réseaux (BiGRe)
Université Libre de Bruxelles
Bruxelles, Belgium
ariane@bigre.ulb.ac.be

Denis Thieffry
Institut de Biologie de l'Ecole
Normale Supérieure (IBENS)
UMR ENS
CNRS 8197
INSERM 1024
Paris, France
thieffry@ens.fr

ISSN 1064-3745 e-ISSN 1940-6029
ISBN 978-1-4939-6156-6 ISBN 978-1-61779-361-5(eBook)
DOI 10.1007/978-1-61779-361-5
Springer New York Dordrecht Heidelberg London

Printed on acid-free paper

Humana Press is part of Springer Science+Business Media (www.springer.com)

Preface

Network-based representations have become pervasive in most fields in biology. Focusing on networks applied to bacteria, this volume provides authoritative descriptions of various experimental and computational methods enabling the characterization and analysis of molecular interaction networks.

Intended primarily for postgraduate students and researchers working in the field of experimental and computational microbiology, this volume combines up-to-date reviews along with detailed protocols written by the developers of bioinformatics resources (databases and software tools). Each protocol emphasizes the crucial steps and the way to set up the parameters in order to obtain the best results.

The first section provides an extensive coverage of various experimental and *in silico* approaches aiming at the characterization of network components. The second section is devoted to the presentation of computational approaches to analyze the topology of molecular networks. The third and last section further introduces a variety of methods and tools enabling to generate qualitative or quantitative dynamical models of molecular processes in bacteria.

Altogether, the volume constitutes a practical guide of methods and tools to characterize, retrieve, visualize, analyze, and manipulate molecular networks.

Bruxelles, Belgium — *Jacques van Helden*
Bruxelles, Belgium — *Ariane Toussaint*
Paris, France — *Denis Thieffry*

Contents

Contributors

CEI ABREU-GOODGER • *European Molecular Biology Laboratory, European Bioinformatics Institute, Hinxton, Cambridge, UK*

STEVEN S. ANDREWS • *Fred Hutchinson Cancer Research Center, Seattle, WA, USA*

MASANORI ARITA • *Department of Computational Biology, Graduate School of Frontier Sciences, The University of Tokyo, Kashiwanoha, Kashiwa, Japan*

GRÉGORY BATT • *INRIA Paris – Rocquencourt, Domaine de Voluceau, Le Chesnay, France*

ANAÏS BAUDOT • *Spanish National Cancer Research Centre, Madrid, Spain*

JÖRN BEHRE • *Theoretical Systems Biology, Institute of Food Research (IFR), Norwich Research Park, Colney Lane, Norwich, United Kingdom*

BRUNO BESSON • *Genostar, Montbonnot, France*

EMMANUELLE BOUVERET • *LISM – CNRS UPR9027, Marseille, France*

SYLVAIN BROHÉE • *ESAT-SCD-SISTA (Bioinformatics group), Katholieke Universiteit Leuven, Belgium*

CHRISTINE BRUN • *TAGC, INSERM U928, Marseille, France*

MICK CHANDLER • *Laboratoire de Microbiologie et Génétique Moléculaire (UMR 5100), Toulouse, France*

CLAUDINE CHAOUIYA • *Instituto Gulbenkian de Ciência, Oeiras, Portugal*

PIERRE-EMMANUEL CIRON • *Genostar, Montbonnot, France*

JULIO COLLADO-VIDES • *Programa de Genómica Computacional, Centro de Ciencias Genómicas, Universidad Nacional Autónoma de México, Cuernavaca, Morelos, Mexico*

LUÍS FILIPE DE FIGUEIREDO • *Chemoinformatics and Metabolism Team, European Bioinformatics Institute (EBI), Wellcome Trust Genome Campus, Hinxton, Cambridgeshire, United Kingdom*

HIDDE DE JONG • *INRIA Rhône-Alpes, Montbonnot, St. Ismier, France*

ESTELLE DUMAS • *INRIA Rhône-Alpes, Montbonnot, St. Ismier, France*

KRISTOF ENGELEN • *Centre of Microbial and Plant Genetics, Katholieke Universiteit Leuven, Heverlee, Belgium*

KAROLINE FAUST • *Research group of Bioinformatics and (eco-)systems biology, Department of Structural Biology; Microbiology Unit (MICR), Department of Applied Biological Sciences (DBIT), Vrije Universiteit Brussel, Bruxelles, Belgium*

ANA CAROLINA FIERRO • *Centre of Microbial and Plant Genetics, Katholieke Universiteit Leuven, Heverlee, Belgium*

MAREN L. FRIESEN • *Department of Molecular and Computational Biology, University of Southern California, Los Angeles, CA, USA*

QIANG FU • *Centre of Microbial and Plant Genetics, Katholieke Universiteit Leuven, Heverlee, Belgium*

JAIR SANTIAGO GARCÍA-SOTELO • *Programa de Genómica Computacional, Centro de Ciencias Genómicas, Universidad Nacional Autónoma de México, Cuernavaca, Morelos, Mexico*

JOHANNES GEISELMANN • *Laboratoire Adaptation et Pathogénie des Microorganismes (CNRS UMR 5163), Bâtiment Jean Roget, Faculté de Médecine-Pharmacie, Université Joseph Fourier, La Tronche, France*
ROBERT GENTLEMAN • *Program in Computational Biology, Fred Hutchinson Cancer Research Center, M2-B876, Seattle, WA, USA*
JOSEPH GERACI • *Ontario Cancer Institute/UHN, Toronto, ON, Canada*
MONIKA HEINER • *Department of Computer Science, Brandenburg University of Technology, Cottbus, Germany*
PIERS J. INGRAM • *Department of Mathematics, Imperial College, South Kensington Campus, London, UK*
EMILY I. JONES • *School of Biological Sciences, Washington State University, Pullman, WA, USA*
IGOR JURISICA • *Ontario Cancer Institute/UHN, Toronto, ON, Canada*
CHRISTOPH KALETA • *Department of Bioinformatics, Friedrich Schiller University Jena, Jena, Germany*
PETER D. KARP • *SRI International, Menlo Park, CA, USA*
WILLIAM P. KELLY • *Theoretical Systems Biology group, Imperial College, London, UK*
WILFRED D. KEPSEU • *Institut Non Linéaire de Nice, CNRS UMR 6618, Université de Nice Sophia Antipolis, Valbonne, France*
ANDREW D. KING • *Department of Industrial Engineering and Operations Research, Columbia University, New York, NY, USA*
MARIO LATENDRESSE • *SRI International, Menlo Park, CA, USA*
NOLWENN LE MEUR • *IRISA, Equipe Symbiose, Université de Rennes I, Rennes, France*
KAREN LEMMENS • *Centre of Microbial and Plant Genetics, Katholieke Universiteit Leuven, Heverlee, Belgium*
GIPSI LIMA-MENDEZ • *Laboratoire de Bioinformatique des Génomes et des Réseaux (BiGRe), Université Libre de Bruxelles, Bruxelles, Belgium*
GEOFFREY LIU • *Ontario Cancer Institute/UHN, Toronto, ON, Canada*
ALEJANDRA LÓPEZ-FUENTES • *Programa de Genómica Computacional, Centro de Ciencias Genómicas, Universidad Nacional Autónoma de México, Cuernavaca, Morelos, Mexico*
KATHLEEN MARCHAL • *Centre of Microbial and Plant Genetics, Katholieke Universiteit Leuven, Heverlee, Belgium*
IRMA MARTÍNEZ-FLORES • *Programa de Genómica Computacional, Centro de Ciencias Genómicas, Universidad Nacional Autónoma de México, Cuernavaca, Morelos, Mexico*
WOLFGANG MARWAN • *Otto von Guericke University & Magdeburg Centre for Systems Biology, c/o Max Planck Institute for Dynamics of Complex Technical Systems, Magdeburg, Germany*
PIETER MEYSMAN • *Centre of Microbial and Plant Genetics, Katholieke Universiteit Leuven, Heverlee, Belgium*
REGIS MONTE • *INRIA Rhône-Alpes, Montbonnot, St. Ismier, France*
PEDRO T. MONTEIRO • *INRIA Rhône-Alpes, Montbonnot, St. Ismier, France*
LUIS MUÑIZ-RASCADO • *Programa de Genómica Computacional, Centro de Ciencias Genómicas, Universidad Nacional Autónoma de México, Cuernavaca, Morelos, Mexico*

AURÉLIEN NALDI • *TAGC, INSERM U928, Marseille, France*
MICHEL PAGE • *INRIA Rhône-Alpes, Montbonnot, St. Ismier, France*
SUZANNE PALEY • *SRI International, Menlo Park, CA, USA*
MATTEO PELLEGRINI • *Department of Molecular, Cell and Developmental Biology, University of California, Los Angeles, CA, USA*
LILIANA PORRÓN-SOTELO • *Programa de Genómica Computacional, Centro de Ciencias Genómicas, Universidad Nacional Autónoma de México, Cuernavaca, Morelos, Mexico*
NATAŠA PRŽULJ • *Department of Computing, Imperial College London, London, UK*
FRANÇOIS RECHENMANN • *INRIA Rhône-Alpes, Montbonnot, St. Ismier, France*
CHRISTIAN ROHR • *Otto von Guericke University & Magdeburg Centre for Systems Biology, c/o Max Planck Institute for Dynamics of Complex Technical Systems, Magdeburg, Germany*
DELPHINE ROPERS • *INRIA Rhône-Alpes, Montbonnot, St. Ismier, France*
HELADIA SALGADO • *Programa de Genómica Computacional, Centro de Ciencias Genómicas, Universidad Nacional Autónoma de México, Cuernavaca, Morelos, Mexico*
AMINAEL SANCHEZ-RODRIGUEZ • *Centre of Microbial and Plant Genetics, Katholieke Universiteit Leuven, Heverlee, Belgium*
STEFAN SCHUSTER • *Department of Bioinformatics, Friedrich Schiller University Jena, Jena, Germany*
HENNING SCHWÖBBERMEYER • *SunGene GmbH, Gatersleben, Germany*
JACQUES-ALEXANDRE SEPULCHRE • *Institut Non Linéaire de Nice, CNRS UMR 6618, Université de Nice Sophia Antipolis, Valbonne, France*
HILDA SOLANO • *Programa de Genómica Computacional, Centro de Ciencias Genómicas, Universidad Nacional Autónoma de México, Cuernavaca, Morelos, Mexico*
OUISSEM SOUIAI • *TAGC, INSERM U928, Marseille, France*
MICHAEL P.H. STUMPF • *Theoretical Systems Biology group, Imperial College, London, UK*
HONG SUN • *Centre of Microbial and Plant Genetics, Katholieke Universiteit Leuven, Heverlee, Belgium*
DENIS THIEFFRY • *Institut de Biologie de l'Ecole Normale Supérieure (IBENS), UMR ENS, CNRS 8197, INSERM 1024, Paris, France*
INGE M. THIJS • *Centre of Microbial and Plant Genetics, Katholieke Universiteit Leuven, Heverlee, Belgium*
ARIANE TOUSSAINT • *Laboratoire de Bioinformatique des Génomes et des Réseaux (BiGRe), Université Libre de Bruxelles, Bruxelles, Belgium*
STIJN VAN DONGEN • *European Molecular Biology Laboratory, European Bioinformatics Institute, Hinxton, Cambridge, UK*
FRÉDÉRIQUE VAN GIJSEGEM • *Laboratoire Interactions Plantes Pathogènes, UMR 217 INRA/AgroParisTech/UPMC Univ Paris 06, Paris, France*
JACQUES VAN HELDEN • *Laboratoire de Bioinformatique des Génomes et des Réseaux (BiGRe), Université Libre de Bruxelles, Bruxelles, Belgium*
SHANNON J. WILLIAMSON • *J. Craig Venter Institute, San Diego, CA, USA*
RÖBBE WÜNSCHIERS • *University of Applied Sciences, Mittweida, Germany*
SHIBU YOOSEPH • *J. Craig Venter Institute, San Diego, CA, USA*

Chapter 1

Bacterial Molecular Networks: Bridging the Gap Between Functional Genomics and Dynamical Modelling

Jacques van Helden, Ariane Toussaint, and Denis Thieffry

Abstract

This introductory review synthesizes the contents of the volume *Bacterial Molecular Networks* of the series *Methods in Molecular Biology*. This volume gathers 9 reviews and 16 method chapters describing computational protocols for the analysis of metabolic pathways, protein interaction networks, and regulatory networks. Each protocol is documented by concrete case studies dedicated to model bacteria or interacting populations. Altogether, the chapters provide a representative overview of state-of-the-art methods for data integration and retrieval, network visualization, graph analysis, and dynamical modelling.

Key words: Protein interactions, Metabolic pathways, Regulatory networks, Graph motifs, Clustering, Dynamical modelling, Bacteria, Metagenomics, Computational biology, Bioinformatics, Systems biology

1. Bacterial Models for Molecular Interaction Networks

Bacteria are everywhere, from the most hospitable to the most hostile environment. They are an important component of the microfauna that supports all biogeochemical (or nutrient) cycles. Besides their well-known role as infectious agents, bacteria are essential for the nutrition of animals (gut microbiome) and plants (nitrogen fixing and other rhizosphere bacteria) and more generally, contribute to the well being of water, soil, and air ecosystems.

For over a century, bacteria have also been fruitful model systems for deciphering fundamental biological mechanisms. In particular, *Escherichia coli*, *Bacillus subtilis*, and a few others have been extensively used to define the basic principles of gene expression and its regulation (1–6).

Jacques van Helden et al. (eds.), *Bacterial Molecular Networks: Methods and Protocols*, Methods in Molecular Biology, vol. 804, DOI 10.1007/978-1-61779-361-5_1, © Springer Science+Business Media, LLC 2012

Nowadays, with over a thousand bacterial genomes sequenced, even greater opportunities have opened for experimental or computational global analysis of metabolism, physiology, and evolution. In particular, the access to comprehensive sets of molecular components (genes, proteins, regulatory signals) is at the basis of the development of novel integrative approaches, aiming at understanding the function of specific sets of these components (operons, regulons, metabolic pathways, protein complexes, etc.) in the context of the whole organism. Biological processes rely on the combined activities of molecules interconnected to form a precisely wired network. The integration of all the cellular processes results in a complex network, comprising several thousands of molecules and interactions. A schematic picture of such networks can nowadays be obtained from the huge data sets gathered from high-throughput technologies ("omics"). These biological networks are usually represented as graphs in the mathematical sense of the term: a set of nodes (representing biomolecules) and edges (describing their interactions). The topological structure of these graphs shapes their dynamical properties, as demonstrated by the pioneering work of René Thomas on the role of positive and negative feedback circuits (7, 8).

The availability of genome-scale molecular networks instigated the swift development of sophisticated software tools to store, query, and visualize the datasets, analyze network structures, identify components, analyze dynamical properties, generate relevant, and testable functional predictions. Today, biologists are confronted to several dozens of software tools that can be combined to address complementary questions about networks of interest. However, using these tools requires to understand not only the biological concepts underlying the datasets, but also the related algorithmic and statistical aspects. It is easy to get lost when confronted to the multiple possibilities for analyzing a given data set: which algorithm should be chosen to answer a particular question? How should the parameters be tuned? How should the results be interpreted? In order to put computer tools in the hands of life sciences researchers, there is a crucial need for well-documented protocols, illustrated by relevant study cases, providing general guidelines as well as more detailed comments about the impact of parametric choices and the potential traps.

This volume targets students and researchers working in the field of experimental or computational biology. Firstly, it offers an overview of the state-of-the-art in network biology and its applications to decipher various types of molecular networks in bacteria. Furthermore, it is a practical guide for computational methods and software tools that can be used to build, retrieve, visualize, analyze, and model biological networks, with concrete microbial applications (Table 1).

Table 1
Software tools covered by the method chapters

Software tool	Website	Interface	Chapter(s)
BioCyc	http://biocyc.org/	Web browser	Latendresse et al.
COLOMBOS	http://ibiza.biw.kuleuven.be/colombos/	Web browser	Fu et al.
DISTILLER	http://ibiza.biw.kuleuven.be/DISTILLER/	Web browser	Fu et al.
GINsim	http://gin.univ-mrs.fr/GINsim/	Java application	Chaouiya et al.
GNA	http://www-helix.inrialpes.fr/gna	Java application	Batt et al.
MAVisto	http://mavisto.ipk-gatersleben.de	Java webstart application	Schwobbermeyer and Wunschiers
MCL	http://micans.org/mcl/	Unix shell prompt	van Dongen and Abreu-Goodger
NeAT	http://rsat.bigre.ulb.ac.be/neat/	Web browser	Brohée Lima-Mendez Faust and van Helden
netZ	http://www3.imperial.ac.uk/theoreticalsystemsbiology/data-software	R package	Kelly et al.
PRODISTIN	http://crfb.univ-mrs.fr/webdistin/	Web browser	Baudot et al.
R graph methods	http://www.r-project.org/ http://www.bioconductor.org/ http://www.graphviz.org/ http://www.boost.org/	R packages	Le Meur and Gentleman
RegulonDB	http://regulondb.ccg.unam.mx	Web browser or MySQL or SOAP	Salgado et al.
RNSC	http://www.cs.utoronto.ca/~juris/data/rnsc/	Unix shell prompt	King et al.
SmolDyn	http://www.smoldyn.org	Unix shell prompt	Andrews
Snoopy	http://www-dssz.informatik.tu-cottbus.de/index.html?/software/snoopy.html	Stand alone application for Mac OSX, Windows and Linux	Marwan et al.

2. Experimental and *In Silico* Approaches to Unravel Interactions, Infer Pathways, and Build Networks (Part I)

The first section of the book provides a general overview of currently available experimental and in silico approaches used to characterize molecular interactions, which constitute the components of biomolecular pathways and networks.

This section starts with a review by *Bouveret and Brun* (Chap. 2), covering the techniques available to microbiologists for studying protein–protein interactions in bacteria, and the relevance of these interactions for high-throughput studies, including the deciphering of various bacterial interactomes.

Beyond description at the level of the organism, recent developments of high-throughput technologies allow for the direct sequencing of DNA from bacterial communities, providing access to the diversity and versatility of microbial populations in various environments and in function of various physicochemical parameters. *Williamson and Yooseph* (Chap. 3) elaborate upon the different steps required for such metagenomic analyses, including DNA extraction, sequencing, and computational analyses.

Microorganisms not only flirt and compete in their natural niches, but they also constantly exchange biomolecules, including DNA. *Toussaint and Chandler* (Chap. 4) summarize the reticulated relationships between prokaryotic species resulting from the exchanges of mobile genetic elements, and propose routes to improve their annotation and to decipher their evolutionary relationships. Directly related to this review, the method chapter by *Lima-Mendez* (Chap. 5) explains how to build a reticulate representation of the relationships between microbial genomes, where genetic material is transmitted by both vertical transmission and horizontal exchanges.

The next chapters deal with in silico resources (software tools and databases) to build, store, and analyze molecular networks. As discussed by *Arita* (Chap. 6), in the case of metabolic networks, the mapping of molecular interactions onto a graph usually implies some loss of information, which can be limited by selecting specific topological transformations that depend on the biological questions addressed. This is further detailed in the following chapter, where *Faust and van Helden* (Chap. 7) present a pragmatic utilization of graph-based representations of metabolic reactions for inferring metabolic pathways from sets of functionally related enzyme-coding genes. The method is illustrated with a study case directly relevant to bacteria, the inferring of metabolic pathways from operons. The same approach can also be applied to sets of functionally related enzyme-coding genes obtained from other sources such as co-expression clusters or phylogenetic profiles. The protocol by *Fu et al.* (Chap. 8) introduces COLUMBUS and DISTILLER, two algorithms dedicated to the mining of public resources for transcriptome data and to infer transcriptional networks by a combined analysis of transcriptional profiles and *cis*-regulatory motifs. *Pellegrini et al.* (Chap. 9) summarize the concepts and methods at the basis of phylogenetic profiling. The approach relies on the analysis of presence/absence of orthologs across a set of phylogenetically related species, in order to predict sets of genes that are likely to be involved in related functions.

The section ends with a presentation of novel interfaces to the most popular databases on bacterial gene regulation (RegulonDB) and metabolism (BioCyc). The protocol by *Salgado et al.* (Chap. 10), describes three ways to retrieve transcriptional networks from RegulonDB, adapted to various types of utilization: simple download, programmatic access via Web services, and direct access to the SQL database management system. The resulting networks can then be used as input to perform different types of computational analyses, e.g., detect over-represented network motifs, identify regulatory modules, or characterize degree distributions. The protocol by *Latendresse et al.* (Chap. 11) introduces a set of flexible tools specifically designed to visualize bacterial metabolic and regulatory networks annotated in the BioCyc database. Although *E. coli* serves as the main reference, the tools described should apply to many other bacterial species, which will be covered by the MetaCyc database once sufficient annotation becomes available about their regulatory networks.

3. Topological Analysis of Bacterial Networks (Part II)

After having collected the components of a given molecular network, biologists soon realize that it is impossible to intuitively grasp its global properties. Irrespective of the method used to obtain the network (case-by-case experiments, high-throughput technologies, computer-based inference), large-scale biomolecular networks are complex, intricate, and often noisy. This complexity stimulated the development of a wide variety of theoretical approaches to analyze network topology (see, e.g., ref. 9, 10) and extract relevant modules. This section presents a set of reviews and method chapters describing a variety of topological approaches: analysis of degree distributions, detection of network motifs, network-based clustering. The chapter by *Geraci et al.* (Chap. 12) introduces the key concepts and describes standard and novel approaches to extract significant network characteristics (e.g., node degree properties, over-represented motifs) in bacterial molecular networks. The following method chapters document specific tools to perform different types of topological analyses. *Kelly et al.* (Chap. 13) present an R library enabling to fit alternative models onto the degree distribution of a given network and to select the most likely model. This type of analysis challenges simplistic models that do not adequately fit the data (see also ref. 11). *Schwoebbermeyer and Wuenschiers* (Chap. 14) describe how to use the software MAVisto to mine networks for recurrent motifs by combining a versatile motif search algorithm with interactive exploration methods and specialized visualization techniques.

The three following chapters present graph-based clustering algorithms that can be used to extract functional modules from biological networks. *van Dongen et al.* (Chap. 15) propose several protocols and study cases where the algorithm MCL is applied to partition interaction networks derived from protein sequence similarities or correlations between gene expression profiles. *King et al.* (Chap. 16) apply the algorithm RNSC to identify densely connected subgraphs that might reveal protein complexes in protein–protein interaction networks. *Baudot et al.* (Chap. 17) illustrate the use of the PRODISTIN tool, which combines hierarchical clustering and analysis of annotated protein functions in order to extract consistent clusters of interacting proteins associated to particular biological processes (Gene Ontology annotations), thereby enabling novel functional predictions.

The two last chapters of this section present generic software packages enabling various types of network analysis. The protocol by *Brohée* (Chap. 18) explains the use of the Network Analysis Tools (NeAT), a Web-based software suite combining a variety of tools to analyze, compare, and classify biological networks (coexpression, transcriptional networks, GO annotations, etc.), with a special emphasis on the definition of robust metrics and control procedures (generation of random networks and network alterations). The chapter by *Le Meur and Gentlemen* (Chap. 19) demonstrates the power of R statistical programming language and packages to analyze statistical and topological properties of graph-based representations of biological networks.

4. Dynamical Modelling of Bacterial Networks (Part III)

A proper understanding of biological processes requires an appreciation of their temporal and spatial behaviour. Although mathematical biologists have been working for decades on the development of dynamical models that recapitulate in silico some essential aspects of the behaviour of living organisms (e.g., cell cycle), the correlation of such models with functional genomic data sets is seldom made. Here again, bacteria offer unmatched opportunities because of their limited complexity compared to multicellular organisms, their amenability to experiments, and the existence of rich data sets (in particular for *E. coli*). The last part of the book presents a series of qualitative and quantitative approaches to model the temporal behaviour of biological networks in bacteria.

Focusing on metabolism, *Behre et al.* (Chap. 20) review established methods to detect and analyze structural invariants, pointing to conserved moieties or elementary fluxes at the basis of asymptotical stationary behaviour. *Marwan et al.* (Chap. 21) present the Petri net formalism, a framework enabling the development of discrete, continuous, or stochastic models, or even combination thereof. Petri nets are particularly suited to the modelling of metabolism and have been recently applied to signal transduction pathways, and, to a lesser extent, to transcriptional regulatory networks. This is illustrated by their case study: the modelling of phosphate regulation in enteric bacteria. *Batt et al.* (Chap. 22) focus on the regulatory network controlling *E. coli* response to carbon starvation, and describe the progressive building of a predictive qualitative dynamical model for this response and its parameterization. The protocol proposed by *Chaouiya et al.* (Chap. 23) uses a logical formalism, that allows for the simulation of bacteriophage λ development, while hinting at the specific roles of feedback circuits found at the core of the phage regulatory network.

Most detailed mechanistic dynamical models are limited to the level of single organisms. In contrast, mathematical ecology and evolutionary studies emphasize the relationships between variants within a population, or between different species. Bacteria are involved in a wide spectrum of interspecies relationships with a wide variety of organisms, from commensalism to mutualism and from symbiosis to parasitism. Analysis and modelling of these relationships is an important emerging field of research illustrated here by two chapters. *Friesen and Jones* (Chap. 24) review eco-evolutionary studies of adaptive dynamics, inclusive fitness, and population genetic models, providing insight into the strengths and weaknesses of each approach and into how current evolutionary methods can contribute to the deciphering of the mechanistic basis of host–symbiont interactions. Focusing on bacterial virulence, *Kepseu et al.* (Chap. 25) review the main mechanisms used by bacteria to trigger the production of virulence factors. The dynamical modelling of these mechanisms leads to the identification of two qualitatively distinct infectious transitions, while an irreversible switch behaviour is at the basis of an efficient onset of virulence.

The studies mentioned above refer to coarse-grain dynamical models of bacterial molecular networks, where variables represent the mean behaviour of population of biomolecules, cells, or species. Justified for large numbers of components, this approximation does not reflect the variety of behaviour that can occur at the microscopic level, in particular when the number of interacting components is small. This consideration has led to the development of stochastic methods for the modelling of the temporal behaviour of individual components (e.g., molecules or cells) in spatially distributed interacting networks. *Andrews* (Chap. 26) presents Smoldyn, a computer program for modelling cellular systems with spatial and stochastic detail. Thanks to a novel rule-based modelling, chemical species and reactions are automatically generated as they arise in simulations, enabling computationally efficient simulations of bacterial systems.

5. Outlook: Bridging the Gap Between Bioinformatics and Dynamical Modelling

During the recent years, sustained by the progress in high-throughput "omics" approaches, our perception of the bacterial world, its dynamics and evolution have undergone a revolution.

Various microbiomes and viromes, sampled from niches as diverse as mammalian gut, oceans or polluted environments, can now be analyzed to answer questions about diseases and environmental equilibrium, which could not even have been formulated a decade ago. Several recently published studies (12–14) illustrate how several high-throughput methodologies can altogether provide a comprehensive picture of transcriptional activity, protein expression and interactions, along with metabolic activity in bacteria such as *Mycoplasma pneumoniae*. The multiplication of such investigations raises a pressing, increasing need to combine methods and expertise in mathematics, informatics, and biology to extract significant insights from the flood of molecular data.

By combining the efficient computational tools such as those presented in this and other volumes of the MiMB series (15–18), researchers will be able to organize, analyze, and visualize large-scale molecular datasets and, most importantly, to assess the statistical relevance of the features of interest. Such tools are essential to infer the molecular behaviour of bacteria in specific conditions (i.e., response to available nutriments, hosts for symbiosis, or pathogenesis, etc.). By and large, high-throughput technologies are providing us with a snapshot of the network state, but a real understanding of the biological systems' behaviour would require a proper formalization of the underlying causal relationships.

This problem is at the core of the domain of mathematical biology since half a century. In the recent years, sophisticated methods and software tools have been developed to model the spatial–temporal behaviour of bacterial networks at different scales and with different levels of details. Arguably, qualitative approaches are better adapted to cope with the current lack of quantitative data. However, with the raise of large-scale quantitative measurements, we can expect more and more opportunities for the application of detailed kinetic or stochastic approaches to the predictive dynamical modelling of bacterial molecular networks. Irrespective of the mathematical formalism, it remains challenging to explicitly articulate functional genomics data analysis and dynamical modelling, even for a single cell. Another recurrent difficulty lies in relating components, interactions, and behaviours, from molecules to microbial populations or ecosystems and that occur at different temporal and spatial scales.

Although microbes have recurrently proved to be much more complex than expected, bacteria still offer excellent opportunities to develop novel concepts, methods, and tools to bridge the gaps between computational biology and dynamical modelling. This is a necessary step to reach the level of understanding commended by "systems biology" and, even more, for the development of synthetic biology, which beyond the characterization and predictive modelling of existing organisms aims at engineering new living systems. Among these, bacteria may be engineered with properties to cope with economic, environmental, or medical challenges (for a recent review on synthetic bacteriology, see ref. 19). Here again, efficient design requires a proper blend of computational and experimental expertise (see e.g., ref. 20).

A first important step in the synthesis of a new bacterial cell was recently reported (21). A Mycoplasma chromosome was synthesized, assembled, and introduced into cells of a different Mycoplasma species, generating cells capable of continuous genome replication and cell division with properties characteristic of the sole synthetic genome. This amazing technical accomplishment does not imply that we can master bacterial reproduction and evolution. Indeed, a systematic exploration of the consequences of rewiring regulatory interactions among global regulators in the best known bacteria, *E. coli*, led to puzzling and unpredictable results in terms of gene expression profiles and global fitness (cf. 22; Isalan, personal communication). Explicit dynamical modelling of the underlying networks may hopefully allow for the understanding of such non intuitive behaviour.

On a longer range, further methodological developments along the lines outlined in this volume will no doubt contribute to go from the understanding and possible manipulation of individual bacterial cells and cellular networks to the management of bacterial populations, of their interactions, as well of their response to surrounding viral and other predators populations.

Acknowledgments

The collaboration between the TAGC (DT) and BiGRe (JvH and AT) laboratories is supported by the Belgian Program on Interuniversity Attraction Poles, initiated by the Belgian Federal

Science Policy Office, project P6/25 (BioMaGNet). The BiGRe laboratory is further supported by the MICROME Collaborative Project funded by the European Commission within its FP7 Programme, under the thematic area "BIO-INFORMATICS – Microbial genomics and bio-informatics" (contract number 222886–2).

References

1. Jacob F, Monod J. (1961) Genetic regulatory mechanisms in the synthesis of proteins. J Mol Biol, 3:318–356.
2. Monod J, Jacob F. (1961) Teleonomic mechanisms in cellular metabolism, growth, and differentiation. Cold Spring Harb Symp Quant Biol, 26:389–401.
3. Brock TD. (1990) *The Emergence of Bacterial Genetics*. Cold Spring Harbor Laboratory Press, Cold Spring Harbor, NY.
4. Beckwith JR, Silhavy TJ. (1992) *The Power of Bacterial Genetics: A Literature-Based Course*. Cold Spring Harbor Laboratory Press, Plainview, NY.
5. Thieffry D. (1996) *Escherichia coli* as a model system with which to study cell differentiation. Hist Philos Life Sci, 18:163–193.
6. Morange M. (1998) *A History of Molecular Biology*. Harvard University Press, Cambridge, MA.
7. Thomas R, D'Ari R. (1990) *Biological Feedback*. CRC Press, Boca Raton, FL.
8. Thieffry D. (2007) Dynamical roles of biological regulatory circuits. Brief Bioinform, 8:220–225.
9. Alon U. (2007) *An Introduction to Systems Biology: Design Principles of Biological Circuits*. Chapman & Hall/CRC, Boca Raton, FL.
10. Martinez-Antonio A, Janga SC, Thieffry D. (2008) Functional organisation of *Escherichia coli* transcriptional regulatory network. J Mol Biol, 381:238–247.
11. Lima-Mendez G, van Helden J. (2009) The powerful law of the power law and other myths in network biology. Mol Biosyst, 5:1482–1493.
12. Guell M, van Noort V, Yus E, Chen WH, Leigh-Bell J, Michalodimitrakis K, Yamada T, Arumugam M, Doerks T, Kuhner S, Rode M, Suyama M, Schmidt S, Gavin AC, Bork P, Serrano L. (2009) Transcriptome complexity in a genome-reduced bacterium. Science, 326:1268–1271.
13. Kuhner S, van Noort V, Betts MJ, Leo-Macias A, Batisse C, Rode M, Yamada T, Maier T, Bader S, Beltran-Alvarez P, Castano-Diez D, Chen WH, Devos D, Guell M, Norambuena T, Racke I, Rybin V, Schmidt A, Yus E, Aebersold R, Herrmann R, Bottcher B, Frangakis AS, Russell RB, Serrano L, Bork P, Gavin AC. (2009) Proteome organization in a genome-reduced bacterium. Science, 326:1235–1240.
14. Yus E, Maier T, Michalodimitrakis K, van Noort V, Yamada T, Chen WH, Wodke JA, Guell M, Martinez S, Bourgeois R, Kuhner S, Raineri E, Letunic I, Kalinina OV, Rode M, Herrmann R, Gutierrez-Gallego R, Russell RB, Gavin AC, Bork P, Serrano L. (2009) Impact of genome reduction on bacterial metabolism and its regulation. Science, 326:1263–1268.
15. Fenyö D. (2010) *Computational Biology, Methods in Molecular Biology. Methods in Molecular Biology*, Vol. 673. Springer.
16. Ladunga I. (2010) *Computational Biology of Transcription Factor Binding. Methods in Molecular Biology*, Vol. 674. Springer.
17. Wu CH, Chen C. (2011) *Bioinformatics for Comparative Proteomics. Methods in Molecular Biology*, Vol. 694. Springer.
18. Hamacher M, Eisenacher M, Stephan C. (2011) *Data Mining in Proteomics. Methods in Molecular Biology*, Vol. 696. Springer.
19. Michalodimitrakis K, Isalan M. (2009) Engineering prokaryotic gene circuits. FEMS Microbiol Rev, 33:27–37.
20. Stricker J, Cookson S, Bennett MR, Mather WH, Tsimring LS, Hasty J. (2008) A fast, robust and tunable synthetic gene oscillator. Nature, 456:516–519.
21. Gibson DG, Glass JI, Lartigue C, Noskov VN, Chuang RY, Algire MA, Benders GA, Montague MG, Ma L, Moodie MM, Merryman C, Vashee S, Krishnakumar R, Assad-Garcia N, Andrews-Pfannkoch C, Denisova EA,

Young L, Qi ZQ, Segall-Shapiro TH, Calvey CH, Parmar PP, Hutchison CA, 3rd, Smith HO, Venter JC. (2010) Creation of a bacterial cell controlled by a chemically synthesized genome. Science, 329:52–56.

22. Isalan M, Lemerle C, Michalodimitrakis K, Horn C, Beltrao P, Raineri E, Garriga-Canut M, Serrano L. (2008) Evolvability and hierarchy in rewired bacterial gene networks. Nature, 452:840–845.

Part I

Experimental and In Silico Approaches to Unravel Network Components

Chapter 2

Bacterial Interactomes: From Interactions to Networks

Emmanuelle Bouveret and Christine Brun

Abstract

In order to ensure their function(s) in the cell, proteins are organized in machineries, underlaid by a complex network of interactions. Identifying protein interactions is thus crucial to our understanding of cell functioning. Technical advances in molecular biology and genomic technology now allow for the systematic study of the interactions occurring in a given organism. This review first presents the techniques readily available to microbiologists for studying protein–protein interactions in bacteria, as well as their usability for high-throughput studies. Two types of techniques need to be considered: (1) the isolation of protein complexes from the organism of interest by affinity purification, and subsequent identification of the complex partners by mass spectrometry and (2) two-hybrid techniques, in general based on the production of two recombinant proteins whose interaction has to be tested in a reporter cell. Next, we summarize the bacterial interactomes already published. Finally, the strengths and pitfalls of the techniques are discussed, together with the potential prospect of interactome studies in bacteria.

Key words: Protein–protein interaction, Tandem affinity purification, Bacterial two-hybrid, Yeast two-hybrid, Protein networks, Interactome

1. Introduction

1.1. From Interactions...

Up to recently, classical biochemical and genetic methods have been used to identify interactions between genes and between proteins at a slow pace. Usually, these experiments follow a protocol established for the elucidation of the functions of one or a few genes/proteins of interest or for a better understanding of a particular biological process. Beyond the success of these approaches, the genomics era has introduced major modifications in the way biologists envisage to get answers to their questioning and as a result, plan experiments at a completely different scale. Genome-wide and more generally global approaches open the possibility to tackle cell functioning as a whole, namely as a system.

Understanding the biological processes at a system-level implies the analysis of interaction networks. The deciphering of these networks first necessitates identifying the relations and

Jacques van Helden et al. (eds.), *Bacterial Molecular Networks: Methods and Protocols*, Methods in Molecular Biology, vol. 804, DOI 10.1007/978-1-61779-361-5_2, © Springer Science+Business Media, LLC 2012

interactions that exist between the parts of the system, i.e., the genes and the proteins. To this aim, large-scale and high-throughput methods have been developed to discover interactions of different nature. Whereas DNA microarray technologies point at related co-expressed genes, ChIP-on-chip and ChIP-seq methodologies reveal protein–DNA interactions occurring *in vivo*, and synthetic genetic array analyses identify genetically interacting genes such as synthetic lethals.

Regarding protein–protein interactions (PPI), two technologies, originally devised to identify specific interactions, protein complex affinity purification and two-hybrid methods, have been optimized to allow for large-scale analyses. The first detects direct and indirect protein–protein interactions occurring within protein complexes by identifying the protein partners of a given protein using affinity purification (1). The second detects direct binary interactions and is generally based on molecular or genetic complementarity principles. For instance, a binary interaction existing between a bait and a prey allows for the reconstitution of a functional transcription factor, which in turn, activates the expression of a reporter gene (2–5).

1.2. ...To Protein–Protein Interaction Networks...

Once identified by these large-scale methodologies, interactions can be assembled into protein–protein interaction maps. These "interactomes" (6) represent a set of possible biophysical interactions between the tested proteins of the studied organism. Their exploration leads either to functional predictions for uncharacterized proteins based on the composition of their network neighbourhood, or provides information about the organization of the biological processes into the cell, their coordination, their actors, etc.

1.3. ...In Bacteria

Whole genome sequencing programs revealed that almost a third of the protein-coding genes in every genome is functionally uncharacterized and occasionally lacks homologs in distant species (7). Even for the deeply studied *Escherichia coli* model bacterium, Gene Ontology annotations (8) for *E. coli* genes (August 2009 release) show that ~30% of the genes cannot be assigned any functional annotations (Fig. 1). Forty-five percent of *E. coli* genes have had either their molecular or cellular function experimentally elucidated, yet both have been proven for only 15%. The extent of our knowledge of gene functions in the model bacterium is thus amazingly limited despite more than half a century of investigations. The situation is obviously even worth for less studied bacterial species. Consequently, it is rather unexpected that despite the relatively small size of bacterial genomes, only few high-throughput protein–protein interaction studies have been performed so far in these organisms. This is quite detrimental to the field of systems biology since, one could reasonably predict that if the functioning of any living organism comes to be fully

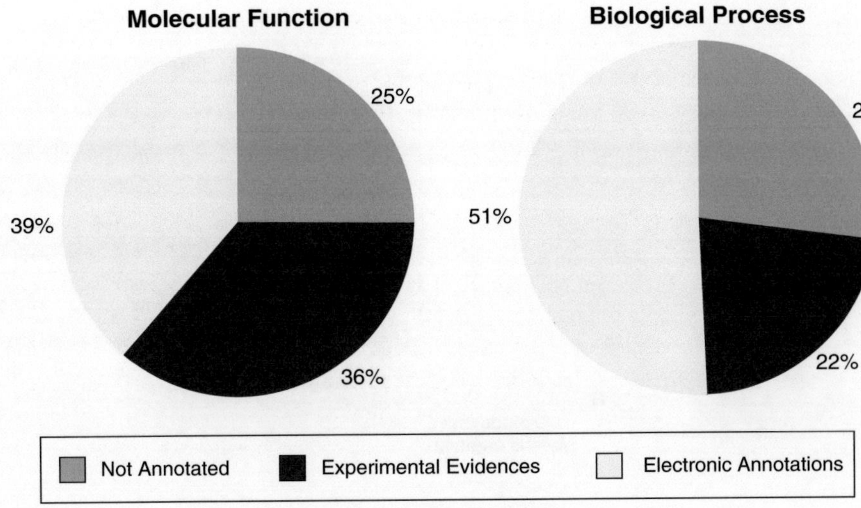

Fig. 1. Extent of the gene ontology annotations for *Escherichia coli* genes.

understood in a near future a bacterium such as *E. coli* or bacteria with smaller genomes appear as the likeliest candidates.

This review focuses first onto the principles of the methods allowing the identification of protein–protein interactions which have already been used or can be used to produce the bacterial interactomes. Second, available resources for bacterial interaction data are presented. Finally, the strengths and weaknesses of the experimental as well as the conceptual approaches are discussed.

2. Methods

2.1. Tandem Affinity Purification

Protein partners in a given complex can be identified through affinity purification. The principle of the methodology consists in (1) fusing a tag to a given protein called the bait, in the organism of interest, (2) producing the protein in physiological conditions that warrant the formation of the natural complexes, and (3) purifying the complexes by affinity. Purified proteins can then be identified by mass spectrometry analyses. If in theory any couple of tag and cognate affinity resin may be chosen, the tandem affinity purification (TAP) has proven overtly successful.

TAP was originally developed in *Saccharomyces cerevisiae* in Bertrand Séraphin's laboratory at EMBL (1). The method is based on a TAP tag composed of two IgG binding domains originating from *Staphylococcus aureus* protein A and the Calmodulin Binding Peptide (CBP) such that the tagged proteins can be purified on IgG and Calmodulin-coated beads, respectively (Fig. 2a). The two parts of the tag are separated by a sequence of seven residues (ENLYFQG), which is specifically

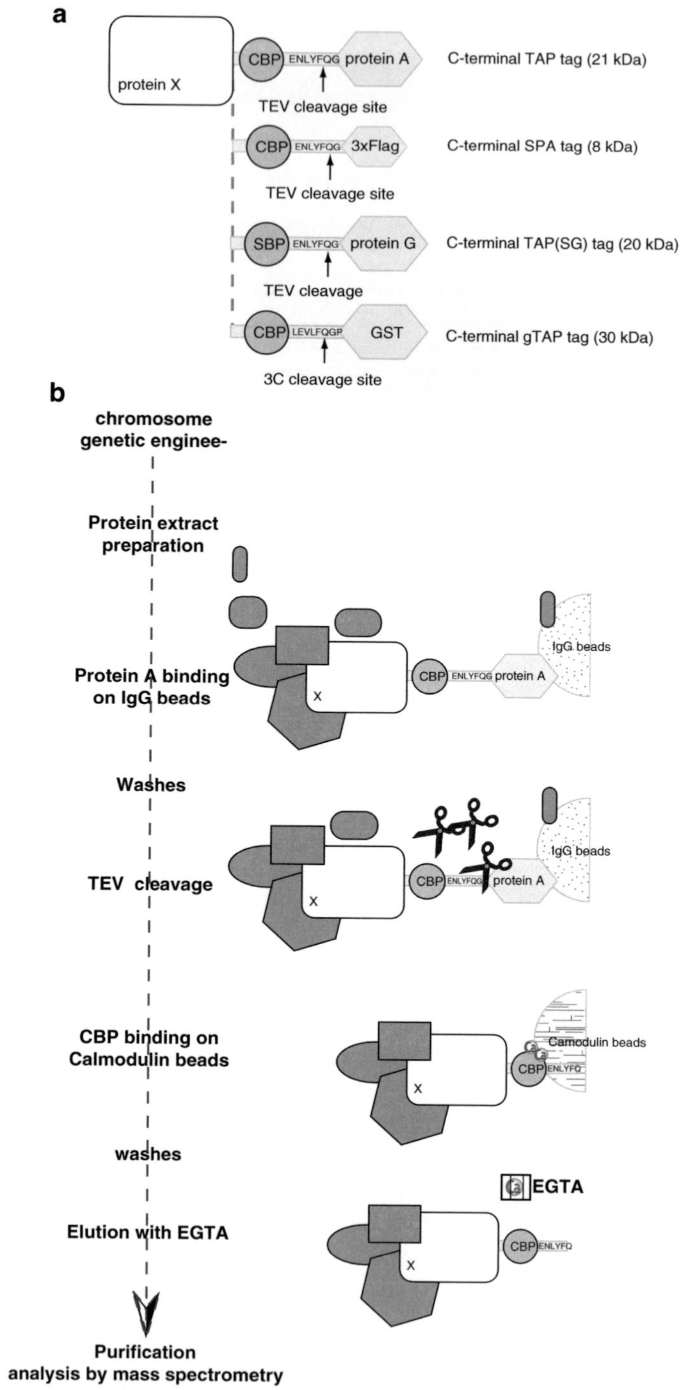

Fig. 2. Principle of TAP-like purifications. (**a**) The organization of different TAP-like tags described in the text is shown. *CBP* calmodulin binding peptide, *SBP* streptavidin binding peptide. TAP, SPA, TAP(SG), and gTAP tags have been described in refs. 1, 12, 22, 23. (**b**) Procedure of the TAP method.

recognized and cleaved by the Tobacco Etch Virus (TEV) protease between the last Q and G residues. In total, the TAP tag is about 20 kDa long and can be placed either at the N terminus or at the C terminus of the bait protein. Purification is performed in two successive steps (Fig. 2b; see ref. 9). First, the protein extract (prepared from a culture of a strain producing the tagged protein) is incubated on IgG-coated beads. After a washing step, the purified material is eluted from the column using the TEV protease. The eluted material is then incubated on Calmodulin-coated beads, the column is washed and the purified material can be eluted by addition of EGTA that chelates the Ca^{2+} ions necessary to the CBP/Calmodulin interaction. The composition of the isolated protein complex is then analyzed by mass spectrometry, with or without prior separation by polyacrylamide gel electrophoresis.

The two successive purification steps rely on different tags and affinity systems and the elution procedures at each step do not modify the buffer characteristics. Hence, this method permits the recovery of a highly pure preparation, while preserving interactions. Yet, the technique not only relies on the purification procedure, but also on the nature of the initial biological material from which the protein extract is prepared. In order to get the best results, the tagged protein must be produced by the cells under physiological conditions. Thus, ideally, the recombinant protein should replace the endogenous wild-type protein and should not be overproduced. When the studied organism is amenable to genetic manipulation, this is the solution of choice. In yeast, the process consists in introducing a cassette coding for the TAP tag and a selection marker at the extremity of the genes of interest by homologous recombination. However, in most organisms, intermediate solutions exist. The tagged proteins can for instance be produced in the corresponding mutants, in an expression system that does not overproduce the tagged protein, or preserves the natural promoter of the gene. Equivalents of the yeast TAP-tagging cassettes have been constructed for *E. coli* (10–12). The TAP method has been applied successfully to low-scale as well as large-scale purification of protein complexes in bacteria as diverse as *E. coli* (13–15), *B. subtilis* (16), *Synechocystis* (17), *Helicobacter pylori* (18), or *Mycoplasma pneumoniae* (19).

2.2. Design of New Tags for TAP-like Methods

Since the first publication of the TAP method, several new tags have been described aiming at improving the binding, the stability, and the functionality of the tagged protein. These methods retain the principle of two successive and independent affinity steps to ensure the physiological relevance of the identified protein interactions.

A wide range of tags is available (20), which could theoretically be used for TAP-like protein complex purification in diverse combinations. We do not describe all of them here, but shall highlight those that have been or can be used in bacteria for systematic protein–protein interaction studies.

For purification in *E. coli*, the sequential peptide affinity (SPA) tag contains three repeats of the Flag epitope instead of protein A used in the original TAP tag (Fig. 2a). The first step of purification is thus performed on an anti-Flag column (12). The SPA tag is much smaller than the TAP tag (5 kDa instead of 20 kDa), and has been shown to increase the stability and the functionality of *E. coli* tagged proteins (12). Very useful and detailed protocols describe the TAP and SPA methods in *E. coli* (21). SPA tag vectors have also been developed for chromosomal tagging in Gram-positive bacteria (22). Furthermore, a new tag (gTAP; Fig. 2a) has been engineered, in which the TEV cleavage site has been replaced by a protease 3C site and the protein A part by the GST tag (22), one advantage being that unlike the TEV protease, protease 3C is active at 4°C.

For purification in mammalian cells, the SG and GS tags have been developed (23). In this case, protein G replaces the protein A part of the TAP tag, and a Strep tag replaces the CBP tag (Fig. 2a). IgG and streptavidin sepharose columns are respectively used for the purification steps and complexes are then eluted with biotin. In addition, the codon usage and the exact sequences of the SG and GS tags for animal cells and other variants in plants (24) have been optimized. These improved tags should work in bacteria where they remain to be tested.

Other combinations of tags have been described and used for protein complex purification in *Salmonella typhimurium* (25, 26), which combine those described above and others such as the 6-histidine tag or the BIO tag (*in vivo* biotinylation signal peptide) (25) in order to replace one or both purification steps.

All the variants described above involve two steps of purification as in the initial TAP method and use elution processes that preserve the interactions. Depending on the organism of interest, the researcher will choose the most adapted constructions.

2.3. Two-Hybrid Approaches

The techniques described above are based on protein co-purification from natural protein extracts prepared from the organism of interest. In contrast, alternative techniques grouped under the general term of "two-hybrid" have been developed in order to test whether two given proteins do or not interact. These techniques only provide information about the existence of binary biophysical interactions regardless of the context in which they naturally occur. In general, these methods rely on the *in vivo* detection of an interaction between two proteins, though often in a heterologous expression system. The two proteins to be tested are each fused to separated domains of a same functional protein and produced together in a reporter cell (Fig. 3). Interaction between the two recombinant proteins is detected *in vivo* in the reporter cell by enabling the reconstitution of either a missing biological activity, or a novel

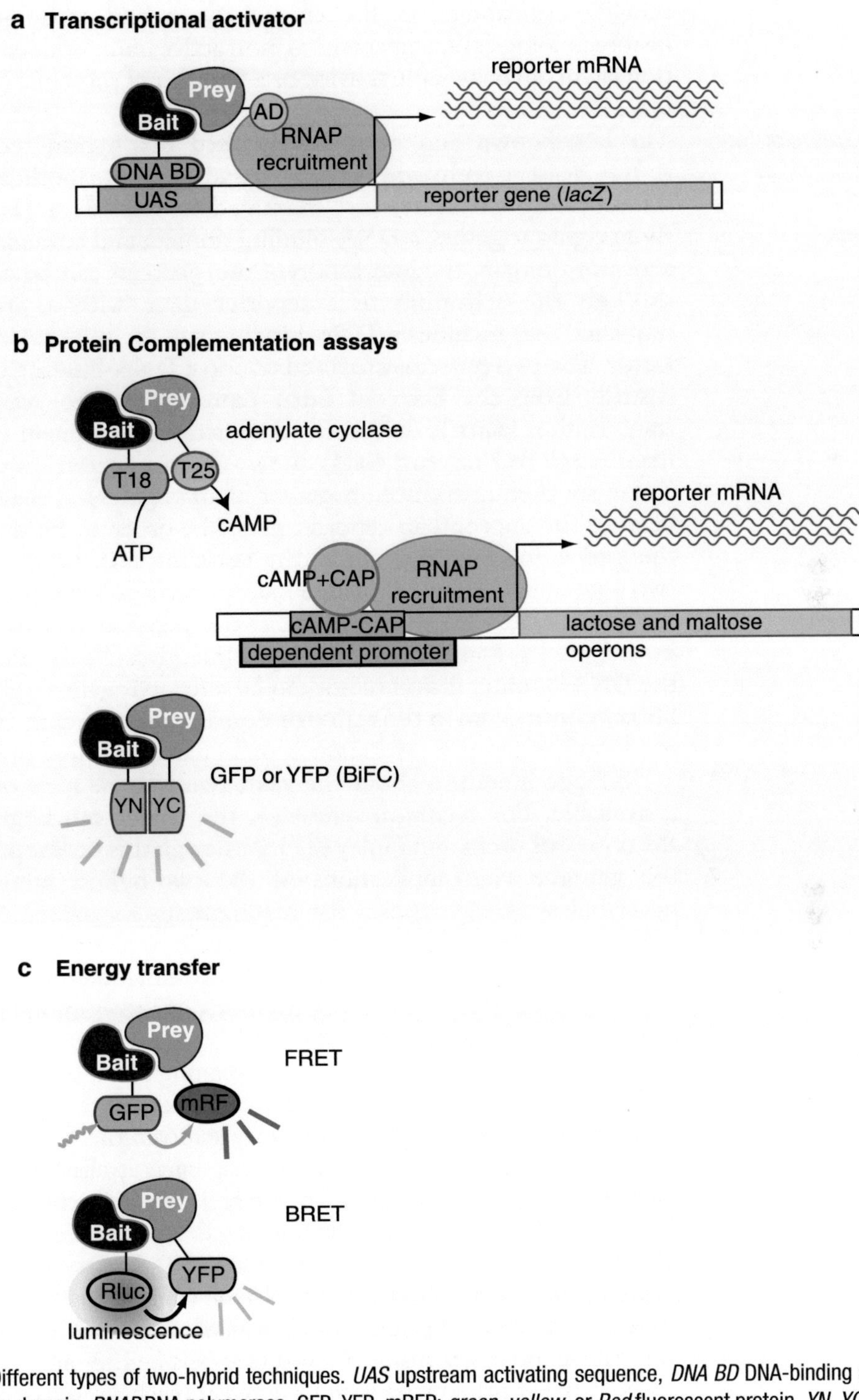

Fig. 3. Different types of two-hybrid techniques. *UAS* upstream activating sequence, *DNA BD* DNA-binding domain, *AD* activation domain, *RNAP* RNA polymerase. GFP, YFP, mRFP: *green*, *yellow*, or *Red* fluorescent protein. *YN*, *YC* N-terminal or C-terminal domains of YFP. *Rluc* Luciferase of *Renilla reniformis*.

activity. Depending on the chosen two-hybrid technique, the restored biological activity consists in an active transcriptional activation factor, an enzymatic activity, or a fluorescent protein.

2.3.1. Classical Yeast Two-Hybrid: Transcriptional Activation Factor Reconstitution

The best-known and most widely used two-hybrid technique is the yeast two-hybrid system based on the modularity of transcription activators, first described in refs. 3, 4 (Fig. 3a). By bringing together a DNA-binding domain and a transcription activator domain, the interaction of two proteins can be detected through the activation of a reporter gene with a promoter sequence that includes a DNA-binding site for the transcription factor. The two proteins are fused one to a DNA-binding domain (usually from the bacterial LexA repressor or the yeast Gal4 transcription factor), the other to an activation domain (usually from *E. coli* B42 or yeast Gal4) on two distinct expression vectors. These are then introduced together in an engineered yeast strain bearing the appropriate reporter gene. In order to be detected, the interaction has to occur in the yeast nucleus. Schematically, two strategies can be adopted. First, a matrix of known proteins for which ORFs are cloned in the two expression vectors can be investigated systematically. Alternatively, a given "bait" cloned in the DNA-binding fusion vector can be screened against a genomic library constructed in the activator domain fusion vector, in order to identify without *a priori* new protein partners called the preys.

A huge literature about the yeast-two hybrids methodology is available. For technical reference, the reader can begin with the review of Brent and Finley (27). Although this review precedes the genome-wide applications of the two-hybrid method, it nevertheless clearly exposes the requirements for validating new interactions, discusses strengths and pitfalls, and the problem of false positives and false negatives, all recommendations that are still fully valid. The issue 328 of Methods of Enzymology (2000) contains the detailed protocols used by various groups. More recent reviews of protocols for screening can be found in specialized journal issues or books such as ref. 28 for array-based screens and ref. 29 for genomic library screening.

The yeast-two hybrid technique has been applied with great success to the identification of interactions between bacterial proteins such as the six proteins of the *E. coli* Tol/Pal system in which interaction domains were mapped (30). It also contributed to identify new partners of the *E. coli* small GTPase Era by genomic library screening (31) and overall was used for the majority of high-throughput screens in bacteria (see below) (32–36).

A two-hybrid system based on the reconstitution of a transcription activation factor has also been designed to detect PPI *in vivo* in *E. coli* cells. It uses the λcI DNA-binding domain and the N-terminal domain of the α subunit of RNA polymerase (RNAP), which is thus directly recruited to the promoter by the interaction (37, 38). This system appears more physiological for studies of bacterial protein interactions and is amenable to genomic library screening.

2.3.2. Enzymatic-Protein Reconstitution (Protein Fragment Complementation Assay)

The second type of two-hybrid approaches consists in the reconstitution of an enzymatic activity or of a protein detectable by its fluorescence emission. This approach called Protein Fragment Complementation Assay (PCA) has been developed using several types of proteins such as adenylate cyclase, DHFR, or GFP (Fig. 3b) (39–41). A variety of systems is available, which all use two distinct expression vectors. Although these methods have not contributed to bacterial network deciphering so far, they should be amenable to high-throughput studies either by cloning the ORFeome of a bacterium, or by screening genomic libraries against a subset of baits.

Bacterial Two-Hybrid

The BACTH (bacterial adenylate cyclase two hybrid) technique has been increasingly used for studies on bacterial proteins in *E. coli* (42). The method is based on the reconstitution of adenylate cyclase activity in an *E. coli cya* mutant, by bringing together the T18 and T25 domains of adenylate cyclase from *Bordetella pertussis* (Fig. 3b). The restored cAMP production is then indirectly detected by its positive effect on the transcription of the lactose and maltose operons, which in turn, can be visualized on reporter colour plates (XGal or Mac Conkey), by β-galactosidase assay, or by selection on lactose or maltose minimal media. Because this technique relies on the reconstitution of an enzymatic activity and a diffusible second messenger, it can be used to detect interactions between membrane protein (13, 39). This method has been increasingly used for systematically testing protein–protein interactions in a subset of proteins of interest from diverse bacteria (39, 42, 43) and its recent adaptation to genomic library screens (44) opens the possibility of high-throughput analysis.

Other PCA Systems Used

In theory, any enzyme that can be split in two non-functional fragments that can re-associate through interaction with a third-part could be used for PCA, as long as the reconstituted activity can be easily detected or selected *in vivo* Murine Dihydrofolate Reductase (mDHFR) has, for instance, been used for genomic library screening in *E. coli* (45). The power of the PCA method based on mDHFR has been highlighted by its transposition to yeast, allowing the construction of an *in vivo* map of the yeast protein interactome (46). Another very successful approach is the Bimolecular Fluorescence Complementation (BiFC) technique, which takes advantage of the refolding of a functional GFP variant upon interaction between two hybrid proteins (Fig. 3b) (47). This technique was used in *E. coli* (48), *Agrobacterium tumefaciens* (49), or *Bacillus* (50). Although none of the bacterial interactomes published so far have used PCA methods, all these methods fulfil the requirements for being used in large-scale studies: they allow for library screening and for an easy detection or selection of the interactions *in vivo*.

2.3.3. Transfer of Fluorescence

A third type of method to detect binary protein–protein interactions *in vivo* is also based on the co-production in the cell of two distinct hybrid proteins. In this case, the interaction brings in close proximity two distinct fluorescent or luminescent proteins that can transfer their energy (Fig. 3c). These FRET and BRET (Fluorescence and Bioluminescence Resonance Energy Transfer, respectively) techniques have been used in bacteria (51, 52). However, interpretation of the data obtained with these techniques is often complex, and hence whether they may be suitable for high-throughput studies in bacteria remains an open question. Nevertheless, FRET and BRET allow for real time analyses (52), and complex technological developments such as combining GFP-based PCA with FRET or BRET in a single assay allow for the *in vivo* detection of multiple interactions, at a time in a given complex (53). This makes fluorescence-based techniques very powerful and promising.

3. Resources

3.1. High-throughput Bacterial Interactomes

Although bacteria, essentially due to the relative small size and low complexity of their genomes, appear as the organisms of choice for systems biology projects, only few and still partial interactomes have been deciphered (Table 1). Two among the techniques described above have been used: affinity purification and the classical yeast two-hybrid methodology.

3.1.1. Affinity Purification Studies

Using the technique developed by Court's laboratory (11) and the SPA approach (12), Emili's and Greenblat's groups constructed a collection of *E. coli* tagged strains to produce an *E. coli* interactome. Strains expressing the TAP or the SPA tags at the C-terminus of ~2,000 proteins were constructed (15, 55) in the recombination inducible DY330 strain. Essential and soluble proteins (15), then orphan proteins (55) were targeted, adding up to ~1/3 of the *E. coli* proteome. The 857 tagged *E. coli* strains described in the first study (15) are now distributed by Open Biosystems (now part of Thermo Fisher Scientific). The global results have been organized as web-accessible resources (http://www.compsysbio.org/bacteriome/ (56) and http://ecoli.med.utoronto.ca (55)). Due to the large number of identified interactions and the high coverage of the *E. coli* proteome, the last study allowed for the prediction of functions for orphan proteins,

Table 1
Bacterial large-scale interaction screens

Bacterial species	Methods	Total ORF number	Number of tested baits	Number of identified preys	Number of identified interactions	References
Helicobacter pylori 26695	Y2H	1,590	261	480 (?)	1,280	Rain et al. (32)
Escherichia coli K12 DY330	TAP-TAG	4,236	648	1,369	5,254	Butland et al. (15)
Escherichia coli K12 W3110	His-TAG	4,236	2,667	2,012	11,511	Arifuzzaman et al. (57)
Escherichia coli K12 W3110	TAP-TAG	4,236	1,241	1,316	5,997	Hu et al. (55)
Mycoplasma pneumoniae M129	TAP-TAG	689	212	411	1,058	Kühner et al. (19)
Campylobacter jejuni NCTC11168	Y2H	1,654	637	1,248	11,687	Parrish et al. (33)
Synechocystis PCC6803	Y2H	3,569	1,037	1,352	3,236	Sato et al. (34)
Mesorhizobium loti	Y2H	7,277	1,542	1,084	3,121	Shimoda et al. (35)
Treponema pallidium	Y2H	1,039	606	357	3,640	Titz et al. (36)
Steptococcus pneumoniae	PING	2,126	28	40	157	Gerber et al. (70)
Steptococcus pneumoniae	Y2H	2,126	–	–	77	MPIDB (61)

thanks to their clustering in protein complexes or functional modules (55).

In a second high-throughput PPI study of *E. coli* using affinity purification (57), all *E. coli* ORFs were tested using a different, less physiological approach. Four thousand three hundred and thirty-nine 6-histine N-terminal tagged proteins were produced on plasmids under the control of an IPTG-inducible T5-lac promoter. It is to note that a single round of purification was used in this study (cf. Subheading 2.1).

Taken as a model for the proteome organization of a genome-reduced bacterium, the interactome of *M. pneumoniae* was recently deciphered by the TAP approach (19). In this case, all TAP-fusions were produced from a genomic integration

Table 2
Main databases for bacterial interactions.

Main databases	Species	Interaction types	Number of bacterial interactions reported	URL
IntAct	All	Binary	38,116	http://www.ebi.ac.uk/intact/
MINT	All	Binary	7,456	http://mint.bio.uniroma2.it/
MPIDB	Bacteria	Binary	23,580	http://www.jcvi.org/mpidb/
Bacteriome	*Escherichia coli*	Direct, indirect and inferred	11,209	http://www.compsysbio.org/bacteriome/

under the control of a unique promoter (456 strains compared to the 689 protein coding genes of the bacterium). Ultimately, the interactome covers approximately 85% of the predicted soluble proteome of *M. pneumoniae.*

3.1.2. Two-Hybrid Screens

Five high-throughput bacterial protein–protein interactions screens using the yeast two-hybrid have been published so far. Here again, two approaches have been followed. First, nearly all the ORFs of the organism were cloned in the yeast two-hybrid vectors and systematically tested in all possible combinations. This array-based approach was performed in *Treponema pallidum* (36) and *Campylobacter jejuni* (33), which both have relatively small genomes (1,039 and 1,654 predicted ORFs, respectively).

Second, in *H. pylori* (32), *Synechocystis* sp. (34), and *Mesorhizobium loti* (35), a subset of "baits" were cloned and screened against a genomic library. In each study, baits were chosen according to specific research interests or traits of the studied bacterium. For example, genes involved in symbiosis or conserved in the host *Arabidopsis thaliana* and genes of unknown function were studied in *M. loti* (35) and *Synechocystis* sp., respectively (34).

3.2. Databases

Interactions identified with high-throughput methods constitute large datasets, which are compiled within interaction databases. Besides general databases such as IntAct (58) or MINT (59) that contain protein–protein interactions detected in all types of species including bacteria (and for which a detailed protocol has been described in ref. 60), two databases are especially dedicated to bacterial protein–protein interactions (Table 2). Whereas Bacteriome.org ((56), http://www.compsysbio.org/bacteriome/) only collects results of systematic proteomic studies on *E. coli* cited above, MPIDB ((61), http://www.jcvi.org/mpidb/about.php) contains 20,000 binary interactions from almost 200 bacterial species and strains. Interestingly, this latter database provides an additional dataset of ~800 interactions identified by classical low-throughput experiments and extracted from publication abstracts. This database complies with the standards for interaction descriptions (62) and participates to international curation efforts (IMEx, http://imex.sourceforge.net/).

Although inferred protein–protein interactions are beyond the scope of this review, orthology information may come as a rescue in face of the paucity of the data. The STRING database ((63), http://string-db.org/), which integrates interaction data from low and large-scale experiments with other types of data (co-expression, genomic context, etc.) allows for inferring interactions by transfer between organisms, where applicable.

3.3. Strengths and Pitfalls

Eukaryotic and bacterial interactomes are far from comprehensive, and their quality and current coverage are difficult to estimate. Many of the available data suffer from (1) the presence of false-positive interactions, i.e., that are detected experimentally but do not exist in the organism in physiological conditions and (2) the absence of a large number of true interactions, the false negatives, which are not identified during the screens. Sources of these pitfalls and ways to overcome them are developed below.

3.3.1. Bias

One caveat of the TAP-like methods is that the procedure is long and uses several washing steps such that only strong interactions are retained (64). The TAP method is thus best suited for the identification of stable and abundant complexes. *In vivo* cross-linking prior the TAP-like purifications has been proposed as a way to overcome this limitation. This additional step does not appear to prevent subsequent protein identification by mass spectrometry. This approach has been successful for protein complex purification from *H. pylori* cells with the TAP tag (18) or with the His-Bio tag on yeast cells (25). TAP-like methods are also biased toward soluble and abundant proteins. Small proteins are missed due in part to the purification procedure, but also to the reliability of mass spectrometry to identify small peptides. On the contrary, the two-hybrid methods are not sensitive to the natural abundance of proteins because these are not expressed in their natural context, and the identification of the partners by sequencing is unequivocal.

3.3.2. Membrane Proteins

The second pitfall of the TAP-like methods is inherent to affinity purification procedures that fail to easily deal with membrane proteins. Some specific protocols for the solubilization of membrane proteins before TAP have been described (65) but these may interfere with the interactions. A combination of cross-linking and membrane solubilization might be the solution to this problem of membrane complex purification. In the case of two-hybrid like techniques, the problem can be circumvented by testing only the soluble domains of membrane proteins. Furthermore, the

BACTH and PCA techniques have been used for the detection of full length integral inner membrane proteins (13, 39).

3.3.3. Recombinant Protein Expression

A common problem in all the methods reported here is that they rely on the production of recombinant fusion proteins (the bait and prey in two-hybrid like methods, only the bait in affinity purifications). These recombinant proteins may either not retain their function, or not be correctly produced or not localize correctly in the cell or a combination of these. This could lead to missing or nonphysiological interactions. In most two-hybrid methods, recombinant proteins are over-produced. Although this may be an advantage when allowing for the detection of interactions involving proteins of low abundance or proteins only produced in specific conditions, it may represent a disadvantage since over-expression can provoke nonphysiological interactions. On the contrary, TAP-like methods are based on the physiological expression of all the partners of the putative protein complexes and are therefore more reliable.

3.3.4. Contaminants

Both TAP-like and two-hybrid-like methods produce false positives, but not for the same reasons. In TAP-like methods, false positives correspond to co-purifying contaminants, mostly abundant enzymes and other proteins, such as ribosomal proteins, but also chaperones that bind to the recombinant bait or its TAP tag region. It is thus important to maintain a list of the usual TAP contaminants found in a given organism following a given procedure. Such a list was published for the TAP study of *E. coli* (55). It contains pyruvate dehydrogenase subunits, chaperones, and ribosomal proteins. The two-hybrid methods are parasitized by (1) the so-called sticky proteins, which are not correctly folded or are produced in a wrong cell compartment and (2) the numerous auto-activating proteins that promote transcription, without any interaction partner. The long experience with yeast two-hybrid in eukaryotes leads to setting up procedures aiming at cleaning up the raw data. Furthermore, any new interaction (whether identified by TAP-like or by two-hybrid-like methods) needs to be validated. The doctrine first described for yeast two-hybrid data remains relevant (27): validation of the interaction by another method, selection of mutations that break and restore the interaction, specificity of the interaction, etc. Noticeably, a tool kit composed of several high-throughput protein interaction assays has been recently benchmarked against true and random interactions in order to calculate a probability of existence for each identified interaction (66). All the methods included in the tool kit being based on the two-hybrid principle (classical yeast two-hybrid, PCA, etc.), one can ideally imagine using it for retesting bacterial interactions.

3.3.5. Sampling Sensitivity

Several large-scale screens may be required for achieving an acceptable coverage of a given interactome (67). The sampling effect is particularly visible when comparing several interaction datasets obtained from the same species. Different affinity purification methods used to generate three sets of data on *E. coli* (cf. Subheading 3.1) identified only 62/17,049 interactions in common (55). Without *a priori* about data quality, this result shows that the interaction space of *E. coli* is still far from being covered.

3.3.6. Confidence Score

All potential problems discussed so far including false-positive and false-negative interactions, stimulated the development of confidence scores for interactions. Besides the experimental confidence score for binary interactions derived from the use of the tool kit evoked previously (66), confidence scores are usually based on other experimental data obtained during the identification process such as the frequency of detection of a given interaction as well as independent data such as co-expression or/and subcellular colocalization of the two interacting partners (32, 33, 68, 69).

The issue of the reliability of the interactions produced by the high-throughput methods is crucial. Although one user may feel more comfortable with interactions that are coherent with prior knowledge on the protein partners, one should remember that a lack of fit may suggest false negatives among the secondary data (taken as prior knowledge) but also that over-validated interactions only point toward already beaten tracks and impair the discovery process.

3.3.7. Feasibility of High-throughput Studies

Finally, the methods may be compared and evaluated regarding their practicability and strengths as high-throughput methods. Two-hybrid methods are clearly favoured in this case, due to their simplicity compared to TAP-like methods that rely on tedious steps of bacterial culture, biochemistry, and mass spectrometry.

4. Outlook

The fact that different methodologies have been used to generate different datasets of interactions that have little in common (55) calls for a comparison of the techniques' sensitivity and coverage. This will enable the estimation of the completeness and quality of the identified interactions, as it was recently done for eukaryotes (66). These high-throughput studies provide a wealth of information to the microbiologist community, especially for the deciphering of gene function at the genomic scale. However, beyond the description of every protein–protein interactions that can occur in a cell, the gain of information about dynamics of protein network is a next step toward a global understanding of cell systems. To understand how the interactome responds and adapts to modification of the cell environment, interactomes obtained in different physiological conditions will have to be compared. This can be achieved experimentally *in vivo* using the TAP-like methods and bioinformatically via the integration of pertinent genomic, transcriptomic, and proteomic data with interactomes deciphered by two-hybrid methods. Hopefully, such comparisons will provide better understanding of the dynamics of the networks, resulting from genetic regulation and changing interactions.

Acknowledgments

Work by E.B. is funded by the CNRS, a young research grant from ANR, and FRM. Work by C.B. is funded by the CNRS, the Inserm, the Université de la Méditerranée and grants from ANR.

References

1. Rigaut G, Shevchenko A, Rutz B, Wilm M, Mann M, Seraphin B. (1999) A generic protein purification method for protein complex characterization and proteome exploration. Nat Biotechnol, 17:1030–1032.
2. Legrain P, Selig L. (2000) Genome-wide protein interaction maps using two-hybrid systems. FEBS Lett, 480:32–36.
3. Chien CT, Bartel PL, Sternglanz R, Fields S. (1991) The two-hybrid system: a method to identify and clone genes for proteins that interact with a protein of interest. Proc Natl Acad Sci USA, 88: 9578–9582.
4. Fields S, Song O. (1989) A novel genetic system to detect protein–protein interactions. Nature, 340:245–246.
5. Walhout AJ, Vidal M. (2001) High-throughput yeast two-hybrid assays for large-scale protein interaction mapping. Methods, 24:297–306.
6. Sanchez C, Lachaize C, Janody F, Bellon B, Roder L, Euzenat J, Rechenmann F, Jacq B. (1999) Grasping at molecular interactions and genetic networks in *Drosophila melanogaster* using FlyNets, an Internet database. Nucleic Acids Res, 27:89–94.
7. Galperin MY, Koonin EV. (2000) Who's your neighbor? New computational approaches for functional genomics. Nat Biotechnol, 18:609–613.
8. Ashburner M, Ball CA, Blake JA, Botstein D, Butler H, Cherry JM, Davis AP, Dolinski K, Dwight SS, Eppig JT, Harris MA, Hill DP, Issel-Tarver L, Kasarskis A, Lewis S, Matese JC, Richardson JE, Ringwald M, Rubin GM, Sherlock G. (2000) Gene ontology: tool for the unification of biology. The Gene Ontology Consortium. Nat Genet, 25:25–29.
9. Puig O, Caspary F, Rigaut G, Rutz B, Bouveret E, Bragado-Nilsson E, Wilm M, Seraphin B. (2001) The tandem affinity purification (TAP) method: a general procedure of protein complex purification. Methods, 24:218–229.
10. Datsenko KA, Wanner BL. (2000) One-step inactivation of chromosomal genes in *Escherichia coli* K-12 using PCR products. Proc Natl Acad Sci U S A, 97:6640–6645.
11. Yu D, Ellis HM, Lee EC, Jenkins NA, Copeland NG, Court DL. (2000) An efficient recombination system for chromosome engineering in *Escherichia coli*. Proc Natl Acad Sci U S A, 97:5978–5983.
12. Zeghouf M, Li J, Butland G, Borkowska A, Canadien V, Richards D, Beattie B, Emili A, Greenblatt JF. (2004) Sequential Peptide

Affinity (SPA) system for the identification of mammalian and bacterial protein complexes. J Proteome Res, 3:463–468.

13. Gully D, Bouveret E. (2006) A protein network for phospholipid synthesis uncovered by a variant of the tandem affinity purification method in *Escherichia coli*. Proteomics, 6:282–293.
14. Gully D, Moinier D, Loiseau L, Bouveret E. (2003) New partners of acyl carrier protein detected in *Escherichia coli* by tandem affinity purification. FEBS Lett, 548:90–96.
15. Butland G, Peregrin-Alvarez JM, Li J, Yang W, Yang X, Canadien V, Starostine A, Richards D, Beattie B, Krogan N, Davey M, Parkinson J, Greenblatt J, Emili A. (2005) Interaction network containing conserved and essential protein complexes in *Escherichia coli*. Nature, 433:531–537.
16. Pompeo F, Luciano J, Galinier A. (2007) Interaction of GapA with HPr and its homologue, Crh: novel levels of regulation of a key step of glycolysis in *Bacillus subtilis*? J Bacteriol, 189:1154–1157.
17. Brown AP, Affleck V, Fawcett T, Slabas AR. (2006) Tandem affinity purification tagging of fatty acid biosynthetic enzymes in *Synechocystis* sp. PCC6803 and *Arabidopsis thaliana*. J Exp Bot, 57:1563–1571.
18. Stingl K, Schauer K, Ecobichon C, Labigne A, Lenormand P, Rousselle JC, Namane A, de Reuse H. (2008) In vivo interactome of *Helicobacter pylori* urease revealed by tandem affinity purification. Mol Cell Proteomics, 7:2429–2441.
19. Kuhner S, van Noort V, Betts MJ, Leo-Macias A, Batisse C, Rode M, Yamada T, Maier T, Bader S, Beltran-Alvarez P, Castano-Diez D, Chen WH, Devos D, Guell M, Norambuena T, Racke I, Rybin V, Schmidt A, Yus E, Aebersold R, Herrmann R, Bottcher B, Frangakis AS, Russell RB, Serrano L, Bork P, Gavin AC. (2009) Proteome organization in a genome-reduced bacterium. Science, 326:1235–1240.
20. Terpe K. (2003) Overview of tag protein fusions: from molecular and biochemical fundamentals to commercial systems. Appl Microbiol Biotechnol, 60:523–533.
21. Babu M, Butland G, Pogoutse O, Li J, Greenblatt JF, Emili A. (2009) Sequential peptide affinity purification system for the systematic isolation and identification of protein complexes from *Escherichia coli*. Methods Mol Biol 564:373–400.
22. Yang X, Doherty GP, Lewis PJ. (2008) Tandem affinity purification vectors for use in gram positive bacteria. Plasmid, 59:54–62.
23. Burckstummer T, Bennett KL, Preradovic A, Schutze G, Hantschel O, Superti-Furga G, Bauch A. (2006) An efficient tandem affinity purification procedure for interaction proteomics in mammalian cells. Nat Methods, 3:1013–1019.
24. Rohila JS, Chen M, Cerny R, Fromm ME. (2004) Improved tandem affinity purification tag and methods for isolation of protein heterocomplexes from plants. Plant J, 38:172–181.
25. Tagwerker C, Zhang H, Wang X, Larsen LS, Lathrop RH, Hatfield GW, Auer B, Huang L, Kaiser P. (2006) HB tag modules for PCR-based gene tagging and tandem affinity purification in *Saccharomyces cerevisiae*. Yeast, 23:623–632.
26. Chowdhury SM, Shi L, Yoon H, Ansong C, Rommereim LM, Norbeck AD, Auberry KJ, Moore RJ, Adkins JN, Heffron F, Smith RD. (2009) A method for investigating protein–protein interactions related to salmonella typhimurium pathogenesis. J Proteome Res, 8:1504–1514.
27. Brent R, Finley RL, Jr. (1997) Understanding gene and allele function with two-hybrid methods. Annu Rev Genet, 31:663–704.
28. Rajagopala SV, Uetz P. (2009) Analysis of protein–protein interactions using array-based yeast two-hybrid screens. Methods Mol Biol, 548:223–245.
29. Serebriiskii IG, Fang R, Latypova E, Hopkins R, Vinson C, Joung JK, Golemis EA. (2005) A combined yeast/bacteria two-hybrid system: development and evaluation. Mol Cell Proteomics, 4:819–826.
30. Walburger A, Lazdunski C, Corda Y. (2002) The Tol/Pal system function requires an interaction between the C-terminal domain of TolA and the N-terminal domain of TolB. Mol Microbiol, 44:695–708.
31. Zhang J, Inouye M. (2002) MazG, a nucleoside triphosphate pyrophosphohydrolase, interacts with Era, an essential GTPase in *Escherichia coli*. J Bacteriol, 184:5323–5329.
32. Rain JC, Selig L, De Reuse H, Battaglia V, Reverdy C, Simon S, Lenzen G, Petel F, Wojcik J, Schachter V, Chemama Y, Labigne A, Legrain P. (2001) The protein–protein interaction map of *Helicobacter pylori*. Nature, 409:211–215.
33. Parrish JR, Yu J, Liu G, Hines JA, Chan JE, Mangiola BA, Zhang H, Pacifio S, Fotouhi F, DiRita VJ, Ideker T, Andrews P, Finley RL, Jr. (2007) A proteome-wide protein interaction map for *Campylobacter jejuni*. Genome Biol, 8:R130.

34. Sato S, Shimoda Y, Muraki A, Kohara M, Nakamura Y, Tabata S. (2007) A large-scale protein–protein interaction analysis in *Synechocystis* sp. PCC6803. DNA Res, 14:207–216.

35. Shimoda Y, Shinpo S, Kohara M, Nakamura Y, Tabata S, Sato S. (2008) A large scale analysis of protein–protein interactions in the nitrogen-fixing bacterium *Mesorhizobium loti*. DNA Res, 15:13–23.

36. Titz B, Rajagopala SV, Goll J, Hauser R, McKevitt MT, Palzkill T, Uetz P. (2008) The binary protein interactome of *Treponema pallidum* – the syphilis spirochete. PLoS ONE, 3:e2292.

37. Dove SL, Hochschild A. (1998) Use of artificial activators to define a role for protein–protein and protein–DNA contacts in transcriptional activation. Cold Spring Harb Symp Quant Biol, 63:173–180.

38. Dove SL, Joung JK, Hochschild A. (1997) Activation of prokaryotic transcription through arbitrary protein–protein contacts. Nature, 386:627–630.

39. Karimova G, Dautin N, Ladant D. (2005) Interaction network among *Escherichia coli* membrane proteins involved in cell division as revealed by bacterial two-hybrid analysis. J Bacteriol, 187:2233–2243.

40. Nyfeler B, Michnick SW, Hauri HP. (2005) Capturing protein interactions in the secretory pathway of living cells. Proc Natl Acad Sci U S A, 102:6350–6355.

41. Remy I, Campbell-Valois FX, Michnick SW. (2007) Detection of protein–protein interactions using a simple survival protein-fragment complementation assay based on the enzyme dihydrofolate reductase. Nat Protoc, 2:2120–2125.

42. Karimova G, Pidoux J, Ullmann A, Ladant D. (1998) A bacterial two-hybrid system based on a reconstituted signal transduction pathway. Proc Natl Acad Sci U S A, 95:5752–5756.

43. Marbouty M, Saguez C, Cassier-Chauvat C, Chauvat F. (2009) ZipN, an FtsA-like orchestrator of divisome assembly in the model cyanobacterium *Synechocystis* PCC6803. Mol Microbiol, 74(2):409–420.

44. Karimova G, Robichon C, Ladant D. (2009) Characterization of YmgF, a 72-residue inner membrane protein that associates with the *Escherichia coli* cell division machinery. J Bacteriol. 191:333–346.

45. Michnick SW, Remy I, Campbell-Valois FX, Vallee-Belisle A, Pelletier JN. (2000) Detection of protein–protein interactions by protein fragment complementation strategies. Methods Enzymol, 328:208–230.

46. Tarassov K, Messier V, Landry CR, Radinovic S, Serna Molina MM, Shames I, Malitskaya Y, Vogel J, Bussey H, Michnick SW. (2008) An in vivo map of the yeast protein interactome. Science, 320:1465–1470.

47. Kerppola TK. (2008) Bimolecular fluorescence complementation (BiFC) analysis as a probe of protein interactions in living cells. Annu Rev Biophys, 37:465–487.

48. Magliery TJ, Wilson CG, Pan W, Mishler D, Ghosh I, Hamilton AD, Regan L. (2005) Detecting protein–protein interactions with a green fluorescent protein fragment reassembly trap: scope and mechanism. J Am Chem Soc, 127:146–157.

49. Cascales E, Atmakuri K, Liu Z, Binns AN, Christie PJ. (2005) *Agrobacterium tumefaciens* oncogenic suppressors inhibit T-DNA and VirE2 protein substrate binding to the VirD4 coupling protein. Mol Microbiol, 58:565–579.

50. Soufo HJ, Graumann PL. (2003) Actin-like proteins MreB and Mbl from *Bacillus subtilis* are required for bipolar positioning of replication origins. Curr Biol, 13:1916–1920.

51. Sourjik V, Berg HC. (2002) Binding of the *Escherichia coli* response regulator CheY to its target measured in vivo by fluorescence resonance energy transfer. Proc Natl Acad Sci U S A, 99:12669–12674.

52. Xu Y, Piston DW, Johnson CH. (1999) A bioluminescence resonance energy transfer (BRET) system: application to interacting circadian clock proteins. Proc Natl Acad Sci U S A, 96:151–156.

53. Sourjik V, Vaknin A, Shimizu TS, Berg HC. (2007) In vivo measurement by FRET of pathway activity in bacterial chemotaxis. Methods Enzymol, 423:365–391.

54. Rebois RV, Robitaille M, Petrin D, Zylbergold P, Trieu P, Hebert TE. (2008) Combining protein complementation assays with resonance energy transfer to detect multipartner protein complexes in living cells. Methods, 45:214–218.

55. Hu P, Janga SC, Babu M, Diaz-Mejia JJ, Butland G, Yang W, Pogoutse O, Guo X, Phanse S, Wong P, Chandran S, Christopoulos C, Nazarians-Armavil A, Nasseri NK, Musso G, Ali M, Nazemof N, Eroukova V, Golshani A, Paccanaro A, Greenblatt JF, Moreno-Hagelsieb G, Emili A. (2009) Global functional atlas of *Escherichia coli* encompassing previously uncharacterized proteins. PLoS Biol, 7:e96.

56. Su C, Peregrin-Alvarez JM, Butland G, Phanse S, Fong V, Emili A, Parkinson J. (2008) Bacteriome.org – an integrated protein

interaction database for *E. coli*. Nucleic Acids Res, 36:D632–D636.

57. Arifuzzaman M, Maeda M, Itoh A, Nishikata K, Takita C, Saito R, Ara T, Nakahigashi K, Huang HC, Hirai A, Tsuzuki K, Nakamura S, Altaf-Ul-Amin M, Oshima T, Baba T, Yamamoto N, Kawamura T, Ioka-Nakamichi T, Kitagawa M, Tomita M, Kanaya S, Wada C, Mori H. (2006) Large-scale identification of protein–protein interaction of *Escherichia coli* K-12. Genome Res, 16:686–691.
58. Kerrien S, Alam-Faruque Y, Aranda B, Bancarz I, Bridge A, Derow C, Dimmer E, Feuermann M, Friedrichsen A, Huntley R, Kohler C, Khadake J, Leroy C, Liban A, Lieftink C, Montecchi-Palazzi L, Orchard S, Risse J, Robbe K, Roechert B, Thorneycroft D, Zhang Y, Apweiler R, Hermjakob H. (2007) IntAct – open source resource for molecular interaction data. Nucleic Acids Res, 35: D561–D565.
59. Zanzoni A, Montecchi-Palazzi L, Quondam M, Ausiello G, Helmer-Citterich M, Cesareni G. (2002) MINT: a Molecular INTeraction database. FEBS Lett, 513:135–140.
60. Chatr-Aryamontri A, Zanzoni A, Ceol A, Cesareni G. (2008) Searching the protein interaction space through the MINT database. Methods Mol Biol, 484:305–317.
61. Goll J, Rajagopala SV, Shiau SC, Wu H, Lamb BT, Uetz P. (2008) MPIDB: the microbial protein interaction database. Bioinformatics, 24:1743–1744.
62. Hermjakob H, Montecchi-Palazzi L, Bader G, Wojcik J, Salwinski L, Ceol A, Moore S, Orchard S, Sarkans U, von Mering C, Roechert B, Poux S, Jung E, Mersch H, Kersey P, Lappe M, Li Y, Zeng R, Rana D, Nikolski M, Husi H, Brun C, Shanker K, Grant SG, Sander C, Bork P, Zhu W, Pandey A, Brazma A, Jacq B, Vidal M, Sherman D, Legrain P, Cesareni G, Xenarios I, Eisenberg D, Steipe B, Hogue C, Apweiler R. (2004) The HUPO PSI's molecular interaction format – a community standard for the representation of protein interaction data. Nat Biotechnol, 22:177–183.
63. Jensen LJ, Kuhn M, Stark M, Chaffron S, Creevey C, Muller J, Doerks T, Julien P, Roth A, Simonovic M, Bork P, von Mering C. (2009) STRING 8 – a global view on proteins and their functional interactions in 630 organisms. Nucleic Acids Res, 37: D412–D416.
64. Jensen LJ, Bork P. (2008) Biochemistry. Not comparable, but complementary. Science, 322:56–57.
65. Gavin AC, Aloy P, Grandi P, Krause R, Boesche M, Marzioch M, Rau C, Jensen LJ, Bastuck S, Dumpelfeld B, Edelmann A, Heurtier MA, Hoffman V, Hoefert C, Klein K, Hudak M, Michon AM, Schelder M, Schirle M, Remor M, Rudi T, Hooper S, Bauer A, Bouwmeester T, Casari G, Drewes G, Neubauer G, Rick JM, Kuster B, Bork P, Russell RB, Superti-Furga G. (2006) Proteome survey reveals modularity of the yeast cell machinery. Nature, 440:631–636.
66. Braun P, Tasan M, Dreze M, Barrios-Rodiles M, Lemmens I, Yu H, Sahalie JM, Murray RR, Roncari L, de Smet AS, Venkatesan K, Rual JF, Vandenhaute J, Cusick ME, Pawson T, Hill DE, Tavernier J, Wrana JL, Roth FP, Vidal M. (2009) An experimentally derived confidence score for binary protein–protein interactions. Nat Methods, 6:91–97.
67. Venkatesan K, Rual JF, Vazquez A, Stelzl U, Lemmens I, Hirozane-Kishikawa T, Hao T, Zenkner M, Xin X, Goh KI, Yildirim MA, Simonis N, Heinzmann K, Gebreab F, Sahalie JM, Cevik S, Simon C, de Smet AS, Dann E, Smolyar A, Vinayagam A, Yu H, Szeto D, Borick H, Dricot A, Klitgord N, Murray RR, Lin C, Lalowski M, Timm J, Rau K, Boone C, Braun P, Cusick ME, Roth FP, Hill DE, Tavernier J, Wanker EE, Barabasi AL, Vidal M. (2009) An empirical framework for binary interactome mapping. Nat Methods, 6:83–90.
68. Zhang B, Park BH, Karpinets T, Samatova NF. (2008) From pull-down data to protein interaction networks and complexes with biological relevance. Bioinformatics, 24:979–986.
69. Han JD, Bertin N, Hao T, Goldberg DS, Berriz GF, Zhang LV, Dupuy D, Walhout AJ, Cusick ME, Roth FP, Vidal M. (2004) Evidence for dynamically organized modularity in the yeast protein–protein interaction network. Nature, 430:88–93.
70. Gerber D, Maerkl SJ, Quake SR. (2009) An in vitro microfluidic approach to generating protein-interaction networks. Nat Methods, 6:71–74.

Chapter 3

From Bacterial to Microbial Ecosystems (Metagenomics)

Shannon J. Williamson and Shibu Yooseph

Abstract

Metagenomics is revolutionizing the field of microbial ecology through techniques that eliminate the prerequisite of culturing. Metagenomic studies of microbial populations in different environments reveal the incredible diversity and adaptive capabilities of these organisms. With the advent of cheaper, high-throughput sequencing technologies, these studies are also producing vast amounts of sequence data. Here, we discuss the different components of a metagenomic study including sample collection, DNA extraction, sequencing, and informatics. We highlight their issues and challenges, and review the solutions that are currently in use. We conclude with examples of metagenomic studies conducted on environments of varying complexities.

Key words: Metagenomics, Microbial ecology, Sample complexity, Shotgun sequencing, Pyrosequencing, Metagenomic assembly, Binning, Gene finding, Metadata

1. Introduction

Metagenomics, or community genomics, is a rapidly emerging discipline that seeks to understand microbial ecosystems by studying the genomic content of the constituent microbes in their natural community setting (1). This is extremely useful and important since microbes that can be cultivated in the laboratory are thought to account for less than 1% of the microbes that exist in many environments (2). In a typical metagenomic study, nucleic acids are extracted from a sample that is collected from an environment of interest and directly sequenced using a random shotgun approach, thus eliminating the need for in vitro culture. We differentiate this from the paradigm (not discussed in this chapter) that uses targeted survey sequencing of marker genes, like the 16S rRNA gene, to interrogate a microbial community. Metagenomic shotgun sequence data are typically used to study two aspects of a microbial community – its taxonomic content (who's there?) and functional content (what are they doing?).

Jacques van Helden et al. (eds.), *Bacterial Molecular Networks: Methods and Protocols*, Methods in Molecular Biology, vol. 804, DOI 10.1007/978-1-61779-361-5_3,

Data from metagenomic studies have provided clues to the roles that constituent microbes play in their communities, and have also pointed to the incredible diversity of these organisms both at the genome and protein levels.

While early metagenomic studies used the Sanger dideoxy sequencing method (3) to generate sequence data, recent ones have switched to the more cost effective next-generation sequencing technologies like the array-based pyrosequencing technology from 454 Inc. (4) and the primer extension with fluorescent reversible terminators method from Illumina Inc. (5). These newer sequencing technologies also produce vastly more data per run (compared to Sanger), and this has resulted in metagenomic projects becoming dominant contributors to sequence databases.

In this chapter, we review various experimental and informatics aspects of metagenomics. We describe sample collection, DNA extraction, library construction, and sequencing. Among the informatics aspects that are addressed are sequence assembly, gene identification and annotation, taxonomic binning, sample comparison, and linking sequence-level observations to the sample metadata that is collected. The complexity of the various analyses is a function of several variables including the diversity of the microbial community, the depth of sequencing, and the choice of sequencing platform. We conclude with examples of metagenomic studies conducted in environments of varying complexity.

2. Experimental Approaches

2.1. Sample Collection and Processing

Metagenomic approaches have been applied to a diverse range of samples originating from a variety of environments; from marine waters to the human gut (6–15). The unifying constant in metagenomic studies is DNA, therefore, the type of sample that is suitable for metagenomic analysis is only limited by the quantity and quality of nucleic acid that can be extracted from it (16). Technological advances in the amplification of very low concentrations of DNA (e.g., linker-adapter amplification and whole genome amplification via the phi29 DNA polymerase) (17, 18) have further expanded the spectrum of samples that are amenable to metagenomic investigation. However, various environments present their own unique challenges regarding the collection of representative microbial communities that must be addressed before DNA extraction can be considered.

The collection of microbes from certain environments, such as aquatic ecosystems, is rather straightforward, as these tend to be fairly homogenous compared to other ecosystems such as soils. Aquatic ecosystems, particularly in the marine environment, typically contain one million bacterial cells per milliliter of surface seawater. Microbial cells are generally concentrated onto membrane filters of predetermined sizes (6, 8, 9) or can be concentrated by the process of tangential flow filtration prior to pelleting by centrifugation or entrapment on physical filters. Particle-associated microbes, such as those found in soils and sediments are more difficult to process for metagenomic library construction and sequencing than their planktonic relatives.

Soils and sediments are complex matrices, subject to both spatial and temporal heterogeneity in addition to possessing complex chemical and biological characteristics (19–21). While sample collection itself is generally an uncomplicated process, the removal of microbial cells from these matrices prior to DNA extraction can pose a significant challenge. In most studies, either direct (i.e., in situ release of DNA from microbial cells) or indirect (i.e., removal of microbial cells from the matrix prior to the release of DNA) extraction techniques have been utilized to investigate microbial community structure and function through metagenomics (19, 22). Each of these approaches can result in a biased microbial community, which can complicate the interpretation of metagenomic information.

2.2. DNA Extraction and Metagenomic Library Construction

Nucleic acids (either DNA or RNA) are extracted from the collected samples either immediately or after short- or long-term cold storage. The analysis of total microbial RNA is more often referred to as "metatranscriptomics," the use of which allows researchers to investigate the expressed complement of genes within a microbial community (23–26). For the sake of the current discussion (i.e., metagenomics), we focus on the extraction of DNA only. It is generally accomplished by lysing microbial cells using various detergents and enzymes (e.g., SDS, lysozyme, and proteinase K), followed by phenol/chloroform purification and alcohol precipitation of DNA, either manually or through the use of prefabricated kits. For samples collected from aquatic environments, DNA extraction is routinely performed directly on membrane or cartridge filters containing concentrated cells (6, 8, 9), as well as cells pelleted through ultracentrifugation. The extraction of DNA from soil and sediment microbes is more difficult than for aquatic microbes due to the wider diversity of cell types (e.g., enrichment of Gram positives and spore formers) and the existence of recalcitrant compounds such as humic acids that can co-purify with DNA and interfere with downstream molecular applications. Therefore, more aggressive cell lysis and DNA purification techniques are often applied to these sample types such as bead-beating and removal of humic acids through treatment with hexadecylmethylammonium bromide (CTAB) (19, 27, 28).

The approach adopted for metagenomic library construction is largely driven by the research hypotheses and goals coupled with the selected sequencing platform (see the following section for additional details on sequencing platforms). Small- and large-insert shotgun metagenomic libraries, representing total community DNA, have been primarily coupled with dideoxy (Sanger) sequencing and have formed the foundation of modern metagenomic studies (6, 8, 9, 29). Small-insert library construction requires the random shearing of purified DNA into small pieces, generally less than 10 kb in length. Fragmentation of DNA can be accomplished through processes such as enzymatic digestion, nebulization, or through the use of specialized shearing instruments produced by HydroShear® and Covaris, Inc. The fragmented DNA is then cloned into a vector and transformed into permissive cells, such as *Escherichia coli*, where the community DNA is replicated prior to plasmid extraction and sequencing. Vectors that are used in large-insert library construction [such as fosmids, cosmids, and bacterial artificial chromosomes (BAC)] can accommodate long, contiguous pieces of DNA as large as 100 kb and beyond. Large-insert vectors are generally maintained at a low-copy number during the cloning process in order to maximize the stability of the inserts and can be induced to high-copy number under specific conditions to increase the yield of target DNA prior to sequencing.

Recently, many researchers have adopted next-generation sequencing techniques for the metagenomic analysis of microbial communities (14, 30). Pyrosequencing, also referred to as 454 sequencing, is a clone-independent technology that has been utilized with increasing frequency since 2005 (31). The preparation of libraries for pyrosequencing is relatively simplistic compared to the preparation of traditional small- and large-insert libraries that are destined for Sanger sequencing, due to the absence of cloning. Similar to small-insert libraries, community DNA is first sheared into small pieces (~300–800 bp). If sufficient quantities of DNA for pyrosequencing (3–5 μg) are available from the microbial community, no primary amplification steps are necessary. Alternatively, the amount of genomic DNA can be augmented through the processes of whole genome amplification (17) or linker-adapted amplification (18) prior to subsequent library construction steps. For fragment libraries, short adaptor sequences (specific for the 3′ and 5′ ends of DNA) are ligated onto the sheared pieces of DNA and are utilized for purification, amplification through emulsion PCR, and sequencing (32). Paired-end libraries can also be produced by (1) circularizing longer spans of genomic DNA (up to 20 kb) with one set of adapter sequences; (2) shearing the circularized DNA; and (3) ligating 3′ and 5′-specific short adaptor sequences to sheared pieces containing the initial adaptor sequences (33).

Table 1
Sequencing statistics for Sanger and next-generation sequencing platforms

Sequencing method	Platform	Required starting material (F/M)[a]	Read length (bp)[b]	Error rate/base	Cost/run[b]	Capacity[b]
Dideoxy (Sanger)	ABI 3730	1–5 μg	~700–1,000	<1.0%[c]	$4–6 K/plate[d]	~576 Kbp/plate[d]
Pyrosequencing (454)	Roche GS FLX	5.0 μg	~200–400	1.0%[a]	$5–6 K	100 Mbp/run
Sequencing by synthesis	Illumina GA II	0.1–1.0 μg/10 μg	~35	<1.5%[b]	$4–6 K	>1 Gbp/run
Ligase mediated	ABI SOLiD system	10 ng–5 μg/ 5 μg–20 μg	~35	0.2%[b]	$6–8 K	>1 Gbp/run

[a] According to the manufacturer's website with the exception of Sanger sequencing. F/M, fragment/mate pair
[b] According to Fox et al. (104) with the exception of Sanger sequencing
[c] According to Blazej et al. (105)
[d] Plate = 384 wells

2.3. Sequencing Technology

There are currently four options when selecting a sequencing platform for metagenomic studies including dideoxy sequencing (Sanger), pyrosequencing (454 – Roche), SOLiD™ (Applied Biosystems), and Illumina® (formerly known as Solexa). Each of these technologies posses pros and cons with respect to metagenomic sequencing performance including overall cost, read length, error rates, and total capacity (Table 1). Sanger sequencing, the only option available when the first metagenomic study was undertaken and published in 2002 (34), still affords the longest read lengths of all available sequencing technologies (35). However, the shear volume of data, increasing read lengths and cost advantage afforded by pyrosequencing has resulted in a sharp decline in Sanger-based metagenomic projects in the past several years.

The 454 (Roche) platform utilizes emulsion PCR to generate millions of copies of DNA fragments from a single unique ssDNA fragment that was initially immobilized on an individual bead. The amplified fragments then undergo massive parallel sequencing in a picotiter plate, producing ~1.2 million total sequences (36). While initial reads lengths were quite short (100–150 bp), recent advances in sequencing chemistries have resulted in reads that are an average of ~450 bp in length. Greater than 100 metagenomic publications currently cite the use of pyrosequencing technology (37) and its capabilities continue to evolve with further increases in read-length, large-insert library, and paired-end sequencing capabilities and improvements to sample barcoding. The utility of the SOLiD™ (*S*equencing by *O*ligo *L*igation and *D*etection) and the Illumina® Genome Analyzer sequencing platforms for metagenomic investigations is still unclear (see Mardis, 2008 for a review of these technologies (38)). Both technologies produce an incredible amount of data in a single run (Table 1), but at the expense of very short read lengths that are largely uninformative for the analysis of metagenomic data. However, improvements are occurring at a rapid pace and it may not be long before the benefits outweigh the detriments and these platforms become the staples of metagenomic sequencing.

3. Bioinformatics Approaches and Challenges

Sequence data generated from a metagenomic sample originate from a mixture of microbes, which, depending on the diversity of the environment being sampled, can have varying genome sizes, % GC content, and codon usage patterns, may belong to taxonomically distinct regions of the tree of life and can have different functional and metabolic capabilities. The taxonomic identity of a metagenomic sequence is typically not known a priori. The various computational steps that are useful in interpreting a metagenomic dataset include assembly, gene finding, annotation, and taxonomic binning. Analysis and interpretation of metagenomic data is also impacted by the choice of sequencing technology that is used. While the newer sequencing platforms produce much larger volumes of data, they also have higher error rates.

3.1. Assembly

Due to the "cosmopolitan" nature of metagenomic samples, recovering complete or near-complete genomes of the constituent organisms via sequence assembly can be a challenge. Factors such as the diversity of the community, sequence coverage of individual genomes and strain or sub-ribotype variation influence the assembly quality of a metagenomic sample. Assemblies of even moderately diverse communities have been reported to be very fragmented, with average contig sizes of only a few thousand base pairs in length and with a large fraction of the reads being unassembled (6). Assemblers used thus far are the same as those used to assemble isolate genomes, with some parameter modifications (39, 40). Metagenomic sequence assembly is nevertheless an active research area. Alternate assembly approaches include strategies that start with a genomic feature/region of interest and greedily extend the existing contig or that use lower read overlap stringencies to build contigs that reflect consensus of groups of related strains rather than individual genomes (6).

3.2. Gene Finding

A consequence of fragmented assemblies is that a large fraction of the gene sequences predicted in these data are fragmentary. Gene finding can be done either on assemblies or on individual sequence reads. The latter is possible because of the high-gene-coding density in prokaryotes. Several methods for identification of protein coding genes in metagenomic data have been proposed. Approaches based on heuristically derived models for gene finding have been applied to detect bacterial and archaeal genes in fragmentary sequence data. These methods use the strong correlation that has been noted between the GC content of an organism and its codon usage frequencies (41, 42). Models of codon and dicodon frequencies are heuristically derived from available complete genomes and incorporated in the gene finding process (43, 44). Evidence-based approaches (using sequence similarity obtained by performing searches against databases of known protein sequences (9, 12, 45)) have also been used. These approaches, however, impose limits on the protein families that can be explored, since truly novel protein families will not be detected. In addition, proteins that are divergent members in known families may also be overlooked, even though there could be other proteins in the metagenomic data that could transitively link them to the known members of the family. Sequence similarity based methods for identification of such families have been proposed as well (46, 47).

3.3. Annotation

Gene finding is followed by functional annotation, a process by which biological information is attached to the predicted protein coding genes. Examples of biological information include biochemical activity, biological process, expression patterns, or regulation and interactions with other genes. All of this tagged information is essential for establishing functional relationships for gene interactions and is fundamental not only for the understanding of the functional organization at the cellular level, but also for establishing the relationship of the cell to its given environment. The functional annotation process typically involves a series of searches of the predicted (often partial) protein sequences against curated and well defined datasets, followed by a set of rules to combine the search results (48, 49). Comparison of functional annotations across metagenomic datasets is not only desirable, but also necessary in order to fully understand the functional capacity of the microbial community under study. Therefore, it is important for investigators to utilize a structured system of controlled vocabularies (known as ontologies) when assigning descriptions to putative gene products.

3.4. Taxonomic Binning

The assignment of taxonomy to sequences is an important component of metagenomic analysis, and allows researchers to address the question of "who is there"? Both phylogenetic and nonphylogenetic methods have been proposed to solve the taxonomic binning problem. Phylogenetic methods (50, 51) typically proceed by searching the query gene (on a metagenomic read sequence) against a reference gene database to identify other members of the gene family and subsequently construct a phylogenetic tree with these sequences. Taxonomic assignment of the query sequence is then inferred based on the clade that it falls within the tree. Horizontal gene transfer events can complicate efforts to assign taxonomy to metagenomics sequences when genes that actually co-occur within a single microbial genome possess different molecular phylogenies (52). This issue can be more difficult to address for metagenomic projects than genome projects since the data consists of a mixture of sequences from a microbial community rather than a single genome. Therefore, strategies that are used to address this problem in genomics, such as the use of consensus trees or comparison of conserved ribosomal trees to trees built from less conserved genes are not as useful with metagenomic data (52). Nonphylogenetic methods refer to those where a phylogeny is not explicitly constructed. Both (semi-) supervised clustering (53) and machine learning (54, 55) approaches are examples of the later. For these methods, sequences are encoded using their k-mer compositions as features. In the clustering based approach, the k-mer signatures are used to compute distances between sequences and these are then used to cluster the sequences into groups. Cluster groups are annotated based on the presence of informative marker genes. In the machine learning based approach, classifier model parameters are "learned" from training sets of sequences of known taxonomic origin. These models are subsequently used to predict taxonomies of metagenomic sequences.

3.5. Fragment Recruitment

The availability of reference genomes from the same environment as a metagenomic sample can be very valuable as these genomes can provide context to the metagenomic data being analyzed. Recruiting metagenomic read sequences to these reference genomes (using a search program like BLAST (56)) has been shown to be very useful (6). Recruitment has been used to identify regions within a genome that are distributed in unusual ways that may be indicative of their role in adapting organisms to particular environments. This technique has also been used to analyze the genome structure of the reference strain relative to its environmental counterparts, and to measure the distribution of an organism, or genes within an organism, across samples.

3.6. Sample Comparison

Once various features (taxonomic and/or functional) of metagenomic samples have been determined, it is possible to use that information to compare samples. Each sample can be represented as a vector. For taxonomic and functional comparisons, the vector dimensions usually, respectively, represent the different categories in a particular taxonomic rank and the protein families or pathways. The value associated with each dimension can be either raw count or normalized data. This encoding of samples allows for the computation of distance measures that serve as input to ordination and clustering methods (57). Methods for direct comparison of metagenomic samples based on overlap of their read sequences have also been proposed (6).

3.7. Metadata

An important aspect of a metagenomics project is the collection of metadata. These consist of various measurements of the environment where the metagenomic sample was collected. For aquatic environments, examples include temperature, pH, salinity, and various nutrient measurements. An important goal of metagenomic investigations is to understand the interactions of the various microbial players in the community with each other and with their environment. It is thus important to correlate the various metadata with the sequence-based observations. Statistical approaches to address this problem include regression of distance matrices (58, 59) and canonical correlation analysis (60).

3.8. Reference Genomes and Simulated Datasets

For many of the computational problems we have discussed here, the availability of relevant reference data sets (genomes or genes) from an environment is very important, and can greatly improve the quality of inference. In addition, simulated data sets play an important role in evaluating the performance of analysis methods (39, 61). Simulations have been especially useful in testing methods on different microbial population distributions and also under different sequence lengths and error rates.

4. Environmental Metagenomics

Metagenomic interrogation of microbial communities has become an increasingly popular method of analysis because of the level of resolution it can provide and the ever-decreasing cost of sequencing. The type of questions that can be addressed regarding microbial community structure and function largely depends on the complexity of the population present. The predicted complexity of an environment will also influence the depth of sequencing and the types of tools that can be used for data analysis. The following section provides some examples of how metagenomic techniques were applied to microbial communities of low, medium, and high complexity and their outcomes. Application of metagenomics to the study of the human microbiome is also included.

4.1. Low Complexity

Metagenomic investigations of low-complexity microbial communities can provide an extraordinary amount of insight into community structure, function, adaptation, and evolution. The study conducted by Tyson and colleagues (62) on a subterranean acid mine drainage (AMD) biofilm (Iron Mountain, CA) is perhaps the best example of applying metagenomic tools to a low-complexity microbial community. AMD is a microbe-mediated process that produces acid as result of sulfide oxidation (62, 63). Initial analysis of the community structure via 16S rRNA sequencing indicated the presence of only six bacterial and archaeal lineages (three each), dominated by bacteria belonging to *Leptospirillum* group II. Shotgun Sanger sequencing and assembly of ~76 Mbp of DNA resulted in the assembly of near-complete genomes of two microbes belonging to *Leptospirillum* group II and *Ferroplasma* type II (Archaea) and partial assembly of bacteria and archaea belonging to *Leptospirillum* group III, *Ferroplasma* type I and G-plasma. Binning of scaffolds by G + C content and sequence coverage was used to separate the assembled genomes and to assign taxonomy based on the presence of 16S rRNA genes.

The assembly of near-complete genomes allowed the authors of this study to perform an in-depth examination of the extent and source of genetic variability of the microbes comprising the dominant populations in the biofilm, a very challenging feat almost impossible for more complex microbial assemblages. The absence of nucleotide polymorphisms present in the 16S and 23S rRNA genes and intergenic region of the *Leptospirillum* group II composite genome suggested that a single strain of this organism has been sampled. A similar lack of polymorphism was detected in the rRNA genes of the assembled *Ferroplasma* type II genome. However, the overall nucleotide polymorphism pattern suggested that the composite genome was the result of homologous recombination between several strains of *Ferroplasma* type II.

Perhaps the most significant aspect of this study in terms of ecology stemmed from the robust metabolic reconstruction of the dominant assembled genomes. The presence of genes involved in the reductive acetyl-CoA pathway in the composite *Leptospirillum* group II genome and the abundance of sugar and amino acid transporters in the composite *Ferroplasma* type II genome suggested that these organisms rely on chemoautotrophic and heterotrophic lifestyles, respectively. Nitrogen fixation was linked to *Leptospirillum* group III, a relatively rare, but important member of the community and genes involved in iron oxidation were linked to both *Leptospirillum* group II and *Ferroplasma* type II. Lastly, important clues as to how biofilms are established and maintained by these microbes, and how these have adapted to the extreme conditions that characterize AMD systems were retrieved from the assembly of the dominant genomes. This included the presence of genes involved in motility, desiccation prevention and floatation of biofilms, metal resistance, and cytoplasm pH stability.

One of the goals of metagenomic sequencing is to unravel the metabolic potential of a microbial community, with the hope that enough information will be gained to make educated decisions about how to culture specific microbes. The metabolic information regarding nitrogen fixation that was extracted from this metagenomic project was subsequently used to isolate *Leptospirillum ferrodiazotrophum* sp. nov., the first representative of *Leptospirillum* group III (64). This achievement was facilitated by the restrained species diversity present within the AMD biofilm and will be much more difficult to accomplish in more complex microbial systems.

4.2. Medium Complexity

Marine microbial communities are considered taxonomically diverse according to both single gene surveys (65) and metagenomic investigations (6, 8, 9). The complexity of microbal populations in marine ecosystems is driven by a variety of environmental factors and can vary both in space and time. However, even the simplest marine microbial community is expected to be more complex than those recovered from extreme environments like AMD. According to the Genomes OnLine Database (GOLD) (66), at least 10 major marine microbial metagenomic studies have been published (6–9, 11, 26, 30, 65, 67–69) and there are an additional 40 projects with publicly available data.

Three landmark studies in marine microbial metagenomics were conducted between 2003 and 2004 including the Sargasso Sea pilot study, the *Sorcerer II* Global Ocean Sampling (GOS) Expedition, and the HOT/ALOHA study with results published in the following years (6, 7, 9, 11, 45). The Sargasso Sea pilot study, conducted by Venter and colleagues (45), established the groundwork for the GOS circumnavigation, a massive undertaking geared toward characterizing the diversity of marine microbes inhabiting the surface waters of the world's oceans (6, 7). The Sargasso Sea was chosen as the initial sampling location due to its long oceanographic sampling history and because, at the time, it was believed that the diversity of microbes inhabiting this marine ecosystem was constrained due to severe nutrient limitation (45). The results from this study were surprising to most marine microbial ecologists, particularly with respect to taxonomic and functional diversity. The analysis of ~1 Mbp of Sanger sequence data revealed at least 1,800 microbial "species" (148 previously unknown phylotypes) and 1.2 million new genes (45). Perhaps one of the most exciting findings was the discovery of 782 new genes coding for proteorhodopsin, a protein that functions as a light-driven proton pump (70). Proteorhodopsin-containing bacteria in the marine environment may be able to use phototrophy as a mechanism to produce energy to support metabolism and cellular growth in nutrient poor, oligotrophic waters (71).

The *Sorcerer II* GOS Expedition was a significant expansion of the Sargasso Sea pilot study that demonstrated the relevance of shotgun sequencing to address questions surrounding marine microbial ecology. Undertaken by the J. Craig Venter Institute, it sampled surface waters around the globe at ~200 mile intervals (6). A subset of 44 samples collected during the circumnavigation constituted "Phase I" of the Expedition. DNA from 0.1 to 0.8 μm size fraction of these samples was sequenced and analyzed producing 7.7 million Sanger reads and a total of 5.9 Gbp of non-redundant data (6). This study revealed an incredible level of microbial diversity, primarily at the intra-ribotype level. While only a handful of dominant microbial ribotypes could be inferred through metagenomic assembly, intra-ribotype diversity manifested itself through the presence of hypervariable genomic islands, orthologous sequence variation, and total gene content (6). These observations, coupled with discovery of "spectral tuning" of proteorhodpsins highlighted how metagenomic sequencing of microbial communities can reveal important taxonomic and functional information of ecological significance.

The first phase of the GOS Expedition also provided a unique opportunity to examine protein family expansion and novelty (7) since the Phase I GOS dataset nearly doubled the number of predicted proteins within public databases at the time of analysis (7). Protein sequences from NCBI nr, NCBI Prokaryotic Genomes, TIGR Gene Indices, and Ensemble were clustered together into families of evolutionarily related sequences using a novel clustering technique in order to determine the influence of increasing protein prediction on the rate of protein family discovery (7). Analysis of protein clusters indicated that the rate of protein discovery was near linear, suggesting that discovery will continue to increase with the addition of new sequences. Furthermore, a significant portion of protein clusters (containing at least 20 members; 43%) demonstrated no homology to known proteins and hence was unique to the GOS dataset. While this observation points to how deep metagenomic sequencing can stimulate more questions than answers, the mapping of 6,000 previously unknown publicly available ORFans to the GOS data simultaneously validated their predictions.

The metagenomic study of microbial communities inhabiting seven different depths of the North Pacific Subtropical Gyre conducted by DeLong and colleagues (9, 11) provided an intriguing contrast to the GOS Expedition study. First, the library construction approaches adopted by these two studies were fundamentally different. Small-insert (<2 kb) shotgun libraries were produced from GOS microbial DNA (6) while large-insert (40–50 kb) fosmid libraries were produced as part of the Hawaii Ocean Time-series (HOT)/ALOHA study (9). In addition, while GOS samples were collected from surface samples around the globe,

HOT samples were collected in one location, but from multiple depths. These differences in sampling strategies not only influenced the questions that could be addressed, but also the conclusions that were ultimately drawn.

Metagenomic analysis of microbial community DNAs (~64 Mbp of Sanger sequence) collected from the surface of the ocean down to the seafloor revealed varying patterns of taxonomic and functional diversity as well as metabolic plasticity (9). While the authors observed the "expected" microbial taxa at various depths in the water column, depth-dependent patterns of gene content and metabolic pathways were more informative of microbial community processes. The functional complement of the surface ocean and upper water column differed dramatically from deeper waters and included predicted proteins involved in light-dependent processes (e.g., photosynthesis and proteorhodopsins) as well as some involved in bacteriophage defense, nutrient transport, motility, and chemotaxis. Deeper waters were characterized by an increase in proteins that potentially facilitate surface attachment such as pilus components, polysaccharide, and antibiotic synthesis as well as marked differences in carbon and nitrogen utilization. The contrast in the gene repertoire between shallow and deep water microbial communities demonstrates the utility of metagenomic sequencing in revealing the adaptive nature of marine microbial communities.

4.3. High Complexity

Soil microbial communities are believed to be the most complex on Earth. The estimated global abundance of soil microbes is staggering (4–5 $\times$ 10^{30} total) (72) with an estimated tens of thousands of genotypes per gram of soil (73, 74). This diversity is believed to be so large that intensive metagenomic sequencing efforts would be required in order to adequately cover the taxonomic and functional breathe of indigenous microbes. Despite the collective acknowledgment that metagenomic investigations of soil are necessary in order to more fully understand the contribution of microbes to global processes such as terrestrial carbon cycling and agricultural sustainability (75), soil metagenomic studies are poorly represented compared to those from other environments. Recently, Vogel and colleagues (75) proposed a massive community-wide effort to deeply sequence a "reference" soil metagenome, otherwise known as the TerraGenome. The goal of the TerraGenome project is to engage researchers from around the world in an effort to completely sequence the soil metagenome from Park Grass, a long-running (150 years) field experiment at the Rothamsted agricultural institute in the United Kingdom (75). This soil metagenome will serve as "reference" or comparative guide for other soil investigations including additional metagenomic, -transcriptomic, and -proteomic studies. This proposal has generally been met with enthusiasm, although it has not been devoid of criticism (72, 76). The ultimate success of the project will depend on several factors including funding, the robustness of cell and DNA extraction methods (19), sequencing technology, and data interpretation.

So far, due to the richness of soil microbiota genotypes, most culture-independent investigations have focused on single gene surveys, primarily 16S rRNA genes (73, 77–80). This type of analysis provides insight into the taxonomic diversity of soil microbes, but not into the functional information that whole genome shotgun sequencing can provide. Rondon and colleagues conducted one of the first metagenomic studies aimed at linking phylogenetic diversity and function of soil microbes collected from the West Madison Agricultural Research Station (81). Large-insert libraries (BAC) were constructed and screened for biological activity and pooled metagenomic DNA (extracted from BAC clones) was used as a template for 16S rRNA amplification. While the full taxonomic and functional potential of the microbial community was not explored through random high-throughput sequencing of clones, the authors of this study were able to select specific clones that displayed interesting enzymatic properties to explore further.

Tringe and colleagues adopted a less selective approach to the metagenomic analysis of an agricultural soil collected from Minnesota (82). Despite the production of 1-Mbp of data, little genomic assembly was accomplished (<1%), confirming the complexity of the community as determined by a corresponding 16S rRNA survey. Functional analyses at the gene (ortholog), operon, and cellular process (KEGG) levels demonstrated enrichments in genes involved in the degradation of plant matter, operons involved in potassium channeling and processes such as plasmid-mediated horizontal gene transfer and antibiotic biosynthesis (82). This more generalized approach represents the minority of soil metagenomic studies. By far, the majority of published reports to date have undertaken a targeted approach to sequencing and analysis that usually involves functional screening (83). Since soil microbes are potentially rich sources of bioactive molecules involved in processes such as antibiotic production and resistance, degradation of xenobiotic compounds, and biofuel production (83, 84), many of the metagenomic investigations of soil have focused on the detection of specific microbial functions (84–89).

4.4. Human Microbiome

The numbers of microbial cells that thrive both on and within the human body are believed to outnumber human cells by an order of magnitude (90). Together, these microbial populations are known as the "microbiome," influencing processes from metabolic efficiency to immunity (90). The goal of the Human Microbiome Project (HMP), an international initiative established in 2007, is to use metagenomic techniques to characterize the distribution and evolution of the human microbiome (90), with the hope of significant scientific advancements going from a better understanding and prevention of disease (through the identification of diagnostic biomarkers) to the discovery of biotechnologically relevant enzymes (90). The HMP is designed as a three-tiered project revolving around the generation of sufficient microbial "reference" genomes and the production of 16S rRNA and metagenomic sequence data from a "moderate number of samples" (90, 91).

Several research groups began the phylogenetic/taxonomic characterization of the human microbiota through 16S rRNA analyses (92–95). More recently, 16S-based and whole genome metagenomic analyses were coupled in an effort to better understand the functional capabilities of the human microbiome (12, 15, 96). As an example, Turnbaugh and colleagues reported on the core gut microbiome in obese and lean twins (15). The authors examined the fecal microbial communities of adult female monozygotic and dizygotic twins and their mothers (154 total subjects) using 16S rRNA and whole community DNA pyrosequencing. Approximately 1.9 million full and partial length 16S rRNA sequences and 2 Gbp of metagenomic data were produced and notable results emerged including (1) a shared familial gut microbiome; (2) a direct relationship between obesity and bacterial diversity; and (3) a "core" microbiome at the gene family (rather than phylotype) level. Despite the overlap in gut microbial communities between family members, each individual's gut microbiome demonstrated variation in the specific bacterial "species" or phylotypes present. Furthermore, the diversity of the microbial community associated with obese individuals was dramatically reduced when compared with lean individuals with the main differences reflected in the abundances of Bacteroidetes (reduced in obese subjects) and Actinobacteria (enriched in lean subjects). Functional analysis of the metagenomic data revealed the presence of "core" gene families encoded by the various gut microbial communities that did not correlate with taxonomic diversity. The pathways represented by the "core" gene families included those involved in carbohydrate and amino acid metabolism, amino-sugar metabolism and *N*-glycan degradation, all processes that are relevant to the functionality of the gut.

Most recently, an international group of researchers were the first to use metagenomic sequencing exclusively to examine the genomic potential of microbial gut flora (97). Qin and colleagues used the Illumina sequencing technology to deeply sequence total DNA extracted from fecal samples collected from 124 European adults. Samples were collected from healthy and obese adults, as well as individuals afflicted with inflammatory bowel disease (IBD). A total of 576.7 Gb of metagenomic sequence data was produced and 3.3 million unique genes were identified. Almost all of the identified genes were attributed to bacteria (~99%), with an estimated 1,000–1,150 total bacterial species represented among the collection of samples. Individuals shared a large number of microbial "species," with a minimum of 160 "species" per person. This "species core" concept conflicts with earlier 16S rRNA-based studies that found substantial diversity at the phylotype level (15), and the authors of the Illumina study suggest that this discrepancy is likely due to undersampling of the community. Notably, despite the "species core" tangible differences were detected in the relative abundance of bacterial "species" in healthy individuals versus those with IBD. Qin et al. also found that a significant fraction of the microbial genes could be mapped to those detected in other human gut microbiome studies as well as 89 gut reference bacterial genomes suggesting that a common gene pool exists among gut microbes. Further analysis revealed that a functional core, or minimal human gut metagenome, was present among the individuals sampled. The core included genes that are likely required for healthy gut function and were involved in processes such as degradation of complex polysaccharides as well as synthesis of amino acids, vitamins, and fatty acids. Collectively, these observations will constitute a model of how metagenomic techniques can be exploited to better understand microbial community interactions, adaptation, and evolution in the human microbiome.

5. Perspectives

Metagenomics is still evolving but has already brought a revolution in the field of microbial ecology in a variety of environments. While cultivation barriers have been side-stepped through the utilization of molecular markers and metagenomic techniques, cultivation efforts are still necessary in order to get better insight into microbial physiology and to gain more reference genome sequences. The advent of single cell genomics has facilitated the production of reference genomes as well as the comparative genomic analysis of microbes (98, 99) and comes as a complement to metagenomic analyses. The production of metatranscriptomic and proteomic data from environmental microbes will also build upon the base of knowledge gained through metagenomics and will provide a dieper understanding of how gene expression and protein synthesis respond to changing environmental conditions.

The capabilities and challenges associated with metagenomics will continue to change and grow as next-generation sequencing platforms are fully adopted and the newest sequencing technologies are put into practice. For example, Single Molecule Real Time DNA sequencing technology (developed by Pacific Biosciences) promises high-quality kilobase-length reads over short run times

(100). This technology would obviously instigate major changes to current metagenomic techniques, as for example, sequence assembly and offer new possibilities to produce large datasets for a fraction of the cost of traditional Sanger sequencing. However, caution will still be necessary during the analysis of these datasets due to the possible production of artifacts generated by intrinsic sequencing error or biases in amplification processes (101, 102). Perhaps the most significant challenge we face in the immediate future is the overwhelming amount of data that is currently being and will be produced in the near future due to the development and implementation of these next-generation sequencing technologies (103). This huge volume of data not only presents substantial computing challenges, but also calls for the development of novel analysis tools to efficiently process the data. The future of metagenomics is therefore only limited by our ability to stay one step ahead of the data deluge.

Acknowledgments

This work was supported by the Office of Science (BER), U.S. Department of Energy, Cooperative Agreement No. De-FC02-02ER63453.

References

1. Handelsman J. (2004) Metagenomics: application of genomics to uncultured microorganisms. Microbiol Mol Biol Rev, 68 (4):669–685.
2. Whitman WB, Coleman DC, Wiebe WJ. (1998) Prokaryotes: the unseen majority. Proc Natl Acad Sci U S A, 95 (12):6578–6583.
3. Sanger F, Nicklen S, Coulson AR. (1977) DNA sequencing with chain-terminating inhibitors. Proc Natl Acad Sci U S A, 74 (12):5463–5467.
4. Margulies M, et al. (2005) Genome sequencing in microfabricated high-density picolitre reactors. Nature, 437:376–380.
5. White RA, 3rd, et al. (2009) Digital PCR provides sensitive and absolute calibration for high throughput sequencing. BMC Genomics, 10:116.
6. Rusch DB, et al. (2007) The Sorcerer II Global Ocean Sampling expedition: northwest Atlantic through eastern tropical Pacific. PLoS Biol, 5(3):e77.
7. Yooseph S, et al. (2007) The Sorcerer II Global Ocean Sampling expedition: expanding the universe of protein families. PLoS Biol, 5(3):e16.
8. Venter JC, et al. (2004) Environmental genome shotgun sequencing of the Sargasso Sea. Science, 304(5667):66–74.
9. DeLong EF, et al. (2006) Community genomics among stratified microbial assemblages in the ocean's interior. Science, 311 (5760):496–503.
10. Beja O, et al. (2000) Bacterial rhodopsin: evidence for a new type of phototrophy in the sea. Science, 289(5486):1902–1906.
11. Konstantinidis KT, et al. (2009) Comparative metagenomic analysis of a microbial community residing at a depth of 4,000 meters at station ALOHA in the North Pacific subtropical gyre. Appl Environ Microbiol, 75 (16):5345–5355.
12. Gill SR, et al. (2006) Metagenomic analysis of the human distal gut microbiome. Science, 312(5778):1355–1359.

13. Palmer C, et al. (2007) Development of the human infant intestinal microbiota. PLoS Biol, 5(7):e177.
14. Turnbaugh PJ, Gordon JI. (2009) The core gut microbiome, energy balance and obesity. J Physiol, 587(Pt 17):4153–4158.
15. Turnbaugh PJ, et al. (2009) A core gut microbiome in obese and lean twins. Nature, 457(7228):480–484.
16. Hugenholtz P, Tyson GW. (2008) Microbiology: metagenomics. Nature, 455 (7212):481–483.
17. Binga EK, Lasken RS, Neufeld JD. (2008) Something from (almost) nothing: the impact of multiple displacement amplification on microbial ecology. ISME J, 2 (3):233–241.
18. Pfannkoch C, et al. (2010) Hydroxyapatite-mediated separation of dsDNA, ssDNA and RNA genotypes from natural viral assemblages and metagenomic library construction. Appl Environ Microbiol, 76 (15):5039–5045.
19. Williamson KE, et al. Optimizing the extraction of prokaryotes from soils for high-throughput metagenomic sequencing. Appl Environ Microbiol, Submitted.
20. Dell'Anno A, Bompadre S, Danovaro R. (2002) Quantification, base composition, and fate of extracellular DNA in marine sediments. Limnol Oceanogr, 47(3):899–905.
21. Williamson KE, Radosevich M, Wommack KE. (2005) Abundance and diversity of viruses in six Delaware soils. Appl Environ Microbiol, 71(6):3119–3125.
22. Robe P, et al. (2003) Extraction of DNA from soil. Eur J Soil Biol, 39:183–190.
23. Warnecke F, Hess M. (2009) A perspective: metatranscriptomics as a tool for the discovery of novel biocatalysts. J Biotechnol, 142 (1):91–95.
24. Shi Y, Tyson GW, DeLong EF. (2009) Metatranscriptomics reveals unique microbial small RNAs in the ocean's water column. Nature, 459(7244):266–269.
25. Poretsky RS, et al. (2009) Analyzing gene expression from marine microbial communities using environmental transcriptomics. J Vis Exp, (24).
26. Gilbert JA, et al. (2008) Detection of large numbers of novel sequences in the metatranscriptomes of complex marine microbial communities. PLoS ONE, 3(8):e3042.
27. Fortin N, et al. (2004) Soil washing improves the recovery of total community DNA from polluted and high organic content sediments. J Microbiol Methods, 56(2):181–191.
28. Carrigg C, et al. (2007) DNA extraction method affects microbial community profiles from soils and sediment. Appl Microbiol Biotechnol, 77(4):955–964.
29. Tyson GW, et al. (2004) Community structure and metabolism through reconstruction of microbial genomes from the environment. Nature, 428(6978):37–43.
30. Dinsdale EA, et al. (2008) Functional metagenomic profiling of nine biomes. Nature, 455(7214):830.
31. Poinar HN, et al. (2006) Metagenomics to paleogenomics: large-scale sequencing of mammoth DNA. Science, 311 (5759):392–394.
32. Products and solutions: System workflow. (2009) [http://www.454.com/products-solutions/how-it-works/index.asp].
33. Products and solutions: Multi span paired end reads. (2009) [http://www.454.com/products-solutions/experimental-design-options/multi-span-paired-end-reads.asp].
34. Breitbart M, et al. (2002) Genomic analysis of uncultured marine viral communities. Proc Natl Acad Sci U S A, 99(22):14250–14255.
35. Wommack KE, Bhavsar J, Ravel J. (2008) Metagenomics: read length matters. Appl Environ Microbiol, 74(5):1453–1463.
36. Blow N. (2008) Metagenomics: exploring unseen communities. Nature, 453 (7195):687–690.
37. Publications and resources. (2009) [http://www.454.com/publications-and-resources].
38. Mardis ER. (2008) The impact of next-generation sequencing technology on genetics. Trends Genet, 24(3):133–141.
39. Mavromatis K, et al. (2007) Use of simulated data sets to evaluate the fidelity of metagenomic processing methods. Nat Methods, 4 (6):495–500.
40. Raes J, Foerstner KU, Bork P. (2007) Get the most out of your metagenome: computational analysis of environmental sequence data. Curr Opin Microbiol, 10(5):490–498.
41. Besemer J, Borodovsky M. (1999) Heuristic approach to deriving models for gene finding. Nucleic Acids Res, 27 (19):3911–3920.
42. Karlin S. (2001) Detecting anomalous gene clusters and pathogenicity islands in diverse bacterial genomes. Trends Microbiol, 9 (7):335–343.
43. Noguchi H, Park J, Takagi T. (2006) MetaGene: prokaryotic gene finding from environmental genome shotgun sequences. Nucleic Acids Res, 34(19):5623–5630.

44. Hoff KJ, et al. (2008) Gene prediction in metagenomic fragments: a large scale machine learning approach. BMC Bioinformatics, 9:217.
45. Venter J, et al. (2004) Environmental genome shotgun sequencing of the Sargasso Sea. Science, 304(5667):66–74.
46. Yooseph S, Li W, Sutton G. (2008) Gene identification and protein classification in microbial metagenomic sequence data via incremental clustering. BMC Bioinformatics, 9:182.
47. Krause L, et al. (2006) Finding novel genes in bacterial communities isolated from the environment. Bioinformatics, 22(14): e281–e289.
48. Markowitz VM, et al. (2008) IMG/M: a data management and analysis system for metagenomes. Nucleic Acids Res, 36(Database issue):D534–D538.
49. Overbeek R, et al. (2005) The subsystems approach to genome annotation and its use in the project to annotate 1000 genomes. Nucleic Acids Res, 33(17):5691–5702.
50. Bowler C, et al. (2008) The Phaeodactylum genome reveals the evolutionary history of diatom genomes. Nature, 456 (7219):239–244.
51. Krause L, et al. (2008) Phylogenetic classification of short environmental DNA fragments. Nucleic Acids Res, 36(7):2230–2239.
52. Swithers KS, Gogarten JP, Fournier GP. (2009) Trees in the web of life. J Biol, 8 (6):54.
53. Chatterji S, et al. (2008) CompostBin: A DNA composition-based algorithm for binning environmental shotgun reads. Res Comput Mol Biol, 17–28.
54. McHardy AC, et al. (2007) Accurate phylogenetic classification of variable-length DNA fragments. Nat Methods, 4(1):63–72.
55. Brady A, Salzberg SL. (2009) Phymm and PhymmBL: metagenomic phylogenetic classification with interpolated Markov models. Nat Methods, 6(9):673–676.
56. Altschul S, et al. (1990) Basic local alignment search tool. J Mol Biol, 215(3):403–410.
57. Legendre P, Legendre L. (1998) *Numerical Ecology.* 2nd ed. Elsevier, Amsterdam.
58. McArdle B, Anderson M. (2001) Fitting multivariate models to community data: a comment on distance-based redundancy analysis. Ecology, 82(1):290–297.
59. Zapala MA, Schork NJ. Multivariate regression analysis of distance matrices for testing associations between gene expression patterns and related variables. Proc Natl Acad Sci U S A, 103(51):19430–19435.
60. Gianoulis TA, et al. (2009) Quantifying environmental adaptation of metabolic pathways in metagenomics. Proc Natl Acad Sci U S A, 106(5):1374–1379.
61. Richter DC, et al. (2008) MetaSim: a sequencing simulator for genomics and metagenomics. PLoS One, 3(10):e3373.
62. Tyson G, et al. (2004) Community structure and metabolism through reconstruction of microbial genomes from the environment. Nature, 428(6978):37–43.
63. Applications, C.o.M.C.a.F. (2007) *The New Science of Metagenomics: Revealing the Secrets of Our Microbial Planet.* National Research Council.
64. Tyson GW, et al. (2005) Genome-directed isolation of the key nitrogen fixer *Leptospirillum ferrodiazotrophum* sp. nov. from an acidophilic microbial community. Appl Environ Microbiol, 71(10):6319–6324.
65. Sogin ML, et al. (2006) Microbial diversity in the deep sea and the underexplored "rare biosphere". Proc Natl Acad Sci U S A, 103 (32):12115–12120.
66. Genomes OnLine Database: Metagenome Projects. (2009) [http://www.genomesonline.org/gold.cgi?want=Metagenomes].
67. Martin-Cuadrado AB, et al. (2007) Metagenomics of the deep Mediterranean, a warm bathypelagic habitat. PLoS ONE, 2(9):e914.
68. Dinsdale EA, et al. (2008) Microbial ecology of four coral atolls in the Northern Line Islands. PLoS ONE, 3(2):e1584.
69. Frias-Lopez J, et al. (2008) Microbial community gene expression in ocean surface waters. Proc Natl Acad Sci U S A, 105 (10):3805–3810.
70. Beja O, et al. (2000) Bacterial rhodopsin: evidence for a new type of phototrophy in the sea. Science, 289(5486):1902–1906.
71. Gomez-Consarnau L, et al. (2007) Light stimulates growth of proteorhodopsin-containing marine Flavobacteria. Nature, 445 (7124):210–213.
72. Singh BK, et al. (2009) Soil genomics. Nat Rev Micro, 7(10):756.
73. Roesch LF, et al. (2007) Pyrosequencing enumerates and contrasts soil microbial diversity. ISME J, 1(4):283–290.
74. Gans J, Wolinsky M, Dunbar J. (2005) Computational improvements reveal great bacterial diversity and high metal toxicity in soil. Science, 309(5739):1387–1390.

75. Vogel TM, et al. (2009) TerraGenome: a consortium for the sequencing of a soil metagenome. Nat Rev Micro, 7(4):252.
76. Baveye PC. (2009) To sequence or not to sequence the whole-soil metagenome? Nat Rev Micro, 7(10):756.
77. Fierer N, Jackson RB. (2006) The diversity and biogeography of soil bacterial communities. Proc Natl Acad Sci U S A, 103 (3):626–631.
78. Schloss PD, Handelsman J. (2006) Toward a census of bacteria in soil. PLoS Comput Biol, 2(7):e92.
79. Fierer N, et al. (2007) Metagenomic and small-subunit rRNA analyses reveal the genetic diversity of bacteria, archaea, fungi, and viruses in soil. Appl Environ Microbiol, 73(21):7059–7066.
80. Kielak A, et al. (2009) Phylogenetic diversity of Acidobacteria in a former agricultural soil. ISME J, 3(3):378–382.
81. Rondon MR, et al. (2000) Cloning the soil metagenome: a strategy for accessing the genetic and functional diversity of uncultured microorganisms. Appl Environ Microbiol, 66 (6):2541–2547.
82. Tringe S, et al. (2005) Comparative metagenomics of microbial communities. Science, 308(5721):554–557.
83. van Elsas JD, et al. (2008) The metagenomics of disease-suppressive soils – experiences from the METACONTROL project. Trends Biotechnol, 26(11):591–601.
84. Pang H, et al. (2009) Identification of cellulase genes from the metagenomes of compost soils and functional characterization of one novel endoglucanase. Curr Microbiol, 58 (4):404–408.
85. D'Costa VM, et al. (2006) Sampling the antibiotic resistome. Science, 311 (5759):374–377.
86. D'Costa VM, Griffiths E, Wright GD. (2007) Expanding the soil antibiotic resistome: exploring environmental diversity. Curr Opin Microbiol, 10(5):481–489.
87. Zhang K, et al. (2009) Identifying natural product biosynthetic genes from a soil metagenome by using T7 phage selection. Chembiochem, 10(16):2599–2606.
88. Morimoto S, Fujii T. (2009) A new approach to retrieve full lengths of functional genes from soil by PCR-DGGE and metagenome walking. Appl Microbiol Biotechnol, 83 (2):389–396.
89. Chen IC, et al. (2009) Isolation and characterization of a novel lysine racemase from a soil metagenomic library. Appl Environ Microbiol, 75(15):5161–5166.
90. Turnbaugh PJ, et al. (2007) The human microbiome project. Nature, 449 (7164):804–810.
91. Gordon JI. (2005) Extending our view of self: the Human Gut Microbiome Initiative (HGMI). [http://www/genome.gov/Pages/Research/Sequencing/SeqProposals/HGMISeq.pdf].
92. Gao Z, et al. (2007) Molecular analysis of human forearm superficial skin bacterial biota. Proc Natl Acad Sci U S A, 104 (8):2927–2932.
93. Eckburg PB, et al. (2005) Diversity of the human intestinal microbial flora. Science, 308(5728):1635–1638.
94. Hyman RW, et al. (2005) Microbes on the human vaginal epithelium. Proc Natl Acad Sci U S A, 102(22):7952–7957.
95. Pei Z, et al. (2004) Bacterial biota in the human distal esophagus. Proc Natl Acad Sci U S A, 101(12):4250–4255.
96. Kurokawa K, et al. (2007) Comparative metagenomics revealed commonly enriched gene sets in human gut microbiomes. DNA Res, 14(4):169–181.
97. Qin J, et al. (2010) A human gut microbial gene catalogue established by metagenomic sequencing. Nature, 464:59–67.
98. Woyke T, et al. (2009) Assembling the marine metagenome, one cell at a time. PLoS ONE, 4(4):e5299.
99. Lasken RS. (2007) Single-cell genomic sequencing using multiple displacement amplification. Curr Opin Microbiol, 10 (5):510–516.
100. Eid J, et al. (2009) Real-time DNA sequencing from single polymerase molecules. Science, 323(5910):133–138.
101. Gomez-Alvarez V, Teal TK, Schmidt TM. (2009)Systematic artifacts in metagenomes from complex microbial communities. ISME J, 3(11):1314–1317.
102. Kunin V, et al. (2010) Wrinkles in the rare biosphere: pyrosequencing errors lead to artificial inflation of diversity estimates. Environ Microbiol, 12(1):118–123.

103. Editorial. (2009) Metagenomics vs Moore's Law. Nat Methods, 6(9):623.

104. Fox S, Filichkin S, Mockler TC. (2009) Applications of Ultra-High-Throughput Sequencing. In *Methods Molecular Biology*, Humana Press, Clifton, NJ, 79–108.

105. Blazej RG, Kumaresan P, Mathies RA. (2006) Microfabricated bioprocessor for integrated nanoliter-scale Sanger DNA sequencing. Proc Natl Acad Sci U S A, 103 (19):7240–7245.

Chapter 4

Prokaryote Genome Fluidity: Toward a System Approach of the Mobilome

Ariane Toussaint and Mick Chandler

Abstract

The importance of horizontal/lateral gene transfer (LGT) in shaping the genomes of prokaryotic organisms has been recognized in recent years as a result of analysis of the increasing number of available genome sequences. LGT is largely due to the transfer and recombination activities of mobile genetic elements (MGEs). Bacterial and archaeal genomes are mosaics of vertically and horizontally transmitted DNA segments. This generates reticulate relationships between members of the prokaryotic world that are better represented by networks than by "classical" phylogenetic trees. In this review we summarize the nature and activities of MGEs, and the problems that presently limit their analysis on a large scale. We propose routes to improve their annotation in the flow of genomic and metagenomic sequences that currently exist and those that become available. We describe network analysis of evolutionary relationships among some MGE categories and sketch out possible developments of this type of approach to get more insight into the role of the mobilome in bacterial adaptation and evolution.

Key words: Mobile genetic elements, Intercellular gene transfer, Recombination, Reticulate classification, Prokaryotes adaptation and ev0olution

1. Introduction

Gene transfer between bacterial strains and species has been recognized for over 50 years. More recently, however, the true extent of lateral gene transfer (LGT) among prokaryotes has become evident through extensive comparative genomics stimulated by the accelerating number of available genomic and metagenomic sequences in the public databases. Prokaryotic genomes are highly mosaic, composed of DNA segments acquired both by vertical descent and by lateral transfer. Recognition of this has given rise to the concept of the species pan-genome: the complete set of genes present in all strains of a given species (1). This is generally much larger than the core-genome shared by all strains of a species.

Jacques van Helden et al. (eds.), *Bacterial Molecular Networks: Methods and Protocols*, Methods in Molecular Biology, vol. 804, DOI 10.1007/978-1-61779-361-5_4, © Springer Science+Business Media, LLC 2012

Prokaryotes are clearly set apart from most eukaryotes which are insulated from the promiscuous exchanges involved in LGT firstly by the constraints of the nucleus and secondly, in "higher" eukaryotes, by their multicellular organization and the specialization between the germ line and somatic cells. Extensive LGT is now recognized to be central to prokaryote evolution. This seriously challenges the notion of species and the traditional view of the evolutionary relationship between organisms described in terms of phylogenetic trees (2). Gene transfer and reorganization are responsible for the appearance of new strains with additional accessory functions leading to the emergence of new pathogenic traits and resistance to antibacterial drugs (3) or novel metabolic pathways, which permit growth on novel substances and facilitate bioremediation (e.g., refs. 4, 5).

Because microorganisms, and in particular prokaryotes, exert a crucial influence over all aspects of life, it is fundamental to understand genome dynamics in prokaryotic populations both quantitatively and qualitatively within the multitude of ecosystems they inhabit. This will allow a full appreciation of the adaptive behavior of prokaryotes in response to environmental changes over the short and long term and of the dynamics of prokaryotic populations in these various ecosystems. The agents of LGT, for example plasmids, phages, and conjugative transposons or integrative conjugative elements (ICEs; reviewed in ref. 6), are central to the mechanisms of DNA acquisition, establishment, and maintenance in new hosts, and to their impact on the host cell cycle, regulatory circuitry, and fitness.

In principle, quantification of laterally acquired DNA in genomes is possible *in silico* based on analyzes such as phylogenetic incongruence and nucleotide composition (GC skew, GC%, tri- and tetranucleotide composition, specific short nucleotide "words," etc.; see ref. 7 for a comparative approach, ref. 8 for a more recent method). However, such approaches remain predictive and reflect changes over ill defined time scales. But most importantly, they make no clear distinction between the genetic agents responsible for transfer and the result of their activities. Hence they provide little insight into the processes involved, the detailed genomic history of a particular organism, or the population and environmental context within which exchanges occur.

Through the years, genetic, biochemical, and structural analyses have elucidated many of the molecular mechanisms responsible for gene transfer and recombination and, to a lesser extent, their regulation. In addition to transfer itself, we also know much about the mechanisms of DNA establishment, maintenance, and, for some, the domestication of the transferred material in the new host. Some of the ways in which cells are in turn armed to avoid intrusion of new DNA (e.g., DNA restriction/modification systems, genomic sequence divergence (9), antiaddiction (10), or the recently discovered "immunity" endowed by CRISPR systems (11–13)) and the way in which the transferred entities have developed mechanisms to counter these defenses (e.g., restriction alleviation mechanisms (14)) have also been identified. A number of quantitative data are available from laboratory experiments but very few have been acquired in natural environments. Our knowledge is clearly not yet sufficient to permit a systemic approach to the complexity of LGT. For this, it is necessary to address the nature and function of the newly acquired DNA, the regulatory circuitry involved, the signals that trigger exchange, the way cells respond to those signals to promote donor/recipient interactions, how recipient cells manage the intrusion of DNA and accommodate the resulting changes in fitness, how incoming DNA is targeted to particular genomic sites or regions.

Martinez and coworkers (15) have discussed many of the gaps in our knowledge that remain to be filled before we can attempt a systemic approach of even a particular case of LGT: the emergence of antibiotic resistance in bacterial populations. Predicting such events would facilitate the establishment of new therapeutic strategies to avoid or at least minimize resistance acquisition from the environment. Similarly, strategies to improve bioremediation would benefit from a better understanding of the mechanisms of acquisition, organization, and expression of degradation pathways for toxic compounds.

In the present review, we shall attempt to extend this perspective to the agents of genetic transfer in general. We shall first briefly review the mechanisms of LGT and the nature of the agents of transfer. Indeed these processes and features are crucial to permit a comprehensive and robust annotation of transferable genomic sequences, a prerequisite to any type of large scale analysis aimed at understanding the dynamics and impact of LGT on prokaryotic genomes and populations for short-term adaptation and long-term evolution.

We shall then describe the few cases where networks have been used to unravel some aspects of LGT. We shall suggest a schema to represent LGT as a set of interacting networks and summarize some of the many questions that remain to be addressed before such a complex network can start to be implemented.

2. Mobile Genetic Elements (MGEs) and the Mobilome

The agents of LGT are the MGEs. These are defined in the SO sequence ontology (http://www.sequenceontology.org/cgi-bin/miso.cgi, SO:0001037) as "A nucleotide region of varying length with either intragenome, intergenome or intercellular mobility, and which often carries the information necessary for transfer and for recombination with the host genome." The definition can be amended to better fit the prokaryotic cell structure by replacing intragenomic by intracellular mobility.

Various intercellular MGEs have been identified over the years including plasmids, bacteriophages (and archaeal viruses), and ICEs. Intracellular MGEs promote movement and reassortment of DNA within a genome. They include insertion sequences (IS), transposons (Tns), and integron cassettes, many of which can sequester extraneous genes and, when combined with the intercellular MGEs, can take part in extensive gene transfer from the genome of one individual to another (refs. 6, 16 and figures therein). MGEs and the DNA they mobilize (see Table 1 for a more detailed description of the major MGE classes) constitute what is also called the "horizontal gene pool" or mobilome (17, 18). All MGEs can be viewed as assemblies of functional modules, each of which contains one or more genes involved in the assembly of multiprotein complexes with one or more enzymatic activities and responsible for a particular biological process. All complete/active MGEs share the processes of establishment, maintenance, and domestication. Intercellular MGEs are in addition capable of intercellular transfer.

2.1. The Processes of DNA Transfer

Three major biological processes ensure LGT: transformation, conjugative transfer and viral infection and the release of viral particles. All were discovered more than half a century ago.

2.1.1. Transformation

Natural transformation (19), the direct assimilation of exogeneous DNA by the cell, is observed in many Gram^{+} and Gram^{-} bacterial species (see e.g., refs. 20–22). The protein complex responsible for DNA assimilation, sometimes called the transformasome, belongs to type IV secretion systems (T4SS), which translocate DNA and protein substrates across prokaryotic cell envelopes (23, 24). Although not MGE-dependent, transformation can, at least in theory, transfer MGEs as part of a transforming DNA segment. These MGEs may then establish in the new host genome.

A new perspective on the possible prevalent role of transformation in LGT in nature has been provided by the recent appraisal of the immense number of phages in the biosphere, in particular in the oceans, and of the effect of their massive infection and lysis of bacterial blooms (25). Lysis would release DNA susceptible to assimilation by neighboring cells in amounts that far exceeds that produced by DNA secretion through a T4SS by some cells (23) or even by natural cell lysis.

2.1.2. Conjugation

Conjugative plasmids and ICEs are intercellular MGEs that transfer between cells by conjugation. In Gram– bacteria this process involves donor–recipient cell contacts through a T4SS (thus related to the transformation protein complex), which here is associated

Table 1
A modular description of some well-defined MGE classes

Category	Length	Characteristic features related to mobility	Baggage
Extrachromosomal			
Plasmid	100 bp- to several kb	Replication, partition, maintenance, site-specific recombinase (TBR, SBR)	Antibiotic resistance, catabolic genes, pathogenicity determinants, others
Conjugative plasmid	Several kb	Same as plasmids, conjugation machinery	Same as plasmids
Phage	5–600 kb	Replication (DDE transposase in transposable phages) lysis and capsid genes, sometimes TBR, SBR, protelomerase	Pathogenicity determinants, photosynthesis genes
Integrated			
Prophage		Same as phage, *att* TBR or SBR at one end, often adjacent *tRNA* gene, often flanking direct repeat	Same as phage
IS	700–3,500 bp	DDE transposase, *IR* and short flanking direct repeat or other "transposase" and cognate end sequences	None
Unitary Transposon	3.5-several kb	DDE transposase, *IR*, SBR, TBR	Resistance
Composite Transposon	5 to several kb	Flanking pair of identical IS in same or opposite orientation, flanking short direct repeat	Antibiotic resistance, catabolic genes, metal resistance
Integrated Conjugative Element (ICE or CTn)	Up to several kb	Site-specific recombinase, *att* conjugation machinery often adjacent *tRNA* gene	Antibiotic resistance, catabolic genes
Pathogenicity Island (PAI)	Up to several kb	Site specific recombinase, *att*, often adjacent *tRNA* gene	Pathogenicity determinants
Gene cassettes	One gene	*attC*, in arrays, proximity of site-specific TBR	Antibiotic resistance, catabolic genes, others
Gene transfer elements	Around 6 kb	Lysis genes, capsid genes	

IR: inverted repeat; include the sites for transposase binding and induced cleavage. *att*: attachment site; site for site-specific recombinase (TBR, SBR see text for details) binding and induced recombination. On transposable phages, the DDE transposase binding sites are usually called *att* as well.

IS transposases belong to various families of enzymes, DDE, Y, Ser and Y2 transposases (93).

Most presently known unitary transposons belong to the Tn3 and Tn7-related families. The first typically encode a DDE transposase and a site-specific SBR or TBR, which catalyze resolution of the cointegrate produced by replicative transposition ((48) and references therein). The second also code for a DDE transposase and a set of accessory proteins that direct transposition to one or more specific or random sites. Tn*7* encodes a restriction-type enzyme, TnsA, which cleaves the second strand of the transposon, resulting in conservative transposition (see also Figure 2). Other transposons in the family encode an SBR instead, and thus likely transpose by the replicative mode as do Tn*7* mutants lacking TnsA activity (94).

(continued)

Table 1 (continued)

Composite transposons include identical IS sequences at their ends, in the same or opposite orientations, which flank the passenger genes. Paradigm elements of that class are *Tn9* (*IS1*, resistance to chloramphenicol), *Tn5* (IS*50*, resistance to kanamycin) and Tn*10* (IS*10*, resistance to tetracycline).

Integration and excision of most ICEs, pathogenicity islands (PAIs) and phages are catalyzed by TBRs, less frequently by SBRs, the same enzymes that catalyze resolution of cointegrates or regenerate plasmid monomers from multimeric forms (maintenance).

Not all ICEs conjugative machineries are related to those commonly found on conjugative plasmids (95) and vice-versa, not all ICEs conjugation machines have been found on plasmids.

PAIs belong to the large class of genomic islands; they are characterized by the presence of pathogenicity determinants. Some are mobile (95) but many need complementation by a complete element to move. Many reported PAIs are ICEs. Others transfer by lysis and infection through complementation by a phage [see for instance (72)].

Gene cassettes are the mobile part of integrons (96). These contain a fixed platform, consisting of a gene coding for aTBR, its cognate binding site *attI* and a promoter for gene cassette expression. The TBR catalyses integration and excision of mobile cassettes composed usually of a single gene and the complementary attachment site, *attC*. Unlike other TBRs those that lead to cassette integration require a single stranded substrate (97).

Although most temperate phages use TBR or SBR to integrate and excise, transposable phages encode a DDE transposase, which ensures integration but also replication (by replicative transposition) of the phage genome.

The term 'capsid' is here meant to include the head and tail of the tailed phages or *Caudovirales*, i.e. *Podo*, *Myo* and *Syphoviridae* (see ICTV classification at http://www.ncbi.nlm.nih.gov/ICTVdb/) as well as the capsids of all other classes of phages.

Gene transfer agents are produced by various bacterial species (98). They are virus-like elements that seem to function only as gene exchange mediators by transduction through packaging host rather than their own DNA.

Note that besides the genes listed here, most MGEs contain a more or less important set of genes with unknown function including "orfans", i.e. genes that have no similarity with other genes in nucleotide/protein sequence databases.

by means of a coupling protein to a helicase/endonuclease (the relaxase) that initiates DNA transfer at the origin of transfer, *oriT* (26). In Gram+ bacteria, the protein components involved are not as well characterized although some are related to the T4SS (27). Some of these transfer machines are efficient in liquid culture while others only function efficiently on solid surfaces. This presumably reflects the types of environment where donor/recipient interactions take place in nature (e.g., fluids or biofilms (28)).

2.1.3. Infection and Lysis

Bacteriophage and archaeal virus are intercellular MGEs that move between cells by infection and lysis. Many phages package host genes instead of their own within the viral capsids at non-trivial frequencies and, upon subsequent infection, these genes are transferred to a new host by the process of transduction (ref. 29 but see also ref. 30).

2.2. The Processes of DNA Establishment in the New Host

Once in a new host, provided it overcomes the cell's defenses against invasion of genetic materials (e.g., restriction and other nucleases and CRISPRs) an intercellular MGE may proceed to establish itself in, and become part of, the new host genome.

Transfer may have occurred using a single stranded intermediate (e.g., most plasmids and ICE DNA entering by conjugation and transforming DNA, which upon entry is converted to single stranded DNA by a membrane bound nuclease) which is converted to its double-strand DNA form before establishing as either an extrachromosomal unit or plasmid, or as an integrated entity in the form of a prophage, an ICE or another genomic island.

2.2.1. Establishing as an Autonomous Replicon

Establishment as a circular or linear plasmid requires that the DNA molecule regains its circular or linear form and that proper control of gene expression limits the burden imposed on the host cell. Regenerating a circular form is often catalyzed by the relaxase that initiated the transfer process (see ref. 31 for a recent review on relaxases). Very few conjugative linear plasmids are known and their mechanism of transfer remains to be characterized. Non conjugative linear plasmids could enter a cell via transformation. Some of these have covalently closed ends (telomeres), which are processed by an enzyme called protelomerase (32). Others, such as recently characterized *Salmonella typhi* linear plasmid (33) or certain linear Streptomyces replicons (e.g., refs. 34, 35) have covalently capped terminal inverted repeats (tirs).

2.2.2. Establishing by Integration

Integration in the host genome (chromosome or plasmid) occurs using one of a limited number of recombination mechanisms catalyzed by MGE and/or host encoded enzymes and accessory proteins (reviewed in ref. 36). Very importantly, these processes are similar to those used by intracellular MGEs to move within and rearrange genomes.

Homologous Recombination

If incoming DNA has the appropriate length and sufficient identity with resident sequences, it can "integrate" by homologous recombination. Unless the entering DNA segment contains insertions or deletions compared to the resident DNA, the process will result in sequence replacement. Transforming DNA normally establishes by homologous recombination and hence should essentially occur in closely related hosts.

All living cells code for homologous recombination enzymes. They fulfill the crucial repair of double stranded (DS) breaks but can also generate genomic rearrangement by acting on multiple copies of identical repeats including multiple copies of the same MGE, especially ISs (see description in Table 1). These serve as "portable" regions of homology and, depending on their relative orientation, recombine to generate a deletion (same orientation) or an inversion (opposite orientation, Fig. 1). Many MGEs encode part or all of their own homologous recombination enzymes, some of which are less constrained by sequence identity (37), broadening the range of integrable sequences and the impact of homologous recombination on genetic change.

Site-Specific Recombination

Site-specific serine-based recombinase (SBR) and tyrosine-based recombinase (TBR) catalyze reciprocal recombination between short specific attachment sites *att*, which have little sequence identity. TBRs and SBRs, which both form a covalent intermediate with their target DNA (for details see ref. 36), appear as interchangeable functional modules on MGEs. TBRs catalyze cassette gain and loss in integrons. Temperate phages and ICEs can use TBR or SBR to integrate into and excise from their host chromosome. Plasmids may encode many TBRs, whose function remains unknown (38) but some of which resolve plasmid multimers as do SBRs (39). In Tns such as Tn*3* that fuse donor and target replicons as part of their transposition pathway (see below), resolution of the resulting cointegrates into the final transposition products can be catalyzed by either SBRs or TBRs.

Transpositional Recombination

Transposition moves a DNA segment from one location to another in a genome, via an RNA or DNA intermediate. Although bacteria carry introns that transpose via an RNA intermediate (40) transposition via DNA intermediates is generally the rule. This is often catalyzed by DDE transposases (so named because of a highly conserved triad of amino acids, Asp, Asp Glu, which coordinate essential Mg^{2+} ions), which specifically recognize and catalyze cleavage of a single phosphodiester bond at each end of the element, generating a single 3′OH which is then often used as a nucleophile in cleavage of the opposite strand (reviewed in ref. 36). As illustrated in Fig. 2, these ends then launch a nucleophilic attack on the target DNA, generating various DNA intermediates, which are processed with or without replication of the element. A common scar left by all these reactions is a direct repeat (DR) of a few base pairs that flank the newly transposed DNA segment.

In the special case where only a single strand is cleaved at each end, transfer to another target molecule and subsequent replication of the Tn results in a replicon fusion, to generate a cointegrate (Fig. 1). Transposition within one replicon generates a deletion or an inversion of a DNA segment adjacent to the transposing element. These outcomes were predicted by a model proposed by Shapiro (41) and first demonstrated for transposable phage Mu (for a review see ref. 42). These transposition-driven rearrangements generate products which are indistinguishable from those described earlier as result of homologous recombination between two identical MGE copies. It is therefore not straightforward to discriminate between the two routes toward a given rearrangement. The nature of the enzyme encoded by the MGE and the scars left by the event(s) may provide a hint into the molecular mechanisms involved (see below).

More recently a second major class of ISs, the IS200/IS605 family, has been identified. These transpose using a single strand

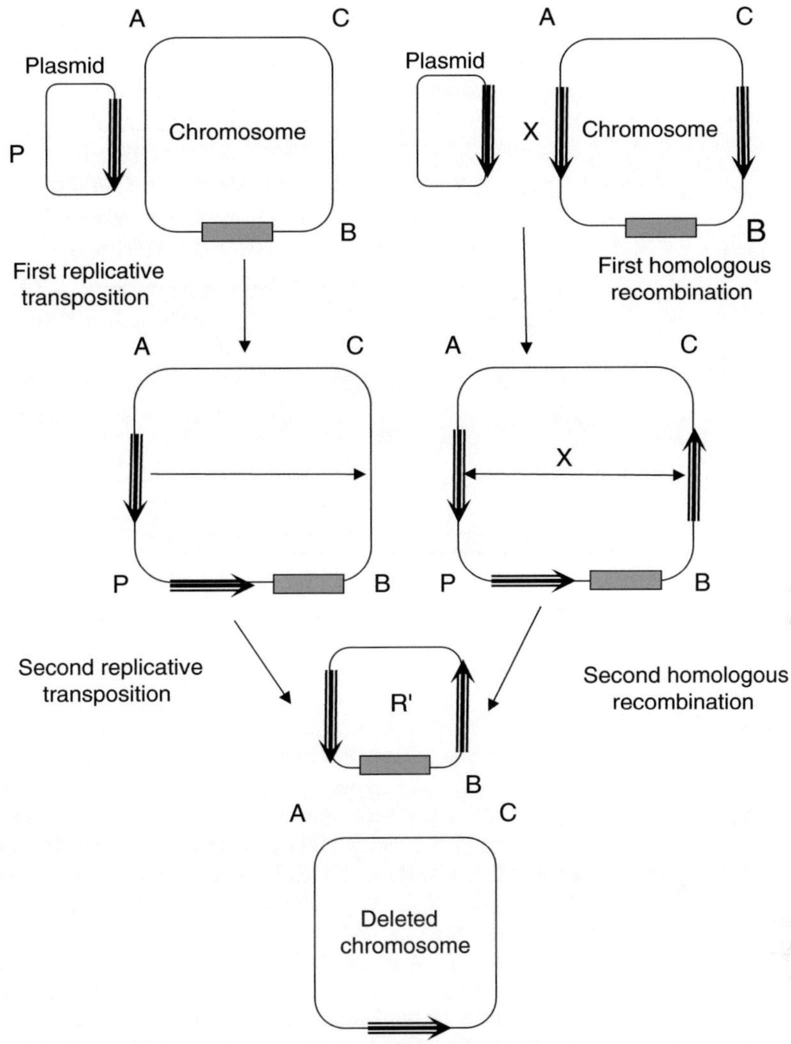

Fig. 1. MGE as reorganizers of genomes. The schematized genome contains one plasmid and one chromosome. On the left side, only the plasmid bears a single MGE coding for a DDE transposase (*large arrow*), which upon a first replicative transposition event into the chromosome generates a cointegrate. The plasmid DNA P is integrated between the chromosome markers A and B and flanked by the two copies of the MGE in the same orientation. One copy of the MGE launches a second replicative transposition on a target site on the same chromosome (*thin arrow*). This can split it into two replicons, one larger plasmid that carries a segment of the chromsome (the R′ with the gray rectangle initially located on the chromosome) and two copies of the MGE, and a chromosome deleted from the segment that moved onto the plasmid and one MGE copy, remnant of the first transposition event. No sequence homology is necessary for these reactions. On the right side, the same final replicons are produced by a sequence of two homologous recombination events (X). Here, sequence homology is mandatory and provided by multiple copies of the same MGE. Three initial copies distributed and oriented as shown could recombine to produce similar R′ and deleted chromosome as above. In both pathways, the second recombination could generate inversion of a chromosomal DNA segment instead of a deletion. In the "transposition" pathway on the left, the same MGE copy is only needed to bind the cleaved target such that instead of each MGE end binding the same strand, they would bind the opposite strand at the target. On the homologous recombination pathway on the right, inversion would have occurred provided the MGE copy between B and C was in the opposite orientation (redrawn from ref. 18).

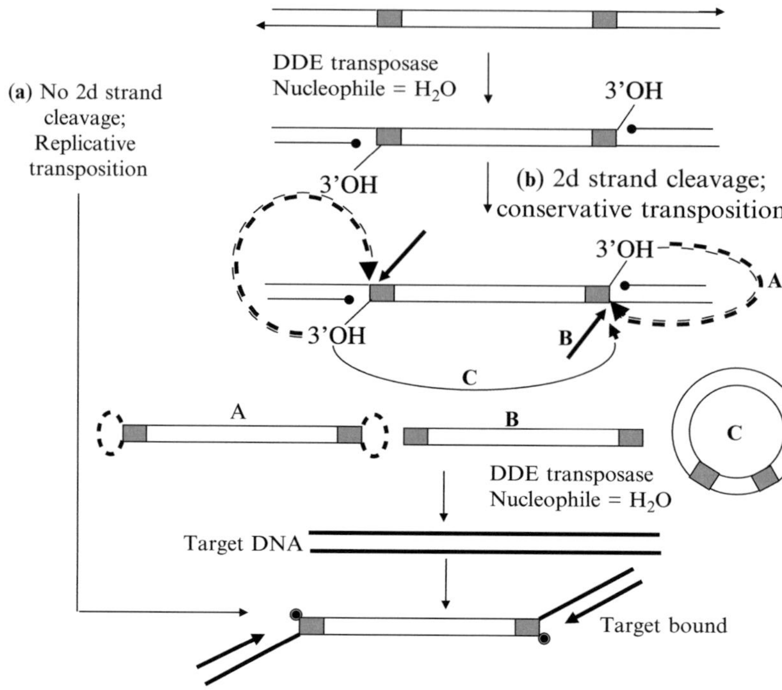

Fig. 2. Transpositional recombination catalyzed by DDE transposases. DDE transposases specifically bind the ends (*gray boxes*) of the cognate MGE. Using H_20 as the nucleophile, the enzyme introduces nicks that generate single reactive 3′OH ends. Depending on the transposase, several reaction pathways are then possible. (**a**) The reactive 3′OH ends directly nick and bind the target DNA on the two strands, a few base pairs apart (staggered cut). The resulting "Shapiro" intermediate (for details on the whole process, see ref. 41) carries 3′OH extremities on the target DNA, where replication by the host DNA polymerase initiates (*arrow* in target bound structure). The gap created by the staggered cut is filled generating a short direct repeat (DR). Replication proceeds along the MGE, which is thus duplicated so that the process is called "replicative transposition." Depending on the relative location of the initial MGE copy and the transposition target, the resulting structure will be a cointegrate (target and MGE on different replicons), a deletion or an inversion (target and MGE on the same replicon, and cleavage/ligation of the same or the opposite strand, see Fig. 1). (**b**) The second strand at the ends of the MGE is cleaved before the attack on the target. A. Some DDE enzymes, as those encoded by the paradigm transposons Tn*5* and Tn*10*, direct the 3′OH reactive ends to cleave the opposite strand of the same MGE end (A) generating a hairpin intermediate. The hairpin ends are again cleaved by the transposase generating 3′OH reactive ends, which attack the target. B. A second enzyme distinct from the transposase cleaves the second strand at the ends before the attack on the target as in transposon Tn*7*. In A and B scenarios, repair replication fills in the gap, generating the flanking DR. The MGE itself is not duplicated and the process is called conservative transposition. C. Some transposases cut the other end on the same strand, generating a Fig. 8 intermediate (not shown), which upon replication of the element by the host machinery generates a circular intermediate again cleaved by the transposase to generate the final transposition products (100, 101).

circular intermediate and target (43) and insert preferentially into the lagging strand template at the replication fork (44).

The set of transfer and recombination processes described so far may not be exhaustive (see for instance ref. 45 for a description of other modes of recombination) but is sufficient to illustrate how well bacteria are equipped to gain and loose genetic material, and the extent to which MGE activities can reassort their host genome (reviewed in ref. 18). Surprisingly, except for homologous recombination this important arsenal of recombination activities has received little attention when considering the role of LGT and the mobilome in the evolution of prokaryotes (46, 47).

2.3. Toward a Comprehensive Annotation of MGEs

2.3.1. Distinguishing Processes and Enzyme Activities

We have purposely evaded the classical terms "integrase," "resolvase," and "transposase" when considering recombinases. This is because any of these proteins can catalyze several processes as illustrated in Table 1. Members of each of the three enzyme families are extremely diverse. They exhibit different degrees of target specificity and directionality in the recombination process (48, 49). Moreover, some members are not involved in recombination (e.g., RNaseH (50)). Low target specificity, as for TBRs of some conjugative Tn (51) allows for reintegration of the element at more or less random locations, which in the case of ICEs can occur after conjugative transfer into a new host. Such events will tend to be considered as transpositions. The DDE transposase family also includes proteins with various target sequence or DNA structure specificity. In addition, retroviral integrases belong to the same DDE enzyme family (reviewed in ref. 36), which has lead to much confusion in annotation of bacterial genomes. For SBRs, there are at least two major groups: long proteins (~500 amino acids) which promote integration/excision of some phages and ICEs; and short (~200 amino acids) ones which catalyze plasmid dimer and cointegrate resolution but also inversion of specific DNA segments (48). SBRs are therefore annotated as integrases, resolvases, or invertases without any consistency.

In all these instances, confusion results from the use of a single term to describe both a process and a molecular activity. This can be easily avoided by dissociating the processes (e.g., integration, resolution, transposition) from the molecular activities that catalyze them. Any process can then be associated with one among several enzymatic activities including TBR, SBR, DDE, or activities required for homologous recombination.

The same ambiguity and solutions apply to the T4SS. This cellular component can be annotated as such but depending on whether the location of the corresponding genes is chromosomal aside from or within pathogenicity islands, or associated with an ICE or plasmid, the T4SS can be involved in one of three processes: transformation, secretion, or conjugation. In addition, some T4SS components have a molecular activity (e.g., ATPase, pilin) that should also be part of a comprehensive annotation.

Appropriate terms for annotating sequence types (plasmid, MGE, chromosome, etc., see above), cellular components, biological processes, and molecular functions are available in structured ontologies such as SO, GO ((52), http://www.geneontology.org/), or the MGE dedicated MeGO ((53), http://aclame.ulb.ac.be/

Classification/mego.html). Such structured ontologies have the additional advantage that all the terms are associated with relationships directly accessible to computational analysis.

Within this annotation frame, MGEs are thus defined as extra-(plasmid) or intrachromosomal mosaic DNA sequences containing genes coding for proteins involved in processes (conjugation, integration, etc.) controlled by cellular components (e.g., T4SS) or molecular activities (e.g., recombinases) that ensure mobility, establishment, and maintenance as well as the acquisition and loss of a variety of passenger genes.

In addition, it is important that the annotation of an integrated MGE include the borders of the element. Some properties of these borders are determined by the type of recombinase that catalyzes integration which is often encoded by the MGE itself and located close to one end. This is generally not included in the annotated prokaryotic genomes since recognition of MGE ends on a genomic sequence is a serious challenge. For instance, one large MGE may include several shorter ones (integrons are often integrated in Tns, IS sequences are often found in genomic islands or plasmids, which also often carry one or more Tns and several Tns often fit into a single large one). We now briefly examine the nature of the sequences that may be expected at the borders of integrated MGEs in relation to the nature of the enzyme involved in their integration.

2.3.2. Recognizing MGE Flanking Sequences

Site-specific recombinases recognize specific *att* sequences located on the MGE and the target replicon (chromosome or plasmid). Recombination between these sites generates DRs flanking the integrated MGE, which may (e.g., 59 base pairs elements between cassettes in integrons, ref. 54) or may not (see for instance Prophinder prophage predictions at http://aclame.ulb.ac.be/Tools/Prophinder/) be reliably identified. TBRs appear to preferentially target tRNA genes (55). Although to our knowledge this preference has never been precisely quantified, the presence of an adjacent tRNA gene is a feature used in several computational methods devoted to the prediction of genomic islands in sequenced genomes (e.g., Islander (56), IslandViewer (57)).

DDE recombinase bind more or less conserved sequences located in an inverted orientation at each end of their cognate MGE (terminal inverted repeats, IRs). In addition, many DDE enzymes generate a staggered cut at the target site which, following DNA repair or replication, generates a short DR flanking the transposed MGE copy (reviewed in ref. 36). The IR and DR sequences are easily identified for some MGEs such as ISs and Tns. In larger Tns, internal sequences similar to the IRs may be present confusing the identification of the ends, which may nevertheless be marked by the DR scar.

2.3.3. Predicting Integrated MGEs

If plasmid and phage genomes are readily accessible for sequencing and annotation, integrated MGEs, including prophages, should be recognized in the host sequences. Ends and characteristic genes described above that are physically associated on the chromosome could be used to identify integrated MGEs, but this is not straightforward due to their heterogeneity. Except for the presence of recombination and transfer proteins on complete active MGEs, the passenger genes they carry can include antibiotic resistance, pathogenicity determinants, catabolic, and other genes as illustrated in Table 1. Integrated MGEs may sometimes be defined by the presence of "empty" sites (appearing as insertions/deletions) in closely related genomes if these sequences are available for comparison. Surprisingly, this property has so far only seldom been exploited in prediction methods.

Several computational methods exist to predict particular types of MGEs such as prophages (58–60), integrons (61), and genomic islands (e.g., MobilomeFINDER (62) that incorporates the pairwise genome alignment method MAUVE (63) and experimental microarrays, MOSAIC (64) or IslandPick (57) just to mention a few). A dedicated site, ISfinder (http://www-is.biotoul.fr (65, 66)) with various analytical tools is available for identification and annotation of complete and partial ISs in prokaryotic genomes. The performance of these methods is yet to be optimized (7). Combining the detection of groups of genes with particular sequence features (67) with the presence of transfer and recombination proteins, and their cognate binding and flanking sequences are features that could contribute to improving integrated MGE prediction methods. Table 1 summarizes some criteria that could be used. It describes the main subclasses of well-defined MGEs as combinations of transfer and establishment functions and the more or less important set of passenger genes they usually carry. The nature of the passenger genes may confer an element with an identity such as phage, pathogenicity/virulence, resistance, catabolic, metabolic, or fitness island (68). However, there are no robust barriers between these classes, because the same sort of passenger genes is carried by various types of MGEs. Examples of these include: antibiotic resistance genes found on plasmids, Tns, and integrons; and virulence factors on plasmids, genomic islands, and phages. This sharing of functions is not limited to passenger genes: temperate phages establish in a cell as prophages either by integrating into the host genome or by forming a circular (69) or a linear plasmid (70); some ISs transpose by a rolling-circle replication mechanism similar to that used for intercellular transfer by plasmids and conjugative Tns and by some phages to replicate (18); and ICEs can combine similar conjugation machineries with a TBR or DDE enzymes to integrate and excise (71, 72).

2.3.4. The Importance of Defective MGEs

Besides complete, active MGEs that encode cell–cell transfer machinery and/or recombinases and maintenance proteins, it is also important to identify the many defective elements lacking part or all of these functional modules but carrying the corresponding target sites for DNA binding and cleavage. The difficulties of MGE annotation are increased when dealing with these incomplete elements, which can be as short as two active ends since in principle they might be mobilized by supplying recombination functions in trans. One striking example is that of the staphylococcal pathogenicity islands (SaPIs), which owe their mobility to their relationship with certain phages (reviewed in ref. 73). Other examples include the Miniature Transposable elements (MITEs) which inhabit many prokaryotic genomes (74). These incomplete elements are also important in understanding the evolutionary history of the host genome.

Overall, the robust annotation of MGE, especially of their ends is a main challenge for the near future. Combining information such as robust enzyme families and end reference sequence sets provided in databases dedicated to well characterized and annotated MGEs (Table 2) with the use of structured ontologies is one way forward, which could cross-feed with more effective and accurate prediction of integrated MGEs.

3. Regulation of LGT

The regulatory circuitry that controls the lysis–lysogeny switch of bacteriophage lambda inspired the founding work in modeling biological systems (e.g., ref. 75). Description of the regulation of the many processes involved in MGE activities (see Subheading 2) and LGT and of the many ways they may be integrated within the host regulatory networks would constitute an entire dedicated review. Most available data relate to a limited set of paradigm processes. These include: natural transformation in *Streptococcus pneumoniae*, *Neisseria*, or *Haemophilus*; conjugative transfer of F, RP4, Ti plasmids, and the ICE elements Tn*1545*, CtnDOT, and SXT; the transposition of several ISs and Tns; and the lysogenic switch of several bacteriophages. Many examples of MGE regulation within host regulatory circuits have been described. Quorum sensing, for instance, regulates competence (22) and Ti plasmid conjugation (76). The SOS response regulates SXT conjugation (77), gene cassette shuffling within integrons (78), and the induction of many prophages. IHF and/or other host

Table 2
Web resources for MGE genomics

ACLAME	http://aclame.ulb.ac.be/
Gene Ontology	http://www.geneontology.org
Integrated Microbial Genomes	http://img.jgi.doe.gov/cgi-bin/pub/main.cgi
IS Finder DB	http://www-is.biotoul.fr
Islander DB	http://cbs.ym.edu.tw/services/islander/islander.cgi
IslandViewer (IslandPick)	http://www.pathogenomics.sfu.ca/islandviewer/query.php
MobilomeFINDER	http://mml.sjtu.edu.cn/MobilomeFINDER/online.htm
Mobyle (integrons)	http://mobyle.pasteur.fr/cgi-bin/portal.py?form=xxr
MOSAIC	http://genome.jouy.inra.fr/mosaic/
PAIDB	http://www.gem.re.kr/paidb
Phage sequences at NCBI	http://www.ncbi.nlm.nih.gov/genomes/static/phg.html
Phage_Finder	http://phage-finder.sourceforge.net
PhiGO	http://aclame.ulb.ac.be/perl/Services/amigo_phage/go.cgi
Plasmid Genome Database	http://www.genomics.ceh.ac.uk/plasmiddb
Plasmid sequences at NCBI	http://www.ncbi.nlm.nih.gov/genomes/static/o.html
Prophage Finder	http://www.bioinfo.de/isb/2006060020/
Prophinder	http://aclame.ulb.ac.be/Prophinder
Sequence Ontology	http://www.sequenceontology.org/
T4-like Phage Genomes	http://phage.ggc.edu/

nucleoid-associated proteins, all of which are global regulators (79), are required for plasmid F transfer (80) and some aspects of transposition catalyzed by DDE enzymes (reviewed in ref. 36). As might be expected, transposition is generally maintained at a low level as a result of regulations occurring at many levels of transposase expression and activity (promoter strength, impinging of transposase transcription, sequestration of transposase translation initiation signals, programmed translational frameshifting, translation termination, transposase stability, stability of the transposase-end complex, preferential *cis*-activity, etc. ref. (81) and references therein). Regulation of MGE activities is clearly a field which is ripe for modeling, which in turn could orient new experimental avenues of investigation and highlight some general motifs/rules/routes that predominate within the complex regulatory interactions involved in MGE-host compatibility.

4. Network Perspectives of the Mobilome

Once reliable ways of identifying MGEs are available and a sufficient understanding of their mobility mechanisms and how these are integrated into host cell physiology has been acquired, we can begin ask about the evolutionary history of present genomic structures. As discussed above, MGE genomes are mosaics, which has given rise to extensive debate concerning bacteriophage classification (82, 83) and have stimulated calls for a reticulate rather than a hierarchical classification for these types of genomes (84). More generally, the contribution of LGT to genomic structure and evolution of prokaryotes has also been the source of controversy (see for instance refs. 85–87) putting into question the limits of traditional phylogenetic approaches (trees, etc.) for analyzing relationships among prokaryotic organisms. Networks and graph analyses emerge as a promising alternative to better capture the dynamics of evolutionary changes (see ref. 2 and references therein for a detailed discussion).

4.1. Toward a Reticulate Classification

A method of edge-weighted networks was proposed to generate a reticulate classification of the more easily manageable phage (88) (see also Chapter 5) and plasmid (89) genomes and to better incorporate the extensive gene exchange that occurs between these genomes. The method represents the web of relationships as graphs where nodes are phage or plasmid genomes, and edges measure phage–phage or plasmid–plasmid similarities in terms of shared genes/proteins. Various measures of graph topology can be applied such as "betweenness," a centrality measure which provides an idea of which nodes hold the network together, or "closeness" which measures the centrality of a node. These approaches bring further insight into the contribution of individual genomes to a gene/protein network (88). The reticulate classification is generated by a two-step clustering system such that each genome is associated with a membership vector, which quantitatively characterizes the membership of the genome in a given node to the set of clusters (see Fig. 3). The authors also proposed an alternative network and reticulate classification based on the sharing, by phage or plasmid genomes, of conserved evolutionary modules, which consist of families of related proteins that share the same phylogenetic profile.

A similar network approach was recently applied to a larger set of DNA and protein sequences from phages, plasmids, chromosomes, and various metagenomes (90). This analysis provides a new perspective on evolutionary changes in relation to MGEs. The networks obtained allowed the authors to propose "genetic worlds," which correspond to isolated clusters of DNA carrying

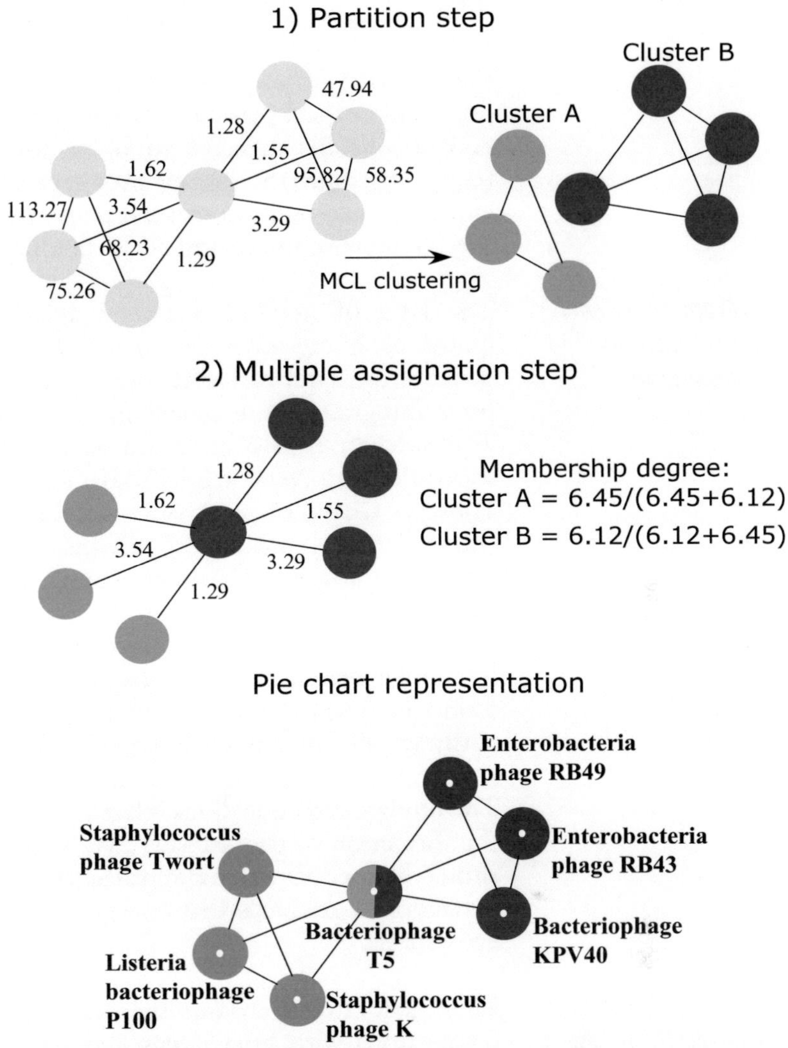

Fig. 3. A two-step clustering procedure for reticulate classification. Phage T5 is used here to illustrate the two-step procedure allowing for the display of its relationships to other phages. Relationships between pairs of phage genomes are "weighted" (numbers on edges) as a measure of the number of protein/protein families shared by these genomes (88). In the first step of partitioning, the graph is clustered. In the present example, clustering produces two clusters A and B (light and dark gray circles/ genomes). In the second "multiple assignation" step, the intercluster connections are recovered and used to calculate the membership degree, that is, the weight (statistical significance) of the intercluster connections relative to the weight of all connections. The node is colored according to the proportion of total weight of the connections to each cluster. The method can be applied to a graph with any number of nodes/genomes (see ref. 88 for more details).

entities (plasmids, phages, or chromosomes) that share DNA principally or exclusively and in which molecular evolution seems to obey a particular mode and tempo. Not unexpectedly, this cell-centered approach suggests a central role for plasmids in the spreading of DNA and contributes to improving current views of the processes and mechanisms responsible for genetic diversity and evolution of different cell lineages. The study does not, however, specifically address the role of integrated MGEs, although one would expect plasmid–chromosome and phage–chromosome connections for instance to point toward integrated plasmids (e.g., Hfr strains) ICEs and prophages, respectively. Closing the circle, the exploitation of such networks would thus appear as a way to improve integrated MGE prediction methods.

4.2. From MGE Gene/Protein Networks to Ecosystems

The type of network outlined above addresses evolutionary questions. Many other aspects of MGE activities could obviously benefit from a network approach, once well-defined sets of MGEs, both integrated and episomal, are available. Projects such as The Sorcerer II Global Ocean Sampling (http://www.sorcerer2expedition.org/version1/HTML/main.htm, Williamson and Yooseph, Chapter 3, and references therein), the Tara Expeditions (http://www.oceans.taraexpeditions.org/), or the International Census of Marine Microbes aim at a global analysis of open microbial ecosystems. These microbes are likely the source of most of the "new" genes acquired by MGEs whether detrimental to human health such as those coding for antibiotic resistance found in hospital environments or potentially useful catabolic pathways allowing microbes to degrade manmade chemicals in waste water treatment plants and other polluted environments. The fundamental questions related to the processes at the onset of the formation of new MGEs, that is, of the mobilome dynamics should be part of systemic approaches aimed toward capturing the dynamics of genomes that compose these various complex open ecosystems.

4.2.1. Connecting MGEs to Cellular and Higher Level Networks

MGE gene/functional modules and regulatory networks establish connections with host genes and regulatory networks, respectively. Such MGE networks could thus become components of a global environmental network. Additional components of such global networks would be needed to describe the flow of MGEs between bacterial cells/populations within ecosystems, population flow (cells and viruses) between ecosystems and the many effects of environmental changes (wind, water currents, temperature, pH, pressure, mineral and organic matter composition, etc.). These factors may influence MGE gene expression, transfer, and/or recombination. Similarly pollutants are expected to challenge populations and incite the spread of specific MGEs and/or their preferred hosts in order to handle the toxic compounds.

Before we can contend with these levels of complexity, simpler fundamental questions about MGEs remain to be answered. Many of these are directly amenable to network analysis. ISs, for instance, form families with representatives in sets of more or less related species of Bacteria and/or Archaea and their plasmids. Simple network analysis based on these families could uncover some of the rules that restrict the ability of a particular IS to persist or amplify in only some hosts. But for this, accurate and robust genome annotation is essential. Other simple questions relate to the similarity/identity between IS located on the chromosome(s) and plasmids of a given strain or to the prevalence of all or some MGEs in intracellular parasite versus host species that thrive in more open environments. A new software package, ISsaga (IS semiautomatic genome annotation) integrated into the ISfinder platform will be capable of performing at least some of these tasks.

5. More Challenges for the Future

We have already listed some of the challenges to be met before complex networks or modeling of MGE activities can be implemented. Aside from improved prediction and annotation of integrated MGEs and more structured data on the regulation of MGE activities, we need more information on the frequencies of lateral transfer, whether by transformation, conjugation, or infection. Although these can be easily measured in the laboratory, measuring "natural" events is more challenging. There are now indications that at least some transfer events occur in nature at frequencies similar if not higher than those measured in the laboratory (91). Studies of LGT in model biofilms might provide a step forward, not only for frequency measurements but also for the study of some regulatory aspects of transfer and for the analysis of the incidence of various types of selection (92).

Identifying MGEs in the very fragmented nucleotide sequences currently accessible in metagenomic data (see Williamson and Yooseph, Chapter 3) is another major challenge. Phages are a special case since they can be separated from the bulk of host cells prior to DNA extraction, allowing for a dedicated sequencing effort (93). This, however, does by no means eliminate phages from the bulk of the cellular DNA, since some phages are adsorbed to their host and some cells are committed to phage production as a result of either infection or induction of a prophage at the time of sampling. In addition, even for phage sequence sets, assembly is likely to generate "in silico" recombinants, difficult to distinguish from recombination occurring in vivo. Related types of "in silico" artifacts can be introduced by repeated sequences such as ISs. They can be the source of genome rearrangements (Fig. 1), and their presence can also result in assembly errors in genomic sequences. These artifacts will of course be exaggerated for metagenomic sequences. Gaining a better understanding of the mobilome within the genomic context will be one step toward the identification and characterization of MGE sequences within metagenomic datasets.

Recent work supports a paradoxical role for phages in maintaining diversity in ocean bacterial populations (30) in addition to their role in DNA transfer. This finding is for sure just the tip of the iceberg of the multitude of unexpected relationships between MGEs and complex ecological systems that can emerge from global analysis of the mobilome and its various components.

Acknowledgments

We wish to thank Hans Geiselmann for his encouragement and useful comments, Raphael Leplae and Gipsi Lima-Mendez for discussion and comments, and Laurence van Melderen and Shannon Williamson for their critical reading of the manuscript. Work on the ACLAME project in the BiGRe laboratory is financed by the European Space Agency (ESA-PRODEX) and the Belgian Science Policy (Belspo) through the EXANAM project (PRODEX agreements No. C90358), the Fonds de la Recherche Scientifique Médicale (FRSM), the Belgian Program on Interuniversity Attraction Poles, initiated by the Belgian Federal Science Policy Office, project P6/25 (BioMaGNet), the BioSapiens Network of excellence funded under the sixth Framework program of the European Communities (LSHG-CT-2003-503265) and the intramural programme of the CNRS (LMGM Toulouse).

References

1. Medini D, Donati C, Tettelin H, Masignani V, Rappuoli R. (2005) The microbial pangenome. Curr Opin Genet Dev, 15:589–594.
2. Bapteste E, O'Malley MA, Beiko RG, Ereshefsky M, Gogarten JP, Franklin-Hall L, Lapointe FJ, Dupre J, Dagan T, Boucher Y, Martin W. (2009) Prokaryotic evolution and the tree of life are two different things. Biol Direct, 4:34.
3. Baquero F. (2009) Environmental stress and evolvability in microbial systems. Clin Microbiol Infect, 15(Suppl. 1):5–10.
4. Nagata Y, Endo R, Ito M, Ohtsubo Y, Tsuda M. (2007) Aerobic degradation of lindane (gamma-hexachlorocyclohexane) in bacteria and its biochemical and molecular basis. Appl Microbiol Biotechnol, 76:741–752.
5. Wackett LP, Sadowsky MJ, Martinez B, Shapir N. (2002) Biodegradation of atrazine and related s-triazine compounds: from enzymes to field studies. Appl Microbiol Biotechnol, 58:39–45.
6. Frost LS, Leplae R, Summers AO, Toussaint A. (2005) Mobile genetic elements: the agents of open source evolution. Nat Rev Microbiol, 3:722–732.
7. Langille MG, Hsiao WW, Brinkman FS. (2008) Evaluation of genomic island predictors using a comparative genomics approach. BMC Bioinformatics, 9:329.
8. Arvey AJ, Azad RK, Raval A, Lawrence JG. (2009) Detection of genomic islands via segmental genome heterogeneity. Nucleic Acids Res, 37:5255–5266.

9. Matic I, Taddei F, Radman M. (1996) Genetic barriers among bacteria. Trends Microbiol, 4:69–72.
10. Saavedra De Bast M, Mine N, Van Melderen L. (2008) Chromosomal toxin-antitoxin systems may act as antiaddiction modules. J Bacteriol, 190:4603–4609.
11. Barrangou R, Fremaux C, Deveau H, Richards M, Boyaval P, Moineau S, Romero DA, Horvath P. (2007) CRISPR provides acquired resistance against viruses in prokaryotes. Science, 315:1709–1712.
12. Beloglazova N, Brown G, Zimmerman MD, Proudfoot M, Makarova KS, Kudritska M, Kochinyan S, Wang S, Chruszcz M, Minor W, Koonin EV, Edwards AM, Savchenko A, Yakunin AF. (2008) A novel family of sequence-specific endoribonucleases associated with the clustered regularly interspaced short palindromic repeats. J Biol Chem, 283:20361–20371.
13. Marraffini LA, Sontheimer EJ. (2010) Self versus non-self discrimination during CRISPR RNA-directed immunity. Nature, 463:568–571.
14. Nekrasov SV, Agafonova OV, Belogurova NG, Delver EP, Belogurov AA. (2007) Plasmid-encoded antirestriction protein ArdA can discriminate between type I methyltransferase and complete restriction-modification system. J Mol Biol, 365:284–297.
15. Martinez JL, Baquero F, Andersson DI. (2007) Predicting antibiotic resistance. Nat Rev Microbiol, 5:958–965.
16. Toussaint A, Merlin C. (2002) Mobile elements as a combination of functional modules. Plasmid, 47:26–35.
17. Thomas CM. (2000) *The Horizontal Gene Pool: Bacterial Plasmids and Gene Spread.* Harwood Academic Publishers, Amsterdam.
18. Merlin C, Mahillon J, Nesvera J, Toussaint A. (2000) Gene Recruiters and Transporters: The Modular Structure of Bacterial Mobile Elements. In *The Horizontal Gene Pool: Bacterial Plasmids and Gene Spread.* Edited by Thomas CM, Harwood Academic Publishers, Amsterdam, 363–409.
19. Avery OT, Macleod CM, McCarty M. (1944) Studies on the chemical nature of the substance inducing transformation of pneumococcal types: induction of transformation by a desoxyribonucleic acid fraction isolated from pneumococcus type Iii. J Exp Med, 79:137–158.
20. Kovacs AT, Smits WK, Mironczuk AM, Kuipers OP. (2009) Ubiquitous late competence genes in Bacillus species indicate the presence of functional DNA uptake machineries. Environ Microbiol, 11:1911–1922.
21. Averhoff B. (2009) Shuffling genes around in hot environments: the unique DNA transporter of *Thermus thermophilus.* FEMS Microbiol Rev, 33:611–626.
22. Claverys JP, Martin B, Polard P. (2009) The genetic transformation machinery: composition, localization, and mechanism. FEMS Microbiol Rev, 33:643–656.
23. Alvarez-Martinez CE, Christie PJ. (2009) Biological diversity of prokaryotic type IV secretion systems. Microbiol Mol Biol Rev, 73:775–808.
24. Hamilton HL, Dillard JP. (2006) Natural transformation of Neisseria gonorrhoeae: from DNA donation to homologous recombination. Mol Microbiol, 59:376–385.
25. Rohwer F, Thurber RV. (2009) Viruses manipulate the marine environment. Nature, 459:207–212.
26. de la Cruz F, Frost LS, Meyer RJ, Zechner EL. (2010) Conjugative DNA metabolism in Gram-negative bacteria. FEMS Microbiol Rev, 34:18–40.
27. Abajy MY, Kopec J, Schiwon K, Burzynski M, Doring M, Bohn C, Grohmann E. (2007) A type IV-secretion-like system is required for conjugative DNA transport of broad-host-range plasmid pIP501 in gram-positive bacteria. J Bacteriol, 189:2487–2496.
28. Ong CL, Beatson SA, McEwan AG, Schembri MA. (2009) Conjugative plasmid transfer and adhesion dynamics in an *Escherichia coli* biofilm. Appl Environ Microbiol, 75:6783–6791.
29. Weinbauer MG, Rassoulzadegan F. (2004) Are viruses driving microbial diversification and diversity? Environ Microbiol, 6:1–11.
30. Rodriguez-Valera F, Martin-Cuadrado AB, Rodriguez-Brito B, Pasic L, Thingstad TF, Rohwer F, Mira A. (2009) Explaining microbial population genomics through phage predation. Nat Rev Microbiol, 7:828–836.
31. Garcillan-Barcia MP, Francia MV, de la Cruz F. (2009) The diversity of conjugative relaxases and its application in plasmid classification. FEMS Microbiol Rev, 33:657–687.
32. Ravin NV. (2003) Mechanisms of replication and telomere resolution of the linear plasmid prophage N15. FEMS Microbiol Lett, 221:1–6.
33. Baker S, Hardy J, Sanderson KE, Quail M, Goodhead I, Kingsley RA, Parkhill J, Stocker

B, Dougan G. (2007) A novel linear plasmid mediates flagellar variation in *Salmonella typhi*. PLoS Pathog, 3:e59.

34. Bentley SD, Brown S, Murphy LD, Harris DE, Quail MA, Parkhill J, Barrell BG, McCormick JR, Santamaria RI, Losick R, Yamasaki M, Kinashi H, Chen CW, Chandra G, Jakimowicz D, Kieser HM, Kieser T, Chater KF. (2004) SCP1, a 356,023 bp linear plasmid adapted to the ecology and developmental biology of its host, *Streptomyces coelicolor* A3(2). Mol Microbiol, 51:1615–1628.
35. Huang CH, Chen CY, Tsai HH, Chen C, Lin YS, Chen CW. (2003) Linear plasmid SLP of *Streptomyces lividans* is a composite replicon. Mol Microbiol, 47:1563–1576.
36. Craig NL, Craigie R, Gellert M, Lambowitz AM. (2003) *Mobile DNA II.* ASM Press, Washington, DC.
37. Martinsohn JT, Radman M, Petit MA. (2008) The lambda red proteins promote efficient recombination between diverged sequences: implications for bacteriophage genome mosaicism. PLoS Genet, 4:e1000065.
38. Leplae R, Lima-Mendez G, Toussaint A. (2006) A first global analysis of plasmid encoded proteins in the ACLAME database. FEMS Microbiol Rev, 30:980–994.
39. Austin S, Ziese M, Sternberg N. (1981) A novel role for site-specific recombination in maintenance of bacterial replicons. Cell, 25:729–736.
40. Toro N, Jimenez-Zurdo JI, Garcia-Rodriguez FM. (2007) Bacterial group II introns: not just splicing. FEMS Microbiol Rev, 31:342–358.
41. Shapiro JA. (1979) Molecular model for the transposition and replication of bacteriophage Mu and other transposable elements. Proc Natl Acad Sci U S A, 76:1933–1937.
42. Symonds N, Toussaint A, van de Putte P, Howe MM. (1987) *Phage Mu.* Cold Spring Harbor Laboratory Press, New York.
43. Guynet C, Hickman AB, Barabas O, Dyda F, Chandler M, Ton-Hoang B. (2008) In vitro reconstitution of a single-stranded transposition mechanism of IS608. Mol Cell, 29:302–312.
44. Ton-Hoang, B., Pasternak, C., Siguier, P., Guynet, C., Hickman, AB., Dyda, F., Sommer, S., and Chandler, M. (2010) *Cell 142*(**3**), 398–408. PMID 069190
45. Ehrlich SD, Bierne H, d'Alencon E, Vilette D, Petranovic M, Noirot P, Michel B. (1993) Mechanisms of illegitimate recombination. Gene, 135:161–166.
46. Wirth T, Falush D, Lan R, Colles F, Mensa P, Wieler LH, Karch H, Reeves PR, Maiden MC, Ochman H, Achtman M. (2006) Sex and virulence in *Escherichia coli*: an evolutionary perspective. Mol Microbiol, 60:1136–1151.
47. Koonin EV, Wolf YI. (2008) Genomics of bacteria and archaea: the emerging dynamic view of the prokaryotic world. Nucleic Acids Res, 36:6688–6719.
48. Smith MC, Thorpe HM. (2002) Diversity in the serine recombinases. Mol Microbiol, 44:299–307.
49. Hallet B, Sherratt DJ. (1997) Transposition and site-specific recombination: adapting DNA cut-and-paste mechanisms to a variety of genetic rearrangements. FEMS Microbiol Rev, 21:157–178.
50. Rice P, Craigie R, Davies DR. (1996) Retroviral integrases and their cousins. Curr Opin Struct Biol, 6:76–83.
51. Rajeev L, Malanowska K, Gardner JF. (2009) Challenging a paradigm: the role of DNA homology in tyrosine recombinase reactions. Microbiol Mol Biol Rev, 73:300–309.
52. Consortium GO. (2010) The Gene Ontology in 2010: extensions and refinements. Nucleic Acids Res, 38:D331–D335.
53. Toussaint A, Lima-Mendez G, Leplae R. (2007) PhiGO, a phage ontology associated with the ACLAME database. Res Microbiol, 158(7):567–571.
54. Joss MJ, Koenig JE, Labbate M, Polz MF, Gillings MR, Stokes HW, Doolittle WF, Boucher Y. (2009) ACID: annotation of cassette and integron data. BMC Bioinformatics, 10:118.
55. Reiter WD, Palm P, Yeats S. (1989) Transfer RNA genes frequently serve as integration sites for prokaryotic genetic elements. Nucleic Acids Res, 17:1907–1914.
56. Mantri Y, Williams KP. (2004) Islander: a database of integrative islands in prokaryotic genomes, the associated integrases and their DNA site specificities. Nucleic Acids Res, 32 (Database issue):D55–D58.
57. Langille MG, Brinkman FS. (2009) IslandViewer: an integrated interface for computational identification and visualization of genomic islands. Bioinformatics, 25:664–665.
58. Fouts DE. (2006) Phage_Finder: Automated identification and classification of prophage regions in complete bacterial genome sequences. Nucleic Acids Res, 34 (20):5839–5851.

59. Bose M, Barber RD. (2006) Prophage Finder: a prophage loci prediction tool for prokaryotic genome sequences. In Silico Biol, 6:0020.
60. Lima-Mendez G, Van Helden J, Toussaint A, Leplae R. (2008) Prophinder: a computational tool for prophage prediction in prokaryotic genomes. Bioinformatics, 24:863–865.
61. Rowe-Magnus DA, Guerout AM, Biskri L, Bouige P, Mazel D. (2003) Comparative analysis of superintegrons: engineering extensive genetic diversity in the Vibrionaceae. Genome Res, 13:428–442.
62. Ou HY, He X, Harrison EM, Kulasekara BR, Thani AB, Kadioglu A, Lory S, Hinton JC, Barer MR, Deng Z, Rajakumar K. (2007) MobilomeFINDER: web-based tools for in silico and experimental discovery of bacterial genomic islands. Nucleic Acids Res, 35: W97–W104.
63. Darling AC, Mau B, Blattner FR, Perna NT. (2004) Mauve: multiple alignment of conserved genomic sequence with rearrangements. Genome Res, 14:1394–1403.
64. Chiapello H, Bourgait I, Sourivong F, Heuclin G, Gendrault-Jacquemard A, Petit MA, El Karoui M. (2005) Systematic determination of the mosaic structure of bacterial genomes: species backbone versus strain-specific loops. BMC Bioinformatics, 6:171.
65. Siguier P, Perochon J, Lestrade L, Mahillon J, Chandler M. (2006) ISfinder: the reference centre for bacterial insertion sequences. Nucleic Acids Res, 34:D32–D36.
66. Kichenaradja P, Siguier P, Perochon J, Chandler M. (2010) ISbrowser: an extension of ISfinder for visualizing insertion sequences in prokaryotic genomes. Nucleic Acids Res, 38:D62–D68.
67. Cortez D, Forterre P, Gribaldo S. (2009) A hidden reservoir of integrative elements is the major source of recently acquired foreign genes and ORFans in archaeal and bacterial genomes. Genome Biol, 10:R65.
68. Hacker J, Hochhut B, Middendorf B, Schneider G, Buchrieser C, Gottschalk G, Dobrindt U. (2004) Pathogenomics of mobile genetic elements of toxigenic bacteria. Int J Med Microbiol, 293:453–461.
69. Ikeda H, Tomizawa J. (1968) Prophage P1, and extrachromosomal replication unit. Cold Spring Harb Symp Quant Biol, 33:791–798.
70. Rybchin VN, Svarchevsky AN. (1999) The plasmid prophage N15: a linear DNA with covalently closed ends. Mol Microbiol, 33:895–903.
71. Toussaint A, Merlin C, Monchy S, Benotmane MA, Leplae R, Mergeay M, Springael D. (2003) The biphenyl- and 4-chlorobiphenyl-catabolic transposon Tn4371, a member of a new family of genomic islands related to IncP and Ti plasmids. Appl Environ Microbiol, 69:4837–4845.
72. Brochet M, Da Cunha V, Couve E, Rusniok C, Trieu-Cuot P, Glaser P. (2009) Atypical association of DDE transposition with conjugation specifies a new family of mobile elements. Mol Microbiol, 71:948–959.
73. Novick RP, Subedi A. (2007) The SaPIs: mobile pathogenicity islands of *Staphylococcus*. Chem Immunol Allergy, 93:42–57.
74. Siguier P, Filee J, Chandler M. (2006) Insertion sequences in prokaryotic genomes. Curr Opin Microbiol, 9:526–531.
75. Thieffry D, Thomas R. (1995) Dynamical behaviour of biological regulatory networks – II. Immunity control in bacteriophage lambda. Bull Math Biol, 57:277–297.
76. Su S, Khan SR, Farrand SK. (2008) Induction and loss of Ti plasmid conjugative competence in response to the acyl-homoserine lactone quorum-sensing signal. J Bacteriol, 190:4398–4407.
77. McLeod SM, Burrus V, Waldor MK. (2006) Requirement for Vibrio cholerae integration host factor in conjugative DNA transfer. J Bacteriol, 188:5704–5711.
78. Guerin E, Cambray G, Sanchez-Alberola N, Campoy S, Erill I, Da Re S, Gonzalez-Zorn B, Barbe J, Ploy MC, Mazel D. (2009) The SOS response controls integron recombination. Science, 324:1034.
79. Dorman CJ. (2009) Nucleoid-associated proteins and bacterial physiology. Adv Appl Microbiol, 67:47–64.
80. Gamas P, Caro L, Galas D, Chandler M. (1987) Expression of F transfer functions depends on the *Escherichia coli* integration host factor. Mol Gen Genet, 207:302–305.
81. Nagy Z, Chandler M. (2004) Regulation of transposition in bacteria. Res Microbiol, 155:387–398.
82. Hendrix RW, Lawrence JG, Hatfull GF, Casjens S. (2000) The origins and ongoing evolution of viruses. Trends Microbiol, 8:504–508.
83. Nelson D. (2004) Phage taxonomy: we agree to disagree. J Bacteriol, 186:7029–7031.
84. Lawrence JG, Hatfull GF, Hendrix RW. (2002) Imbroglios of viral taxonomy: genetic exchange and failings of phenetic approaches. J Bacteriol, 184:4891–4905.

85. Doolittle WF. (1999) Phylogenetic classification and the universal tree. Science, 284:2124–2129.
86. Doolittle RF. (2000) Searching for the common ancestor. Res Microbiol, 151:85–89.
87. Gogarten JP, Doolittle WF, Lawrence JG. (2002) Prokaryotic evolution in light of gene transfer. Mol Biol Evol, 19:2226–2238.
88. Lima-Mendez G, Van Helden J, Toussaint A, Leplae R. (2008) Reticulate representation of evolutionary and functional relationships between phage genomes. Mol Biol Evol, 25:762–777.
89. Leplae R, Lima-Mendez G, Toussaint A. (2010) ACLAME: a CLAssification of Mobile genetic Elements, update 2010. Nucleic Acids Res, 38:D57–D61.
90. Halary S, Leigh JW, Cheaib B, Lopez P, Bapteste E. (2010) Network analyses structure genetic diversity in independent genetic worlds. Proc Natl Acad Sci U S A, 107:127–132.
91. Kay E, Chabrillat G, Vogel TM, Simonet P. (2003) Intergeneric transfer of chromosomal and conjugative plasmid genes between *Ralstonia solanacearum* and *Acinetobacter* sp. BD413. Mol Plant Microbe Interact, 16:74–82.
92. Molin S, Tolker-Nielsen T. (2003) Gene transfer occurs with enhanced efficiency in biofilms and induces enhanced stabilisation of the biofilm structure. Curr Opin Biotechnol, 14:255–261.
93. Casas V, Rohwer F. (2007) Phage metagenomics. Methods Enzymol, 421:259–268.
94. Curcio MJ, Derbyshire KM. (2003) The outs and ins of transposition: from mu to kangaroo. Nat Rev Mol Cell Biol, 4:865–877.
95. May EW, Craig NL. (1996) Switching from cut-and-paste to replicative Tn7 transposition. Science, 272:401–404.
96. Juhas M, Power PM, Harding RM, Ferguson DJ, Dimopoulou ID, Elamin AR, Mohd-Zain Z, Hood DW, Adegbola R, Erwin A, Smith A, Munson RS, Harrison A, Mansfield L, Bentley S, Crook DW. (2007) Sequence and functional analyses of Haemophilus spp. genomic islands. Genome Biol, 8:R237.
97. Mazel D. (2006) Integrons: agents of bacterial evolution. Nat Rev Microbiol, 4:608–620.
98. Bouvier M, Demarre G, Mazel D. (2005) Integron cassette insertion: a recombination process involving a folded single strand substrate. EMBO J, 24:4356–4367.
99. Lang AS, Beatty JT. (2007) Importance of widespread gene transfer agent genes in alpha-proteobacteria. Trends Microbiol, 15:54–62.
100. Polard P, Ton-Hoang B, Haren L, Betermier M, Walczak R, Chandler M. (1996) IS911-mediated transpositional recombination in vitro. J Mol Biol, 264:68–81.
101. Duval-Valentin G, Marty-Cointin B, Chandler M. (2004) Requirement of IS911 replication before integration defines a new bacterial transposition pathway. EMBO J, 23:3897–3906.

Chapter 5

Reticulate Classification of Mosaic Microbial Genomes Using NeAT Website

Gipsi Lima-Mendez

Abstract

The tree of life is the classical representation of the evolutionary relationships between existent species. A tree is appropriate to display the divergence of species through mutation, i.e., by vertical descent. However, lateral gene transfer (LGT) is excluded from such representations. When LGT contribution to genome evolution cannot be neglected (e.g., for prokaryotes and mobile genetic elements), the tree becomes misleading. Networks appear as an intuitive way to represent both vertical and horizontal relationships, while overlapping groups within such graphs are more suitable for their classification.

Here, we describe a method to represent both vertical and horizontal relationships. We start with a set of genomes whose coded proteins have been grouped into families based on sequence similarity. Next, all pairs of genomes are compared, counting the number of proteins classified into the same family. From this comparison, we derive a weighted graph where genomes with a significant number of similar proteins are linked. Finally, we apply a two-step clustering of this graph to produce a classification where nodes can be assigned to multiple clusters. The procedure can be performed using the Network Analysis Tools (NeAT) website.

Key words: Reticulate classification, Evolutionary network, Lateral gene transfer

1. Introduction

Charles Darwin described the evolution of living species as the tree of life (1). The tree became the standard representation of evolutionary relationships between living organisms, and discrete taxa were used to further classify them into related groups. This representation is undoubtedly very handy and "easy to interpret," but it is not suitable when hybridization between species, lateral gene transfer (LGT) or recombination within a population have taken place. In bacteria, the contribution of LGT has been represented as "vines" connecting the tree branches (2, 3). Mobile genetic elements, e.g., bacteriophages, evolve by exchanging genetic information at such an extent that it is hard to distinguish

Jacques van Helden et al. (eds.), *Bacterial Molecular Networks: Methods and Protocols*, Methods in Molecular Biology, vol. 804, DOI 10.1007/978-1-61779-361-5_5, © Springer Science+Business Media, LLC 2012

vertical from lateral gene transmission. Hence, networks appear as a more appropriate way to display and analyze this sort of relationships and overlapping groups within those graphs are more suitable for classification. Hence, we have developed a methodology to classify bacteriophage genomes into multiple classes, thereby complying with their polyphyletic origin (4). Here, we describe that protocol for classifying mosaic genomes, using bacteriophages as a study case. All the steps can be performed using the Network Analysis Tools (NeAT) suite (5).

1.1. Outline of the Method

The method is based on the comparison of genomes regarding their gene/protein content and thus rests on the initial grouping of related proteins into families.

The input of the procedure is a two-column file describing the composition of the genomes in terms of protein families. The first column specifies the protein families and the second the organisms that harbor them.

1.1.1. Network Building

Given two genomes A and B, containing a and b protein families respectively, the probability to observe at least c protein families in common is estimated using the hypergeometric formula:

$$P\text{ - val} = P(X \geq c) = \sum_{i=c}^{\min(a,b)} \frac{C_a^i C_{n-a}^{b-i}}{C_n^b}, \quad (1)$$

where n is the total number of protein families identified in all the genomes. The P-value (P-val) is the probability to consider the number of common families as significant when it results from chance. To correct for multiple-testing, this P-value is multiplied by the number of comparisons, thus turning it into an E-value, which is in turn converted into a significance index sig:

$$\text{Sig} = -\log(E - \text{value}) = -\log(P - \text{value} \cdot T) \quad (2)$$

Pairs of genomes having a sig value above a given threshold are joined by an edge with a weight equal to the sig value.

1.1.2. Network (Fuzzy) Clustering

The reticulate classification is established using a two-step clustering procedure. First, the Markov clustering (MCL) algorithm (6) is used to partition the graph into clusters of similar genomes. Each genome then belongs to a single cluster, which does not reflect the evolutionary history of mosaic genomes. Indeed, many genomes are related to genomes outside their assigned cluster. Since all these intercluster links are present in the genome graph, the second step of the clustering procedure calculates the *membership* of a node g to cluster c as the proportion of edge weights w linking node g to nodes in cluster c:

$$B_{g,c} = \frac{\sum_{i \in c} w_{g,i}}{\sum_{j \in N} w_{g,j}}, \tag{3}$$

where w is the weight of an edge, i and j are respectively, the indexes for the nodes of cluster c and the set of nodes (genomes) N.

In other words, for each genome/node, the weights of the connections made to a given cluster are summed up and divided by the sum of the weights of all connections for that genome, thereby quantifying the extent of membership of the genome/node to a particular cluster. The matrix $B = \{B_{g,c}\}$ contains the *membership* values for every node–cluster pair. Hence, we refer to the matrix B as the *membership matrix*.

1.1.3. Visualization

The membership matrix can be obtained in text format, and/or visualized as a heatmap, where a gray gradient represents the degree of membership of the genomes to the clusters.

1.2. Study Case: Reticulate Classification of Bacteriophages

We chose to apply the method to the classification of phage genomes, because of the well-documented mosaic nature of their genomes. However, the method can be applied to any set of genomes (specially prokaryotes, for which the LGT contribution to their evolution cannot be neglected).

Double stranded (ds) DNA phages have been proposed to share a common gene pool within which any phage can acquire a given gene or set of genes from another phage (7). Based on the nature of the genetic material, morphology, and some other molecular features such as packaging mechanism or the presence of RNA polymerase, the reference phage taxonomy proposed by the International Committee on the Taxonomy of Viruses (ICTV) does not properly reflect these genetic exchanges. Hierarchical clustering of phages relying on overall sequence similarity (8) has the same limitation because it blends modules with different evolutionary histories.

Hereafter, we apply the reticulate classification procedure to the set of phage genomes stored in the ACLAME database (9).

2. Materials

2.1. Software

This protocol can be executed using a standard web browser with an Internet connection providing access to tools available as a part of the NeAT website (http://rsat.ulb.ac.be/neat/).

2.2. File Formats

The starting point is a file describing the composition of genomes in terms of protein/gene families (input class file). The file ought to be in the text format, with two columns separated by the tab character. The first column contains the protein/gene family identifiers (ID). The second column corresponds to the organism's ID or name. For example:

```
# Family ID  MGE ID

family:vir:1 mge:101

family:vir:1 mge:102

family:vir:1 mge:104

family:vir:2 mge:165

family:vir:2 mge:1656

family:vir:2 mge:166

family:vir:2 mge:1661

...
```

Results of the protocol are provided in two formats: text and graphical format (png). The text file can be opened with a text editor or with a spreadsheet application while the graphics file can be opened with image viewers.

2.3. Input File

Because the protocol can be applied to any type of (mosaic) genome, we do not consider here the procedures to build the protein families from the genomes to be analyzed. For the study case, the families can be downloaded from a dedicated data repository. Protein families can be built using an all versus all sequence comparison algorithm on the set of proteins encoded by the genomes of interest, followed by clustering (10) or can be downloaded from dedicated databases, e.g., COGs (http://www.ncbi.nlm.nih.gov/COG/), STRING (http://string-db.org/) (11), or ACLAME (9) (http://aclame.ulb.ac.be) (see Note 1).

2.3.1. Sample File

The sample file aclame_fam-mge_viruses_inputFile.*txt* used in this protocol must be downloaded from the sample repository at http://rsat.bigre.ulb.ac.be/rsat/data/published_data/MiMB_BMN_2011/.

3. Methods

3.1. Graph Inference

3.1.1. Pair-Wise Genome Comparison

This section is devoted to all-vs.-all genome comparisons using the tool *Compare classes* from the NeaT website (http://rsat.ulb.ac.be/neat/).

1. In the NeAT menu (left frame), click on *Compare classes/clusters*.
2. Locate the previously downloaded file aclame_fam-mge_viruses_inputFile.txt by clicking on the *Browse* button on the *Query classes* section.
3. Select *Compare query classes to query classes* (do not specify reference classes).
4. Check the boxes *Prevent self-comparison* and *Prevent reciprocal comparisons.*
5. In the box *Output options*, select *Class pairs*; in the *Return fields* menu, check *Hypergeometric probability* and *Occurrences* and leave unchecked the other fields (i.e., *Jaccard Index* and *Sorting criterion*).
6. Keep all default parameters of the *Threshold* menu, i.e., lower threshold set to 1 for *Query size*, *Reference size* and *Intersection size*, lower threshold of *Significance* set to 0, all other thresholds set to "none."
7. Click *GO* and wait for the results (about a few seconds for this application, see Note 2).
8. The results page provides access to the output in text (tab-delimited) and HTML format. Save the text file (*.tab*) by right-clicking over the link and then select the *Save Link as* option. Provide a relevant file name (e.g., compare-classes_output.txt).
9. This tab-delimited result file can be opened with any text editor or with Excel, to check the column contents of the *compare-classes* result. The header of the field should indicate that the first column contains reference classes, the second column query classes. In our case, query and reference classes are identical, since we compared each mobile genetic element to each other. Check the column containing the *significance of the intersection* (*sig* column). In principle, this should be the 12th column.

3.1.2. Building and Visualizing the Graph

Using the tool *Convert-graph*, the results of *Compare-classes* can be converted into a graph through the following steps.

1. From the *Compare-classes* result page, select *Convert table to graph* in the *Next step* box. This will open the *Convert-graph*

form and automatically upload the *Compare-classes* output table. Beside *Column containing the weight or label*, enter the number of the *sig* column (value 12), as identified in step 9.

2. Select *tab-delimited* as output format.
3. Click *GO* and wait for the results (a few seconds).
4. Save the tab-delimited result file by right-clicking on top of the link and selecting *Save link as*. Provide a relevant file name (e.g., genomes_graph.txt).
5. To visualize the graph click the button *Display the graph* in the Next step box (see Note 3A).
6. Select the option Edge width proportional to weight, change the output format to PNG and keep the output format layout with the default values (see Note 3B).
7. Click the *GO* button and wait for the image (a few seconds).
8. Save the result file by right-clicking on top of the link and selecting *Save link as* (name it for example genomes_graph.png).

3.2. Clustering

3.2.1. Partition

This section describes the steps for partitioning the genome graph into disjoint clusters using the MCL algorithm (6) (see also Van Dongen and Abreu-Goodger, Chapter 15).

1. In the NeAT menu (left frame), click on MCL clustering. Alternatively, you can navigate two pages back to the *Convert-graph* result page and there click the button *MCL Graph clustering* in the *Next Step* box. This will automatically upload the graph in the MCL form, so you can immediately pass to step 4.
2. Locate the graph file downloaded in step 4 of previous section (genomes_ graph.txt) by clicking on the *Browse* button on the *Graph* section.
3. Keep *tab* as input format. Keep the values 1 as *source*, 2 as *target* and introduce 3 as *weight*, column respectively.
4. Select 2 as *Inflation value.*
5. Click *GO* and wait for the results (a few seconds).
6. The result page displays a graph showing the distribution of cluster sizes, and a list of result files in different formats. Among the result files, save the *Clusters (tab format)* file by right-clicking and selecting *Save link as*. Provide a relevant file name (e.g., mcl_infl2.txt). Save also the file that provides the cluster size distribution (see Note 4).
7. Do not leave this page as it provides direct access to the next step of the protocol.

3.2.2. Fuzzy Clustering

The last section of this protocol takes the genome graph and the clusters obtained in the previous section as input. The clusters are mapped back onto the graph in order to assign the genomes to multiple clusters, based on inter- and intracluster weighted links. The output of this section is the membership table, in text and graphic format.

1. From the MCL result page, select *Cluster membership* as next step. This will open the *graph-cluster-membership* form and automatically upload the genome similarity graph and the MCL clusters as input files.
2. In the section *Membership matrix*, set the *Stat* option to *weight* (default).
3. Set the *number of decimals* to *2* for the membership values.
4. Click *GO* and wait for the results (a few seconds).
5. Save the membership table (name it, e.g., mge_cluster_membership.txt) as well as the high- and low-resolution membership heat-maps by right-clicking over the respective links and selecting *Save link as.*

3.3. Biological Interpretation of the Results

The network of bacteriophage genomes generated in section 3.1.2 is represented in Fig. 1. The image shows groups of genomes with no significant evolutionary links, while most of the genomes generated in section 3.1.2 are found in a main component. This component is rather heterogeneous; i.e., a cluster structure is apparent. Following this protocol, these clusters are discovered but their interconnections are kept. The resulting membership matrix is returned in two formats, a text file and an image file. The matrix contains 59 columns, each representing a cluster, and 425 rows, each representing a bacteriophage genome. Table 1 and Fig. 2 display a section of the text and image output files, respectively. They contain the results of the classification for *Pseudomonas aeruginosa* bacteriophages DMS3 (mge:1649), D3112 (mge:1522), and MP22 (mge:1867), *Escherichia coli* phages Mu (mge:320) and P27 (mge:11), *Salmonella enterica* bacteriophage ST64B (mge:96) and *Shigella flexneri* phage V (SfV, mge:97). These phages are temperate, i.e., they are capable of co-existing with the host in an integrated prophage state. However, they separate in two groups according to the mechanism they use to integrate their genome into that of their host. Phages P27, SfV and ST64B use site-specific recombination, while D3112, DMS3, MP22, and Mu are transposable phages (for more details about these mechanisms, see Toussaint and Chandler, Chapter 4).

The group distributes over 10 clusters out of the 59 obtained for the whole dataset. Phages DMS3, D3112, and MP22 belong mainly to cluster 31 with memberships close to 90%. Phage Mu distributes mainly between clusters 31 and 32 with memberships

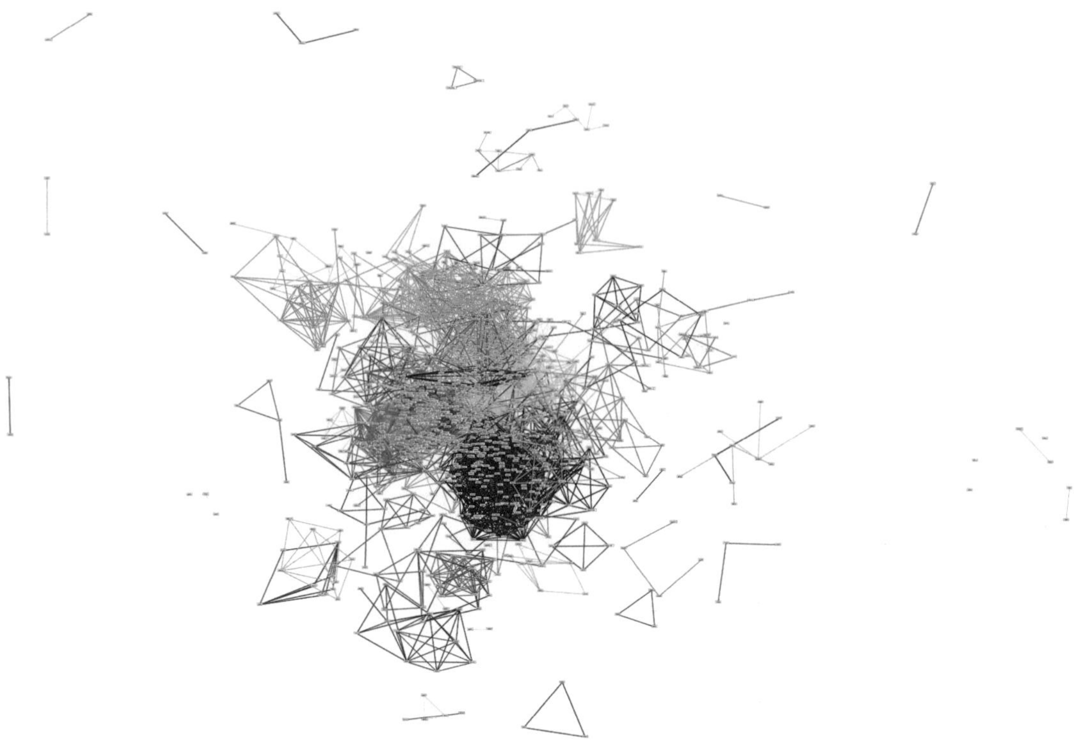

Fig. 1. Graphical representation of the bacteriophage genome network. Genomes with no significant similarity are disjoint in separate components of the graph.

of 0.32 and 0.41, respectively, making a bridge with phages P27, ST64B, and SfV. The gene/protein families encoded by these phages can be searched in the ACLAME database to explain this complex grouping. DMS3, D3112, MP22, and Mu share a very similar genetic structure, but the first three have a long noncontractile tail and thus belong to the *Siphoviridae* ICTV family, while Mu has a contractile tail and hence belongs to the *Myoviridae* family. Phages SfV, ST64B, and P27 share tail genes with phage Mu, while the rest of their genomes relates them with the group of lambdoid phages (clusters 3 and 17). Previously documented (12–14), the "chimeric" structure of these phages is captured with our reticulate classification method, since these three phages are placed in cluster 32 with Mu (membership 0.2, 0.38, 0.31 for P27, ST64B, and SfV, respectively) and belong to various degrees to the clusters 1, 3, and 4, where important groups of integrative temperate phages are grouped (i.e., lambdoid phages and *Staphylococcus* and *Streptococcus* infecting phages).

In summary, this methodology proposes a graph-based classification of genomes for which there is a complex web of connections with their evolutionary neighbors. It can be applied to prokaryotic genomes for which network representations have

Table 1
Section of the membership table

	Node–Cluster membership									
Nodes\clusters	**cl_1**	**cl_3**	**cl_4**	**cl_6**	**cl_8**	**cl_17**	**cl_24**	**cl_31**	**cl_32**	**cl_33**
mge:1649	0.00	0.00	0.00	0.00	0.00	0.00	0.00	0.86	0.01	0.00
mge:1522	0.00	0.00	0.00	0.00	0.00	0.00	0.00	0.88	0.00	0.00
mge:320	0.00	0.00	0.00	0.00	0.00	0.00	0.00	0.32	0.41	0.00
mge:1867	0.00	0.00	0.00	0.00	0.00	0.00	0.00	0.87	0.00	0.00
mge:1574	0.06	0.55	0.02	0.06	0.00	0.05	0.00	0.08	0.18	0.00
mge:11	0.20	0.12	0.20	0.01	0.14	0.08	0.00	0.02	0.20	0.01
mge:96	0.10	0.13	0.10	0.00	0.10	0.13	0.00	0.03	0.38	0.02
mge:97	0.15	0.23	0.09	0.02	0.06	0.09	0.01	0.03	0.31	0.01

Values for the genomes (rows) membership to the clusters (columns) are shown. Clusters with no genome membership value greater than zero have been discarded

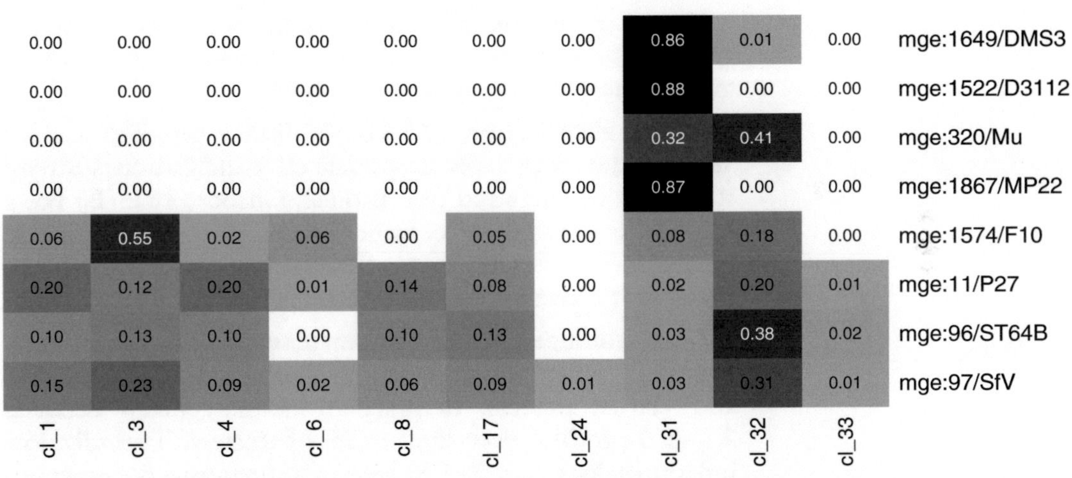

Fig. 2. Reticulate classification of bacteriophages. The section of the heatmap represents the cluster memberships for a subset of genomes. For clarity, only clusters for which the membership of one of the genomes is larger than zero are shown. The transposable phages DMS3, D3112, and MP22 belong almost exclusively to cluster 31. Phage Mu, which is also transposable but has a different tail type, further belongs to cluster 32, where phages SfV, ST64B, and P27 are also affiliated.

been also recently proposed (15). By partitioning the graph in the first step, it is possible to assign the genomes "mainly" to one cluster, while the second step recovers the intercluster connections. To understand the basis of the clustering assignment, it is possible to examine the protein families shared between genomes allocated to a set of clusters as done in the study case.

4. Notes

1. Input file: For applications to mobile genetic elements, the composition of genomes in protein families can be directly obtained from the download section of the ACLAME database (http://aclame.ulb.ac.be/perl/Aclame/Tools/exporter.cgi). The last release of the database is the default version (version 4 at the time of writing this protocol). Make sure to select the category of mobile genetic element you wish to classify, e.g., viruses, prophages, plasmids, or a combination of those. Go in the *download ACLAME families* section and click on the download button next to *Table with ACLAME family IDs and non-redundant MGE IDs*. Save and uncompress this file to make it suitable for the first step of the protocol.
2. Building the network: The amount of genomes being compared will influence the feasibility and time of the all vs. all genome comparisons. The default parameters are set to allow for comparing all pairs of genomes sharing at least one protein in the same family. When many more genomes are compared so that the execution time is long, the threshold on the intersection (number of families shared) can be set at a higher value. It is also crucial to uncheck the *Sorting criterion* box to avoid unnecessary long execution time.
3. Network visualization: (A) Graph display provides a static view of the graph. More specialized visualization software tools may be preferred (see Brohée, Chapter 18). (B) For a high-quality image, select rather postscript (ps) as the image output format. Beware that in this case you may not be able to visualize it directly from the web interface.
4. MCL clustering: The main parameter of MCL is the inflation factor. It influences clustering granularity, i.e., the number and size of clusters. In practice, inflation values between 2 and 3 produce very similar sets of clusters. Typically, low inflation values (e.g., 1–1.5) keep many genomes together in a large component, which is partitioned at larger inflation values. To select an appropriate inflation value, the user should compare the results over an inflation range (e.g., 1.5–4) and select a value of inflation leading to a reasonable scattering of genomes. This assessment can be done from the cluster size distribution file, which can be saved from the result page of MCL.

Acknowledgments

I wish to thank Raphael Leplae, Sylvain Brohée and Jacques van Helden for the maintenance of the ACLAME database and NeAT services, respectively, as well as Ariane Toussaint and Denis Thieffry for critical reading of the manuscript and testing the protocol.

Work on the ACLAME project in the BiGRe laboratory was financed by the European Space Agency (ESA-PRODEX) and the Belgian Science Policy (Belspo) through the EXANAM project (PRODEX agreements No. C90358), the Fonds de la Recherche Scientifique Médicale (FRSM), the Belgian Program on Interuniversity Attraction Poles, initiated by the Belgian Federal Science Policy Office, project P6/25 (BioMaGNet), and the BioSapiens Network of excellence funded under the sixth Framework program of the European Communities (LSHG-CT-2003-503265).

References

1. Darwin C. (1859) *On the Origin of Species by Means of Natural Selection, or the Preservation of Favoured Races in the Struggle for Life*. John Murray, London.
2. Doolittle WF. (1999) Phylogenetic classification and the universal tree. Science, 284:2124–2129.
3. Kunin V, Goldovsky L, Darzentas N, Ouzounis CA. (2005) The net of life: reconstructing the microbial phylogenetic network. Genome Res, 15:954–959.
4. Lima-Mendez G, Van Helden J, Toussaint A, Leplae R. (2008) Reticulate representation of evolutionary and functional relationships between phage genomes. Mol Biol Evol, 25:762–777.
5. Brohee S, Faust K, Lima-Mendez G, Sand O, Janky R, Vanderstocken G, Deville Y, van Helden J. (2008) NeAT: a toolbox for the analysis of biological networks, clusters, classes and pathways. Nucleic Acids Res, 36: W444–W451.
6. van Dongen S. (2000) Graph Clustering by Flow Simulation. In *Centre for Mathematics and Computer Science*. Utrecht, Amsterdam, 173.
7. Hendrix RW, Smith MC, Burns RN, Ford ME, Hatfull GF. (1999) Evolutionary relationships among diverse bacteriophages and prophages: all the world's a phage. Proc Natl Acad Sci U S A, 96:2192–2197.
8. Rohwer F, Edwards R. (2002) The Phage Proteomic Tree: a genome-based taxonomy for phage. J Bacteriol, 184:4529–4535.
9. Leplae R, Lima-Mendez G, Toussaint A. ACLAME: a CLAssification of Mobile genetic Elements, update 2010. Nucleic Acids Res, 38:D57–D61.
10. Enright AJ, Van Dongen S, Ouzounis CA. (2002) An efficient algorithm for large-scale detection of protein families. Nucleic Acids Res, 30:1575–1584.
11. Jensen LJ, Kuhn M, Stark M, Chaffron S, Creevey C, Muller J, Doerks T, Julien P, Roth A, Simonovic M, Bork P, von Mering C. (2009) STRING 8 – a global view on proteins and their functional interactions in 630 organisms. Nucleic Acids Res, 37: D412–D416.
12. Allison GE, Angeles D, Tran-Dinh N, Verma NK. (2002) Complete genomic sequence of SfV, a serotype-converting temperate bacteriophage of *Shigella flexneri*. J Bacteriol, 184:1974–1987.
13. Recktenwald J, Schmidt H. (2002) The nucleotide sequence of Shiga toxin (Stx) 2e-encoding phage phiP27 is not related to other Stx phage genomes, but the modular genetic structure is conserved. Infect Immun, 70:1896–1908.
14. Mmolawa PT, Schmieger H, Heuzenroeder MW. (2003) Bacteriophage ST64B, a genetic mosaic of genes from diverse sources isolated from *Salmonella enterica* serovar typhimurium DT 64. J Bacteriol, 185:6481–6485.
15. Dagan T, Artzy-Randrup Y, Martin W. (2008) Modular networks and cumulative impact of lateral transfer in prokaryote genome evolution. Proc Natl Acad Sci USA, 105(29), 10039–10044.

Chapter 6

From Metabolic Reactions to Networks and Pathways

Masanori Arita

Abstract

Enzymatic reactions form a hypergraph structure and their translation into a graph structure accompanies an information loss. This chapter introduces well-known topological transformations from metabolic reactions to a graph, and discusses their advantages and disadvantages. Also discussed is the legitimacy of defining cofactors or currency metabolites, and suitable application area of each representation.

Key words: Metabolism, Network, Pathway, Modularity, Database

1. Introduction

After the publication of Barabási and Albert's (1) inspiring work, computing the degree exponent of a network became a standard process in analyzing biological networks. Irrespective of the network under investigation, researchers routinely discuss the evolutionary implication(s) of its hubs, motifs, or modules. They also associate characteristics of the network, such as the average path length or clustering coefficient with biological implications (Table 1). These properties, called network metrics, are actively applied to characterize and classify the network topology while ignoring dynamics (see Chapter 13) (2). Metabolism was the first such application (3) and ever since, many variations have been proposed.

The notable characteristic of network biology, i.e., metric-based understanding of biology, is its topological representation and associated discretization. In real metabolism, each reaction includes multiple substrates and products. Enzymes are regulated by multiple components and protein modules, and their reaction mechanism is characterized by various physicochemical quantitative parameters, such as the Gibbs energy and enzyme velocity. In the analysis of network biology, on the other hand, quantitative factors are reduced to a simple graph structure and the main focus is on the degree distribution, reachability between nodes, or modularity.

Jacques van Helden et al. (eds.), *Bacterial Molecular Networks: Methods and Protocols*, Methods in Molecular Biology, vol. 804, DOI 10.1007/978-1-61779-361-5_6, © Springer Science+Business Media, LLC 2012

Table 1
Network metrics: well-known parameters used to characterize the topology of networks

Degree, density, and assortativity	*Degree* is the number of edges for each node. High-degree nodes are called hubs *Density* is the proportion of edges for each node to the maximum degree. Hubs are high-density nodes *Clustering coefficient* (CC) is the average of the edge density of immediate neighbors for a given node. Natural networks are known to have high (ca.0.3 and up) CCs *Assortativity* is a correlation coefficient between the degree distributions of node neighbors. When many nodes are connected with nodes of different degrees, the network is called *dissortative*
Distance and path	*Distance* between nodes is the least number of edges to connect them, i.e., the length of their shortest path *Path* is the actual sequence of edges. *Average path length* (APL) is the average of distances between all pairs of nodes. Many natural networks are known to have small APLs
Centrality	*Centrality* is the criterion to determine the "center" of the network in an undirected graph. (In a directed graph, the equivalent notion is called *prestige*.) *Degree-centrality* chooses the most-connected hub as the center. *Closeness centrality* chooses the node whose average distance to all other nodes is the minimum *Betweenness centrality* chooses the node which are most used in shortest paths between all pairs of nodes
Motif	*Motif* is a small "building block" or frequently appearing pattern in the network. Typically, all patterns up to five nodes are exhaustively searched in the network. Well-known directed motifs include feed-forward and feed-back loops (see Chapters 12 and 14)
Modularity	*Modularity* is a degree of separation for sub-networks. Many natural networks are known to have high modularity. The study on modularity is immature and no clear definition exists

There are advantages and disadvantages in the discretized representation. One important benefit is its simplicity. Very large-scale networks are amenable to algorithmic analyses as a simple graph but not as a more complicated representation (e.g., differential equations). Also, facile computation and simplistic interpretation could attract many computational biologists into this research field. On the backstage, however, many biological aspects are overlooked; as will be explained, even the original network topology may be sacrificed. In this chapter, we introduce topological transformations from metabolic reactions to graph representation and discuss their advantages and disadvantages. We also discuss the legitimacy of defining cofactors or currency metabolites.

2. Strategies for Graph Transformation

The graph structure is a collection of directed/undirected binary relations defined as the paired set of nodes and edges. Each relation, or edge, has only one source and one target node. (The source and target are not distinguished in the case of undirected graphs.) Originally, standard metabolic reactions catalyze multiple substrates into products. Instead of a graph, they form a hypergraph where each edge may connect any number of nodes, as represented by a box in Fig. 1a. Information on metabolite

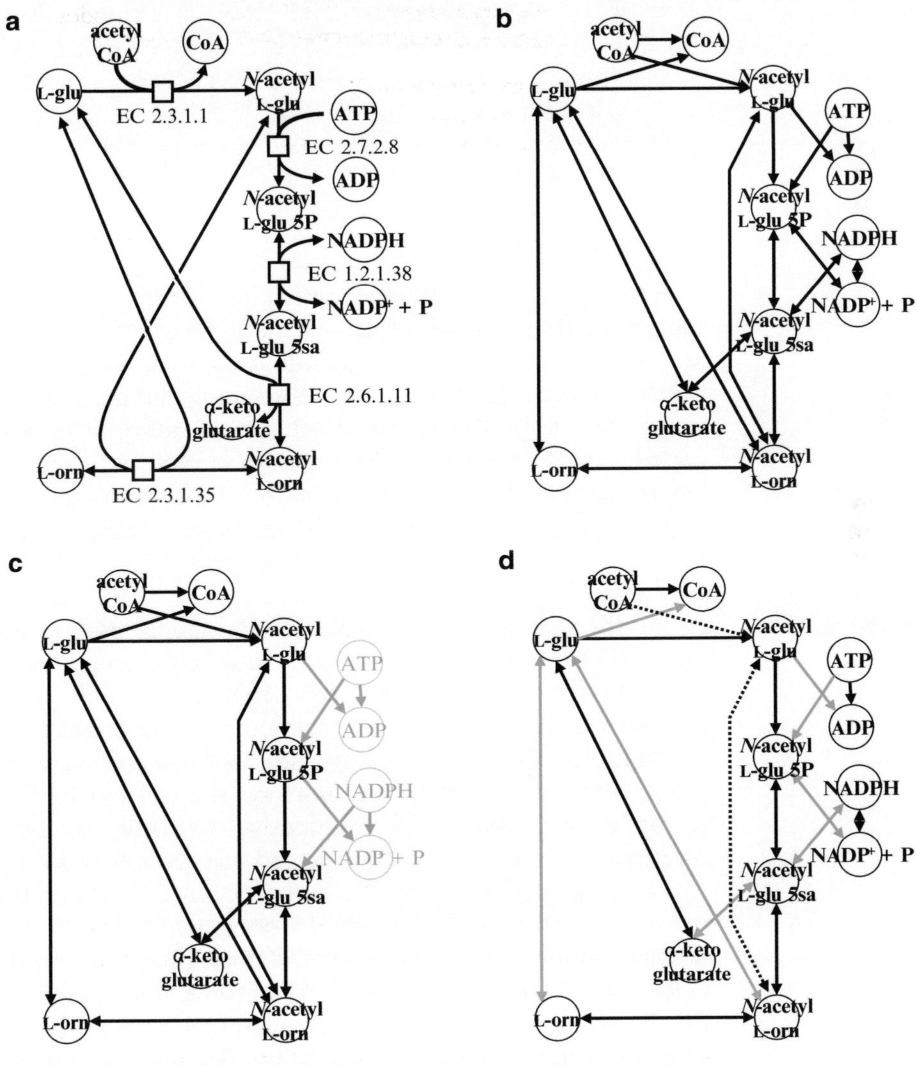

Fig. 1. Three network representations (**b–d**) for the L-ornithine biosynthetic pathway (**a**). (**a**) *Bipartite representation*; (**b**) *fully connected substrate graph*; (**c**) *substrate graph* without currency metabolites; and (**d**) substrate–product *graph*. *Dotted* and *solid lines* indicate the acetyl moiety and the carbon backbone of L-glutamate, respectively. *Dotted lines* are missed in the *substrate graph* with approximate structural relationships.

Table 2
Network hubs in different graph representations for *Escherichia coli* metabolism

Jeong et al. (3)	Fell and Wagner (4)	Ma and Zeng (9)	Arita (12)
H_2O	L-Glutamate	Glycerate 3P	CO_2
ADP	Pyruvate	D-Ribose 5P	Pyruvate
P	Coenzyme A	Acetyl CoA	Acetyl CoA
ATP	α-Ketoglutarate	Pyruvate	ATP
L-Glutamate	L-Glutamine	D-Xylulose 5P	D-Glucose
$NADP^+$	L-Aspartate	D-Fructose 6P	L-Glutamate
PP	Acetyl CoA	5P-D-ribose 1PP	D-Galactose
NAD^+	Phosphoribosyl PP	L-Glutamate	Coenzyme A
NADPH	Tetrahydrofolate	D-Glyceraldehyde 3P	*S*-adenosyl L-methionine
NADH	Succinate	L-Aspartate	5P-D-ribose 1PP

The list of top ten hub metabolites for each graph transformation. The list is a reproduction from Arita (12). *P* (ortho)phosphate, *PP* pyrophosphate

structures is implicitly represented in this hypergraph format, and the structural information can be easily lost in the reduction of metabolism into a graph. Depending on their purposes, several transformation strategies exist for mapping the metabolic hypergraph into a graph. Information loss is unavoidable to some extent, and no transformation strategy is perfect. To explain the respective benefit(s) of transformation, we introduce major strategies one by one, using L-ornithine biosynthesis, the pathway from L-glutamate to L-ornithine, as an example (Fig. 1).

2.1. Fully Connected Substrate Graph

As in the traditional metabolic chart, graph nodes represent metabolites in the substrate graph. Nodes are linked if they participate in the same reaction (Fig. 1b). In the fully connected representation, substrates in a reaction are fully linked with their products, i.e., each reaction is transformed to a complete bipartite graph. The network therefore portrays the relation by "reaction membership." Apparently, frequently occurring metabolites in reactions become the most connected metabolites or hubs.

In their pioneering study, Jeong et al. (3) constructed fully connected substrate graphs for 43 organisms representing three domains of life. They "unexpectedly" found that the average path length of the network (APL; see Table 1 for definition) was identical for all organisms (around 3–4 steps), and the ranking of the most connected metabolites was practically identical. In order of frequency, the hubs were H_2O, ADP, orthophosphate (P), ATP, L-glutamate, $NADP^+$, pyrophosphate (PP), NAD^+, NADPH, and NADH (Table 2). These metabolites are usually referred to as inorganics or cofactors. Jeong et al. (3) argued that "the same highly connected substrates may provide the connections between modules responsible for distinct metabolic functions." Indeed, these metabolites function as a shortcut to achieve the small APL, as L-glutamate is directly linked with L-ornithine in Fig. 1b.

2.2. Substrate Graph Without Currency Metabolites

For biochemists, inorganics and cofactors are minor players. In the traditional metabolic chart, they are depicted in smaller fonts and their linkage is often unattended. Some cofactors have been called currency metabolites because of their ubiquity. To exclude such currency metabolites and inorganics from the substrate graph, Fell and Wagner (4) manually specified hub metabolites to be excluded in their network reconstruction in *Escherichia coli* (Fig. 1c). Their removal list included ATP, ADP, NAD^+, $NADP^+$, NADH, NADPH, CO_2, ammonia, sulfate, thioredoxin, P, and PP. That is, except for L-glutamate, all top ten hubs in the work of Jeong et al. were removed and the remaining hub metabolites were brought forward to form the new ten hubs: L-glutamate, pyruvate, Coenzyme A (CoA), α-ketoglutarate, L-glutamine, L-aspartate, acetyl CoA, phosphoribosyl PP, tetrahydrofolate, and succinate.

The removal of currency metabolites is the most popular procedure in metabolic analysis; importantly, it does not change the topological statistics that are found in the fully connected, bipartite graph representation (4). In other words, currency metabolites are carefully chosen so that the resulting network retains the network distance, modularity, and other important properties of the metabolism (5, 6). In this sense, the decision is inevitably artificial and context-dependent. For example, Wagner and Fell (7) argued that L-glutamate is an important hub based on its ranking, but its high degree originated from many amino-transfer reactions, where L-glutamate and α-ketoglutarate function as amino-group donor and acceptor, respectively. They not only play important roles in amino acid metabolism, but they also function as cofactors in the amino-group transfer. Likewise, currency metabolites should not be removed from their own biosynthesis/degradation pathways. Clearly, the dichotomy of metabolites between currency and commodity (noncurrency) is unacceptable. Recent trends focus on network modularity rather than hubs (6, 8), but the same observation applies. It is a fundamental problem in the modern network analysis.

2.3. Substrate Graph with Structural Relationship

One way to ensure the proper treatment of currency metabolites is to verify each transformation manually. Ma and Zeng (9) manually specified edges between substrates and products for over 3,000 reactions so that edges connect structurally related nodes only. Their rule of thumb was to link moieties that share more than two carbons, thus overlooking the methyl and acetyl transfers and noncarbon atoms (Fig. 1d). Later, Rahman et al. (10) proposed the use of structural similarity to automatically obtain the same effect in their Pathway Hunter Tool. The good news is that this method does not raise the dichotomy between currency and commodity metabolites. Its disadvantage however, lies in the bias originating from the number of reactions used. Because of this bias, the top ten hubs Ma and Zeng identified in the *E. coli* network included D-xylulose 5P and D-ribose 5P (Table 2), which play relatively minor roles from the metabolic perspective.

The direct reflection of the number of reactions as node degrees represents another fundamental problem regardless of the assignment of currency metabolites. Metabolic reactions for genes are often incomplete and ambiguous. Many enzymes exhibit broad substrate specificity and they are sometimes found to perform alternative functions. Such ambiguity is exemplified in the over 500 reassignments of the EC numbers (denoted as transferred entries in the EC hierarchy) and the frequent use of generic terms, such as “amino acid,” “alcohol,” “acyl,” or “long-chain,” in the systematic names of enzymes. Since such generic terms correspond to multiple substrates, the number of reactions corresponding to a given set of EC numbers may vary depending on the interpretation. Therefore, direct counting of the number of reactions is not a robust way for estimating metabolic hubs. For example, EC 2.3.1.85 (acyl-CoA:malonyl-CoA *C*-acyltransferase) is associated with as many as 30 reactions in the KEGG RPAIR database (in fact, many of them are not acyltransferase reactions and are automatically associated with this EC number according to an attached comment by the IUBMB) (11). If all reactions are automatically transformed to a graph representation, the common cofactors for this EC reaction, CoA and NADPs, receive too many graph edges. Still, the molecular structure of the main substrate, acyl CoA, remains ambiguous. The resulting graph structure depends on how these reactions are curated for each target species (including the removal of inappropriate assignments). In general, such curation problem has been ignored, especially by computational biologists, possibly because it is time consuming and its discussion is of little computational interest.

2.4. Substrate–Product Graph

To reduce the bias originating from the interpretation of multiple reactions, a different abstraction has been proposed; we call it the substrate–product graph (12). Instead of counting the number of reactions, this method counts the number of structural transformations by decomposing each reaction into a set of atomic mappings, and is independently employed to construct the KEGG RPAIR database (11). Atomic mapping comprises a set of atomic position pairs for a substrate–product relationship corresponding to a migrating moiety in the reaction. For example, atomic mapping between ATP and ADP represents a loss of orthophosphate from ATP (kinase) and mapping between L-glutamate and α-ketoglutarate represents an exchange of an amino and oxo group (aminotransferase). In the substrate–product graph, nodes correspond to metabolites as in the substrate graph, but edges correspond to atomic mappings. No matter how often each atomic mapping appears in the reaction set (e.g., the atomic mapping between ATP and ADP, or NAD and NADH, appears quite frequently), only one edge is drawn for each structural correspondence. Since a cofactor is always involved in the same reaction pattern, its function corresponds to only one atomic mapping and the node degree is kept small. Most importantly, this representation can correct network statistics. Specifically, the small APL of the metabolic network was lost in this representation (13). The top ten hubs became CO_2, pyruvate, acetyl CoA, ATP, D-glucose, L-glutamate, D-galactose, CoA, *S*-adenosyl L-methionine, and phosphoribosyl PP. CO_2 was the topmost hub because of the various decarboxylation reactions. Likewise, the high rank of acetyl CoA and CoA resulted from many acyl transfers. In this representation, we reconfirmed the importance of L-glutamate pointed out by Wagner and Fell (7), and the statistics also implies the ancientry of the purine-related pathways from the plasticity of the adenosine moiety. Based on protein folds, the purine pathway was reported to be one of the most ancient subnetworks (14).

By focusing on the number of structural changes, the substrate–product graph concisely represents the structural plasticity in the metabolic network (Fig. 1d). The reduction to atomic mappings has both merits and demerits. First, it is robust to the ambiguity in genome annotations, such as the interpretation of general reactions. On the other hand, it is computationally expensive and loses information derived from redundancies in the reaction network. Nonetheless, the method is slowly gaining popularity. Blum and Kohlbacher (15) proposed a degree-weighted version of the substrate–product graph. Pitkanen et al. (16) presented a new, practical method that can detect branched pathways at atomic resolution. Faust et al. (17) used the RPAIR database to reconstruct the substrate–product graph and optimized weighting schemes for metabolic pathway finding (also see Chapter 7). All approaches reported significant improvement in metabolic analyses through the consideration of structural relationships.

2.5. Reaction Graph

The reaction graph, or enzyme-centric graph, is a yet another representation of a metabolic network (7). In this graph, each reaction corresponds to a graph node; an edge is drawn if two reactions share a substrate or product. The representation is a dual image of the substrate graph: nodes and edges in the reaction graph correspond to edges and nodes in the substrate graph. Since there is a one-to-one correspondence between reaction- and substrate graphs, both representations contain the same amount of information. Therefore, the benefits and disadvantages of the substrate graph also apply to the reaction graph and it is not worthwhile to make comparisons as long as the graphs are dual images.

2.6. Bipartite Graph

Bipartite graph is a graph with two types of nodes, where either type is linked to the other type only. A metabolic network can be naturally mapped to a bipartite representation with metabolite- and reaction nodes. It represents association of metabolites with their reactions, and is often used for network visualization (Fig. 1a). Strictly speaking, the bipartite representation does not constrain that all substrates in a reaction should be together consumed to produce all its products. Also, the representation does not distinguish incoming edges or outgoing edges (in this sense, Fig. 1a is not an accurate bipartite view because it uses two kinds of edges, straight and curved, to denote coenzymes). In summary, the representation does not interfere with the above mentioned network interpretations because their advantages and disadvantages are directly applicable to the bipartite case.

3. Available Data Resources and Their Usage

In the previous section, we reviewed six major graph representations for enzymatic reactions. Reaction data are available on many Web sites (Table 3). The original authoritative source is the IUBMB enzyme nomenclature, where EC numbers are assigned and updated (18). A reliable source of kinetic information is the BRENDA database, where detailed kinetic parameters and substrate specificity are available for each organism (19). Most metabolic information is based on these databases, and the network reconstruction usually undergoes organism-specific modification/curation. Discretized metabolic data should be applied with care because each representation has its original purpose. Here, we introduce a few important tips to be considered.

3.1. Different Databases Use Different Metabolite Names

Unfortunately, metabolite names are no more standardized than protein names. Standardization of metabolite names must precede network reconstruction. For example, α-ketoglutarate can be written as 2-oxoglutarate, oxoglutaric acid, 2-ketoglutarate, or α-ketoglutaric acid. The LIGAND section in the KEGG database

Table 3
Freely available data for reaction information

IUBMB Enzyme Nomenclature (http://www.chem.qmul.ac.uk/iubmb/enzyme/)	Enzyme names and reactions for each EC number with references
BRENDA (http://www.brenda-enzymes.org/)	Organism, Km value, isozyme, and substrate specificity information with references
MetaCyc (http://biocyc.org/metacyc/index.shtml)	Species-specific enzyme and pathway information for over 350 organisms with references
KEGG (http://www.genome.jp/kegg/reaction/)	Substrate–product information for EC reactions without resolving general names for over 1,100 organisms
BiGG (http://systemsbiology.ucsd.edu/In_Silico_Organisms/)	Genome-scale metabolic reconstruction for over 30 organisms

is a good source of such synonyms (20). It must be noted that many enzymatic reactions also use generic names, such as "alcohol" or "amino acid." Many reactions do not specify stereochemical information (e.g., serine instead of L-serine). Such discrepancies must be manually resolved by referencing genomic information in network reconstruction. Already curated data may be available for some model organisms (Table 3), but otherwise, manual curation is a necessary step. Not a few computational studies bulk-download reaction data from the KEGG- and other databases to reconstruct metabolic networks. It is not easy to verify the extent of curation from the brief description of their reconstruction policy, but without standardization, the results are not sufficiently accurate for an estimation of the global network properties.

3.2. Not All Connected Reactions form Pathways

A metabolic pathway consists of a sequence of reactions through which some structural moiety must be passed down. This prerequisite is not always satisfied by a graph path in the metabolic network. In Fig. 1b, for example, the link between L-glutamate and CoA, or between *N*-acetyl L-glutamate 5-semialdehyde and α-ketoglutarate is not a valid pathway because there is no carbon or nitrogen transfer between them. However, many studies computed the shortest paths in the substrate graph to identify the metrics of metabolic networks. It must be noted that such computation does not reflect true metabolic pathways although it may capture some general characteristics.

The directionality of reactions is another issue that causes headaches. There are many discrepancies between databases on directionality, and it is difficult to estimate directionality from enzyme structures or catalytic formulas. Kümmel et al. (21) proposed a heuristic algorithm to check reversibility based on the Gibbs energy, but Maskow and von Stockar (22) argued that the Gibbs energy does not facilitate determination of feasibility without considering intracellular ionic strength, the pH value, and the concentration of abundant metabolites. For now, it is safer to assume that most reactions are reversible; irreversible reactions are, for example, decarboxylation and cyclization reactions.

3.3. Kinetic Parameters Are Measured In Vitro, Not In Vivo

Kinetic parameters reported in the BRENDA and other databases were measured in vitro assuming the classic model of Michaelis and Menten (23). These parameters are good indicators for estimating the relative efficiency of enzymes; however, their true catalytic velocity within cells remains unknown. This is why not every kinetic parameter is directly applicable to simulation studies. Likewise, these values alone cannot indicate the reversibility of reactions. These observations justify the importance of qualitative studies because currently available parameters are not sufficient to fully depict intracellular conditions.

4. Network Applications: Constraint-Based or Graph-Theoretic Approaches

There are two mainstreams in utilizing reconstructed metabolic networks: the constraint-based- and the graph-theoretic approach. The former focuses on the metabolic capability of an organism in terms of metabolite fluxes and attempts to find the basis of the possible flux space. In contrast, the graph approach focuses on the metrics of metabolic networks and tries to understand flux in terms of pathways.

4.1. Constraint-Based Approach

This approach is best represented by flux balance analysis (FBA) and the elementary flux mode (EFM) (24, 25). In these strategies, the metabolic hypergraph is implemented as a matrix to solve its linear constraints, and the basis of the flux space is called extreme pathways or elementary modes, respectively (see Chapter 20) (26). The main constraints are the steady-state hypothesis for metabolites (no increase or decrease in their amounts) and the capacity constraints for reactions (directionality and the maximum value for flux). The elementary mode is more constrained in the formalization of a metabolic pathway and therefore harder to compute. The lack of scalability has been a bottleneck in the application of the EFM, but a new scalable approach was recently presented by the EFM group (27). A detailed comparison between FBA and EFM is available elsewhere (28).

The resolution of generic names and the gap filling of major pathways are crucial steps for both strategies. Currently available reconstructions need substantial updates to be called complete. In the "complete" in silico strain (iJR904) of *E. coli*, for example, biosynthesis for 64 metabolites is reported to be flawed (29, 30). While gaps in many biosynthetic/degradation pathways remain, manual verification of completeness is difficult. The graph-theoretic representation of pathways is needed to fill these pathway gaps.

4.2. Graph-Theoretic Approach

To find metabolic pathways, the approach concentrates on the topological aspects of the network. A commonly used method is finding the shortest paths by tracking node connectivity (10, 12, 15–17) (also see Chapter 7). Since the direct application of the shortest path algorithm to the substrate graph produces a huge number of false-positive pathways, the substrate–product graph or its alternative with atomic information is the proper choice. This requirement is obvious especially when nitrogen or sulfur metabolism is considered. Another often overlooked problem is the consideration of molecular symmetry in pathway finding algorithms (11, 16). Since many molecules have symmetric atomic positions, simple connection of atomic mappings, such as the RPAIR information may miss important pathways. For example, simplistic connection cannot reproduce the carbon fate in the TCA cycle, where many symmetric molecules are involved. However, the consideration of symmetry information is closely related with the stereo-specificity of enzymes (e.g., pro-chirality) and is not easily solvable.

4.3. Identifying the Network Modularity

Some recent works focused on network modularity and its evolutionary implications in addition to pathways (14, 31). Strangely, the substrate–product graph has not been used to identify the modularity of metabolic networks; this may be attributable to the ambiguous definition of modularity.

A common pitfall in assessing network modularity is the use of KEGG metabolic maps as functional categories. These maps are drawn with aesthetic considerations; e.g., the same pathways are drawn multiple times for an easier understanding, and there is no well-defined criterion for omitting currency metabolites. Consequently, most maps contain multiple functional modules in metabolism. This observation is evident when the KEGG maps are compared with the MetaCyc pathway repository, the other comprehensive pathway resource for all organisms (see Chapter 11). The MetaCyc contains over 1,400 pathway modules in contrast to 160 KEGG maps (32). The lack of atom-resolved guidelines for the modularity of metabolites hampers a proper comparison between KEGG maps (substrate graphs with duplication), MetaCyc pathways (bipartite graphs in LISP format), and the substrate–product representation for computing pathways. Therefore, the study of modularity remains at an immature stage and provides a number of interesting research topics.

5. Future Perspectives

Many researchers fail to recognize the proper representation of metabolic networks and their existing variations. It is crucial, especially for computational biologists, to understand the purpose of network representation and its appropriate use in a relevant context. To fully utilize the power of network analysis, more biological understanding is necessary. In a short term, we need to define biologically justifiable network modules. The use of KEGG pathway maps or similar classifications as the modularity standard is scientifically indefensible. Furthermore, a public database for atomic mappings that is easily understandable and applicable must be established to facilitate application of the substrate–product graph with atomic information by nonexperts. It should be noted that the substrate graph can be easily biased by the treatment of generic reactions and by the definition of currency metabolites (the KEGG RPAIR database does not recognize this problem). In view of the lack of standard names or identifiers even for basic compounds, however, it would be a challenging task to establish a common repository for metabolic pathways and associated information in a single institution. In this respect, a publicly shared portal may be a solution so that we can establish a community-based repository for metabolic information (33). In a longer term, the use of artificial chemical network to study the evolution of metabolism provides an interesting perspective because the fortuity of the real world renders evolutionary analysis possible only with such theoretical simulations (34). However, these trials should incorporate, at least, elemental information, such as carbon, nitrogen and sulfur, because their balance is crucial in understanding the metabolism.

Acknowledgments

The author thanks the editors and reviewers for valuable comments. This work is supported by JST-BIRD and Grant-in-Aid for Scientific Research on Priority Areas "Systems Genomics" from the Ministry of Education, Culture, Sports, Science and Technology, Japan.

References

1. Barabási A-L, Albert R. (1999) Emergence of scaling in random networks. Science, 286(5439):509–512.
2. Liu B. (2007) *Web Data Mining – Exploring Hyperlinks, Contents and Usage Data*. Springer Series on Data-Centric Systems and Applications.
3. Jeong H, Tombor B, Albert R, Oltvai ZN, Barabási A-I. (2000) The large-scale organization of metabolic networks. Nature, 407(6804):651–654.
4. Fell DA, Wagner A. (2000) The small world of metabolism. Nat Biotechnol, 18(11): 1121–1122.
5. Huss M, Holme P. (2006) Currency and commodity metabolites: their identification and relation to the modularity of metabolic networks. IET Syst Biol, 1(5):280–285.
6. Parter M, Kashtan N, Alon U. (2007) Environmental variability and modularity of bacterial metabolic networks. BMC Evol Biol, 7:169.
7. Wagner A, Fell D. (2001) The small world inside large metabolic networks. Proc R Soc Lond B, 268(1478):1803–1810.
8. Kreimer A, Borenstein E, Gophna U, Ruppin E. (2008) The evolution of modularity in bacterial metabolic networks. Proc Natl Acad Sci U S A, 105(19):6976–6981.
9. Ma H, Zeng A-P. (2003) Reconstruction of metabolic networks from genome data and analysis of their global structure for various organisms. Bioinformatics, 19(2):270–277.
10. Rahman SA, Advani P, Schunk R, Schrader R, Schomburg D. (2005) Metabolic pathway analysis web service (Pathway Hunter Tool at CUBIC). Bioinformatics, 21(7):1189–1193.
11. Kotera M, Hattori M, Oh M-A, Yamamoto R, Komeno T, Yabuzaki J, Tonomura K, Goto S, Kanehisa M. (2004) RPAIR: a reactant-pair database representing chemical changes in enzymatic reactions. Genome Informatics, 15:P062 (poster abstract)
12. Arita M. (2003) In silico atomic tracing by substrate–product relationships in *Escherichia coli* intermediary metabolism. Genome Res, 13:2455–2466.
13. Arita M. (2004) The metabolic world of *Escherichia coli* is not small. Proc Natl Acad Sci U S A, 101(6):1543–1547.
14. Caetano-Anollés G, Kim HS, Mittenthal JE. (2007) The origin of modern metabolic networks inferred from phylogenomic analysis of protein architecture. Proc Natl Acad Sci U S A, 104(22):9358–9363.
15. Blum T, Kohlbacher O. (2008) Using atom mapping rules for an improved detection of relevant routes in weighted metabolic networks. J Comput Biol, 15(6):565–576.
16. Pitkanen E, Jouhten P, Rousu J. (2009) Inferring branching pathways in genome-scale metabolic networks. BMC Syst Biol, 3(1):103.
17. Faust K, Croes D, van Helden J. (2009) Metabolic pathfinding using RPAIR annotation. J Mol Biol, 388(2):390–414.
18. IUBMB Enzyme Nomenclature. [http://www.chem.qmul.ac.uk/iubmb/enzyme/]
19. Chang A, Scheer M, Grote A, Schomburg I, Schomburg D. (2009) BRENDA, AMENDA and FRENDA the enzyme information system: new content and tools in 2009. Nucleic Acids Res, 37(Database issue):D588–D592.
20. Kanehisa M, Araki M, Goto S, Hattori M, Hirakawa M, Itoh M, Katayama T, Kawashima S, Okuda S, Tokimatsu T, Yamanishi Y. (2008) KEGG for linking genomes to life and the environment. Nucleic Acids Res, 36(Database issue):D480–D484.
21. Kümmel A, Panke S, Heinemann M. (2006) Systematic assignment of thermodynamic constraints in metabolic network models. BMC Bioinformatics, 7:512.
22. Maskow T, von Stockar U. (2005) How reliable are thermodynamic feasibility statements of biochemical pathways? Biotechnol Bioeng, 92(2):223–230.
23. Symbolism and Terminology in Enzyme Kinetics (IUBMB Recommendation 1981). Enzyme Reactions & Inhibition. [http://www.chem.qmul.ac.uk/iubmb/kinetics/ek4t6.html].
24. Price ND, Reed JL, Palsson BØ. (2004) Genome-scale models of microbial cells: evaluating the consequences of constraints. Nat Rev Microbiol, 2(11):886–897.
25. Schuster S, Fell DA, Dandekar T. (2000) A general definition of metabolic pathways useful for systematic organization and analysis of complex metabolic networks. Nat Biotechnol, 18(3):326–332.
26. Schilling CH, Schuster S, Palsson BØ, Heinrich R. (1999) Metabolic pathway analysis: basic concepts and scientific applications in the post-genomic era. Biotechnol Prog, 15(3):296–303.
27. Kaleta C, de Figueiredo LF, Schuster S. (2009) Can the whole be less than the sum

of its parts? Pathway analysis in genome-scale metabolic networks using elementary flux patterns. Genome Res, 19(10):1872–1883.
28. Klamt S, Stelling J. (2003) Two approaches for metabolic pathway analysis? Trends Biotechnol, 21(2):64–69.
29. Reed JL, Vo TD, Schilling CH, Palsson BØ. (2003) An expanded genome-scale model of *Escherichia coli* K-12 (iJR904 GSM/GPR). Genome Biol, 4:R54.
30. Satish Kumar V, Dasika MS, Maranas CD. (2007) Optimization based automated curation of metabolic reconstructions. BMC Bioinformatics, 8:212.
31. Guimerà R, Nunes Amaral LA. (2005) Functional cartography of complex metabolic networks. Nature, 433(7028):895–900.
32. Caspi R, Altman T, Dale JM, Dreher K, Fulcher CA, Gilham F, Kaipa P, Karthikeyan AS, Kothari A, Krummenacker M, Latendresse M, Mueller LA, Paley S, Popescu L, Pujar A, Shearer AG, Zhang P, Karp PD. (2010) The MetaCyc database of metabolic pathways and enzymes and the BioCyc collection of pathway/genome databases. Nucleic Acids Res, 38(Database issue):D473–D479.
33. Arita M. (2009) What can metabolomics learn from genomics and proteomics? Curr Opin Biotechnol, 20(6):610–615.
34. Hintze A, Adami C. Evolution of complex modular biological networks. PLoS Comput Biol, 4(2):e23.

Chapter 7

Predicting Metabolic Pathways by Sub-network Extraction

Karoline Faust and Jacques van Helden

Abstract

Various methods result in groups of functionally related genes obtained from genomes (operons, regulons, syntheny groups, and phylogenetic profiles), transcriptomes (co-expression groups) and proteomes (modules of interacting proteins). When such groups contain two or more enzyme-coding genes, graph analysis methods can be applied to extract a metabolic pathway that interconnects them.

We describe here the way to use the Pathway extraction tool available on the NeAT Web server (http://rsat.ulb.ac.be/neat/) to piece together the metabolic pathway from a group of associated, enzyme-coding genes.

The tool identifies the reactions that can be catalyzed by the products of the query genes (seed reactions), and applies sub-graph extraction algorithms to extract from a metabolic network a sub-network that connects the seed reactions. This sub-network represents the predicted metabolic pathway.

We describe here the pathway prediction process in a step-by-step way, give hints about the main parametric choices, and illustrate how this tool can be used to extract metabolic pathways from bacterial genomes, on the basis of two study cases: the isoleucine–valine operon in *Escherichia coli* and a predicted operon in *Cupriavidus (Ralstonia) metallidurans*.

Key words: Metabolic pathways, Pathway prediction, Pathway discovery, Sub-network extraction, NeAT

1. Introduction

Since decades, bacteria such as *Escherichia coli* have served as model organisms for characterizing metabolic reactions and pathways. The visual representation of this cumulated knowledge (e.g., the famous Boehringer-Mannheim metabolic wall chart (1)) reveals the complexity of the multiple interconnections between pathways. Exhaustive repositories have been organized and made available in several pathway databases, such as BioCyc (2) (see Chapter 11), or KEGG (3), enabling algorithmic approaches for reconstructing metabolism from a genome and analyzing pathway conservation and divergence between organisms.

Jacques van Helden et al. (eds.), *Bacterial Molecular Networks: Methods and Protocols*, Methods in Molecular Biology, vol. 804, DOI 10.1007/978-1-61779-361-5_7,

Genome and transcriptome data permit to identify groups of enzyme-coding genes that are associated by a variety of criteria: co-localization (e.g., operons), co-regulation (*cis*-regulatory motifs), co-expression (transcriptome data), and co-occurrence (phylogenetic profiles). Such associations suggest the involvement of the corresponding enzymes in some common biochemical process.

The simplest approach to interpret such sets of associated genes (the *query genes*) is to identify the reactions catalyzed by their products (*query reactions*) and map them onto pathways stored in specialized databases (e.g., refs. 4–7). However, this mapping approach has several shortcomings.

1. It relies on a collection of predefined pathways, and does not allow the detection of organism-specific variants.
2. It is unable to uncover novel pathways, even if those pathways are built from known "building blocks" (reactions and compounds).
3. When the genes match multiple pathways, the mapping approach fails to establish cross-pathway relationships.

Alternatively, pathways can be discovered by interconnecting the query reactions within the complete metabolic network, built from all the reactions and compounds from some metabolic database. The difficulty of this approach is to select, among many possible pathways, the ones that are most relevant for understanding the functional relationships between the query genes. Indeed, a naive approach that would link the query reactions by finding the shortest path between the catalyzed reactions would merely return irrelevant pathways, because it would follow invalid shortcuts through "hub" metabolites (H_2O, ATP, H^+, etc.) (8).

To circumvent this problem, we previously developed a method relying on weighted path finding, where a specific penalty is attached to each compound according to its degree in the metabolic network (number of reactions in which it participates) (9, 10). We further showed that weighted path finding can be improved (11) by building a network from KEGG reactant pairs (RPAIR) (12, 13), where relationships between reaction substrates and products are explicitly labeled according to their role in the reaction (main, trans, cofac, ligase, or leave, see Note 3).

The "2-ends" path finding problem (finding paths between two query reactions/compounds) can be extended to obtain pathways from multiple query reactions/compounds. We developed such a method (14, 15) that takes as input a set of *seed nodes* (genes, compounds, and reactions) and a metabolic network, and attempts to return a sub-network that interconnects the seed nodes at a minimal "cost," where the cost is a weighted sum of the intermediate compounds and reactions used to link the seed nodes.

Our path finding and sub-network extraction tools can be accessed via a user-friendly Web interface on the NeAT Web site (16).

In this article, we explain how pathways can be predicted by sub-network extraction, and illustrate the step-by-step procedure with two realistic examples in the bacteria *E. coli* and *Cupriavidus (Rastonia) metallidurans*, respectively.

2. Outline of the Pathway Extraction Method

The sub-network extraction procedure takes two inputs:

2.1. Input

1. A set of *seed nodes* (genes, reactions, or compounds) are provided by the user. Typical seed nodes are groups of co-expressed genes, or genes belonging to the same operon.
2. A metabolic network, built from a comprehensive collection of reactions and compounds. The collection can either be organism-specific (i.e., only contain the reactions known to be catalyzed in that organism) or generic (i.e., contain all reactions from a metabolic database). NeAT proposes several precomputed metabolic networks, and permits to upload a custom network.

2.2. Metabolic Networks

2.2.1. Preloaded Networks

The Web form of the *Pathway extraction* tool proposes several precomputed networks, which were built from exhaustive collections of reactions obtained from metabolic databases KEGG (3) and MetaCyc (17).

A metabolic network is basically built from all the small molecules and reactions stored in the reference database, irrespective of their involvement in annotated pathways. A first filtering step is applied by removing polymers (DNA, RNA, and peptides) and glycans. Furthermore, reactions involving polymers were removed as well. Additional filter steps may improve the pathway prediction accuracy (see Note 1), but curation of the networks is out of scope of this work.

2.2.2. Custom Networks

The *Pathway extraction* tool also permits to upload a custom metabolic network. A complementary tool, the *KEGG network provider*, permits to build a specific network from the KEGG PATHWAY database by selecting all the reactions potentially catalyzed by the enzyme-coding genes identified in one or several user-selected organisms.

The *Pathway extraction* tool also allows to upload any other type of network (e.g., protein interactions).

2.2.3. Reaction Reversibility

In the MetaCyc and KEGG RPAIR networks available on the NeAT Web site, all reactions are treated as reversible. This choice is justified by the fact that, under physiological conditions, irreversible reactions can be traversed in the opposite direction due to mass action.

It is however possible to specify irreversible reactions in the custom networks.

2.2.4. Weight Policies

The pathway extraction algorithms are able to deal with a network where a specific weight is assigned to each node. This weight is indicative of a penalty: high-weighting nodes will tend to be avoided when establishing the links between seed nodes.

For nonmetabolic networks (e.g., protein interactions), the pathway extraction tool assigns to each node a weight corresponding to its degree. For metabolic networks, only the weight of a compound node is set to its degree and the weight of a reaction node is set to one. We previously showed that this is a sensible weight policy for large metabolic networks (10, 11).

Alternatively, the unit weight policy (each node receives a weight of one) is available but not recommended because it does not penalize the hub metabolites.

Finally, the user may assign custom weights, according to any criterion considered as relevant.

2.3. Sub-network Extraction Algorithms

The pathway prediction tool supports several algorithms for extracting sub-networks from the input network.

Iterative path finding algorithms, such as Takahashi–Matsuyama (18) and iterative recursive enumeration algorithm (REA) (19) attempt to extract a sub-network of minimal weight, by searching the lightest paths between seed nodes and the growing sub-network (Takahashi–Matsuyama) or any pair of seed nodes (iterative REA), and combining them in order to build the sub-network that minimizes the total weight. The *lightest path* is defined as the path that minimizes the sum of node weights.

The *random walks* approach (termed kWalks) (20, 21) seeks to maximize the relevance of the sub-network it extracts. The *relevance* of a node or an edge is defined as the expected number of times it is crossed during random walks, where each random walk starts and ends in a seed node. Once a list of edge relevances is obtained, the sub-network is built by adding edges in order of decreasing relevance to the seed nodes. The algorithm stops if either all seed nodes are connected or no more edges can be added.

An important feature of the kWalks algorithm is its time efficiency, which outperforms by far the iterative path finding algorithms. The kWalks algorithm is not specifically designed to extract the optimal sub-network between multiple seed nodes, but it can be used in a *hybrid approach*, where kWalks is first used to strongly reduce the network size by selecting a relevant sub-network around the seed nodes, before applying an iterative path finding algorithm.

We have evaluated these algorithms for a number of different parameters on 71 metabolic pathways from *S. cerevisiae* (14, 15). This evaluation showed that a combination of kWalks with the Takahashi–Matsuyama algorithm yielded the highest average accuracy.

3. Study Cases: Predicting Metabolic Pathways from Operons

We present two study cases that illustrate the prediction of metabolic pathways from operons in bacterial genomes.

1. The first study case deals with an operon from *E. coli K12*, whose function is already known and which therefore serves as a positive control. For this example, we use a metabolic network built from all the reactions of the MetaCyc database.
2. In the second study case, we extract a pathway for genes predicted to belong to an operon of unknown function, predicted in the genome of *Cupriavidus (Ralstonia) metallidurans* by a simple distance-based method. We extract a pathway by interconnecting seed reactions in a network built from all reactant pairs (RPAIRS) of the KEGG metabolic database.

In bacteria, the grouping of genes in polycistronic transcription units (operons) can provide an important clue for understanding the organism-specific response to metabolic changes in the environment. In model organisms, enzymes belonging to a common operon have often been shown to catalyze successive steps of a metabolic pathway. For example, the histidine operon of *E. coli K12* contains eight genes, six of which code for enzymes involved in the histidine biosynthetic pathway. From the list of genes of this operon, we should thus be able to reconstruct the entire histidine biosynthesis pathway.

Grouping of pathway-related enzymes in operons is a quite common situation. On this basis, we could consider that, if a set of enzyme-coding genes are found in the same operon, they are possibly involved in the same metabolic process. By extracting from the complete metabolic network a sub-network linking those enzymes, we can predict the metabolic pathway they are likely to catalyze. In some cases like the His operon, all enzymes are located in the same operon, and the extraction would simply put the catalyzed reactions in the correct order. In other cases, the operon may contain only a subset of the enzymes, and the extraction permits to find the missing links.

We should however introduce some restrictions here.

- Firstly, there is no one-to-one relationship between metabolic pathways and operons. Actually, the His operon is an ideal case, but some other pathways may be split into several operons, or even scattered over the whole genome (e.g., the methionine biosynthesis pathway in *E. coli*). In such cases, other criteria may be used to regroup genes: similarity between *cis*-regulatory motifs, expression profiles, etc.
- Secondly, we have to be aware that for most bacterial genomes, the grouping of genes in operons is unknown. Operons can however be predicted with a reasonable level of accuracy, using distance-based (22) or synteny-based methods (23).

3.1. Study Case 1: Extraction of a Metabolic Pathway from the Isoleucine–Valine (ilv) Operon in E. coli and the MetaCyc Network

We illustrate metabolic pathway discovery first on a group of query genes taken from the isoleucine–valine (ilv) operon in *E. coli*. This operon contains seven genes, six of which are coding for enzymes involved in the biosynthesis of isoleucine and valine. Information on this operon was obtained from RegulonDB (24).

Starting from these genes, we extract a sub-network from a generic metabolic network built from all reactions and compounds in MetaCyc (version 11) (17), the well-annotated tier of BioCyc. This network consists of 4,893 compounds and 5,358 reactions. In order to allow reactions to be traversed in both directions, we instantiate two nodes per reaction (see Note 2). The complete network thus contains 15,609 nodes in total.

3.2. Study Case 2: Extraction of a Metabolic Pathway from a Predicted Operon in Cupriavidus (Ralstonia) metallidurans and the KEGG RPAIR Network

Cupriavidus metallidurans (formerly named *Ralstonia metallidurans*) is known for its capability to survive in environments with high concentrations of heavy metals (25). Its metabolism was so far reconstructed only by automatic procedures (e.g., in BioCyc with PathoLogic (26)).

The program *infer-operon* of the RSAT package (27) applies a simple distance-based method (derived from ref. 2), where consecutive genes oriented in the same direction and separated by a distance shorter than 55 bp are predicted to belong to the same operon. Given the 6,176 protein-coding genes of *Cupriavidus metallidurans*, infer-operons predicted 1,220 operons (i.e., transcription units containing at least two genes), 58 of which contain three or more enzyme-coding genes.

As a study case, we selected an operon that consists of three enzyme-coding genes (Rmet_4878, Rmet_4877 and Rmet_4875) and an additional gene that codes for an uncharacterized protein (Rmet_4876). This operon is hereafter referred to as "operon Rmet_4878", after the name of its leader gene.

Given the three enzyme-coding genes, we extract a sub-network from a generic metabolic network built from KEGG RPAIR (version 49), which consists of 16,826 nodes (11,066 reactant pairs and 5,760 compounds) and 44,236 edges (see Note 3). In contrast to the MetaCyc network, the KEGG RPAIR network is undirected (see Note 2). As for a justification of the network choice, see Note 4.

4. Materials

4.1. Software

The only software required to carry out this protocol is a Web browser with a connection to the Internet.

The *Pathway extraction* tool used here is publicly available as a part of the NeAT Web interface (http://rsat.ulb.ac.be/neat/) (16) (see also Chapter 18).

4.2. File Formats

4.2.1. Input Network (Graph)

Several metabolic networks (or graphs, see Note 5) are supported on the NeAT Web server. Alternatively, custom networks can be uploaded in the following formats: tab-delimited (node and edge list, see detailed description on the NeAT Web site), GML (Graph Modeling Language (28)), and KGML (KEGG Markup Language (29)).

4.2.2. Seed Nodes

For metabolic networks from KEGG or MetaCyc, the following seeds are accepted: gene names, (partial) enzyme names (MetaCyc only), UniProt identifiers (MetaCyc only), RefSeq identifiers (MetaCyc only), (partial) compound names, EC numbers and identifiers of compounds, reactions, and reactant pairs (reactant pairs are accepted for KEGG only). For any other network, seeds have to be identifiers of nodes present in the network.

4.2.3. Output Graph

The extracted "pathway" is a subgraph of the input network (see Note 6). The pathway can be exported in various formats: tab-delimited, GML, DOT (Graphviz graph drawing language (30)), or VisML (VisANT XML). The resulting files can be loaded into various visualization tools, such as VisANT (31), Cytoscape (32) or yEd (33).

5. Methods

5.1. Study Case 1: Extraction of a Metabolic Pathway from the MetaCyc Network (Ilv Operon)

In this section, we show step-by-step how to extract a metabolic pathway from the MetaCyc network. The network is provided by the *Pathway extraction* Web application.

5.1.1. Pathway Extraction Parameters

1. In the NeAT menu (http://rsat.ulb.ac.be/neat/), open the title box *Path finding and pathway extraction* (click on the white triangle to expand the tool list). Click *Pathway extraction*.
2. In the pop-up menu *Preloaded network*, select the metabolic network "MetaCyc".
3. Select "degree" as *Weighting scheme*.
4. Paste the following genes into the seed node text field (one gene name by line):

   ```
   ilvL
   ilvG_1
   ilvG_2
   ilvM
   ilvE
   ilvD
   ilvA.
   ```
5. Next to the *seed node* text field, select "Genes/Enzymes" as *identifier type*.
6. For the option *Metacyc network: select source organism of genes/enzymes*, select "*Escherichia coli* K-12 substr. MG1655".
7. For the option Algorithm, select "kWalks + Takahashi–Matsuyama" (see Note 7).
8. Leave the options *Preprocess and Postprocess* unchecked (see Notes 9 and 10).
9. Set *Iteration number* to 1 (see Note 11).
10. Set the *Percentage of input network* to 0.05 (see Note 12).
11. Leave unchecked the option *Compute weights with kWalks* (see Note 13).
12. On the bottom of the form, click the "GO" button.

5.1.2. Reaction Grouping Rules

1. After a few seconds, a page will display the result of the association of genes to MetaCyc reaction identifiers (see Note 14). Figure 1 reproduces this result.

NeAT - Pathway extraction seed node selection

The table below lists information on each of the given seed node identifiers.
Green means that the given seed node identifier has been found in the database.
Orange means that there are several matches for the given seed node identifier and that you have to choose the appropriate one. For instance, if you give "lysine" as seed node identifier, should it be L-Lysine or D-Lysine?
Red means that the given seed node identifier could not be found in the database and will be neglected.

You provided invalid identifier(s)!

Provided identifier	Name in MetaCyc	Associated EC numbers	Seeds used for pathway prediction	Group of seed	Identifier type	Problem with identifier
ilvD	IlvD	4.2.1.9	[DIHYDROXYISOVALDEHYDRAT-RXN/DIHYDROXYMETVALDEHYDRAT-RXN]	ilvD_group6	Gene	
ilvL			[]	ilvL_group1	Gene	Gene is not present in the database for the selected organism.
ilvA	IlvA	4.3.1.19	[THREDEHYD-RXN]	ilvA_group7	Gene	
ilvE	IlvE	2.6.1.42, 2.6.1.5, 2.6.1.57	[PHEAMINOTRANS-RXN/BRANCHED-CHAINAMINOTRANSFERVAL-RXN/BRANCHED-CHAINAMINOTRANSFERLEU-RXN/TYROSINE-AMINOTRANSFERASE-RXN/BRANCHED-CHAINAMINOTRANSFERILEU-RXN]	ilvE_group5	Gene	
ilvG_2	IlvG_2	2.2.1.6	[ACETOLACTSYN-RXN/ACETOOHBUTSYN-RXN]	ilvG_2_group3	Gene	Given gene is annotated as pseudogene!
ilvM	IlvM	2.2.1.6	[ACETOLACTSYN-RXN/ACETOOHBUTSYN-RXN]	ilvM_group4	Gene	
ilvG_1			[]	ilvG_1_group2	Gene	Gene is not present in the database for the selected organism.

Fig. 1. First result page of the *pathway extraction* tool, displaying the links from seed genes (isoleucine–valine operon) to EC numbers and reactions in MetaCyc.

2. For the group treatment, select "Group reactions by EC number" (see Note 15).
3. Click on "GO".
4. After 2 or 3 min (depending on server load), the result should appear on the screen.

5.1.3. Interpretation of the Pathway Extracted from the Ilv Operon of E. coli

The first step of pathway extraction is to establish the links between seed genes and the reactions they can catalyze. MetaCyc allows linking genes directly to reactions. Reactions are annotated with EC numbers, which systematically classify enzymes according to their catalytic mechanisms.

The mapping of genes to reactions shows that, among the seven genes of the *ilv* operon, five are associated with some enzymatic function (Fig. 1). Several genes (*ilvE*, *ilvM*, and *ilvD*) are associated to multiple reactions having the same EC number. This reflects the fact that the EC number notation is less precise than the reaction identifiers indicated by MetaCyc because one EC number may describe a reaction mechanism that can act on different substrates. For instance, EC number 2.6.1.42 is associated to three reactions in MetaCyc, each involving a different

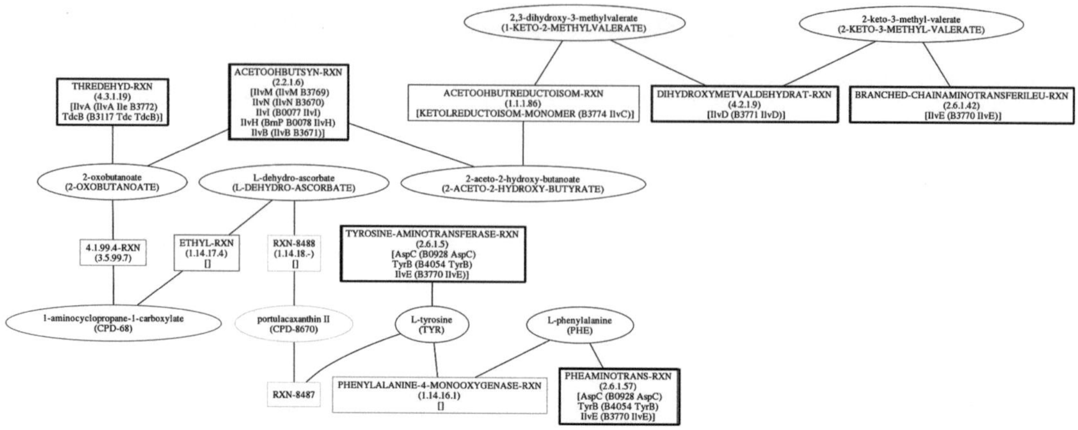

Fig. 2. Pathway returned for study case 1 (isoleucine–valine operon in *Escherichia coli* and MetaCyc network). *Shape code*: *ellipse*: compound (labeled with its name and MetaCyc identifier); *rectangle*: reaction (labeled with its MetaCyc identifier, associated EC numbers and enzymes in MetaCyc). *Color code*: *thick black border*: seed node and *gray border*: linking nodes.

amino acid (leucine, valine, and isoleucine). However, each enzyme generally acts on only one of its potential substrates, the one produced by the preceding reaction of the pathway. Our pragmatic approach is thus to group reactions by EC numbers: the set of reactions associated with a given EC number are considered as an equivalence group, and as soon as one of them has been included in the solution, the other ones are neglected.

The pathway extracted from MetaCyc given the genes of the ilv operon is shown in Fig. 2. Alternative pathways may be returned by the tool, but all feature the branch from reaction THREDEHYD-RXN to reaction BRANCHED-CHAINAMINOTRANSFERILEU-RXN. The pathway depicted in Fig. 2 contains 12 reactions, six of which are catalyzed by the seed genes.

Below the image of the predicted pathway, the result page lists correspondences between this predicted pathway and a list of reference pathways (those annotated in the MetaCyc database). Matching pathways are sorted according to the significance of the overlap (Fig. 3). The pathway predicted for the *ilv* operon matched most significantly the MetaCyc pathway "isoleucine biosynthesis I (from threonine)": the intersection between the 16 nodes (compounds and reactions) of the reference pathway (column *R*) and the 21 nodes of the predicted pathway (column *Q*) covers 10 nodes (column *QR*). Given the total node number of the MetaCyc network (10,251), the probability to observe such an overlap by chance is 8*e*–25 (column *P_val*).

This nominal probability can however be misleading, since the predicted pathway was compared with a large number of reference pathways (1,752): the chances to get a spurious match are multiplied by the number of comparisons between

Significance of overlap between predicted subnetwork and reference pathways

ref = Reference pathway
query = Predicted pathway
R = Number of nodes in reference pathway.
Q = Number of nodes in predicted pathway.
QR = Number of nodes in the intersection of the reference and predicted node set.
QvR = Number of nodes in the union of the reference and predicted node sets.
R!Q = Number of nodes present in the reference but not in the predicted node set.
Q!R = Number of nodes present in the predicted but not in the reference node set.
jac_sim = Jaccard similarity. For 2 node sets A and B: jac_sim = |A intersection B| / |A union B|)
P_val = P-value of the intersection, calculated with he hypergeometric function. Pval = P(X >= QR). The population size corresponds to the node number in the input network (10251).
E_val = E-value of the intersection. E_val = P_val * number_of_tests. The number of tests corresponds to the number of reference pathways in the selected metabolic database (1752).
sig = Significance of the intersection. sig = -log10(E_val)

ref	query	R	Q	QR	QvR	R!Q	Q!R	jac_sim	P_val	E_val	sig
isoleucine_biosynthesis_I_(from_threonine)	predicted	16	21	10	27	6	11	0.37037	8e-25	1.4016000000000001E-21	20.853
superpathway_of_leucine,_valine,_and_isoleucine_biosynthesis	predicted	35	21	11	45	24	10	0.24444	4.4e-23	7.7088E-20	19.113
isoleucine_biosynthesis_I	predicted	27	21	10	38	17	11	0.26316	8.3e-22	1.45416E-18	17.837
isoleucine_biosynthesis_IV	predicted	7	21	7	21	0	14	0.33333	4.9e-20	8.584800000000001E-17	16.066
superpathway_of_threonine_metabolism	predicted	42	21	10	53	32	11	0.18868	1.4e-19	2.4528E-16	15.61
isoleucine_biosynthesis_II	predicted	22	21	7	36	15	14	0.19444	8.3e-15	1.45416E-11	10.837
valine_biosynthesis	predicted	14	21	6	29	8	15	0.20690	1e-13	1.752E-10	9.756
isoleucine_biosynthesis_III	predicted	10	21	5	26	5	16	0.19231	5.4e-12	9.460799999999999E-9	8.024
alanine_biosynthesis_I	predicted	11	21	3	29	8	18	0.10345	1.2e-06	0.0021024	2.677
superpathway_of_betalain_biosynthesis	predicted	47	21	4	64	43	17	0.06250	2.2e-06	0.0038544	2.414
superpathway_of_alanine_biosynthesis	predicted	14	21	3	32	11	18	0.09375	2.7e-06	0.0047304	2.325
tyrosine_biosynthesis_IV	predicted	2	21	2	21	0	19	0.09524	4e-06	0.0070079999999999995	2.154
leucine_biosynthesis	predicted	19	21	3	37	16	18	0.08108	7e-06	0.012264	1.911
4-hydroxyphenylpyruvate_biosynthesis	predicted	3	21	2	22	1	19	0.09091	1.2e-05	0.021024	1.677
aspartate_biosynthesis	predicted	5	21	2	24	3	19	0.08333	4e-05	0.07008	1.154
isoleucine_biosynthesis_V	predicted	5	21	2	24	3	19	0.08333	4e-05	0.07008	1.154
superpathway_of_phenylalanine,_tyrosine,_and_tryptophan_biosynthesis	predicted	41	21	3	59	38	18	0.05085	7.5e-05	0.1314	0.881

Fig. 3. The *pathway extraction* tool generates a list of reference pathways that overlap with the predicted pathway by calling the RSAT/NeAT tool compare-classes. The pathway predicted for the isoleucine–valine operon most significantly overlaps with the "isoleucine biosynthesis I (from threonine)" pathway in MetaCyc.

predicted and reference pathways. This multitesting effect is compensated by computing the *E*-value (column *E_val*), which indicates the number of matches expected by chance: $E\text{val} = P\text{val} \times N_{\text{R}}$, where N_{R} is the number of reference pathways. The rows having an *E*-value >1 in Fig. 3 should thus not be considered as statistically significant.

As pointed out by the hypergeometric test, the predicted pathway correctly reproduces the annotated isoleucine pathway but contains additional reactions and compounds. Figure 4 shows a comparison of the predicted and annotated pathway.

The isoleucine–valine operon is a particular case of parallel pathways: the same set of enzymes catalyzes two independent pathways (isoleucine and valine biosynthesis), by applying the same succession of chemical operations on two different substrates (pyruvate for valine biosynthesis and threonine for isoleucine biosynthesis).

The predicted pathway fails to cover the valine pathway because seed reactions were grouped according to their EC numbers. An (inclusive) OR relationship exists between reactions of an EC number group, which means that only one reaction needs to be contributed to the predicted pathway to consider the EC number group as connected. However, the valine and the isoleucine reference pathway contains four EC numbers associated to multiple reactions (2.2.1.6, 1.1.1.86, 4.2.1.9, and 2.6.1.42). Since the four EC number groups could be connected with the intermediates of the isoleucine pathway only, the valine pathway was not predicted.

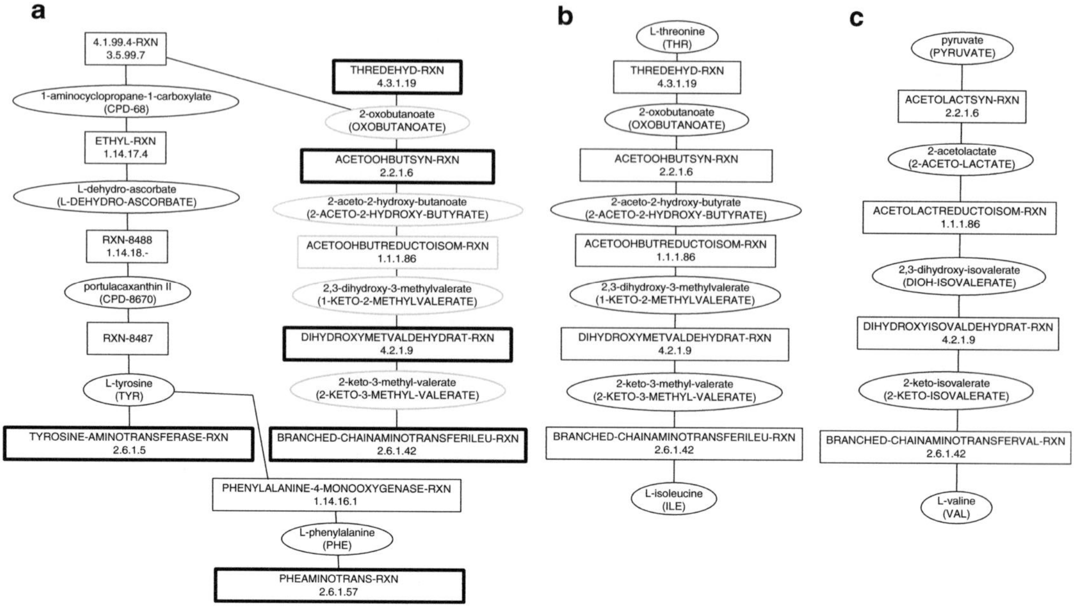

Fig. 4. Comparison of the pathway predicted from the *ilv* operon (**a**) and two reference pathways: isoleucine (**b**) and valine (**c**) biosynthesis. Non-seed nodes present in the predicted and the reference pathway are counted as true positives, those present in the predicted but absent from the reference pathway as false positives. *Shape code*: *ellipse*: compound (labeled with its name and database identifier); *rectangle*: reaction/reactant pair (labeled with MetaCyc identifier and associated EC numbers). *Color code a*: *thick black border*: seed, *gray border*: true positive, *black border*: false positive

5.2. Study Case 2: Extraction of a Metabolic Pathway from the KEGG RPAIR Network

For the second study case, we test a different network, built from the complete set of reactant pairs (RPAIRS) of the KEGG database.

5.2.1. Pathway Extraction Parameters

1. In the NeAT menu (left frame), click *Pathway extraction.*
2. In the pop-up menu *Preloaded network*, select the metabolic network “KEGG RPAIR”.
3. Make sure that the Weighting scheme is set to “degree”.
4. Paste the following genes in the seed node text field:

   ```
   Rmet_4878
   Rmet_4877
   Rmet_4876
   Rmet_4875.
   ```

5. For the option *Identifier type*, check the radio button “Genes/Enzymes”.
6. Next to the seed node text field, click on the pop-up menu KEGG *networks: select source organism of genes/enzymes*, and select “*Cupriavidus metallidurans* (rme)”.

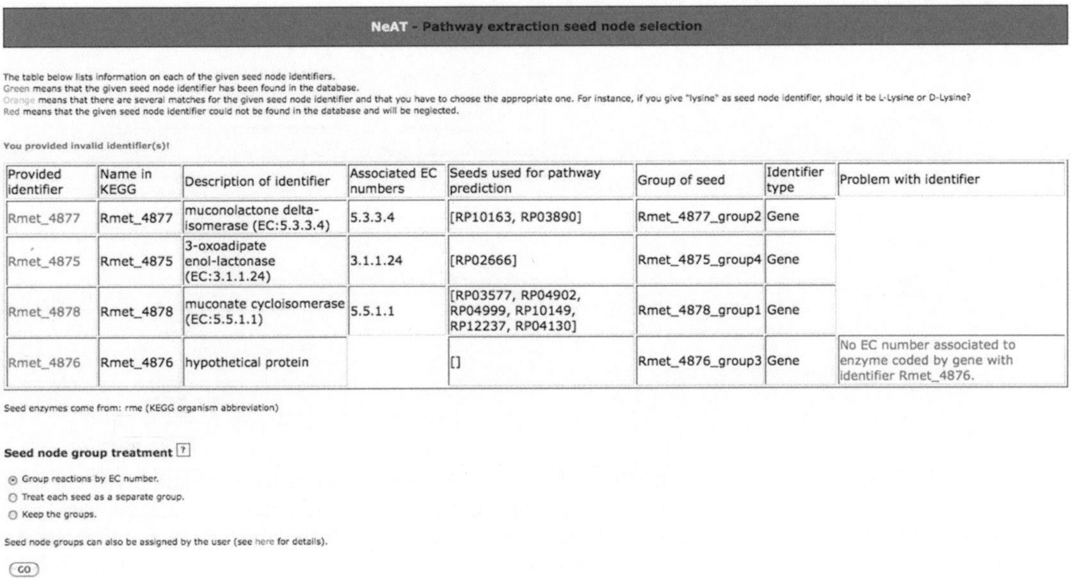

NeAT - Pathway extraction seed node selection

The table below lists information on each of the given seed node identifiers.
Green means that the given seed node identifier has been found in the database.
Orange means that there are several matches for the given seed node identifier and that you have to choose the appropriate one. For instance, if you give "lysine" as seed node identifier, should it be L-Lysine or D-Lysine?
Red means that the given seed node identifier could not be found in the database and will be neglected.

You provided invalid identifier(s)!

Provided identifier	Name in KEGG	Description of identifier	Associated EC numbers	Seeds used for pathway prediction	Group of seed	Identifier type	Problem with identifier
Rmet_4877	Rmet_4877	muconolactone delta-isomerase (EC:5.3.3.4)	5.3.3.4	[RP10163, RP03890]	Rmet_4877_group2	Gene	
Rmet_4875	Rmet_4875	3-oxoadipate enol-lactonase (EC:3.1.1.24)	3.1.1.24	[RP02666]	Rmet_4875_group4	Gene	
Rmet_4878	Rmet_4878	muconate cycloisomerase (EC:5.5.1.1)	5.5.1.1	[RP03577, RP04902, RP04999, RP10149, RP12237, RP04130]	Rmet_4878_group1	Gene	
Rmet_4876	Rmet_4876	hypothetical protein		[]	Rmet_4876_group3	Gene	No EC number associated to enzyme coded by gene with identifier Rmet_4876.

Seed enzymes come from: rme (KEGG organism abbreviation)

Seed node group treatment ?

- ◉ Group reactions by EC number.
- ○ Treat each seed as a separate group.
- ○ Keep the groups.

Seed node groups can also be assigned by the user (see here for details).

GO

Fig. 5. First result page of the pathway extraction tool, displaying the links from seed genes (the predicted *Cupriavidus metallidurans* operon) to EC numbers and reactant pairs in KEGG.

7. Scroll down and activate the option *Preprocess* (see Note 9).
8. Leave the option *Postprocess* unchecked.
9. Set the other options in this form as described in steps 7, 9, 10 and 11 for the first study case.
10. Click *GO*. After a few seconds, you should obtain a table linking the seed genes to reactions.
11. For the option *Seed node group treatment*, select "Group reactions by EC number".
12. Click GO. After some minutes, the result should be displayed.

5.2.2. Interpretation of the Pathway Extracted from the Cupriavidus metallidurans Operon of Unknown Function

Figure 5 displays the mapping of seed genes to KEGG reactant pairs. In KEGG, genes have to be linked to reactions via EC numbers. Three of the four genes of the predicted operon are associated with enzymes. The link between enzymes and reactant pairs via EC numbers introduces some ambiguities: Rmet_4877 is associated to two reactant pairs via EC number 5.3.3.4, whereas Rmet_4878 is associated to six reactions via the EC 5.5.1.1.

Figure 6 depicts the pathway extracted for the enzyme-coding genes of the predicted operon Rmet_4878 in *C. metallidurans*. All the reactant pairs used in this pathway are highlighted in bold, showing that they all correspond to seed genes. From 2-Oxo-2,3-dihydrofurano-5-acetate, the pathway shows two parallel branches

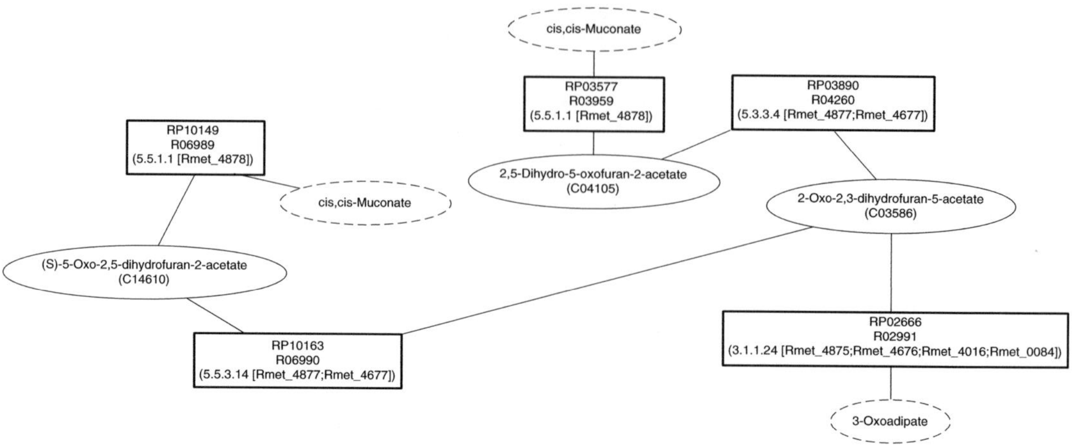

Fig. 6. Pathway extracted for study case 2 (operon predicted in *Cupriavidus metallidurans*, the KEGG RPAIR network). *Shape code*: *ellipse*: compound (labeled with name and KEGG identifier); *rectangle*: reactant pair (labeled with RPAIR identifier, associated reactions, EC numbers and enzymes-coding genes). *Color code*: same as Fig. 2. In addition *dotted border*: compounds adjacent to seed reactant pairs added after prediction.

Significance of overlap between predicted subnetwork and reference pathways

ref = Reference pathway
query = Predicted pathway
R = Number of nodes in reference pathway.
Q = Number of nodes in predicted pathway.
QR = Number of nodes in the intersection of the reference and predicted node set.
QvR = Number of nodes in the union of the reference and predicted node sets.
R!Q = Number of nodes present in the reference but not in the predicted node set.
Q!R = Number of nodes present in the predicted but not in the reference node set.
jac_sim = Jaccard similarity. For 2 node sets A and B: jac_sim = |A intersection B| / |A union B|)
P_val = P-value of the intersection, calculated with he hypergeometric function. Pval = P(X >= QR). The population size corresponds to the node number in the input network (16826).
E_val = E-value of the intersection. E_val = P_val * number_of_tests. The number of tests corresponds to the number of reference pathways in the selected metabolic database (145).
sig = Significance of the intersection. sig = -log10(E_val)

ref	query	R	Q	QR	QvR	R!Q	Q!R	jac_sim	P_val	E_val	sig
Benzoate_degradation_via_hydroxylation	predicted	98	11	7	102	91	4	0.06863	5.9e-14	8.555E-12	11.068

Fig. 7. The pathway predicted for the *Cupriavidus metallidurans* operon overlaps significantly with the "Benzoate degradation via hydroxylation" KEGG reference map.

passing by the same succession of EC numbers (5.3.3.4 then 5.5.1.1), but with different substrates. The two branches merge because the two last RPAIRS (RP10149 and RP03577) produce the same molecule (*cis,cis*-Muconate). The pathway extraction thus suggests two alternative paths for producing *cis,cis*-Muconate from 2-Oxo-2,3-dihydrofuran-5-acetate, but this apparent branching is due to the presence of stereoisomers in KEGG. Indeed, the two KEGG compound entries 2,5-Dihydro-5-oxofuran-2-acetate (C04105) and (S)-5-Oxo-2,5-dihydrofuran-2-acetate (C14610) refer to the same compound, namely muconolactone. The first describes the compound without taking into account stereoisomers, whereas the second refers to a stereoisomer.

The predicted pathway maps to only one KEGG map, namely the "Benzoate degradation via hydroxylation" reference map (see Fig. 7). Among the eight nodes of the predicted pathway, five are

Properties of predicted subnetwork

Node number	11
Edge number	10
Compound number	6
Reaction number	5
Number of components	1
Number of branching points (branching points are nodes having more than 2 neighbours, direct and reverse direction of reaction are counted as one neighbour)	1
Number of orphan seed groups	0
Orphan seed groups	[]
Maximal distance between two seed nodes (distance = number of edges)	2.0
Mean distance between two seed nodes	2.0
Seed node coverage (Ratio of seed nodes versus all nodes (direct and reverse direction of a reaction count as one node))	0.455
Seed reaction coverage (Ratio of seed reactions versus all reactions (direct and reverse direction of a reaction count as one node))	1.0
Known pathways overlapping with predicted pathway	[Benzoate_degradation_via_hydroxylation]
Compounds occurring in known pathways	[C02480, C00846, C14610, C03586]
Reactions occurring in known pathways	[]

Fig. 8. Basic properties of the pathway predicted for the *Cupriavidus metallidurans* operon.

found on this map. The three predicted nodes not in the map are the stereoisomer of muconolactone and its adjacent reactions.

In summary, the discovered pathway converts *cis,cis*-muconate (C02480) into 2-Oxo-2,3-dihydrofuran-5-acetate (C03586), via muconolactone. These are steps of the well-known catechol ortho-cleavage pathway. In MetaCyc, this pathway is stored as catechol degradation III (identifier: PWY-5417).

Thus, the pathway extraction tool predicts that *C. metallidurans* is capable of degrading catechol. This prediction agrees with biochemical knowledge on *C. metallidurans* (34).

5.3. General Guidelines for Interpreting Pathway Extraction Results

After having discussed the particular interpretation of two study cases, we provide hereafter some general guidelines for interpreting pathways predicted by sub-network extraction.

Beyond the visual inspection of the pathway diagrams (Figs. 2 and 6), the reliability of the predicted pathways can be evaluated on the basis of some statistics (e.g., Fig. 8) and by a detailed analysis of the nodes used to link the seeds.

- Seed node coverage. The seed node coverage is defined as the fraction of seeds among all nodes of the predicted pathway. A high seed node coverage indicates that the predicted pathway is strongly supported by the seed nodes.
- *Distance between seed nodes.* The reliability of the predicted pathway decreases with the distance between seeds. For instance, a long branch not supported by any seed node except the terminal one very likely represents a false positive. The part of the pathway predicted for the first study case going from reaction THREDEHYD-RXN to reaction TYROSINE-AMINOTRANSFERASE-RXN is a good example for such a branch.
- *Presence of enzymes catalyzing predicted reactions in the query organism.* The resulting pathway consists of seed nodes and predicted nodes connecting the seeds. A pathway is more reliable if all its predicted reactions are likely to be catalyzed in the organism of interest. However, reactions not associated

to any gene in the query organism might also be spontaneous. Also, some enzymes may be missing due to errors in the genome annotation process.

- *Presence of predicted nodes in the source data set.* Closer inspection of the source data set may indicate that seed nodes and some predicted nodes belong to the same group. For instance, in a microarray data set, genes may have been removed from a cluster because their expression ratios were below the given threshold. However, these genes may be part of the pathway connecting the seed genes.
- *Reproducibility.* If several data sources exist for the same kind of data, pathway discovery can be repeated in a network constructed from an alternative source. For example, *Pathway extraction* offers metabolic networks from databases assembled by two independent groups, namely the KEGG and MetaCyc teams. The reproduction of a pathway discovered in the KEGG network with the MetaCyc network may increase the confidence in this pathway.
- *Known pathways.* Known pathways overlapping with the extracted pathway may help to interpret the latter one. For instance, the extracted pathway may be a combination of known pathways or may combine a known pathway with a novel one. To ease interpretation, *Pathway extraction* displays links to the overlapping pathways annotated in KEGG or MetaCyc (depending on the network used).
- *Layout. Pathway extraction* makes use of Graphviz to compute the pathway layout. Graphviz does not deal well with large networks, making interpretation of large pathways difficult. We recommend to export the resulting pathway into a tool offering various graph layouts, such as VisANT, yEd, or Cytoscape. The *Pathway extraction* result page includes a section called "Next steps", which offers options to visualize the predicted pathway with VisANT or the NeAT program *display-graph* or to export it into another format with the NeAT tool *convert-graph.*
- *Subsequent analysis. Pathway* extraction may return large pathways, which are impossible to interpret. These pathways may be submitted to other graph analysis tools, e.g., to MCL (see Chapter 15), RNSC (see Chapter 16) (both integrated into NeAT), or to the NeAT tools *graph-neighbors*, *Pathfinder* or even *Pathway extraction* itself (all available via the "Next steps" section) to divide the sub-network into clusters, to extract neighbors of query nodes, or to connect nodes of interest.

Notes 16 and 17 discuss errors and weaknesses of pathway extraction. Note 10 can be consulted for the treatment of disconnected result pathways.

6. Notes

1. *Network quality.* The accuracy of predicted pathways strongly depends on the quality of the input network. If a specific node is absent from the input network, a pathway containing this node cannot be predicted. On the other side, a wrong connection between two nodes might lead to the extraction of an erroneous pathway. KEGG LIGAND is known to contain problematic entries (e.g., refs. 35, 36), which were not treated in NeAT.
2. *Construction of metabolic networks.* In the MetaCyc network, a reaction may have several substrates and/or products. To prevent path-based graph algorithms to go from one substrate to another substrate or to go from one product to another product, the network is directed. The two directions of each reaction are represented by two different, mutually exclusive nodes (9). Since each reactant pair has by definition only one substrate and one product (see Note 3), the KEGG RPAIR network can be undirected. The metabolic networks supported on the NeAT Web site are symmetric, meaning that the path from A to B has the same weight as the path from B to A ($w(p_{A \to B}) = w(p_{B \to A})$). Note that sub-network extraction is much quicker in symmetric than in asymmetric networks.
3. *Reactant pairs in KEGG.* In KEGG, reactions are decomposed into reactant pairs, which establish a link between a single substrate and a single product of a reaction (12, 13). Different roles are assigned to reactant pairs. For example, in the reaction R00480:

   L-Aspartate + ATP ↔ 4-Phospho-L-aspartate + ADP,

   the reactant pair L-Aspartate/4-Phospho-L-aspartate (RP00932) has the role "main", whereas the reactant pair ATP/4-Phospho-L-aspartate (RP06173) has the role "trans". This annotation helps to improve the accuracy of pathway discovery (11).
4. *Choice of metabolic network.* The two metabolic databases used by the *Pathway extraction* tool have different strengths and weaknesses.
   - *Gene – reaction mapping.* MetaCyc links genes directly to their reactions. When querying KEGG via its Web site or via its API, such a linking cannot be achieved. Instead, genes have to be linked to reactions via EC numbers, i.e., one has to retrieve all EC numbers associated to an enzyme-coding gene and then all reactions of each EC

number, thus loosing precision in the gene-reaction mapping.

- *Predicted versus annotated pathways.* BioCyc makes a clear distinction between annotated and computationally predicted pathways (MetaCyc pathways are experimentally confirmed, BioCyc tier 2 and 3 pathways are not), which is not the case for KEGG.
- *Organism coverage.* MetaCyc does not cover as many organisms as KEGG does. For instance, MetaCyc version 11 did not contain *Ralstonia metallidurans*, but this organism was present in KEGG version 49.0.
- KEGG offers reactant pair annotations, which increase pathway prediction accuracy (see Note 3).
  For all these reasons, the KEGG RPAIR network was selected for the second study case.
  The choice of the metabolic network depends on the application in mind.

5. *Networks and graphs.* The concept of "biological network" refers to a set of entities (e.g., genes, proteins, metabolites, and reactions) linked by relationships (e.g., genetic interaction, protein interaction, and substrate/product relationship). This corresponds to the mathematical definition of a graph: a set of *nodes* (also called *vertices*) linked by *edges* (undirected graphs) or *arcs* (directed graphs). In this article, the terms "network" and "graph" are considered as synonyms. We prefer "network" for referring to the biological concept (e.g., metabolic network), and "graph" for its formal representation (e.g., graph file format).
6. *Pathways and paths.* Biochemists use the term "pathway" to indicate a set of interconnected reactions involved in some biological function. A biochemical pathway can be branched and contain cycles. This differs from the graph theory definition of pathway, which refers to a succession of adjacent edges with no branch or cycle. In this article, we use the term "pathway" for the biochemical concept, and "path" for the graph theory concept.
7. *Choice of algorithm.* By default, Pathway extraction executes a combination of kWalks with the Takahashi–Matsuyama algorithm (18). This combination, termed hybrid algorithm, yielded the highest accuracy in our evaluation (15) and runs quicker than the Takahashi–Matsuyama algorithm alone. In principle, the Takahashi–Matsuyama algorithm may return different solutions upon repeated launches. However, in our experience, this is rarely the case. Repetitive REA returns the same solution for repeated launches (except for the case where not all paths were collected due to a huge number of

alternatives, see Note 16). Note that the computational complexity of repetitive REA squares that of Takahashi–Matsuyama, while its accuracy in our evaluation was below the accuracy of the latter algorithm. However, since repetitive REA computes paths for all seed node pairs, it will connect non-orphan seeds even in the presence of orphan seeds (see Note 8). In contrast, Takahashi–Matsuyama will fail to connect remaining seeds when it encounters an orphan seed.

8. *Orphan node.* An orphan node is not connected to any other node in the network, i.e., it has no edges.

9. *Preprocessing.* Preprocessing only applies to RPAIR networks. It consists in linking pairs of adjacent seeds before running the extraction algorithm. Two types of adjacency are treated.

   (a) If the seed set contains a RPAIR and one of its compounds, the compound and the RPAIR are joined.

   (b) If the seed set contains two main RPAIRS (see Note 3) sharing the same compound, they are directly connected, irrespective of the compound weight.

   In most cases, joining main reactant pairs increase the accuracy of extraction. However, there are annotated pathways that contain nonmain reactant pairs (see ref. 11). Preprocessing is restricted to RPAIR networks because in networks of reactions we do not know if a compound shared by two seed reactions is a main or a side compound. Applying such a preprocessing to reaction networks would thus link reactions via irrelevant compounds, such as H_2O or ATP.

10. *Postprocessing.* Pathway extraction may return a pathway consisting of several components. In this case, the extraction may be repeated with the option Postprocess enabled. During postprocessing, each component of the result network is treated as a seed node group. Postprocess attempts to connect these groups using the selected pathway extraction algorithm. Consider for example three seed groups A, B, and C, containing the seed nodes A1, A2 (in seed group A), B1, B2 (in seed group B), and C1 (in seed group C). During pathway prediction, A1 may be connected to B1, and B2 to C1. This results in a predicted pathway consisting of two components. In order to connect these components, we may launch pathway prediction on rearranged seed groups, namely A1 + B1 as one group, and B2 + C1 as the other group. If we find a connection between the two rearranged seed groups, a pathway consisting of one component results.

11. *Iteration number.* The iteration number determines how often the kWalks algorithm is executed. Calling kWalks iteratively enhances the difference between relevant and irrelevant nodes and edges. However, each iteration comes at the cost of

an increased runtime and above a certain threshold does no longer improve pathway prediction accuracy. For metabolic networks, our evaluation showed that 3 is a reasonable iteration number. In hybrid algorithms, iteration of kWalks does not have a significant impact on accuracy, and can in some case reduce the accuracy; therefore, the iteration number should be set to three when using kWalks alone, and to one when using the hybrid algorithm.

12. *Percentage of input network.* The parameter "percentage of the input network" determines the size of the sub-network extracted by kWalks in the first step of a hybrid algorithm. Reducing the input network percentage may increase the accuracy for large networks. However, if the network is not sufficiently large, it may lead to a disconnected sub-network in the first step of the hybrid algorithm, preventing the extraction of a pathway. The default value of this parameter (5%) is tailored to the supported metabolic networks and may need adjustment for other networks.

13. *Weight discovery.* The kWalks algorithm computes relevances on nodes and edges. If neither a good weight policy nor custom weights are available, these relevances can be used as weights.

14. *Associating genes to reactions.* Seed genes and EC numbers have to be associated to reactions to connect them in a metabolic network consisting only of reactions and compounds. The more precise this association is, the more accurate is the extraction of pathways. An enzyme-coding gene may be associated to several EC numbers (e.g., yeast gene aro1 codes for a pentafunctional enzyme with EC numbers 1.1.1.25, 2.5.1.19, 2.7.1.71, 4.2.3.4, and 4.2.1.10) and each EC number in turn may be associated to several reactions. Vice versa, a reaction may be associated to several genes (e.g., in the case of isoenzymes). Thus, an N:N relationship between genes and reactions exists. For the KEGG network, the current KEGG database is queried to obtain EC numbers associated with the query gene. These EC numbers are linked to their reactions using a local database containing KEGG data. For the KEGG RPAIR network, each reaction is in addition associated to its main reactant pair(s) (see Note 3). For instance, gene ilvD is annotated with EC number 4.2.1.9, which is in KEGG version 49.0 associated to reactions R01209 (with main reactant pairs RP01363 and RP10127) and R04441 (RP04080). MetaCyc annotation allows to link genes directly to reactions without going via EC numbers. For instance, in MetaCyc version 11, gene ilvM is associated to two reactions with identifiers ACETOOHBUTSYN-RXN and ACETOLACT-SYN-RXN.

15. *Group treatment*. When genes or enzymes are given as queries, a single query can correspond to multiple seed nodes for different reasons: (1) enzymes are characterized by an EC number, which may correspond to multiple reactions, reflecting either a broad substrate spectrum (e.g., alcohol dehydrogenase), or the capability of the enzyme to use alternative co-factors (e.g., $NADPH^+$ or $NADH^+$); (2) a single gene/enzyme may contain several enzymatic domains catalyzing different steps of a same pathway (e.g., aspartokinase/homoserine dehydrogenase I from *E. coli*). In the first case, when a single EC is linked to multiple reactions, we would generally consider those reactions as alternative choices, among which only one has to be used in the pathway. In the second case (multidomain enzymes), it is generally relevant to incorporate the reactions catalyzed by the different domains as separate seeds.

    The pathway extraction tool offers three seed group treatments: (1) groups may be preserved while merging overlapping groups ("Keep the groups"), (2) groups may be separated in as many sub-groups as unique seed nodes exist ("Treat each seed as a separate group"), or, in the case of genes or enzymes (3) seed reactions may be regrouped according to their EC numbers ("Group reactions by EC number"). The third choice is recommended to deal with seed genes/enzymes. On the one hand, it allows multidomain enzymes to contribute all their EC numbers to the pathway (AND relationship between EC numbers) and, on the other hand, it restricts the number of alternative reactions contributed by one EC number in case of ambiguities (inclusive OR relationship between reactions belonging to one EC number).

16. *Errors*. The following errors may occur:

    (a) Seeds not present in the input network. Obviously, a pathway cannot be predicted for the given seed nodes if some of them are absent from the input network.

    (b) Pathway extraction failed to extract a subgraph. This may have two reasons:

        - No path exists between any of the given seed nodes.
        - Paths exist, but have been filtered out because they contain mutually exclusive reactant pairs (does not apply to kWalks).

    (c) *Missed paths*. The choice of unit weight as a weight policy may result in such a high number of alternative paths that paths-based algorithms fail to collect all of them. In this case, the resulting pathway is incomplete. In addition, for each repetition of the extraction, a different part of the result pathway might be missed.

17. *Weaknesses.* The accuracy of pathway prediction decreases with a reduction of the seed node number and an increase of possible alternative pathways. Therefore, highly connected pathways in the center of metabolic networks, such as glycolysis, can only be predicted correctly with a sufficiently high number of seed nodes. Concerning metabolic networks, stoichiometry is ignored, which may result in the extraction of erroneous pathways (37). However, see ref. 38 for a discussion of stoichiometry in metabolic pathway discovery. In addition, paths-based approaches can only partly predict cyclic or spiral-shaped metabolic pathways (such as fatty acid biosynthesis).

Acknowledgments

KF was supported by Actions de Recherches Concertées de la Communauté Française de Belgique (ARC grant number 04/09-307). The BiGRe Laboratory is supported by the Belgian Program on Interuniversity Attraction Poles, initiated by the Belgian Federal Science Policy Office, project P6/25 (BioMaGNet), and by the MICROME Collaborative Project funded by the European Commission within its FP7 Programme, under the thematic area "BIO-INFORMATICS – Microbial genomics and bio-informatics" (contract number 222886–2).

References

1. Michal G. (1998) *Biochemical Pathways: An Atlas of Biochemistry and Molecular Biology.* Wiley-Spektrum.
2. Caspi R, Foerster H, Fulcher CA, Kaipa P, Krummenacker M, Latendresse M, Paley S, Rhee SY, Shearer AG, Tissier C, Walk TC, Zhang P, Karp PD. (2008) The MetaCyc Database of metabolic pathways and enzymes and the BioCyc collection of Pathway/Genome Databases. Nucleic Acids Res, 36:D623–D631.
3. Kanehisa M, Araki M, Goto S, Hattori M, Hirakawa M, Itoh M, Katayama T, Kawashima S, Okuda S, Tokimatsu T, Yamanishi Y. (2008) KEGG for linking genomes to life and the environment. Nucleic Acids Res, 36:D480–D484.
4. Dahlquist KD, Salomonis N, Vranizan K, Lawlor SC, Conklin BR. (2002) Gen-MAPP, a new tool for viewing and analyzing microarray data on biological pathways. Nat Genet, 31:19–20.
5. Thimm O, Bläasing O, Gibon Y, Nagel A, Svenja Meyer PK, Selbig J, Müller LA, Rhee SY, Stitt M. (2004) MAPMAN: a user-driven tool to display genomics data sets onto diagrams of metabolic pathways and other biological processes. Plant J, 37:914–939.
6. Paley SM, Karp PD. (2006) The Pathway Tools cellular overview diagram and Omics Viewer. Nucleic Acids Res, 34:3771–3778.
7. Adler P, Reimand J, Jäanes J, Kolde R, Peterson H, Vilo J. (2008) KEGGanim: pathway animations for high-throughput data. Bioinformatics, 24:588–590.
8. van Helden J, Wernisch L, Gilbert D, Wodak S. (2002) *Graph-Based Analysis of Metabolic Networks.* Springer-Verlag, Volume 38 of Ernst Schering Res Found Workshop, 245–274.
9. Croes D, Couche F, Wodak S, van Helden J. (2005) Metabolic PathFinding: inferring relevant pathways in biochemical networks. Nucleic Acids Res, 33:W326–W330.

10. Croes D, Couche F, Wodak S, van Helden J. (2006) Inferring meaningful pathways in weighted metabolic networks. J Mol Biol, 356:222–236.
11. Faust K, Croes D, van Helden J. (2009) Metabolic pathfinding using RPAIR annotation. J Mol Biol, 388:390–414.
12. Kotera M, Hattori M, Oh MA, Yamamoto R, Komeno T, Yabuzaki J, Tonomura K, Goto S, Kanehisa M. (2004) RPAIR: a reactant-pair database representing chemical changes in enzymatic reactions. Genome Informatics, 15:P062.
13. Kotera M, Okuno Y, Hattori M, Goto S, Kanehisa M. (2004) Computational assignment of the EC numbers for genomic-scale analysis of enzymatic reactions. J Am Chem Soc, 126:16487–16498.
14. Faust K, Callut J, Dupont P, van Helden J. (2008) Inference of Pathways from Metabolic Networks by Subgraph Extraction. In *Proceedings of the Second International Workshop on Machine Learning in Systems Biology (MLSB)*, 13–14 September 2008, Brussels. Edited by Wehenkel L, d'Alché Buc F, Moreau Y, Geurts P, 27–38.
15. Faust K, Dupont P, Callut J, van Helden J. (2010) Pathway discovery in metabolic networks by subgraph extraction. Bioinformatics, 26:1211–1218.
16. Brohée S, Faust K, Lima-Mendez G, Sand O, Janky R, Vanderstocken G, Deville Y, van Helden J. (2008) NeAT: a toolbox for the analysis of biological networks, clusters, classes and pathways. Nucleic Acids Res, 36:W444–W451.
17. Krieger CJ, Zhang P, Mueller LA, Wang A, Paley S, Arnaud M, Pick J, Rhee SY, Karp P. (2004) MetaCyc: a multiorganism database of metabolic pathways and enzymes. Nucleic Acids Res, 32:D438–D442.
18. Takahashi H, Matsuyama A. (1980) An approximate solution for the Steiner problem in graphs. Math Japonica, 24:573–577.
19. Jimenez V, Marzal A. (1999) Computing the K Shortest Paths: A New Algorithm and an Experimental Comparison. In *Proceedings of the Third International Workshop on Algorithm Engineering (WAE 1999)* 1999. Vol. 1668, 15–29.
20. Dupont P, Callut J, Dooms G, Monette JN, Deville Y. (2006) Relevant subgraph extraction from random walks in a graph. Research Report UCL/FSA/INGI RR 2006–07.
21. Callut J. (2007) First passage times dynamics in Markov models with applications to HMM induction, sequence classification, and graph mining. PhD thesis, Université Catholique de Louvain.
22. Moreno-Hagelsieb G, Collado-Vides J. (2002) A powerful non-homology method for the prediction of operons in prokaryotes. Bioinformatics, 18(Suppl. 1):S329–S336.
23. Ermolaeva MD, White O, Salzberg SL. (2001) Prediction of operons in microbial genomes. Nucleic Acids Res, 29(5):1216–1221.
24. Gama-Castro S, Jiménez-Jacinto V, Peralta-Gil M, Santos-Zavaleta A, Peñaloza Spinola MI, Contreras-Moreira B, Segura-Salazar J, Muñiz Rascado L, Martínez-Flores I, Salgado H, Bonavides-Martínez C, Abreu-Goodger C, Rodríguez-Penagos C, Miranda-Ríos J, Morett E, Merino E, Huerta AM, Treviño Quintanilla L, Collado-Vides J. (2008) RegulonDB (version 206.0): gene regulation model of *Escherichia coli* K-12 beyond transcription, active (experimental) annotated promoters and Text-presso navigation. Nucleic Acids Res, 36.
25. Mergeay M, Monchy S, Vallaeys T, Auquier V, Benotmane A, Bertin P, Taghavi S, Dunn J, van der Lelie D, Wattiez R. (2003) Ralstonia metallidurans, a bacterium specifically adapted to toxic metals: towards a catalogue of metal-responsive genes. FEMS Microbiol Rev, 27:385–410.
26. Karp P, Paley S, Romero P. (2002) The Pathway Tools software. Bioinformatics, 18:S225–S232.
27. Thomas-Chollier M, Sand O, Turatsinze JV, Janky R, Defrance M, Vervisch E, Brohée S, van Helden J. (2008) RSAT: regulatory sequence analysis tools. Nucleic Acids Res, 36:W119–W127.
28. GML(Graph Modelling Language). [http://www.infosun._m.unipassau.de/Graphlet/GML/gml-tr.html].
29. KGML (KEGG Markup Language). [http://www.genome.jp/kegg/xml/].
30. GraphViz (Graph Visualization Software). [http://www.graphviz.org/].
31. Shannon P, Markiel A, Ozier O, Baliga NS, Wang JT, Ramage D, Amin N, Schwikowski B, Ideker T. (2003) Cytoscape: a software environment for integrated models of biomolecular interaction networks. Genome Res, 13:2498–2504.
32. Hu Z, Ng D, Yamada T, Chen C, Kawashima S, Mellor J, Linghu B, Kanehisa M, Stuart J, DeLisi C. (2007) VisANT 3.0: new modules for pathway visualization, editing, prediction and construction. Nucleic Acids Res, 35: W625–W632.
33. yEd (Java Graph Editor). [http://www.yworks.com/en/productsyed about.html].

34. Sauret-Ignazi G, Gagnon J, Béguin C, Barrelle M, Markowicz Y, Pelmont J, Toussaint A. (1996) Characterisation of a chromosomally encoded catechol 1,2-dioxygenase (E.C. 1.13.11.1) from Alcaligenes eutrophus CH34. Arch Microbiol, 166:42–50.
35. Ott M, Vriend G. (2006) Correcting ligands, metabolites, and pathways. BMC Bioinformatics, 517.
36. Poolman M, Bonde B, Gevorgyan A, Patel H, Fell D. (2006) Challenges to be faced in the reconstruction of metabolic networks from public databases. EE Proc Syst Biol, 153:379–384.
37. de Figueiredo LF, Schuster S, Kaleta C, Fell DA. (2008) Can sugars be produced from fatty acids? A test case for pathway analysis tools. Bioinformatics, 24:2615–2621.
38. Planes F, Beasley J. (2008) A critical examination of stoichiometric and path-finding approaches to metabolic pathways. Briefings Bioinformatics, 9:422–436.

Chapter 8

Directed Module Detection in a Large-Scale Expression Compendium

Qiang Fu, Karen Lemmens, Aminael Sanchez-Rodriguez, Inge M. Thijs, Pieter Meysman, Hong Sun, Ana Carolina Fierro, Kristof Engelen, and Kathleen Marchal

Abstract

Public online microarray databases contain tremendous amounts of expression data. Mining these data sources can provide a wealth of information on the underlying transcriptional networks. In this chapter, we illustrate how the web services COLOMBOS and DISTILLER can be used to identify condition-dependent coexpression modules by exploring compendia of public expression data. COLOMBOS is designed for user-specified query-driven analysis, whereas DISTILLER generates a global regulatory network overview. The user is guided through both web services by means of a case study in which condition-dependent coexpression modules comprising a gene of interest (i.e., "directed") are identified.

Key words: Regulatory network, Motif, Microarray compendium, Transcriptomics, Expression profile, Coexpression module, Bi-cluster, Condition annotation, Web service

1. Introduction

Omics-based approaches are increasingly being used to uncover underlying mechanisms of bacterial behavior (1). The obligate deposit of high-throughput experiments in public databases upon publication has tremendously increased the amount of publicly available experiments. Mining these data sources helps in gaining a global condition-dependent view on bacterial gene regulation (2, 3), in expanding the current knowledge on transcriptional interactions with novel reliable predictions (4, 5), and in comparing transcriptional networks across species (1, 6). It also offers molecular biologists the possibility to see their own dedicated analysis in light of what is already available. Inferring transcriptional networks from these public data usually requires complex normalization procedures (7) and computationally

Jacques van Helden et al. (eds.), *Bacterial Molecular Networks: Methods and Protocols*, Methods in Molecular Biology, vol. 804, DOI 10.1007/978-1-61779-361-5_8, © Springer Science+Business Media, LLC 2012

intensive algorithms. To enhance the usability of these tools, they are often wrapped in a web service.

In this book chapter, we illustrate how compendia of public expression data can be used to identify condition-dependent coexpression modules, in which a particular gene of interest is involved ("directed" module detection), by means of two web services: COLOMBOS (8) and DISTILLER (the latter in combination with the visualization tool ViTraM). Figure 1 illustrates the difference between both approaches. COLOMBOS only relies on expression data to retrieve condition-dependent modules, implying that there is a functional relationship between the module genes, but not necessarily that they are regulated by the same (set of) transcription factor(s). DISTILLER in contrast, incorporates additional motif data and the constraint that module genes should share the same regulatory program, implying that the

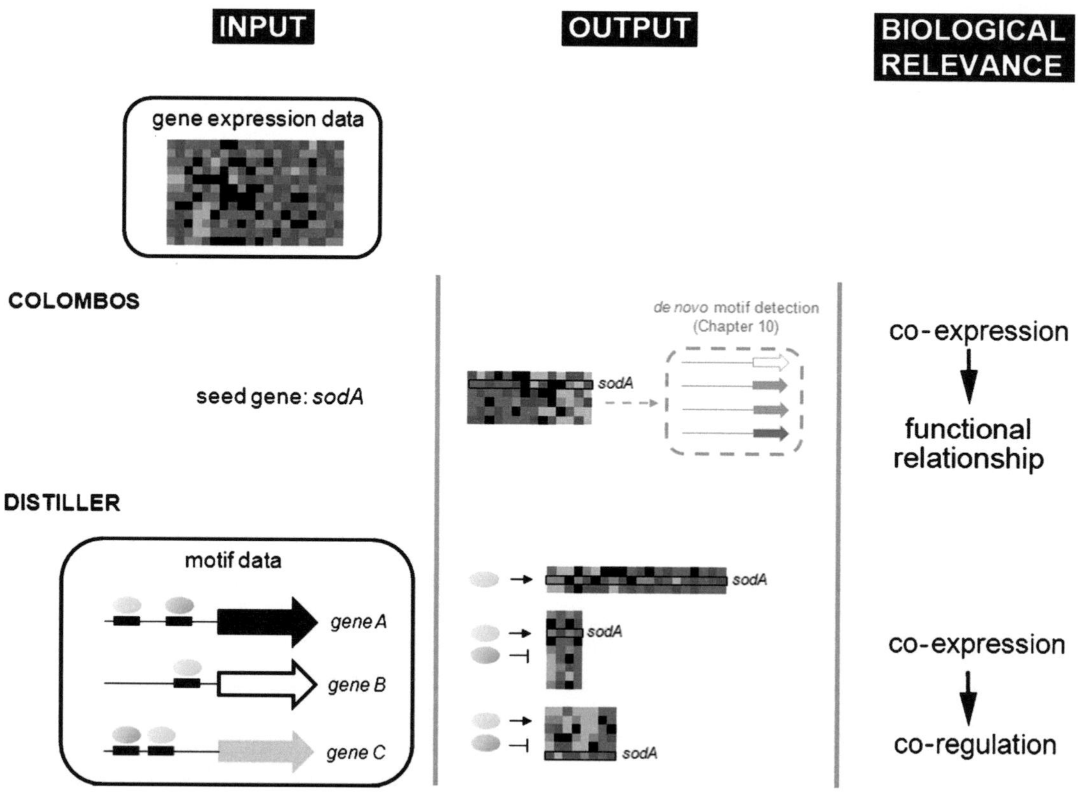

Fig. 1. A large-scale expression compendium is required as input for both methods. Coexpression modules are detected by COLOMBOS starting from a user provided seed gene (*sodA*, see case study). DISTILLER searches for coexpression modules in a global fashion (information on seed genes is not a prerequisite) and is constrained by regulator-to-gene assignments. Coexpression patterns retrieved by COLOMBOS point toward a functional relationship among the module genes but do not necessarily imply coregulation. To gain insights on the regulatory program of those genes, de novo motif detection tools could be applied to analyze their promoter regions. Inherent to the regulator-to-gene constraints implemented in DISTILLER, there is a direct link between coexpression and coregulation for the genes in the modules that it retrieves.

module's condition-dependent coexpression can be directly linked to transcriptional coregulation. To clearly demonstrate the functionalities of these web services, the gene *sodA* is used as a case study. This gene encodes for a protein with superoxide dismutase activity, which reduces harmful free radicals of oxygen formed during normal metabolic cell processes (9–11). It is known to be regulated by several regulators, such as, Fur, SoxS, and MarA. As such its expression is coupled to different biological conditions: multiple antibiotic resistances (MarA), superoxide resistance (SoxS), and the intracellular iron pool (Fur). By applying different analysis approaches (COLOMBOS and DISTILLER), we will show how to identify *sodA* containing modules, i.e., genes that behave similarly to *sodA* in a condition-dependent way from large-scale expression compendia.

2. Materials

2.1. Cross-Platform Expression Compendium

The cross-platform expression compendium of *Escherichia coli* used in these analyses consists of a large amount of publicly available microarrays (1,747 microarrays) derived from different platforms (e.g., cDNA microarrays and Affymetrix). Data were normalized to remove cross platform differences and experimental conditions were annotated in a formal manner (see below). The expression data are presented as expression log ratios of measurements in a test condition versus a reference condition. Because of this inherent comparison in each log ratio, the condition annotation makes use of the concept of "(condition) contrasts." A *condition contrast* is the set of all genes' expression log ratios, each of which represents the log ratio of a gene's expression values obtained in one particular experiment for which a test is compared to a reference sample. The condition annotations of each contrast thus indicate which aspect(s) (experimental conditions) has/have been changed between the test and reference condition.

The condition annotation is specified using a set of formal condition properties (representing for instance mutations, compounds in the growth medium, treatments, and general growth conditions) and associated absolute values for these properties (for instance, concentration of a certain compound in the growth medium). As such, we can specify the condition annotation of each condition contrast rigidly as a vector representing the differences between these property values in test and reference condition for each of the relevant condition properties (for instance, the concentration difference between test and reference condition). This enables the mathematical comparison of contrast annotations. For instance, a contrast condition annotation specified as "$FeSO_4$: 50 μM" describes that for this condition contrast, the concentration of $FeSO_4$ has changed by a quantity of "50 μM," in the medium of the test versus the reference sample. Due to the large number of different properties and their distinct nature, it may not be easy for a user to find those condition properties he or she is interested in. To facilitate this, we have provided a condition ontology similar to Gene Ontology (GO) (12). This ontology maps the condition properties used to annotate the condition contrasts to the biological processes or functionalities they most likely affect. Each property can be mapped to one or more ontology terms and thus ensures an intuitive, higher level view of the condition contrasts. As an example, the condition ontology term "amino acid metabolic process" includes condition properties that are linked to amino acid metabolism, such as the addition of arginine to the medium, *trpR* mutation (tryptophan transcriptional repressor), etc. The condition ontology superimposed on the condition annotation thus represents a broader, more intuitive view of the altered stimuli that are responsible for the observed expression changes (i.e., log ratios) of the condition contrasts.

2.2. Coexpression Modules

Genes that have the same regulatory program (here defined as having the same transcription factor binding motifs in their promoter regions) behave similarly: their expression responds in a similar manner to certain alterations in the organism's intra- or extracellular environment. This is called *coexpression*. The evidence of coexpression points to a functional relationship between coexpressed genes, and might sometimes be an indication of shared regulatory programs. Such a functional relatedness is only enforced when coexpression occurs across multiple different conditions. Since each value in our data set does not represent an absolute measure of expression, but rather a relative one associated to a condition contrast (test versus reference), coexpression in this case translates to genes showing log ratios that are significantly different from zero for at least one condition contrast, and for each of the involved contrasts show coherent changes in the same direction (either up or down). Both methods that we apply in this chapter, COLOMBOS and DISTILLER, employ different computations to score this coexpression, but essentially look for the same phenomenon.

Notice that genes are only coexpressed under certain condition contrasts. This combined set of genes and condition contrasts, where the coexpression pattern appears, is what we refer to here as a *module*. Extra constraints can be used to tune the module concept for specific purposes. For example, the requirement could be added that genes in the module should have some (known) motifs in common.

2.3. Software

2.3.1. COLOMBOS

COLOMBOS (http://ibiza.biw.kuleuven.be/colombos/), an acronym for *COLections Of Micrcroarrays for Bacterial OrganismS*, is a web service (Fig. 2) for interactively exploring, querying,

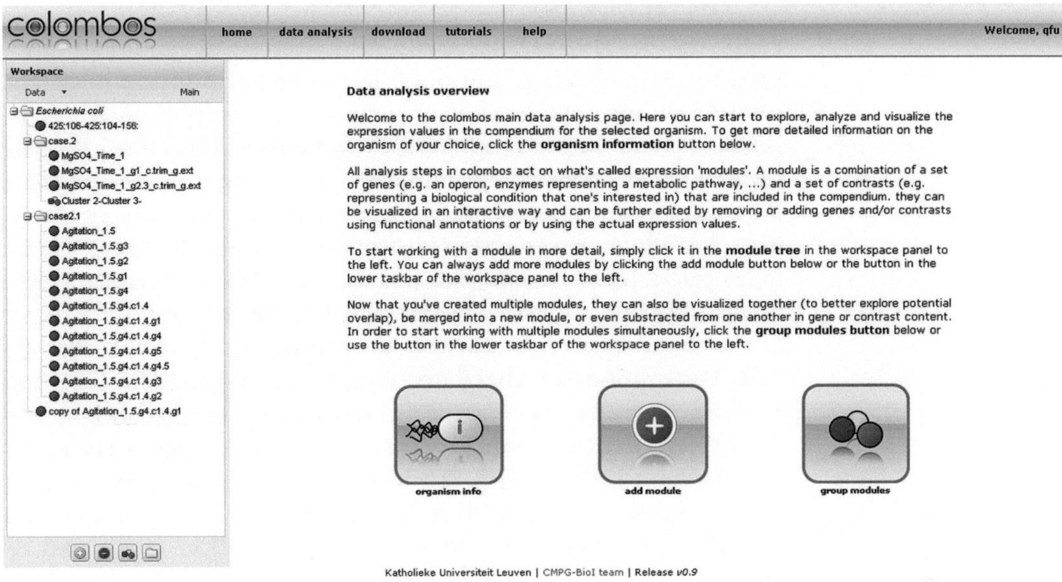

Fig. 2. The interface of the COLOMBOS web service is composed of three parts: the website navigation header at the *top*, the workspace (showing a module tree) at the *left-hand side*, and the operating space (here showing the "Data analysis overview" panel) filling the *center*.

analyzing and visualizing data from cross platform expression compendia through an intuitive interface. Currently, it provides the possibility to query expression compendia based on the annotation of their experimental condition contrasts and/or genes. COLOMBOS can be used for instance to search for genes being coexpressed with one or more genes of interest under a prespecified set of condition contrasts, or to search for the condition contrasts in an expression compendium under which a prespecified gene set is coexpressed. As such, it can identify coexpression modules based on user input. In COLOMBOS, modules can be visualized as interactive heatmaps, showing the annotation of genes and condition contrasts as obtained from public databases and corresponding literature. See Note 1 for version and supported browsers of the website.

2.3.2. DISTILLER

DISTILLER (2) stands for *Data Integration System To Identify Links in Expression Regulation*. It is a data integration framework that searches for condition-dependent transcriptional modules by combining expression data with information on the interaction between a regulator and its corresponding target genes, for instance based on motif screening. A module is here defined as a set of genes that are both coexpressed across a sufficiently large number of columns in the expression matrix and that share motif instances for the same regulator(s). Note that the expression data used by DISTILLER could be either absolute expression values or log ratios.

DISTILLER searches the modules in a global fashion over the entire dataset through three steps: identification candidate modules, module filtering, and module extension. Built on advanced itemset mining approaches, it first exhaustively enumerates all possible valid *closed* modules as candidates (see Note 2). The candidates are partially redundant as they might share the same genes or condition contrasts. Therefore, in the second step (filtering), a module prioritization is provided in which an interest score is assigned to each module. The score takes into account the probability that a module of the same size can be found by chance in the input dataset, but at the same time, penalizes the overlaps with those modules that have already been reported. The top ranking modules are statistically significant yet distinctive, and thus selected as the output of this step. These initial modules obtained by DISTILLER are usually called "seeds" as they are obtained under stringent thresholds. This has the advantage that all identified interactions between a transcription factor and its target genes are highly reliable (high precision), but that true interactions might be missed (lower sensitivity). Hence, in the last step (module extension), the obtained modules can be further extended with additional genes by relaxing the used thresholds on the required minimal coexpression level and motif score. Note that it is not required to execute all three steps. The result of each step can directly be analyzed and visualized by the user.

For this case study, the data consist of the same compendium as the one accessed through the COLOMBOS website. Hence, the columns of the expression matrix, across which DISTILLER defines coexpression, correspond to the condition contrasts as defined in the compendium (see Subheading 2.1). We will run DISTILLER through the web service at http://ibiza.biw.kuleuven.be/DISTILLER/ (Fig. 3). For compatibility issues, See Note 1.

2.3.3. ViTraM

For DISTILLER output analysis, we use the module visualization software ViTraM (VIsualization of TRAnscriptional Modules) to visualize overlapping transcriptional coexpression modules together with the motif information in an interactive way. In this chapter, we only briefly discuss this tool. For more details we refer to ref. 13.

Note that for this case study, we used ViTraM 2.0. It can be downloaded at http://homes.esat.kuleuven.be/~kmarchal/ViTraM/Index.html. See Note 1 for compatibility issues and memory requirements of the tool.

2.4. Input File Formats

Different file formats are used by COLOMBOS, DISTILLER, and ViTraM. Depending on their functions, the input files are explained in this section and the output files in the next one. Sample files that contain the data used for our case study are also provided (see Subheading 2.6).

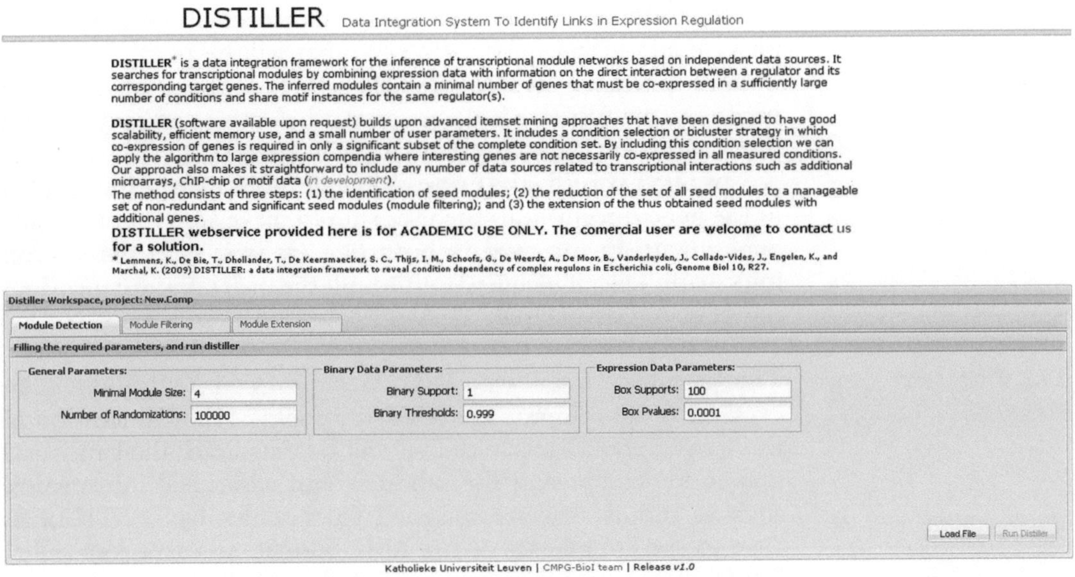

Fig. 3. The interface of the DISTILLER web service is separated into two sections vertically. The *upper part* contains the text that briefly explains the functionality of DISTILLER. The *lower part* is the operating area, where user interacts with the website to run DISITLLER. The *operating area* shown here is the process running interface that is reached after a user has logged in and selected a project. For each step of DISTILLER (see Subheading 2.3.2) a separate tab is provided to run it. In the figure, the first tab for running module detection step is shown in *front*, the tabs for module filtering and module extension is hidden.

2.4.1. DISTILLER Input File Format

DISTILLER requires two input files: an expression data file and a motif data file. In the case study described here, the expression data file contains the log ratio expression values for each gene under each condition contrast. Two possible input formats are allowed for the expression data file. The first one is a matrix-based format, in which each row represents the data of one gene; each column represents a condition contrast in the expression compendium. The only exception is the first row, where the identifiers of each condition contrast are specified. The gene names are specified in the first column of each row. The second format that is permitted is the format of the expression data file that can be obtained for each user created module from COLOMBOS (see Subheading 2.5.1).

The motif data file contains the results of a motif screening analysis (14). Each motif screening score indicates the probability that a motif instance is present in the promoter region of a gene, with the maximum score 1 and the minimum score 0. These data are also imported in a matrix format file, where the rows in the file represent the genes and the columns represent the motifs for which the screening was performed. Similarly to the expression data, the first row in the file contains the motif information with one motif name in each column, and from the second row on, each row represents one gene's motif scores with the corresponding gene name specified in the first column.

In addition to the two files mentioned above, users can provide an operon information file that will be used during the module extension step of DISTILLER. It contains two items in each row, the first one is the gene that belongs to an operon, and the second one is the first gene of that operon.

In DISTILLER, the expression and motif matrix are coupled in the gene direction, so the user should make sure that the same gene identifiers are used in both files (as well as in the operon information file, if included), although the order of the genes does not have to be the same.

2.4.2. ViTraM Input File Format

The visualization software ViTraM requires an XML file and an expression file as input. The XML file contains the information of the transcriptional modules that will be visualized: the genes and conditions composing the modules and additional information such as motifs that are assigned to modules by DISTILLER. Due to the complexity of the XML format, an extra tool called XMLCreator is available at the ViTraM website to automatically generate the XML file from the ".m" output file and the supplementary expression data file of DISTILLER (see Subheading 2.5.2). For the purpose of this study, we do not discuss the XML file format directly used by ViTraM. Interested users can find more detailed information in ref. 13. Conveniently, the tool also generates a smaller expression data file containing only the relevant expression values from those genes and condition contrasts existing in the result modules contained in the ".m" file. Moreover, the tool can also incorporate extra information into XML file, for example, the motif data contained in the binary supplementary file mentioned in Subheading 2.5.2.

2.5. Output File Formats

Below, the different output file formats are explained in each subsection.

2.5.1. COLOMBOS Expression Data Format

As COLOMBOS provides an intuitive browser interface, no input files are required. It does provide a tab delimited text output file containing information on the user created modules consisting of a particular set of genes and the corresponding set of condition contrasts under which these genes are coexpressed. There are two sections in this file. The first one describes the condition information of the module. It specifies for each condition contrast the name and description of its test and reference sample, the corresponding experiment identifier, the online database from which the experiment was retrieved, the microarray platform used to measure the sample hybridizations, and finally, the annotated contrast condition changes between the test and the reference sample. The second section contains three parts: the corresponding condition contrast identifiers in the first row; the gene information (gene locus tag, gene name, and internal gene identifier) in the first three columns; and the log ratio expression data from the expression matrix. Note that starting from the fourth column, each column in the file corresponds to one condition contrast. Each of the rows represents one gene and its expression values for different condition contrasts.

2.5.2. DISTILLER Output File Format

For all three steps of DISTILLER (see Subheading 2.3.2), the outputs are of the same format: ".m" file. The file contains the information of the identified modules, with each module described in a section, separated by an empty line. Each section contains four data fields, each representing a part of the module information. Given a section specifying the information for a module *M*, the field "Significances" holds the *p*-values that were assigned to *M* by the itemset mining algorithm (15) used by DISTILLER. The field "items" contains the information of the genes that belong to *M*, each of which is represented by its row number (see Note 3) in the expression data input file. The field "boxtidset1" specifies the contrasts of *M*, in which the genes in *M* are coexpressed, each of which is represented by its column number in the expression data input file. Finally, the field "tidset1" contains the motifs of *M*, which are shared by the genes in *M*, specified by their column numbers in the motif data input file. Each module is represented by a number indicated between the "{}" brackets after the name of each data field.

Besides the main output file, two supplementary files required to use the visualization software ViTraM, are also provided. They are grouped in a compressed file that can be downloaded from the DISTILLER website. The file whose name starts with "expdata_" contains the DISTILLER preprocessed input expression data, and the file whose name starts with "binary_" contains the DISTILLER preprocessed input motif data. The file formats of these expression data and motif data files are the same as the corresponding input files described in Subheading 2.4.1.

2.6. Sample Files

The following four sample files are provided as example input files for this case study. The file expdata_COLOMBOS_module_information.txt is an example of the exported module data file of COLOMBOS, which, as one of the input formats of the expression data for DISTILLER, contains both the gene and condition contrast information, and the log ratio expression data. The file expdata_DISTILLER_Expression.txt is an example of the other expression data format accepted by DISTILLER. The file binary_DISTILLER_Motif.txt is the motif data containing motif scores for each gene. And finally, the file operon_DISTILLER_OperonGenes.txt is the example operon information file.

Furthermore, three example output files, corresponding to the output generated after each step of DISTILLER, are provided. DISTILLER_Output.m is the output of the seed module identification step, DISTILLER_FilteringOutput.m is the output of the module filtering step, and DISTILLER_output Extended.m is the output of the module extension step. Two files, expdata_DISTILLER_supple.txt and binary_DISTILLER_Motif_supple.txt, are provided as the supplementary files used for visualizing modules contained in those sample output files.

Finally, ViTraM_Modules.XML is provided as the XML file describing the modules, and ViTraM_Expression.txt is provided as the expression data file used by ViTraM.

From http://homes.esat.kuleuven.be/~kmarchal/Supplementary_Information_Fu_2010/sample_files/, one can download all the example files.

3. Methods

We describe the steps required to use two web services, COLOMBOS and DISTILLER, to identify coexpression modules around a (set of) gene(s) of interest (query genes). This will be illustrated by applying them to the query gene *sodA* as a case study.

The first analysis flow is based on the web service COLOMBOS. We show how COLOMBOS can be used to search for genes that are coexpressed with a (set of) query gene(s). In contrast to COLOMBOS, the second approach using DISTILLER detects coexpressed gene sets and their regulatory program across the whole data set in a single run in an unsupervised way. It searches for gene sets that are coexpressed and contain the same motif(s) in their promoter region by combining expression and motif data.

3.1. Identifying Coexpression Modules Containing sodA Using COLOMBOS

To search for coexpressed genes using COLOMBOS, we first create what we call an expression *module*. An initial module can be created by specifying a set of genes of interest (query genes), and use COLOMBOS to extract the most relevant condition contrasts, where the chosen genes are differentially expressed. Next, the module is extended with genes that are coexpressed with the initial gene set under the preselected set of condition contrasts. We explain the workflow based on our case study with a single query gene, *sodA*. The flow consists of the following steps:

Step 1 Go to the COLOMBOS website at http://ibiza.biw.kuleuven.be/colombos/.

Create the Initial Module

Step 2 Create the initial module based on the known gene set of interest (query genes) and the biological conditions that are known to affect these genes when changed.

Step 2.1 To start the analysis, click on "data analysis" in the title bar (Fig. 2) to bring up the data operation interface in the center part of the website. It is separated horizontally into two parts. At the left-hand side, there is a "workspace" information panel that lists all the modules created in a tree structure. The workspace panel is always visible when residing on the data analysis page. (Note that you can always move away from the data analysis page to any of the other COLOMBOS pages, such as the help sections, at any time; the state of the data analysis page will be preserved for whenever you switch back.) Since no module has been created, the workspace panel is currently empty. The other part of the page, referred as the operating space in this text, is where the visualization and analysis of the expression data takes place and is currently occupied by the "Data analysis overview" panel. This panel serves as an intuitive guide showing the user the available data analysis options depending on the current state of the workspace along with some brief explanations. The "Main" button at the top right of the workspace will always bring up the "Data analysis overview" panel again in the operating space.

Since we just started a new session, there is only one button "select organism" available. Click it, and select the organism "*Escherichia coli*," the one that this case study focuses on, from the dropdown box that appears. Notice now this species appears in the "workspace" panel at left as the root of the tree.

Step 2.2 After the species is picked, the "Data analysis overview" panel's content is also chanced accordingly. The button "select organism" is replaced by two other buttons "organism info" and "add module." The workspace panel now also shows that *E. coli* data is loaded. Clicking on the *E. coli* entry in the workspace panel has the same function as clicking the "organism info" button in the operating space: it will bring up detailed information about the organism, its compendium, and associated additional data sources.

We proceed to create our initial module of the gene *sodA*. The "Add module overview" panel appears in the operating space after clicking the "add module" button. In this panel, there are three buttons "select gene only," "select contrasts only," and "select gene and contrast." As explained on the page, each corresponds to a different method for creating a module.

For this case study, we manually select the gene *sodA*, but we do not specify any relevant conditions ourselves. Instead, we have COLOMBOS automatically identify the most relevant condition contrasts for *sodA* based on its expression values. This function is provided by the leftmost button "select gene only." Clicking it brings up the gene selection option panel in the operating space. Four available options are listed here: "By gene name/locus tag," "By transcriptional regulation," "By pathway," and "By transcription unit." We want to select a specific gene (*sodA*), so the first option is our choice. Click the yellow box surrounding the text "By gene name/locus tag," fill in "sodA" in the appeared text box (one gene per row), and then click the "Done" button located at the bottom to save your input. If one or more of the query genes is not recognized or included in the compendium, an error message will appear listing the unresolved entries.

Step 2.3 As mentioned in previous step, we chose the option "select genes only" to create the module. After specifying the genes, a "ranked contrast selection" panel (Fig. 4) will appear in the operating space, which allows the user to select contrasts ranked by a score that prioritizing those show the highest magnitude of change and most coherent coexpression for the selected set of genes (see Note 4). The panel is separated into two parts horizontally. At the left, the condition contrasts are ranked according to the score from highest till the lowest. A cutoff value can be specified in the box above the list to select only those

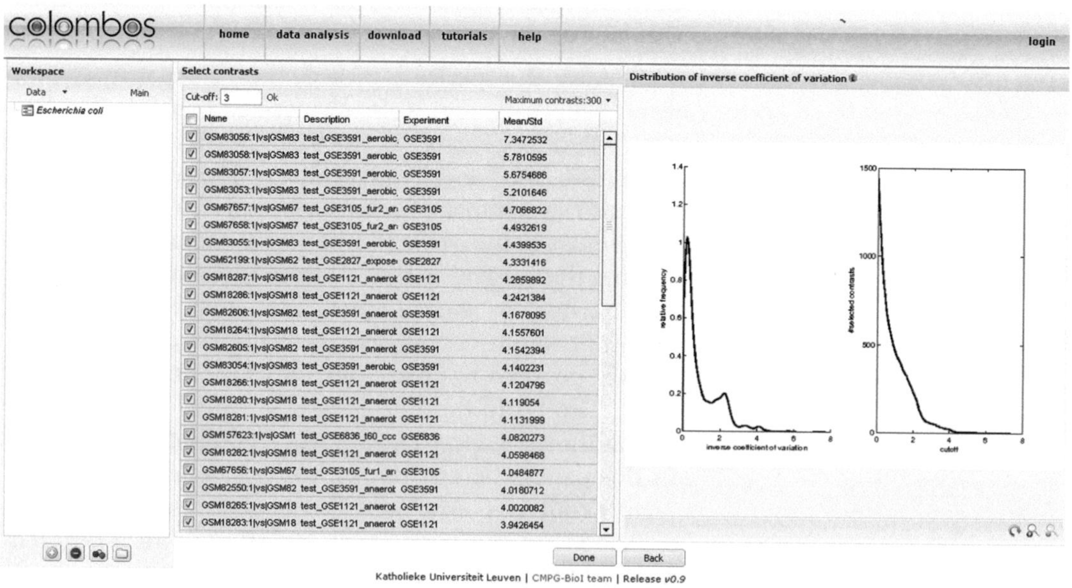

Fig. 4. The "ranked contrast selection" panel is shown occupying the operating space of COLOMBOS webpage. At the *center*, the condition contrasts are ranked according to the score (see Note 4) from highest to lowest. The figures to the *right* are meant to provide additional helpful information and show, from *left* to *right*, a *density plot* showing the distribution of the contrast scores across the whole compendium and a *plot* showing the number of contrasts that will be added if a given cutoff is selected.

contrasts of score higher than it. The figures to the right are meant to provide additional helpful information and show, from left to right, a density plot showing the distribution of the contrast scores across the whole compendium and a plot showing the number of contrasts that will be added if a given cutoff is selected.

For our initial module, we chose cutoff value 3. There is no absolute guideline to choose the cutoff. Several try-outs may be needed to identify the optimal one. Input 3 in the box above the list at the left, and click "Ok" button at the side. The contrast list is then filtered, and all contrasts left in the list are automatically selected to be added to the module. Click "Done" button at the bottom, and fill in "sodA_c3" as the module name to create our initial module.

The module "sodA_c3" is now created and will appear in the module tree in the workspace at the left side as a node directly under the root node (presenting the selected organism "*Escherichia coli*").

Inspect the Existing Modules

Step 3 After a module is created, one can review its information through various available options.

Step 3.1 Click the module you want to check in the workspace at the left to show its module overview tab (Fig. 5)

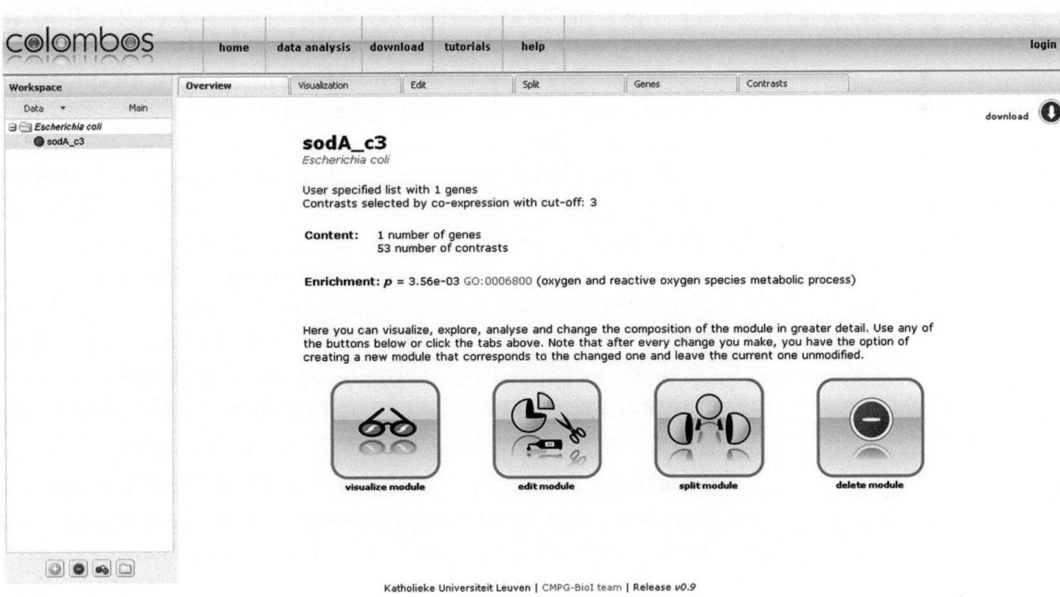

Fig. 5. The module overview tab of COLOMBOS is presented in this figure. In the tab, some brief information on the selected module "sodA_c3" is listed. It is followed by the enrichment section lists the Gene Ontology (GO) terms enriched by the genes in the module together with their enrichment scores. At the *bottom*, there are four buttons that lead to various functions to *visualize*, *edit*, *split*, or *delete* the module.

in the operating space at right. It provides a brief summary of the current module, and serves as a helping guide to analyze and visualize the module's composition. For each module, the summary section shows its name followed by its description at the top. Below the number of included genes and contrasts are indicated, as well a list of GO enrichment scores (see Note 14) for the genes in the module. Only the GO terms of the class biological process are calculated for their enrichment scores, and only those with *p*-values <0.1 are shown.

At the bottom of this panel, there are four buttons to visualize, edit, split, or delete the module. The function of visualization and delete is straightforward. The edit function allows user to modify either the genes or the contrasts of the module. Instead, the split function breaks the current module into several new ones by separating either its genes or its contrasts into distinctive groups and creating a module for each selected group.

Step 3.2. Each module can be visually presented as a 2D heatmap. In the heatmap, each square represents one gene's expression log ratio for a certain condition contrast. A red color indicates over-expression while a green one shows under-expression. The intensity of the color corresponds to the magnitude of the change. The brighter the color, the higher the absolute ratio value.

To view the heatmap, click "Visualization" button in a module's overview tab to switch to heatmap tab. In the tab, the heatmap is shown at the center with an information panel at the right hand side. Hovering over the heatmap, the information of the gene and the condition contrast under the cursor will be shown in that information panel. Various options exist to group the condition contrasts in the heatmap according to their experiments, condition properties, and the condition ontology terms that their associated condition properties belong to. It can be selected from the dropdown box located at the left corner of the bar below the heatmap. The heatmap of our module sorted based on condition ontology (default sorting is based on expression value) is shown in Fig. 6.

Checking our module "sodA_c3," it shows that the top 53 condition contrasts, where *sodA* is most differentially expressed, have been selected. Among those, 34 contrasts belong to the condition ontology term "response to oxygen levels." This ontology term includes condition properties that are linked to cellular processes that are dependent on oxygen availability, such as *fnr* mutations (a global oxygen responsive transcriptional regulator), NO_2 (electron transport decoupler), agitation of

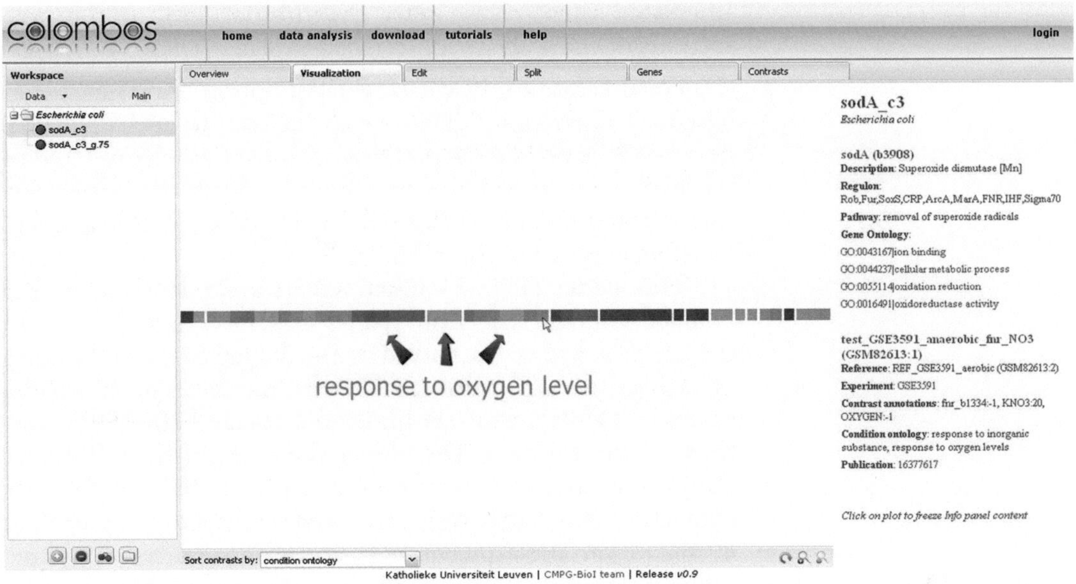

Fig. 6. The COLOMBOS visualization tab in the operating space showing the heatmap of module "sodA_c3." In the heatmap, *each row* corresponds to one gene, and *each column* to one condition contrast. The *information panel* at the *right hand side* shows information of the gene *sodA* and the condition contrast under the cursor. Using the *dropdown box* located at the *left corner of the bar* below the heatmap, one can select various options to group the condition contrasts of the module. In this heatmap, the condition contrasts are grouped by the ontology term they belong to. The condition contrast groups belonging to the ontology term "response to oxygen level" are marked in the figure.

the growth medium, actual oxygen levels, etc. Since the function of *sodA* itself is linked to processes related to oxygen availability, one can expect its expression level to change prominently under these 34 contrasts, and indeed COLOMBOS successfully identifies them as the most relevant contrasts of *sodA*. There are also 17 contrasts belong to ontology term "growth," with three in common with the 34 contrasts mentioned before. The ontology term "growth" is very general, grouping experimental conditions that trigger various biological processes simultaneously at a global cellular scale. Based on this, it is not unexpected to find *sodA* differentially expressed under contrasts belonging to this ontology term.

Extend the Genes of a Module

Step 4 To extend the module with additional genes, we use the edit function. Click "edit module" button to bring the "Edit" tab in front, where three options are available: edit the name/description, the genes, or the contrasts. Click the "edit genes" since we plan to modify the genes of this module. A gene modification panel appears in the "Edit" tab with all options to edit genes. The options in green boxes indicate ways to add genes to the module, whereas those in red boxes indicate ways to remove genes. We do not explain all the options here.

Interested users please refer the tutorial and/or the help sections of the website for detailed explanations.

For this case study, we use the last option "Add new genes based on expression." This allows the user to add additional genes, whose expression patterns are most (anti-)correlated to the expression pattern of the module. Click the green box surrounding the text of this option; a "ranked gene selection" panel will appear in the "Edit" tab.

This panel (Fig. 7) functions in a similar way to the "ranked contrast selection" panel explained in step 2.3. On the left side, a gene list ranked by the degree of the correlation of each gene's expression profile with the mean profile of the current module's gene set under the selected condition contrasts of this module. The higher the score, the more similar the profiles are. A cutoff value can be specified in the box above the list to select only those genes that score higher than it. Based on the (uncentered) correlation, there are three different options available to rank the genes. They can be chosen from the dropdown box that will appear by clicking the text "correlation" in the bar on top of the ranked gene list. Here, we use the default one "correlation." See Note 5 for more information about these correlation scores and how to calculate them.

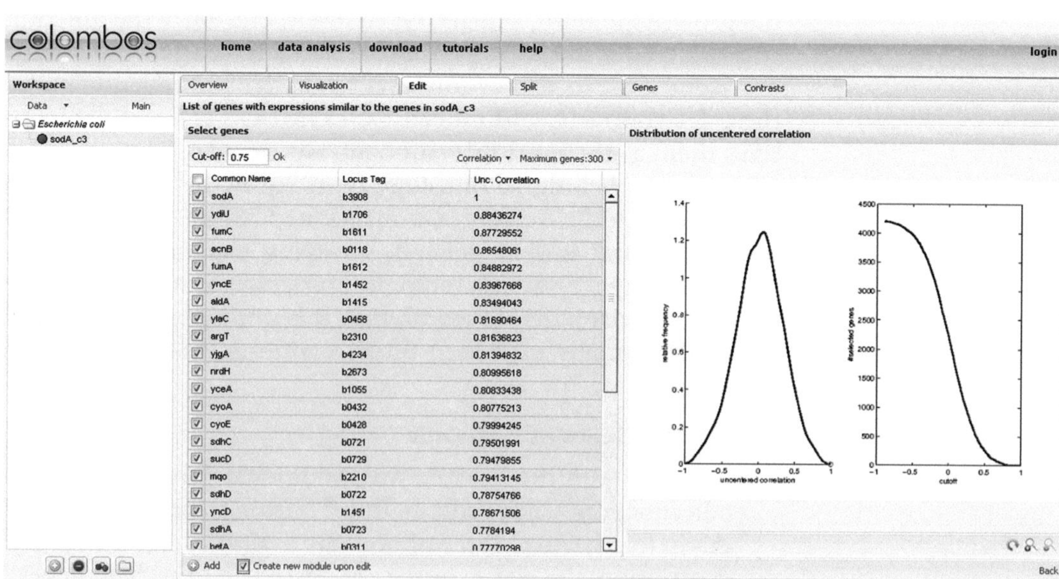

Fig. 7. The COLOMBOS "ranked gene selection" interface. Genes are ranked in the *center* by their correlation to the genes in the module, which at this stage was only *sodA* (correlation scores are calculated as described in Note 5). Different ranking options are available, and can be selected from the *dropdown box* that appears after clicking the text "correlation" above the ranked gene list. The figures in the *right* provides additional helpful information and show, from *left* to *right*, a *density plot* showing the distribution of the correlation scores across all genes in the compendium and a *plot* showing the number of genes that will be added if a given cutoff is chosen. The *circle* at one indicates the gene *sodA* that already exists in the current module, and that is of course perfectly correlated with itself (hence the correlation score of 1).

The figures to the right are meant to provide additional helpful information and show, from left to right, a density plot showing the distribution of the correlation scores across all genes in the compendium and a plot showing the number of genes that will be added if a given cutoff is selected. The circles connected by a red vertical line in the density plot indicate the gene(s) that already exists in the current module and its correlation score.

To extend our module, we use a cutoff value of 0.75. Input 0.75 in the box above the gene list at the left, and click "Ok" button at the side. The list is then filtered, and all genes to be added are automatically selected (checked) in the list to the left. With the option "Create new module upon edit" (located below the gene list), the original module is kept unchanged, and the extended module is saved as a new module. Chose this option, then click the "Add" button at left, and specify "sodA_c3_g.75" (in the popup window that appears) as the name of the new module to create it. It now appears as a node below the original module "sodA_c3" in the workspace.

The resulting module now contains 35 genes with an expression profile similar to that of *sodA* under the selected contrasts (see heatmap Fig. 8). Among the genes in the module are *cyoA*, *cyoB*, *cyoC*, and *cyoE*, subunits of the cytochrome *b* terminal oxidase

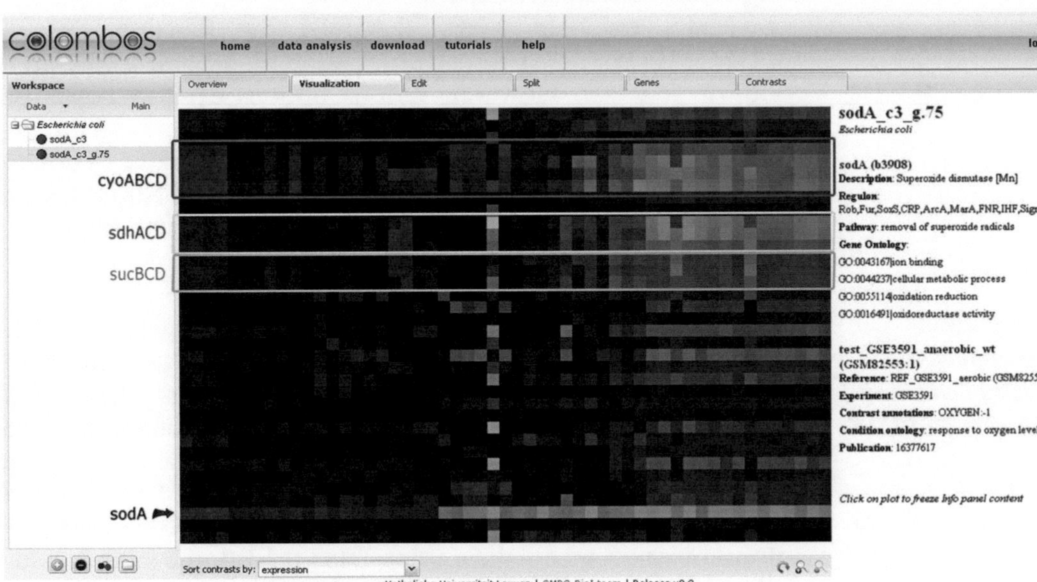

Fig. 8. The heatmap of module "sodA_c3_g.75" is shown here.The row of expression data that corresponds to *sodA* is marked by the *arrow at the bottom*. The expression data of the gene groups that are prominently coexpressed with *sodA* are marked by the *named boxes*. The genes *cyoA*, *cyoB*, *cyoC*, and *cyoE* are in the top one named *cyoABCD*, the genes *sdhA*, *sdhC* and *sdhD* are in the middle one named sdhACD, and the genes *sucB*, *sucC*, and *sucD* are in the bottom one named *sucBCD*.

complex involved in aerobic respiration (16), *sdhA*, *sdhC*, *sdhD* and encoding the succinate dehydrogenase (SdhCDAB) active during Krebs cycle catalyzing the oxidation of succinate to fumarate under aerobic conditions (17) and *sucA*, *sucB*, *sucC*, and *sucD* involved in generating succinyl-CoA, one of the reactants in the Krebs cycle (18). Considering the functions of these gene products (see also enriched GO terms in Table 2), it is not at all unexpected that their expression levels are influenced by oxygen availability. Moreover, many genes in module "sodA_c3_g.75" also share common regulators with *sodA*, such as ArcA, Fur, FNR, CRP, etc. So *sodA* (encoding a superoxide dismutase activity) being coexpressed with the genes involved in aerobic respiration might be essential to protect the cell against oxidative stress.

The fact that we obtained module "sodA_c3_g.75" clearly show the strength of the simple and straightforward method implemented in COLOMBOS. Although this COLOMBOS coexpression module does not necessarily imply coregulation of its genes, this could be investigated further (see Fig. 1). The promoter regions of the genes in this module could now be used as input for motif detection methods. This task could be achieved either by screening the promoter regions with Position-Specific Scoring Matrices (PSSMs) of previously characterized motifs, or by using de novo motif detection methods (19, 20). Note that some of the genes in it are unannotated, the COLOMBOS web service can also serve as a tool to help biologists identify interesting future research candidates.

Here, we only briefly described the functionalities of COLOMBOS related to constructing modules based only on a known gene set. However, various options exist to build the module based on different types of information. Interested users can see Note 6 for more information. It is also possible to download the data of a module to further analyze it (see Note 7). One can also create a user account that enables you to save the workspace and load previously generated data (see Note 8).

3.2. Identifying Transcriptional Modules Containing sodA with DISTILLER

As described above, DISTILLER (2) is a data integration framework that can automatically identify condition-dependent transcriptional modules by combining expression data with information on the interaction between a regulator and its corresponding target genes (here through motif data). Here, we show how to run each of the three steps of DISTILLER through our web service and setup the related parameters. After transcriptional modules are obtained using DISTILLER, the module visualization software ViTraM is used to visualize overlapping transcriptional coexpression modules together with the motif information in an interactive way. At the end, we briefly discuss those modules containing *sodA* in DISTILLER results.

3.2.1. Running DISTILLER

Every search for regulatory modules targeting one specific gene or a set of genes starts from the expression and motif data of all genes of the species. DISTILLER then identifies all possible modules. From the resulting modules, the user can filter out those involving his or her query genes. As a case study, we use DISTILLER to identify the regulatory modules containing *sodA*.

To reduce the running time of DISTILLER, the motif data can be reduced by removing all the motifs that are irrelevant to the specific query gene set. Here, we omitted motifs that were not present in the *sodA* promoter region from our motif dataset so that only seven motifs (ArcA, Rob, SoxS, Fur, IHF, FNR, MarA, and CRP) remained.

Create a User Profile and a Project

Since DISTILLER can be time consuming, a user account needs to be created and a project should be defined for each dataset.

Step 1 Go to the DISTILLER web service at: http://ibiza.biw.kuleuven.be/DISTILLER/. An existing user can login through the front page. New users can create a user profile by clicking the "New User" button and completing the required information. Subsequently, click the "Create User" button. Note that the user's email address is required for sending an email notification of the results (see Note 9).

Step 2 After log in, the project selection interface will appear, where one can directly select an existing project from the combo box to continue, or create a new one by clicking "Create New Project" and giving a name and a brief description of the project. Here, we create a new project named "Ecoli_sodA."

Step 3 After choosing or creating a project, the DISTILLER panel will appear where the user can run the three main steps of DISTILLER, each of which has its own separate tab. As our project is a new one, the user will see the "Module Detection" tab, while the "Module Filtering" and "Module Extension" tabs are disabled.

Identifying Seed Modules

Step 4 The module detection tab allows the user to identify seed modules from the source data. For a new project, no source data files have been uploaded, hence only the "Load file" button at the lower right corner is enabled, and the "run Distiller" button is disabled. Users must first upload the expression data file (use sample file expdata_DISTILLER_Expression.txt) and the motif data file (binary_DISTILLER_Motif.txt). Upload the required data files by clicking the "Load File" button. Note that the web service automatically recognizes the data type of a file according to the prefix of its name. See Note 10 for the mandatory file name prefix of each input file type.

Step 5 After uploading the data files, users can now prepare to run DISTILLER by specifying the required parameters. Three groups of parameters exist: the binary data parameters related to the motif data (see step 6), the expression data parameters (see step 7), and the general parameters (see step 8).

Step 6 Binary data parameters: (1) Binary Support specifies the minimal number of motifs that the genes in a module should have in common and (2) Binary Thresholds is the minimal score a motif instance in the promoter region of a gene should have to be considered as present in that promoter region (see Note 11). For this case study, we choose 1 for the Binary Support and 0.999 as the Binary Threshold. Hence the genes belonging to a module must share at least one motif and a gene is considered to have a specific motif if the corresponding value in the motif data is equal or higher than 0.999.

Step 7 Expression data parameters: (1) Box Support specifies the minimal number of condition contrasts under which the genes in a module should be coexpressed and (2) Box *p*-value is used to generate the threshold bandwidth sequence that is needed to test whether a module passes the constraints on the expression data, i.e., to test whether the genes in the module are sufficiently coexpressed within the selected condition contrasts (see ref. 2 for details). Here, we choose 100 for the Box Support and 0.0001 for the Box *p*-values.

Step 8 General parameters: (1) minimal module size specifies the minimal number of genes that should be in a module and (2) number of randomizations specifies the number of random modules that will be generated for computing a threshold bandwidth sequence for the condition contrast selection. Here, we specify the minimal module size as 4, and the number of randomizations as 10,000. The algorithm uses these parameters to generate 10,000 random modules consisting of four genes to compute the threshold bandwidth sequence.

Step 9 When all parameters are specified, click "Run Distiller." DISTILLER now enumerates all initial modules, each of which contain at least four genes coexpressed in at least 100 condition contrasts and sharing at least one motif. Note these contrasts are selected based on comparing the ordered contrasts bandwidth sequences of given genes with the threshold bandwidth sequence (see ref. 2 for details).

As DISTILLER is computationally expensive, depending on the specified parameters and the dataset size the user will receive a notification email when the process is finished. The email contains the link to the output file holding the information on all identified modules, and the link to the supplementary file bundle (see Note 9) that is required to visualize modules using ViTraM.

Filter the Raw DISTILLER Output

Step 10 After receiving the notification email for the first step of DISTILLER, the user can go back to the DISTILLER website to proceed with the other steps. Open the DISTILLER website, log in with your user account, and select the previously created project. Since the first step of DISTILLER has finished successfully, the tabs "Module Filtering" and the tab "Module Extending" are enabled. Note that the user can directly proceed to the module extending step, but due to reasons specified in Subheading 2.3.2, performing the filtering step first is highly recommended.

Step 11 Click the "Filtering" tab. Specify the number of modules to search in this step. Click on "Filter Result" to proceed. In our case study, we choose 20 modules. The result will be the top 20 modules ranked by their scores computed by DISTILLER (see Subheading 2.3.2). The filtering step is also computationally intensive. Hence, the end of the process will again be notified by an email sent to the user.

Extend DISTILLER Seed Modules

Step 12 When the notification email for the filtering step has been received, the user can now go back to the DISTILLER website to proceed with the "Module Extension" tab.

Step 13 The first parameter required by the module extension step is the Extended MotifThreshold. This is the same type of parameter as the Binary Threshold specified in the module detection step (see step 6 of this section). By choosing a less stringent (lower) value for the Extended MotifThreshold than was chosen for the Binary Threshold, a gene will have more motif candidates. As a result, more genes can satisfy the Binary Support parameter. In the case study, we use 0.95 for the Extended Motif Threshold (as compared to 0.999 for the Binary Threshold).

Step 14 The second parameter, Correlation Percentage, is specified as the correlation threshold applied to select candidate genes for module extension. Given a Correlation Percentage of *p*, only those genes whose expression profile correlation, calculated based on the mean expression profile of a module, is higher than *p*, will be considered as the candidates to extend that module. Here, we choose *p* as 0.95.

Step 15 Candidate extension genes have to comply with both parameters in order to be included in the extended module. However, the genes of a module can also be further extended with operon information if this is available. Candidate extension genes belonging to an operon whose first gene is present in a seed module, only need to satisfy the criterion for the expression profile (Correlation Percentage parameter) to be included into the module. The operon information (i.e., which genes belong to which operon) is included in the analysis by uploading an operon information file (see Subheading 2.4.1) with the prefix "operon_" in its file name.

Step 16 After specifying the parameters and optionally uploading an operon data file, the user can click "Extend Modules" to run the module extending step. A notification email is sent when the process is finished.

After receiving the final output file of DISTILLER, the user can then specifically select the modules that contain his or her query genes, in our case the gene *sodA*. Note that here we only discussed how to use DISTILLER to identify transcription networks. However, a versatile tool like DISTILLER has many other application domains. See Note 12 for more details.

3.2.2. Visualization of DISTILLER Modules with ViTraM

Here, we show how to use ViTraM to analyze modules discovered by DISTILLER. First, as mentioned in Subheading 2.4.2, the output file of DISTILLER needs to be converted into ViTraM readable format. Then, the generated files are used in ViTraM to visualize the DISTILLER results. We discuss both steps separately.

Note that due to the randomization, DISTILLER is nondeterministic. Thus, it is possible that when running DISTILLER in exactly the same way as outlined above, the recovered modules might be different from the ones in this example (Subheading 3.2.1). If the user wants to proceed with exactly the same results as in Subheading 3.2.1 he can use the "DISTILLER_outputExtended.m" sample file.

Prepare ViTraM Readable Files

Step 1 Download the XMLCreator from the ViTraM website at http://homes.esat.kuleuven.be/~kmarchal/ViTraM/Index.html and unzip the file.

Step 2 Run the XMLCreator by typing the command "java-jar-Xms256m-Xmx512m XMLCreator.jar" in a terminal. Windows users can click file "XMLCreator_Windows_512M.bat" to run the XMLCreator. Choose "DISTILLER" as module detection tool in the popup window and click on "OK". A new window appears where user can specify the input and output files. The required input files are the DISTILLER output data file of ".m" format and the "Expression Data" file (see Subheading 2.4.2). In addition, other files can provide extra information for each module that can be visualized by ViTraM, for example, the motif information of each gene. Finally, the names of the output files need to be specified.

Two output files will be generated: the module XML file and the corresponding expression data file consisting of the expression values for those genes and condition contrasts present in any of the modules. When ready, click the button "Make XML File" to generate these output files. If files are generated successfully, a window will appear to notify user.

As an example, we will use sample files provided with this chapter to show how to run XMLCreator. First, specify the file DISTILLER_outputExtended.m as the "DISTILLER output data," and the file expdata_DISTILLER_supple.txt as the "Expression Data". This is the minimal input that is required to run XMLCreator. In addition, we also provide the file binary_DISTILLER_Motif_supple.txt as the "Motif Data" providing motif information to be visualized together with the modules (see Note 13). Specify "ViTraM_Modules.XML" and "ViTraM_Expression.txt" as the output file names in data field "XML Data" and "Expression Data." Click "Make XML File" should generate the output file successfully. To run XMLCreator on the DISTILLER output, pleases always use the supplement files specified in the same notification email of an ".m" output file (see Note 9).

Run ViTraM

From the ViTraM website (http://homes.esat.kuleuven.be/~kmarchal/ViTraM/Index.html), the user can download ViTraM v2.0 and unzip the file. Note, a registration is required by providing us some basic information, and the user must accept our software license.

Step 3 To run ViTraM under windows, click on "ViTraM_Windows_512M.bat" or "ViTraM_Windows_2G.bat" provided. The latter can handle larger datasets, but does require more than 2Gb physical memory in the computer. To run ViTraM under linux or mac, call "ViTraM_Linux&Mac.sh" file.

Step 4 Load "ViTraM_Modules.XML" and "ViTraM_Expression.txt" generated in step 2 by clicking on "Open Module XML File" and "Open Expression File," respectively. After the data are loaded, they can be visualized by clicking on "Load All Modules" in the "Module Selection" panel. This extra loading step enables visualizing a subset of modules in the large dataset, where directly visualizing all the modules in it after data is loaded generally causes memory issues. In such a case, user can use Filter Panel first to select only the relevant modules, then to visualize them. This will significantly reduce the loading time to visualize modules.

Step 5 Subsequently clicking on "View Modules" in the "Module Display" panel to show the modules in the main window "Module Display." In this window, the genes are listed at the left, and the conditions are listed at the top. Each module is represented as one or a group of boxes with its id shown at the top left corner of each box. Note, each module uses a distinctive color for its boxes.

Step 6 Click "View Gene Properties" in "Gene Properties Display" panel, the genes' extra properties can be visualized together with the modules in an extra window located at the left side of the "Module Display" window. Note since we provided motif data while generating XML file in step 2 of this section, they will now appear in this window. The color of each cubic represents the value of the motif score explained in Subheading 2.4.1. The red color corresponds to the high value and the green the low ones. If a score is higher than a threshold, a cross will appear in the corresponding cubic. For more information, we refer user to the manual of ViTraM.

Step 7 The "View Experiment Properties" in the "Experiment Properties Display" panel can open up one extra window that shows the relevant condition data if loaded. In this case study, no such data is available.

Step 8 By clicking on "Favorite Genes" in the Filter Panel at the right hand side, it is possible to display only those modules containing a particular gene. The pop-up window shows on the left a list of all genes. Select the gene "b3908," corresponding to *sodA*, and click the arrow button "->" to add it to the favorite gene list at the right. When clicking "OK", only modules containing the *sodA* gene will be displayed.

Step 9 To optimize the visualization of modules that overlap, click on "Automatic ordering" in the "Module ordering" panel and subsequently "Run Overlap Index" to layout the currently displayed modules in the most optimal way (for details on the algorithms that identify the optimal display of overlapping modules, we refer to ref. 13).

Step 10 Click on "View Modules" and subsequently "Refresh Modules" in the "Overview & Heatmap Display." Then click on "Adding Heatmap." The expression values of the genes for the condition contrasts in the currently displayed modules will be shown by means of a heatmap.

3.2.3. Analyzing Resulting sodA Modules

In the case study we used DISTILLER to search for modules of coexpressed genes sharing at least one motif instances for the same regulator. Resulting modules including the motif instances shared among genes within each module are displayed in Table 1. Since we are interested in *sodA*, transcriptional modules containing this gene were selected and visualized using ViTraM. They are modules 2, 3, 8, and 12 (Fig. 9). The numbering of the modules corresponds to their ranks in the filtered results. Note that DISTILLER works as a global method and thus identifies all possible regulatory

Table 1
Overview of the 20 transcriptional modules identified by DISTILLER

Module	Motifs	# Genes	# Conditions	*sodA*	Genes in modules containing *sodA*
1	Fur	22	97		
2	MarA and SoxS	4	77	Yes	b0871, b1852, b3908, and b3924
3	ArcA and CRP	7	79	Yes	b0118, b0720, b0721, b1276, b1415, b3908, and b4376
4	FNR and IHF	4	226		
5	ArcA and FNR	4	123		
6	CRP	308	107		
7	CRP and FNR	4	87		
8	Fur and CRP	4	146	Yes	b0432, b0721, b2393, and b3908
9	SoxS	4	181		
10	Fur	18	86		
11	MarA	4	206		
12	CRP and FNR	5	90	Yes	b0113, b0432, b1415, b3417, and b3908
13	IHF	8	87		
14	IHF	53	123		
15	FNR	38	102		
16	CRP	76	125		
17	ArcA and CRP	4	148		
18	SoxS	4	95		
19	Fur	4	100		
20	Fur	22	76		

For each of the 20 modules, the number of genes and condition contrasts comprising the module, the motifs that are present in the upstream region of each gene of the module, and the presence of *sodA* is indicated. For each of these modules containing *sodA*, its genes are also listed at the end

modules in the dataset (so not only those with *sodA*). This explains why the numbering of the modules containing *sodA* is not consecutive. The lower the number, the more significant a module is.

Module 2 contains the genes regulated by both SoxS and MarA. The identified condition contrasts in this module belong to the ontology terms “carbohydrate metabolic process,” “growth,” and “response to oxidative stress.” SoxS and MarA are known to participate in the removal of superoxide and nitric oxide and protection from organic solvents (21). One can expect that the condition contrasts belonging to “response to oxidative stress” to

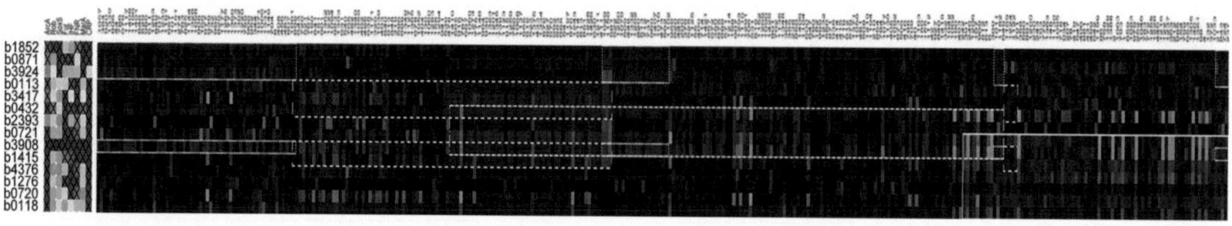

Fig. 9. The ViTraM output shows the four modules containing *sodA*. In this figure, *each row* represents a gene and *each column* represents either a motif (*left hand columns*) or a condition contrast (*right hand columns*). Modules are indicated with a *square* and expression values are visualized as a heatmap. In the heatmap module 2 is indicated by *dotted lines*, module 3 by *solid line*, module 8 by *dash-dotted lines*, and module 12 by *dashed lines*.

be suitable for inducing their expression and therefore the expression of genes regulated by them. We already highlighted the link between the removal of superoxide species and metabolic pathways. It explains why the condition contrasts under "carbohydrate metabolic process" are found in this module. As discussed after step 3 of Subheading 3.1, to observe condition contrasts belonging to "growth" is not unreasonable in this case. In addition to *sodA*, the other genes found in this module are *fpr* (Ferredoxin-NADP reductase), *poxB* (Pyruvate dehydrogenase), and *zwf* (Glucose-6-phosphate 1-dehydrogenase). The enriched GO terms of this gene set (see Table 2) are highly consistent with the observed condition ontology terms.

The genes in module 3 are regulated by both ArcA and CRP. Its condition contrasts are highly related to the ontology terms "growth" and "response to oxygen levels." CRP is a global regulator involves in the degradation of any nonglucose carbon sources and also an antagonist of catabolite repression. On the other hand, ArcA participates specifically in a signal transduction system sensing particular aerobic and anaerobic growth conditions (22). The functions of the genes found in this module (*acnA*, *acnB*, *aldA*, *gltA*, *osmY*, *sdhC*, and *sodA*) are an intersection between global metabolic pathways and specific processes responding to oxygen level changes (see Table 2).

Genes in module 8 are both regulated by CRP and Fur. As previously mentioned, CRP is a global regulator that facilitates bacterial fitness in function of the availability of different carbon sources. Fur is a sensor of intercellular iron concentration, and also participates in the response to reactive nitrogen species (23). The condition contrasts of this module mainly involves "carbohydrate metabolic process," "growth," "detoxification of nitrogen compound," and "lactose catabolic process." Except for *sodA*, the genes found in this module are *cyoA* (Ubiquinol oxidase subunit 2), *nupC* (Nucleoside permease nupC), and *sdhC* (Succinate dehydrogenase cytochrome b556 subunit). Altogether the condition contrasts and the genes selected in this module reflect the linkage between basic metabolism changes induced by

Table 2
GO enrichment of *sodA* modules12

Module	*p*-Value	Gene ontology term	GO term description
COLOMBOS			
sodA_c3_g.75	2.41e–12	GO:0006091	Generation of precursor metabolites and energy
	9.23e–02	GO:0006800	Oxygen and reactive oxygen species metabolic process
	3.17e–02	GO:0006869	Lipid transport
	2.74e–07	GO:0022900	Electron transport chain
	4.22e–02	GO:0042592	Homeostatic process
	1.85e–02	GO:0045454	Cell redox homeostasis
DISTILLER			
module2	7.07e–02	GO:0005996	Monosaccharide metabolic process
	7.11e–03	GO:0006800	Oxygen and reactive oxygen species metabolic process
DISTILLER			
module3	3.11e–05	GO:0006091	Generation of precursor metabolites and energy
	1.77e–02	GO:0006800	Oxygen and reactive oxygen species metabolic process
	7.11e–02	GO:0044262	Cellular carbohydrate metabolic process
DISTILLER			
module8	4.09e–03	GO:0006091	Generation of precursor metabolites and energy
	1.06e–02	GO:0006800	Oxygen and reactive oxygen species metabolic process
	4.09e–03	GO:0022900	Electron transport chain
DISTILLER			
module12	1.42e–02	GO:0006800	Oxygen and reactive oxygen species metabolic process

For the COLOMBOS module and each DISTILLER modules that were found to contain *sodA*, the most significantly enriched Gene Ontology terms [12] in each module are listed together with their enrichment *p*-value, which is calculated based on a hypergeometric distribution (see Note 14). The GO terms are ordered by their GO accession ID

external C-source availability (e.g., glucose concentration) and intracellular energy production through the electron transfer chain (see also enriched GO terms in Table 2).

In module 12, genes are regulated by CRP and FNR, both master-regulators. FNR activates genes involved in anaerobic metabolism and represses genes involved in aerobic metabolism. It also regulates the transcription of many genes with different functions, such as acid resistance, chemotaxis, cell structure, and molecular biosynthesis, among others. The selected condition contrasts of this module constitute a very broad set of ontology terms, including "carbohydrate metabolic process," "growth," "dna repair," "sos response," "lactose catabolic process," and "detoxification of nitrogen compound." This is in line with the global regulation exerted by CRP and FNR.

3.3. Comparison of COLOMBOS and DISTILLER

The COLOMBOS web service provides very straightforward and intuitive ways for analyzing and exploring large-scale expression compendia. It is intended for specific queries based on users' preknowledge on genes or conditions of interest. In each step, this knowledge and user involvement is crucial to explore various functionalities in order to generate good quality results. It is a deterministic approach, which means that when executing exactly the same operations, COLOMBOS will generate exactly the same result. Furthermore, the analysis methods implemented in COLOMBOS are fast, generally taking only seconds or less to generate results for each step.

DISTILLER, on the other hand, is a semiautomatic method. Except for a limited set of parameters, it requires very little user involvement. Guided by both motif information and expression data, the method tries to compose a global regulatory network that covers the whole input dataset (see Table 1). It is nondeterministic due to the randomization involved in generating a threshold bandwidth sequence for the condition selection (see step 8 in Subheading 3.2.1).

In our case study, we used COLOMBOS and DISTILLER to gain insights in the functional processes, in which *sodA* is involved, and the regulatory programs that coordinate them. This resulted in one module indentified by COLOMBOS and four modules identified by DISTILLER. As discussed in previous sections, each module does reflect relevant biological processes, in which genes including *sodA* participate. This does not necessarily require the modules to overlap, as each might represent different processes involving different genes. When these two approaches are used to address the same biological question as was done here, the distinct features of each approach leads to different but meaningful and thus complementary results.

The module "sodA_c3_g.75" identified by COLOMBOS contains several genes involved in various biological processes (see Fig. 8 and the discussion at the end of Subheading 3.1) with very high and coherent expression values for most condition contrasts (in both cases where up- or down-regulation occurred). In contrast, each of the DISTILLER modules contains in general much less genes than module "sodA_c3_g.75." Genes in these modules show less extreme expression levels indicating a weaker coexpression signal compared to "sodA_c3_g.75."

When evaluating the overlap between modules found by these two methods, module 3 and "sodA_c3_g.75" share four genes (*aldA*, *acnB*, *sdhC*, and *sodA*) and 19 condition contrasts ("response to oxygen level" and "growth"). These contrasts represent approximately 25 and 35% of all condition contrasts of each module, respectively. Module 8 and "sodA_c3_g.75" have-three genes (*cyoA*, *sdhC*, and *sodA*) and eight condition contrasts (response to oxygen level) in common, which represent approximately 6 and 15% of all condition contrasts in each module. This kind of similarity can be expected, since we applied both methods to answer same biological question based on the same expression data.

On the other hand, no significant overlap is observed between module 2 and "sodA_c3_g.75". Module 12 shares three genes (*aldA*, *cyoA* and *sodA*) with "sodA_c3_g.75" though, but the two modules have no condition contrasts in common. Upon closer inspection of the expression data, genes in DISTILLER module 2 are coexpressed, but the expression value of *sodA* is not very high for these condition contrasts, when compared with those of "sodA_c3_g.75." Hence, those condition contrasts will not be easily picked up by COLOMBOS in step 2.3 of Subheading 3.1 (unless a much lower selection threshold is used). Furthermore, genes appearing only in the DISTILLER modules show a different expression behavior from that of *sodA* under those condition contrasts selected by COLOMBOS. As a result, when extending genes of a module based on expression profiles (see step 4 of Subheading 3.1), those genes will be ranked less relevant. Consequently, they will hardly be considered as prime candidates for module extension.

In summary, COLOMBOS focuses on using expression values as its main criterion for building modules, resulting in clear expression changes in tight coexpression. It can easily extract prominent coexpression behavior in the expression data, but might miss modules with less significant coexpression patterns. The coexpression patterns retrieved by COLOMBOS can be broadly interpreted as genes being functionally related as their abundance are altered in similar ways in response to various stimuli. This functional relationship could imply coregulation (which might be identified using motif detection algorithms, see also Fig. 1), but it is not a necessary prerequisite. Instead, it cannot be excluded that several regulatory programs might be responsible for the observed coexpression patterns. On the other hand, DISTILLER, as a global method, tries to recover distinctive modules that can be directly linked to a shared regulatory program, i.e., of which the genes are coregulated by the same (set of) regulator(s). Combining motif information with expression data, it successfully retrieves less prominent coexpression patterns with biological significance by utilizing extra information regarding the presence (or prediction, depending on the nature of the motif input data) of transcription factor binding sites in its genes promoter regions. The case study of *sodA* used in this chapter illustrates this complementarity of both approaches for retrieving biologically relevant results.

4. Notes

1. Currently, COLOMBOS (v0.9) runs on three browsers: Firefox, Google Chrome, and Opera. Be sure that you are using the latest version of the browser to have the best compatibilities. Please check the website for updated information. The DISTILLER website (v1.0) is guaranteed to work properly on Firefox and efforts are undertaken to make it compatible with other browsers as well. Please check the website for updated information. ViTraM (v2.0) is written in JAVA; hence, it can run under any platforms with JAVA support. On Windows ViTraM comes in two versions: vitram_Windows_512M.bat or vitram_Windows_2G.bat. The vitram_-Windows_512M.bat version has a minimum memory requirement of 1Gb RAM, and we advise to use a computer with at least 3Gb RAM for running vitram_Windows_2G.bat.
2. A module is considered *closed* if it cannot be further extended by any other gene without reducing the number of motifs shared by all of its genes. To search such a candidate, it starts with the smallest module that contains only one gene, and gradually extends them by merging with other modules. A candidate is found when any extension violates the conditions implied by "*closed.*" Only closed modules that contain the minimal number of motifs and condition contrasts are considered valid (meet the required minimal Supports on the motifs and condition contrasts).
3. When counting the row number of a given gene, there are certain caveats. First, the title line should not be counted. Second, the count should start from 0. Hence the first row is of row number 0 instead of 1. The same applies to the *column number* referred in the ".m" file. There, the *first column* containing gene information should not be counted. And it also starts at 0.
4. The ranking score c of a condition contrast for a group of genes is calculated as the inverse coefficient of variation of those genes' expression values in this contrast. It is defined as the mean divided by the standard deviation of the genes' expression values:

$$c = \frac{\mu}{\sigma}.$$

On the one hand, for expression values of the same mean, the higher the score, the less sparse the values are. It prioritizes the contrasts where genes' expression values are more consistent. On the other hand, for expression values of the same standard deviation, the higher the score, the higher the mean. It prefers the contrasts where genes are highly expressed. The score thus serves as a measure that values both magnitude of expression change in response to a condition contrast, as well as coherence of expression within that contrast.

In case of one gene, as described in step 2.3 of Subheading 3.1, the score of a condition contrast is then degraded to the absolute expression value of that gene under it.

5. The correlation score is the *Uncentered Pearson's correlation* calculated based on the formula:

$$rv = \frac{1}{n}\sum_{i-1}^{n}\left(\frac{xi}{\sigma_x^{(0)}}\right)\left(\frac{yi}{\sigma_y^{(0)}}\right),$$

where $\sigma_x^{(0)} = \sqrt{(1/n)\sum_{i=1}^{n} x_i^2}$; and $\sigma_y^{(0)} = \sqrt{(1/n)\sum_{i=1}^{n} y_i^2}$. Here xi represents the candidate gene expression data at condition contrast i, whereas yi represents a module's mean expression value at contrast i. $\sigma_x^{(0)}$ and $\sigma_y^{(0)}$ are both uncentered standard deviations assuming zero mean of the population, hence they are marked with superscript "(0)." The higher the rv, the more similar the expression profile of a gene is to a module's mean expression profile.

For ranking genes, there are three options provided based on the (uncentered) correlation score calculated. First one is "correlation" which uses rv directly as final score. The second option "Absolution correlation" takes $|rv|$. It ranks both correlated and anticorrelated genes based solely on their similarities. Instead, the third option "Anticorrelation" take $-rv$ as score to favor only the anticorrelated genes.

6. COLOMBOS provides rich functionalities to help the user create coexpression modules. In this case study, we specify the query gene and letting COLOMBOS automatically identify relevant condition contrasts. Instead, one can also specify the conditions in interests and letting COLOMBOS automatically identify the genes that are over-expressed or under-expressed in them. Furthermore, it is also possible to specify both the genes and the conditions in interests in order to check the behavior logged in the existing experimental data.

When specifying genes of interests, apart from manually inputting gene information, one can also select genes based on other annotations obtained from public databases, such as the transcription factor or sigma factor that regulates the gene's transcription (24), the pathway a gene belongs to ref. 25, or the transcription unit a gene belongs to ref. 25. These functionalities are available as options in the gene selection section (left hand side) of the module creation panel.

Additionally, several alternative functions are provided to choose the condition contrasts besides the condition selection based on expression values as used here. In case one is interested in specific experiments, one can select them directly. When a user is interested in specific types of condition properties, a condition hierarchy is available. The properties are grouped into four major categories: genomic, growth, medium, and treatment. Subcategories exist under each major category to further classify them. This functionality can be reached through option "By annotation" in the condition selection section (right-hand side) of the module creation panel. One can also use the condition property ontology in a similar way by selecting "By ontology" in the condition selection section.

Even more options are available when the user modifies an existing module. To check all available functionalities, we refer the online manual of COLOMBOS.

7. Download function: once a final module is identified, it can be exported in the format described in Subheading 2.5.1. To do so first click the module in the workspace to show its overview tab. Then click the download button located at the up right corner of this tab to show the download window. Next, follow the instruction in this window to click the button "Request" at the top left corner. This sends a request to the server to generate the data file for the current module. When ready, a link to the generated data file is shown in the same window. User can click it to save the module data. For this case study, the data of the last module "sodA_c3_g.75" is exported into file expdata_COLOMBOS_module_information.txt (available as a sample file, see Subheading 2.6).
8. An anonymous user can use the analysis interface without logging in, but the data generated will be lost when the webpage is closed. A registered user has the advantage of being able to save the workspace, and load it again any other time when logged in. It takes only two steps to create a new user. After logging in, the "Data" menu items "Load," "Save," and "Save As" will be enabled.
9. Since each step of DISTILLER may take hours or even days to finish, the result of each step is sent to the user by email. This notification email is of a standard format, with subject "DISTILLER process result notification email." On the first line, the information about the finished process and the corresponding user project is presented, followed by the link to the DISTILLER result file. It is followed by the link to the supplementary file bundle required for ViTraM visualization.
10. Note the file uploading interface can load multiple files at one time. For system to recognize the different input files

uploaded, some prespecified name convention is necessary. We require that the name of the expression data file should start with "expdata_," and the name of the motif data file should start with "binary_."

11. The parameter Binary Threshold is used to convert the continuous values in the uploaded motif data into a binary one. When applied to the data, any value which is equal or higher than the threshold will be converted into value 1, and 0 otherwise, resulting in a binary matrix. This binary matrix is then used by DISTILLER to identify modules.
12. DISTILLER can also use other data sources as input, such as for instance information on the binding of regulators as obtained from ChIP-chip experiments (2, 26). Furthermore, DISTILLER can be used together with data source other than motif data to discover other type of networks, as long as connections exist between the data source and the gene expression profile. For instance, protein–protein interaction data or synthetic lethality data can also be included. Binary data is preferred; for other types of data, a preprocessing step to convert them into binary data is required.
13. The XMLCreator allows loading additional data files such as information on the function of the genes, the binding of a regulator to the promoter region of a gene, or information on what was measured in each experimental conditions (see also manual of the XMLCreator at ViTraM's website). When provided, this information can be displayed along with the visualized modules inside ViTraM.
14. The GO enrichment *p*-value is calculated based on a hypergeometric distribution. The lower the *p*-value, the lower the chance of a GO term appearing in a random gene set having the same number of genes as the module in comparison, and the more significant a GO term is enriched in the module.

Acknowledgments

This work is supported by (1) K.U. Leuven: GOA/08/011, CoE EF/05/007 SymBioSys, CREA/08/023; (2) IWT: SBO-BioFrame; (3) IUAP: P6/25 (BioMaGNet); (4) FWO: IOK-B9725-G.0329.09; and (5) HFSP-RGY0079/2007C.

References

1. Fierro AC, Vandenbussche F, Engelen K, Van de Peer Y, Marchal K. (2008) Meta analysis of gene expression data within and across species. Curr Genomics, 9:525–534.
2. Lemmens K, De Bie T, Dhollander T, De Keersmaecker SC, Thijs IM, Schoofs G, De Weerdt A, De Moor B, Vanderleyden J, Collado-Vides J, Engelen K, Marchal K. (2009) DISTILLER: a data integration framework to reveal condition dependency of complex regulons in *Escherichia coli*. Genome Biol, 10:R27.
3. Fadda A, Fierro AC, Lemmens K, Monsieurs P, Engelen K, Marchal K. (2009) Inferring the transcriptional network of *Bacillus subtilis*. Mol Biosyst, 5:1840–1852.
4. Faith JJ, Hayete B, Thaden JT, Mogno I, Wierzbowski J, Cottarel G, Kasif S, Collins JJ, Gardner TS. (2007) Large-scale mapping and validation of *Escherichia coli* transcriptional regulation from a compendium of expression profiles. PLoS Biol, 5:e8.
5. Ernst J, Beg QK, Kay KA, Balazsi G, Oltvai ZN, Bar-Joseph Z. (2008) A semi-supervised method for predicting transcription factor-gene interactions in *Escherichia coli*. PLoS Comput Biol, 4:e1000044.
6. Babu MM, Lang B, Aravind L. (2009) Methods to reconstruct and compare transcriptional regulatory networks. Methods Mol Biol, 541:163–180.
7. Faith JJ, Driscoll ME, Fusaro VA, Cosgrove EJ, Hayete B, Juhn FS, Schneider SJ, Gardner TS. (2008) Many microbe microarrays database: uniformly normalized affymetrix compendia with structured experimental metadata. Nucleic Acids Res, 36:D866–D870.
8. Engelen K, Fu Q, Meysman P, Sánchez-Rodríguez A, De Smet R, Lemmens K, Fierro AC, Marchal K. (2011) COLOMBOS: access port for cross-platform bacterial expression compendia. PLoS One 6(7): e20938.
9. Fawcett WP, Wolf RE. (1995) Genetic definition of the *Escherichia-coli* Zwf Soxbox, the DNA-binding site for Soxs-mediated induction of glucose-6-phosphate-dehydrogenase in response to superoxide. J. Bacteriol. 177:1742–1750.
10. Tardat B, Touati D. (1993) Iron and oxygen regulation of *Escherichia coli* Mnsod expression – competition between the global regulators fur and arca for binding to DNA. Mol Microbiol, 9:53–63.
11. Jair KW, Martin RG, Rosner JL, Fujita N, Ishihama A, Wolf RE. (1995) Purification and regulatory properties of mara protein, a transcriptional activator of *Escherichia-coli* multiple antibiotic and superoxide resistance promoters. J Bacteriol, 177:7100–7104.
12. Ashburner M, Ball CA, Blake JA, Botstein D, Butler H, Cherry JM, Davis AP, Dolinski K, Dwight SS, Eppig JT, Harris MA, Hill DP, Issel-Tarver L, Kasarskis A, Lewis S, Matese JC, Richardson JE, Ringwald M, Rubin GM, Sherlock G. (2000) Gene ontology: tool for the unification of biology. The Gene Ontology Consortium. Nat Genet, 25:25–29.
13. Sun H, Lemmens K, Van den Bulcke T, Engelen K, De Moor B, Marchal K. (2009) ViTraM: visualization of transcriptional modules. Bioinformatics, 25:2450–2451.
14. Hertzberg L, Zuk O, Getz G, Domany E. (2005) Finding motifs in promoter regions. J Comput Biol, 12:314–330.
15. Zaki MJ, Hsiao C. (2002) CHARM: An Efficient Algorithm for Closed Itemset Mining. In *Proceedings of the Second SIAM International Conference on Data Mining*, 457–473.
16. Cotter PA, Gunsalus RP. (1992) Contribution of the fnr and arcA gene products in coordinate regulation of cytochrome o and d oxidase (cyoABCDE and cydAB) genes in *Escherichia coli*. FEMS Microbiol Lett, 70:31–36.
17. Wilde RJ, Guest JR. (1986) Transcript analysis of the citrate synthase and succinate dehydrogenase genes of *Escherichia coli* K12. J Gen Microbiol, 132:3239–3251.
18. Buck D, Guest JR. (1989) Overexpression and site-directed mutagenesis of the succinyl-CoA synthetase of *Escherichia coli* and nucleotide sequence of a gene (g30) that is adjacent to the suc operon. Biochem J, 260:737–747.
19. Tompa M, Li N, Bailey TL, Church GM, De Moor B, Eskin E, Favorov AV, Frith MC, Fu Y, Kent WJ, Makeev VJ, Mironov AA, Noble WS, Pavesi G, Pesole G, Regnier M, Simonis N, Sinha S, Thijs G, van Helden J, Vandenbogaert M, Weng Z, Workman C, Ye C, Zhu Z. (2005) Assessing computational tools for the discovery of transcription factor binding sites. Nat Biotechnol, 23:137–144.
20. Storms V, Claeys M, Sanchez A, De Moor B, Verstuyf A, Marchal K. The effect of orthology and coregulation on detecting regulatory motifs. PLoS One, 5:e8938.

21. Touati D. (2000) Sensing and protecting against superoxide stress in *Escherichia coli*–how many ways are there to trigger soxRS response? Redox Rep, 5:287–293.
22. Compan I, Touati D. (1994) Anaerobic activation of arcA transcription in *Escherichia coli*: roles of Fnr and ArcA. Mol Microbiol, 11:955–964.
23. Mukhopadhyay P, Zheng M, Bedzyk LA, LaRossa RA, Storz G. (2004) Prominent roles of the NorR and Fur regulators in the *Escherichia coli* transcriptional response to reactive nitrogen species. Proc Natl Acad Sci U S A, 101:745–750.
24. Gama-Castro S, Jimenez-Jacinto V, Peralta-Gil M, Santos-Zavaleta A, Penaloza-Spinola MI, Contreras-Moreira B, Segura-Salazar J, Muniz-Rascado L, Martinez-Flores I, Salgado H, Bonavides-Martinez C, Abreu-Goodger C, Rodriguez-Penagos C, Miranda-Rios J, Morett E, Merino E, Huerta AM, Trevino-Quintanilla L, Collado-Vides J. (2008) RegulonDB (version 6.0): gene regulation model of *Escherichia coli* K-12 beyond transcription, active (experimental) annotated promoters and Textpresso navigation. Nucleic Acids Res, 36:D120–D124.
25. Karp PD, Ouzounis CA, Moore-Kochlacs C, Goldovsky L, Kaipa P, Ahren D, Tsoka S, Darzentas N, Kunin V, Lopez-Bigas N. (2005) Expansion of the BioCyc collection of pathway/genome databases to 160 genomes. Nucleic Acids Res, 33:6083–6089.
26. Lemmens K, Dhollander T, De Bie T, Monsieurs P, Engelen K, Smets B, Winderickx J, De Moor B, Marchal K. (2006) Inferring transcriptional modules from ChIP-chip, motif and microarray data. Genome Biol, 7:R37.

Chapter 9

Using Phylogenetic Profiles to Predict Functional Relationships

Matteo Pellegrini

Abstract

Phylogenetic profiling involves the comparison of phylogenetic data across gene families. It is possible to construct phylogenetic trees, or related data structures, for specific gene families using a wide variety of tools and approaches. Phylogenetic profiling involves the comparison of this data to determine which families have correlated or coupled evolution. The underlying assumption is that in certain cases these couplings may allow us to infer that the two families are functionally related: that is their function in the cell is coupled. Although this technique can be applied to noncoding genes, it is more commonly used to assess the function of protein coding genes. Examples of proteins that are functionally related include subunits of protein complexes, or enzymes that perform consecutive steps along biochemical pathways. We hypothesize the deletion of one of the families from a genome would then indirectly affect the function of the other.

Dozens of different implementations of the phylogenetic profiling technique have been developed over the past decade. These range from the first simple approaches that describe phylogenetic profiles as binary vectors to the most complex ones that attempt to model to the coevolution of protein families on a phylogenetic tree. We discuss a set of these implementations and present the software and databases that are available to perform phylogenetic profiling.

Key words: Phylogenetic profiles, Coevolution, Functional associations, Comparative genomics, Coevolving proteins

1. Introduction

The remarkable improvements in sequencing technology that have occurred over the past few decades have made the sequencing of genomes an ever more routine task. To date about 1,000 bacterial genomes have been sequenced along with dozens of eukaryotic ones. Along with the genome sequences themselves, annotation efforts have also progressed so that most new genome sequences are accompanied by detailed descriptions of the positions of the genes encoded within the genome, and the functions of

Jacques van Helden et al. (eds.), *Bacterial Molecular Networks: Methods and Protocols*, Methods in Molecular Biology, vol. 804, DOI 10.1007/978-1-61779-361-5_9,

the proteins and noncoding RNAs that are encoded in the genes. One of the fundamental challenges for computational biologists and bioinformaticians is the inference of interaction networks between these genes that enhance our ability to understand the function of the gene products. To this end, here we discuss the phylogenetic profiling technique, and its uses for probing functional association between proteins.

Sequence homology is the primary tool used to assign a function to a protein. If two proteins have significant similarities between their sequences, then they likely descended from a common ancestor and share a common function. As our experimental knowledge of protein functions increases, this approach allows us to pass functional annotation from a characterized protein to its uncharacterized homologs, and thus annotate an ever-growing number of sequenced proteins. Nonetheless, gaps in our knowledge still remain, which lead to lack of any meaningful functional annotation for many protein families. The function of these orphan proteins may not be studied using homology-based computational approaches, and therefore a different class of nonhomology algorithms must be used in these cases. Among these approaches, phylogenetic profiling is one of the primary tools.

As genomes evolve, genes are deleted or are horizontally transferred from one species to another. The intermingling of genetic material between species, which is particularly common among bacteria, makes the reconstruction of species phylogenetic trees very challenging. At the same time, the abundance of genetic exchanges between organisms offers a unique opportunity to study the coupling of genes within genomes. If at the simplest level we view genomes as bags of genes, and we have access to hundreds of bags, then we can begin to identify pair-wise and higher associations between the elements of the bags. In other words, if genes are often transferred between organisms, we can identify which sets of genes appear to transfer together. Genes that are coupled in this manner are necessarily present or absent within the same organisms, and it is therefore unlikely to find one without the other. The identification of these couplings allows us to infer that the products of the two genes likely function together to achieve a common biological function. They may be subunits of a protein complex, or sequential steps in metabolic pathways. In either case, an organism needs both genes to carry out their function, and having only one of them likely decreases its fitness.

The search for co-occurring protein families across organisms is often referred to as phylogenetic profiling. Here, we review various implementations of phylogenetic profiling, and also discuss databases that use this approach to study protein function. Finally, we discuss extensions of phylogenetic profiling that involve higher order associations between genes.

2. Structure of Phylogenetic Profiles

The first implementation of phylogenetic profiles consisted of binary vectors that captured the presence or absence of homologs of a reference protein across organisms (1). To construct these profiles, first a reference genome was selected (e.g., *E. coli*). Each *E. coli* protein was then aligned to the proteome of each fully sequenced bacteria using BLAST (2). If a hit to one of the proteins in another organism is identified by a significant BLAST threshold (e.g., $E < 1e\text{-}6$), then a 1 is inserted in the corresponding position of that organism in the phylogenetic profile. If no significant hit is found, then a zero is inserted. Using this approach, the entire length of the phylogenetic profile vector is populated. A schematic representation of the construction of phylogenetic profiles is shown in Fig. 1.

When using reference genomes to construct phylogenetic profiles, one generates a separate phylogenetic profile matrix for each organism. However, instead of computing the homologs between a reference genome and other organisms, it is also possible to use orthologous protein families to populate a general phylogenetic profile that does not require a reference genome. The definitions of orthologous protein families vary across implementations, but one example is that provided by the clusters of orthologous groups (COGs) database (3, 4).

Representations of phylogenetic profiles other than binary vectors have also been developed. For example, in one case,

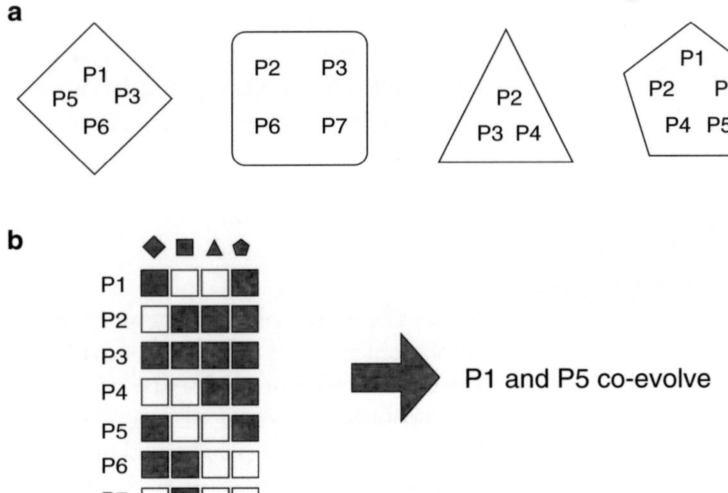

Fig. 1. Schematic representation of the phylogenetic profile method. In panel (**a**), we see four sample genomes with their respective proteins. Panel (**b**) shows phylogenetic profiles that capture the presence or absence of these protein families across genomes. We note that protein family one and five have identical profiles, and are thus coevolving.

instead of a binary value, the logarithm of the BLAST expectation score is used (5). Another approach used matrices instead of vectors to represent each profile (6). The entries of the matrix are the evolutionary distances between orthologous proteins, and the matrix has dimension $N \times N$, where N is the number of organisms being considered, and one would thus generate $N \times N \times P$ profiles for a genomes, where P is the number of protein families. These distance matrices may be compared by computing an internal product that measures the similarity of the corresponding cells across two matrices.

3. Metrics for Comparisons Between Phylogenetic Profiles

Once the phylogenetic profiles have been constructed, the next step consists of systematically comparing all pairs of profiles to detect protein families that have coupled evolutionary patterns. The first approach developed to compare profiles used the Hamming distance, or simply the number of positions where two binary profiles have different values (1). However, this metric does not provide a statistical framework for evaluating the likelihood of observing a specific Hamming distances. To this end, subsequent approaches used the hypergeometric distribution to estimate the probability of observing a certain number, or greater matches between two profiles (7). This metric was further refined in other work to account for the different proteome sizes of the organisms that are used to populate the phylogenetic profiles. A weighted hypergeometric distribution P value was developed to address this limitation (8).

When nonbinary phylogenetic profiles are constructed, then other metrics have to be used. Various studies have used mutual information to measure the similarities between profile vectors (5).

4. Accounting for the Phylogenetic Tree of Organisms

Since the goal of phylogenetic profiling is the identification of protein families that have coevolved, it is important to account for the underlying phylogeny of the organisms used to construct the profiles. The tree of organisms may be used to infer what loss or acquisition events explain the profile of a specific protein family. When comparing two profiles, it is then possible to use parsimony to estimate the smallest number of differences that explain the evolution of two protein families (9). This approach allows us to separate pairs of profiles that have identical scores when non-tree-based metrics, such as the ones described in the previous section, are used.

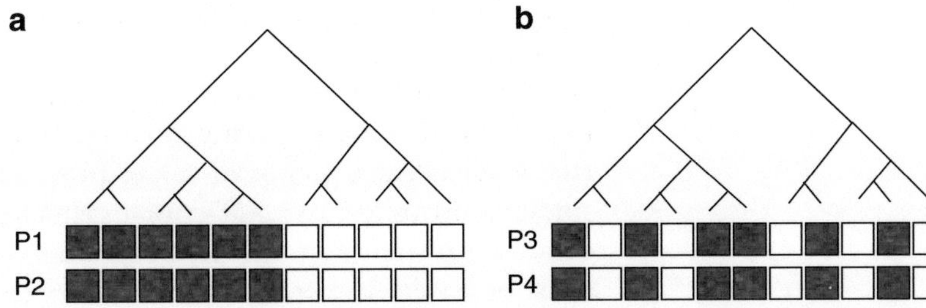

Fig. 2. Tree-based analysis of phylogenetic profiles allows us to identify the most significant coevolutionary events. In panel (**a**), we see two protein families that co-occur across genomes, but are only found in one branch of the species tree, thus we can explain the pattern with a single loss event at the right branch of the first bifurcation, or a single gain in the left branch of the first bifurcation. In contrast, panel (**b**) shows a more interesting co-occurrence pattern that arose due to multiple coordinated loss events across the species surveyed. We conclude that P3 and P4 are more likely coevolving, and hence functionally related, than P1 and P2.

The simple example shown in Fig. 2 illustrates the advantage of considering the underlying tree of organisms when comparing two polygenetic profiles. In panel A, we see two profiles that are identical, but can be explained by the acquisition of the protein in one branch of the tree. In contrast, in panel B we see two proteins that have repeatedly been lost and acquired in multiple branches of the tree. Our hypothesis is that the latter pair is far more likely to be functionally coupled than the former, since their evolutionary pattern requires more coordinated loss events.

Pagel et al. provided a formal model for the estimation of the likelihood of coevolution of two protein families on a tree of organisms (10, 11). They explicitly estimate the coupling parameters of two protein families and are thus able to derive an estimate of their coevolution. However, this formalism requires significant computational resources compared to the simpler metrics.

Recently, we introduced an intermediate approach that partially accounts for the tree of organisms, but is able to do so more efficiently than the Pagel et al. approach (12). This method does not consider the full tree of organisms, but only the ordering of organisms within the tree. Dynamic programming may be used to estimate the optimal ordering or organisms on a tree, by minimizing the distance between adjacent organisms represented by the leaves of the tree (13). Armed with this information, it is possible to implement a simple probabilistic model for the likelihood of observing runs of matching ones in two binary profiles. This allows us to separate pairs of profiles that have matches in only one branch of the tree, from those that are co-occurring in diverse branches of the tree. It also allows us to partially compensate for the fact that certain branches are more populated than others in cases, where a protein family is conserved across the entire branch, since both large and small branches would contribute a single run of occurrences.

The simplest approach to partially account for the tree of organisms when comparing phylogenetic profiles is to prune the tree so that each group of related organisms is represented by a single individual (14). Of course, there is some ambiguity associated with the definition of a related group of organisms, as this must be based on a somewhat arbitrary level of taxonomic similarity. However, once this parameter has been established, this method reduces the biases in the analysis that may arise from the over-representation of one group with respect to another, but does not explicitly correct the remaining biases that are addressed by the two previous approaches.

5. Assessing the Functional Relationships of Coevolving Protein Families

Our hypothesis is that coevolving protein families likely share a related function. A variety of methods have been developed to test this theory. The simplest is to use existing pathway annotation, such as that provided by the Gene Ontology Consortium (15). This annotation allows us to group together proteins that act within the same biological process. Examples of biological processes are the cell cycle, specific metabolic pathways or large protein complexes, such as the flagella. In order to demonstrate that coevolving pairs of protein are functionally related, we simply demonstrate that these pairs are more often members of the same biological process than random pairs. If pairs of proteins within the same process are considered true positives, and those in different processes true negatives, then using a receiver operator characteristic (ROC) curve allows us to measure the enrichment for true positives versus false positives in a ranked list. These types of analyses also enable the comparison of different approaches for the identification of phylogenetic profile pairs, as the methods that yields the largest area under the ROC curve performs best (12, 16).

Beyond the demonstration that phylogenetic profiling allows one to identify functionally related proteins, it is also possible to use this approach to assign a function to a previously uncharacterized protein family. That is, if nothing is known about the function of protein family A, but using phylogenetic profiling one identifies that it is coevolving with functionally characterized protein family B, then it is likely that the function of A is similar to that of B. A recent review demonstrates that over the years a variety of examples of this "guilt by association" technique have been used to predict and then verify the functions of previously uncharacterized protein families (17).

6. Higher-Order Relationships Among Phylogenetic Profiles

The analysis discussed so far has been restricted to the identification of pair-wise relationships between phylogenetic profiles. However, it is also possible that higher order relationships may be identified. Here, we present two examples of approaches that have been used to identify relationships between triplets and larger groups of phylogenetic profiles.

The first approach searches for logic combinations of pairs of triplets that match a third profile (4). For example, protein family C may be present across organisms only when both protein families A and B are present (see Fig. 3). Thus, C is not correlated with the presence of A or B, but is only correlated with the combined presence of families A and B. Bowers et al. developed an implementation of this approach and were able to uncover a large number of examples of logic triplets. They analyzed all eight possible logic relationships and developed a metric for scoring the significance of logic triplets.

A second method that uses higher-order relationships between phylogenetic profiles was developed to identify protein complexes that are duplicated within genomes (18). In this approach, phylogenetic profiles are first compared to identify significant pair-wise relationships. This leads to the creation of a $P \times P$ binary matrix, where P is the number of profiles, whose entries are one if the corresponding (i,j) pair is found to be significantly related and zero otherwise. This matrix is then clustered so that groups of profiles that are significantly related form clusters along the matrix diagonal. These clusters might correspond to components of pathways or protein complexes.

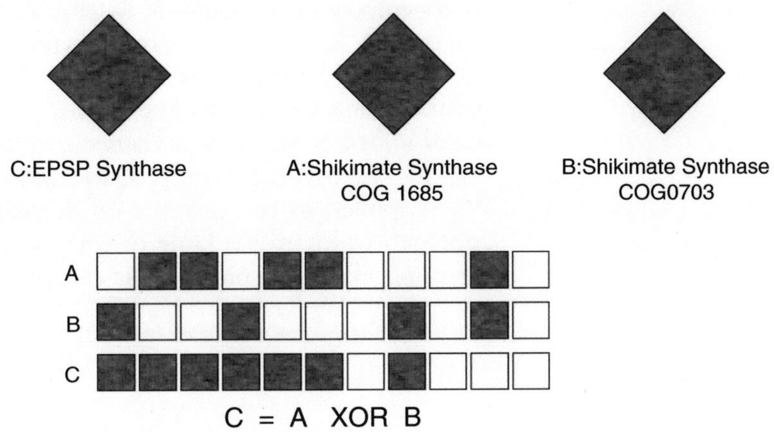

Fig. 3. Logic analysis of phylogenetic profiles. Protein family C (EPSP synthase) is only present in genomes that contain either protein families A (COG1685 Shikimate Synthase) or B (COG0703 Shikimate Synthase), but not both (exclusive or, XOR).

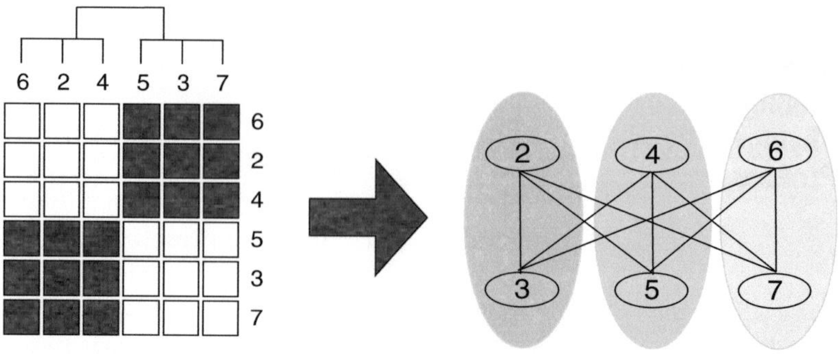

Fig. 4. Phylogenetic profiles may be used to identify duplicated complexes. *On the left*, we see a matrix of significant pair-wise relationships between the phylogenetic profiles of the two corresponding proteins (*black indicates significance and white lack of significance*). These proteins are clustered and ordered in the same manner on the *x* and *y* axes. A block of significant relationships occurs off the matrix diagonal (*upper left to lower right*). This pattern is due to coevolutionary relationships between the subunits of the three complexes shown on the right. The 2 subunits of the blue complex coevolves with the 3, 5, and 7 subunits, which are all homologous to each other. The 4 and 6 subunits show identical co-occurrence patterns. By using additional criteria, such as the presence within an operon, it is possible to separate the three complexes from each other.

Alternatively, one can also cluster the nonbinary version of the profile similarity matrix, whose entries represent the distance between two profiles.

Using this approach, Li et al. also identified off-diagonal clusters. These arise when protein complexes are duplicated within the genome. Consider, for example, a complex that is formed of two subunits (2 and 3), and two other homologous complexes which are formed of two subunits (4 and 5, 6 and 7), as illustrated in Fig. 4. We assume that subunit 2 is homologous to 3, 5, and 7. In this case, the phylogenetic profile of protein 2 would be found to coevolve with proteins 3, 5, and 7. Similarly, protein 3 coevolves with 2, 4, and 6. This pattern of relationships leads to an off-diagonal cluster that contains the pair-wise interactions between the six subunits. Separating these complexes using additional information, such as co-occurrence within operons, allows one to identify the correct association between the subunits. When applied to the genome of *R. palustris*, Li et al. used this approach to identify a large off-diagonal cluster that arose from the duplication of a nitrogenase complex.

7. Available Software

Several databases are available that utilize phylogenetic profiles for functional annotation or network reconstruction. The most widely used is the STRING database (19). This database is constructed using COGs and implements a continuous version of phylogenetic profiles, whose pair-wise similarity is evaluated using mutual information. The underlying phylogenetic tree of organisms is accounted for by collapsing into a single node those taxa in which the presence or absence of a specific gene pair is in agreement in all the species.

Another implementation of phylogenetic profiles is found in the Prolinks database (20). Here, orthology is determined using a simple BLAST cut-off criterion, and a reference organism must be specified. The phylogenetic profiles are encoded as binary vectors and their similarity is determined using the hypergeometric distribution probability. In this implementation, the phylogenetic relationships between organisms are not accounted for, and each genome is treated as an independent random variable (and therefore some closely related genomes can lead to a bias in the analysis). However, future implementations will also include the "runs" correction discussed above to partially compensate for this effect. The PLEX (21) and ViSANT (22) databases implement phylogenetic profiling in a very similar manner.

8. Strength and Pitfalls

As we have discussed, the phylogenetic profile approach treats genomes as bags of genes, and does not consider the order of genes within the genome. This is clearly an oversimplification, and a variety of methods have been developed to search for pairs or larger groups of genes that maintain their proximity across varied organisms (23). Because transcriptional units in bacteria are operons and not individual genes, the evolutionarily conserved sets of genes typically correspond to operons, or fragments of operons. In fact, the conservation of proximity of groups of genes is a strong indication that the group is a part of a bacterial operon.

Many of the databases we have discussed above incorporate not only phylogenetic profiles in their analyses, but also some form of gene proximity conservation measure. Since the pairwise predictions produced by these two approaches do not always overlap, they may be combined using Bayesian techniques to arrive at more accurate coevolutionary networks.

9. Perspectives

We have already seen that as the number of organisms in phylogenetic profiles grew from tens to hundreds, the ability of this technique to identify biologically interesting coevolution events has dramatically increased. We expect that these trends will continue as the number of genomes continues to increase. However, along with these improvements the challenge will be the development of efficient implementations of the approach that can handle the ever-growing number of organisms with fully sequenced genomes. The all versus all alignments of proteins across thousands of genomes is a very computationally intensive task. Furthermore, all the approaches that incorporate the tree of organisms when computing the significance of two profiles are also computationally demanding. Over the next few years, these methodologies will need to be optimized to handle the thousands of bacterial genomes that will be sequenced. These increases in efficiency will undoubtedly be accompanied by ever more useful functional predictions.

Acknowledgments

The author wishes to acknowledge the UCLA-DOE Institute for Genomics and proteomics for support.

References

1. Pellegrini M, Marcotte EM, Thompson MJ, Eisenberg D, Yeates TO. (1999) Assigning protein functions by comparative genome analysis: protein phylogenetic profiles. Proc Natl Acad Sci U S A, 96:4285–4288.
2. Altschul SF, Madden TL, Schaffer AA, Zhang J, Zhang Z, Miller W, Lipman DJ. (1997) Gapped BLAST and PSI-BLAST: a new generation of protein database search programs. Nucleic Acids Res, 25:3389–3402.
3. Tatusov RL, Fedorova ND, Jackson JD, Jacobs AR, Kiryutin B, Koonin EV, Krylov DM, Mazumder R, Mekhedov SL, Nikolskaya AN, Rao BS, Smirnov S, Sverdlov AV, Vasudevan S, Wolf YI, Yin JJ, Natale DA. (2003) The COG database: an updated version includes eukaryotes. BMC Bioinformatics, 4:41.
4. Bowers PM, Cokus SJ, Eisenberg D, Yeates TO. (2004) Use of logic relationships to decipher protein network organization. Science, 306:2246–2249.
5. Date SV, Marcotte EM. (2003) Discovery of uncharacterized cellular systems by genome-wide analysis of functional linkages. Nat Biotechnol, 21:1055–1062.
6. Pazos F, Valencia A. (2001) Similarity of phylogenetic trees as indicator of protein-protein interaction. Protein Eng, 14:609–614.
7. Wu J, Kasif S, DeLisi C. (2003) Identification of functional links between genes using phylogenetic profiles. Bioinformatics, 19: 1524–1530.
8. Kharchenko P, Chen L, Freund Y, Vitkup D, Church GM. (2006) Identifying metabolic enzymes with multiple types of association evidence. BMC Bioinformatics, 7:177.
9. Liberles DA. (2001) Evaluation of methods for determination of a reconstructed history of gene sequence evolution. Mol Biol Evol, 18:2040–2047.
10. Barker D, Pagel M. (2005) Predicting functional gene links from phylogenetic-statistical analyses of whole genomes. PLoS Comput Biol, 1:e3.
11. Barker D, Meade A, Pagel M. (2007) Constrained models of evolution lead to improved prediction of functional linkage from correlated gain and loss of genes. Bioinformatics, 23:14–20.
12. Cokus S, Mizutani S, Pellegrini M. (2007) An improved method for identifying functionally linked proteins using phylogenetic profiles. BMC Bioinformatics, 8(Suppl 4):S7.
13. Bar-Joseph Z, Gifford DK, Jaakkola TS. (2001) Fast optimal leaf ordering for hierarchical clustering. Bioinformatics, 17(Suppl 1): S22–S29.
14. Sun J, Li Y, Zhao Z. (2007) Phylogenetic profiles for the prediction of protein-protein interactions: how to select reference

organisms? Biochem Biophys Res Commun, 353:985–991.

15. Ashburner M, Ball CA, Blake JA, Botstein D, Butler H, Cherry JM, Davis AP, Dolinski K, Dwight SS, Eppig JT, Harris MA, Hill DP, Issel-Tarver L, Kasarskis A, Lewis S, Matese JC, Richardson JE, Ringwald M, Rubin GM, Sherlock G. (2000) Gene ontology: tool for the unification of biology. The Gene Ontology Consortium. Nat Genet, 25:25–29.
16. Jothi R, Przytycka TM, Aravind L. (2007) Discovering functional linkages and uncharacterized cellular pathways using phylogenetic profile comparisons: a comprehensive assessment. BMC Bioinformatics, 8:173.
17. Kensche PR, van Noort V, Dutilh BE, Huynen MA. (2008) Practical and theoretical advances in predicting the function of a protein by its phylogenetic distribution. J R Soc Interface, 5:151–170.
18. Li H, Pellegrini M, Eisenberg D. (2005) Detection of parallel functional modules by comparative analysis of genome sequences. Nat Biotechnol, 23:253–260.
19. Jensen LJ, Kuhn M, Stark M, Chaffron S, Creevey C, Muller J, Doerks T, Julien P, Roth A, Simonovic M, Bork P, von Mering C. (2009) STRING 8 – a global view on proteins and their functional interactions in 630 organisms. Nucleic Acids Res, 37: D412–D416.
20. Bowers PM, Pellegrini M, Thompson MJ, Fierro J, Yeates TO, Eisenberg D. (2004) Prolinks: a database of protein functional linkages derived from coevolution. Genome Biol, 5:R35.
21. Date SV, Marcotte EM. (2005) Protein function prediction using the Protein Link EXplorer (PLEX). Bioinformatics, 21:2558–2559.
22. Hu Z, Hung JH, Wang Y, Chang YC, Huang CL, Huyck M, DeLisi C. (2009) VisANT 3.5: multi-scale network visualization, analysis and inference based on the gene ontology. Nucleic Acids Res, 37:W115–W121.
23. Dandekar T, Snel B, Huynen M, Bork P. (1998) Conservation of gene order: a fingerprint of proteins that physically interact. Trends Biochem Sci, 23:324–328.

Chapter 10

Extracting Regulatory Networks of *Escherichia coli* from RegulonDB

Heladia Salgado, Irma Martínez-Flores, Alejandra López-Fuentes, Jair Santiago García-Sotelo, Liliana Porrón-Sotelo, Hilda Solano, Luis Muñiz-Rascado, and Julio Collado-Vides

Abstract

RegulonDB contains the largest and currently best-known data set on transcriptional regulation in a single free-living organism, that of *Escherichia coli* K-12 (Gama-Castro et al. Nucleic Acids Res 36:D120–D124, 2008). This organized knowledge has been the gold standard for the implementation of bioinformatic predictive methods on gene regulation in bacteria (Collado-Vides et al. J Bacteriol 191:23–31, 2009). Given the complexity of different types of interactions, the difficulty of visualizing in a single figure of the whole network, and the different uses of this knowledge, we are making available different views of the genetic network. This chapter describes case studies about how to access these views, via precomputed files, web services and SQL, including sigma–gene relationships corresponding to transcription of alternative RNA polymerase holoenzyme promoters; as well as, transcription factor (TF)-genes, TF–operons, TF–TF, and TF–regulon interactions. 17

Key words: Bacterial, Regulatory networks, RegulonDB, *Escherichia coli*, Transcription factor, Gene, Operon, Sigma factor, Regulatory interactions, Database, Web services

1. Introduction

1.1. Genetic Network in Bacteria

Escherichia coli is a reference organism because of the rich knowledge available about its biology. The group of Collado-Vides has been curating information on regulation of transcription initiation and operon organization during the last 10 years, making this corpus of knowledge available both in RegulonDB (1–7) and EcoCyc databases (8).

Gene regulation operates at many levels, from transcription initiation to activity of the protein and gene products, including various intermediate levels, such as elongation, termination, mRNA stability, initiation of translation, attenuation, etc. It is true, though,

Jacques van Helden et al. (eds.), *Bacterial Molecular Networks: Methods and Protocols*, Methods in Molecular Biology, vol. 804, DOI 10.1007/978-1-61779-361-5_10, © Springer Science+Business Media, LLC 2012

that the best-known level is that of regulation of transcription initiation (9–11). Indeed, for 50 years, transcriptional regulation has been studied in bacteria by hundreds of research groups, leading to knowledge on thousands of interactions between transcription factors (TFs) and their target genes (12). This complete set of regulatory interactions of gene expression allows bacteria to adapt to environmental changes, minimizing its energy requirements (13). It enables the use of the genotypic potential into specific phenotypic expression profiles; thus, the cell uses interactions under certain conditions (11, 14, 15). For instance, during the stationary phase, changes in gene expression imply a metabolic reorganization, because during starvation there is an increase in cellular resistance to various damaging physical and chemical agents. Through this reorganization, the cell integrity is maintained (13, 16, 17).

Transcriptional regulation also involves interactions between a s factor and the genes expressed by it. The different s factors direct the binding of RNA polymerase to a specific promoter site, initiating transcription (18, 19). In E. coli, the most abundant factor is σ 70; in addition, there are six alternative sigma factors: σ 32, σ 54, σ 28, σ 24, σ 38, and σ 19. Many genes expressed during growth are transcribed by means of σ 70, while the others sigma factors transcribe genes necessary for additional specific conditions, enabling *E.coli* to adapt to different environmental stresses (20–22).

1.2. Outline of the Regulatory Network

Gene regulation involves the participation of fragments in the chromosome – genes and their regulatory regions – and, on the other hand, macromolecules and metabolites in the cytosol.

An essential regulatory region in the chromosome is the promoter; in bacteria, it is defined as the region for binding of the RNAP where transcription initiates. The recognition of RNAP to DNA depends on the sigma factor associated. The sigma 70-RNAP recognizes the well-known −10 and −35 boxes, whereas other motifs have been identified for alternative sigma factors (23, 24). Regulation of transcription initiation also depends on binding sites for TFs (see Note 2) which activate or repress the initiation process (12, 25). Promoters and binding sites are located upstream of the beginning of the regulated gene or groups of genes. The organization of genes in a genomic context makes the regulation process more efficient (12).

In bacteria, those genes whose functions are related are usually adjacent, forming syntenic groups, and their regulation is coordinated (26). In most cases, they are co-transcribed in operons (two or more contiguous genes transcribed from a promoter in the same direction to a downstream terminator (12)); it is also relatively common to find operons with several promoters, sometimes located internally.

It is important to keep in mind that all this machinery has to be sensitive to changes via certain signals that may originate in either the interior or the exterior of the cell. This is achieved either by the activation or inactivation of regulatory elements such as for instance some global TFs like IHF or HNS, and in a more rapid mechanism, by having TFs with active and inactive conformations depending on the binding–unbinding of specific allosteric metabolites, or covalent modifications such as phosphorylation of TFs as in the case of the two-component systems. Additional mechanisms involve secondary messengers implicated in cellular signaling processes: metabolites that directly affect RNAP, proteins that physically alter TFs and their regulating activity, other TFs (27), as well as regulatory sRNAs that affect transcription or translation.

As a consequence of these changes, the active network or total set of active interactions at a given moment or growth condition, will be a fraction of the whole set of interactions. It is important to keep this in mind in the analysis of the data. The dynamic capacity of these networks also implies a temporary condition; because different conditions harbor varying number of active TFs and further analysis can help to understand the association of TFs with different conditions and to elucidate their dynamic interplay with other TFs (28–32).

1.3. Networks and RegulonDB

Curation by our team of regulation of transcription initiation and operon organization has for years gathered knowledge form the original literature as described in RegulonDB and EcoCyc. Contrary to the total number of genes or metabolic pathways, which have remained rather constant in the last years, the number of elements involved in the interactions of gene regulation keeps constantly increasing through the years. For instance, the number of promoters and binding sites for TFs in version 6.7 is close to 1,800 for each of them, as compared to around 300 by the year of the publication of the *E. coli* genome (see Note 3). RegulonDB is constantly updated, with new versions generated three to four times every year. In this database, all the objects are supported by at least one source of evidence, which have been classified as weak or strong depending on the confidence of the methods that support the information (see Note 4).

Gene regulation can be described as a network where nodes are genes, TFs or sigma genes, and the connections or edges (see Note 1) describe the regulatory interaction of an element on another, such as a TF affecting a regulated gene. However, the connections at this abstract level do not consider any additional details of the intricate mechanisms of genetic regulation (33–41).

RegulonDB provides access to the following views of genetic network:

1. The network between TFs and their regulated genes. The two elements of this network are TFs and genes. This view enables

users to describe the network as either regulatory gene to affected gene, or TF to regulated gene. In most cases the graph will be similar, except for autoregulation of TFs where the interaction is a loop in the gene-to-gene network, but an arrow in the TF to gene graph. Note that autoregulation is far from an exceptional case. In the case of heterodimers things get more complex since there will be one node for the protein (i.e., IHF), but two nodes for each gene (i.e., *himA* and *himB*), and a full complex module if a heterodimer were autoregulated.

2. The network between TFs and TUs. Each outgoing edge describes regulation of a TF affecting the group of genes forming a transcription unit, making it a much smaller network than the previous one. The interactions of this network happen at a single promoter by definition. Users again can select representing either the regulatory genes or their TF gene products.
3. The network between TFs and operons. In this network type, the number of interactions and nodes is even smaller than the one of TF–TUs, since an operon may contain multiple TUs. The prize of this simplification is that an edge is not directly related to a single mechanism or regulated promoter. For instance, a single CRP site participates in activating *galp1* whereas it represses *galp2*, both transcribing the *galETK* operon. A single edge with a dual function represents the dual effect of CRP on the operon. This single edge corresponds to two interactions in the TF–TU network.
4. The network limited to regulation among TFs. This is a subset of the network of TFs and regulated genes, containing only those interactions describing how TFs are regulated.
5. The network between sigma factors and their corresponding regulated genes. Note that here interactions refer to transcription not strictly to gene regulation; however, we assigned a positive function to the interactions since they are necessary for transcription.
6. The network between sigma factors and their regulated TUs.
7. The network of sigma factors. The elements of this small network describe which sigma factor transcribes which genes coding for sigma factors.

1.4. Network Applications

Fundamental new discoveries and concepts have emerged from the analysis of this network: for instance, the notion of motifs in a network with feed-forward and bifan modules by the group of Uri Alon (42) as well as topological analyses of the network (43). These are examples of modeling and topology of the network applications; in addition the network has also been used in gene expression analyses (microarray), analysis of ChIP-chip data, analysis of gene expression dynamics, metagenomics, regulatory-network prediction, promoter and TF binding site predictions, orthologous regulatory identification, architecture of operons, promoters, or DNA-binding sites, evolutionary analyses of transcription, and inference of the role of TFs. See an extensive list of these types of applications in Collado-Vides et al. (33).

As mentioned before, the graphical abstractions of the network do not necessarily map directly to mechanisms of regulation. Nonetheless, the tables we provide enable users to extract sets of regulated genes (44) to get sets of upstream regions for pattern matching and pattern discovery associated to the mechanisms of regulation in these networks.

This book contains several chapters of tools that apply to this TRN. For instance, Colombos and Distiller have been used to identify condition-dependent transcriptional networks based on regulator–target gene interaction information in RegulonDB (Fu et al., Chapter 8). There are also Web-based tools for browsing regulatory networks within BioCyc; these tools display regulatory networks of genes from the data in the EcoCyc and RegulonDB databases (Latendresse et al., Chapter 11). The NeAT toolbox compares networks to networks, clusters to clusters, and network to clusters; with this tool researchers can visualize whether groups of nodes belonging to the same class are grouped together in the work and for co-regulated genes (regulons) where the data have been obtained from RegulonDB (Brohée, Chapter 18). The capacity of the MAVisto tool (a tool for biological network motif analysis) to find motifs in biological networks is a well-analyzed transcription regulatory network of information on *E. coli*, and its data were extracted from RegulonDB (Schwöbbermeyer and Wünschiers, Chapter 14).

1.5. Case Studies: Obtaining a Transcriptional Regulatory Network

We describe three case studies that illustrate how to obtain a transcriptional regulatory network for *Escherichia coli* K-12.

In the first case study, we see the way to obtain the precomputed genetic networks from the RegulonDB website. The second and last case uses a Web service and SQL commands, respectively. In the three cases, we get the transcriptional regulatory network between TFs and their regulated genes, but the same protocol applies with the other networks.

The result tables include the type of evidence associated to the interactions irrespective of the level of the network. This is a subtle detail, since the evidence for the existence of the different groups (TUs or operons) is not included in these results; interested users can get these evidences directly from TUs or operons datasets (see Note 5).

1.6. Case Study 1: Downloading Precomputed Genetic Networks (Flat Files and Images) from the RegulonDB Website

This protocol is aimed at people in the field of biology who have no knowledge in programming. The download menu from the RegulonDB website provides direct access to precomputed genetic networks.

1.7. Case Study 2: Obtaining Genetic Networks Through Web Services

The second protocol is aimed at users with an interest in bioinformatics process automation. This case study describes how to obtain information on genetic networks through Web services by using a SOAP protocol. RegulonDB can be accessed via a Web service called "NetWork," which allows extraction of information from the networks described above.

The methods of the Network Web service are the following:

1. getSigmaGene. Get the sigma regulatory network with its target genes.
2. getSigmaTU. Get the sigma regulatory network with its target operons.
3. getTFTF. Get the transcriptional regulatory network, the TF with its target TFs.
4. getTFGene. Get the transcriptional regulatory network with its target genes.
5. getTFOperon. Get the transcriptional regulatory network with its target operons.
6. getTFGeneInOutDegree. Get the complete transcription regulatory interactions for a specific TF: the regulators of the specific TF ("in degree," the degree to which they regulate the TF) and its target genes ("out degree," the degree to which the TF regulate the gene).
7. getColumnNames. Get the column names for the results of the previous methods.

1.8. Case Study 3: Obtaining Genetic Networks Through SQL Queries

For bioinformaticians with expertise in SQL, they have access directly to the genetic networks of RegulonDB via MySQL (see Note 6). In this study case, we describe the procedure to get the TF-gene network via DbVisualizer and CytoScape tools.

2. Materials

2.1. Software

This protocol requires a connection to the Internet and a Web browser.

RegulonDB database is available from http://regulondb.ccg.unam.mx, and the examples used in this work are based on version 6.4.

For the second study case, to execute the Perl client it is necessary to have installed Perl and the SOAP::Lite library. The perl clients were developed using version 5.8.8 and the SOAP:: Lite library version 0.60.

Also, to test the Web services it is necessary to install the SoapUI tool (http://www.eviware.com/soapUI/soapui-products-overview.html). The example shown is based on version 3.0.1.

The last study case requires to have installed DbVisualizer (http://www.dbvis.com) and CytoScape including the CytoSQL plugin (http://www.cytoscape.org, http://www.ptools.ua.ac.be/CytoSQL). We use DbVisualizer version 7.0.5 and CytoSQL version 0.9.

3. Methods

3.1. Case Study 1: Downloading Precomputed Genetic Networks (Flat Files and Images) from the RegulonDB Website

In this study case we show how to download the precomputed network of TFs and its regulated genes in flat file, the meaning of the columns and its respective image of the network. The results of this study case are shown in Fig. 1.

3.1.1. Downloading Flat Files

1. Open a browser and type in the following URL: http://regulondb.ccg.unam.mx/.
2. At the main menu, in the *Downloads* section, select the option *Data Sets*. This displays a table with available data sets organized by section.
3. Go to the section *Regulatory Network Interactions* and select *TF-gene interactions*.

3.1.2. Understanding the Flat File

The file contains the following columns:

1. Transcription Factor (TF) Name. This is the name of the protein that acts as a regulator. Some of these proteins are homo- or heteromeric complexes.
2. Gene coding for the TF (regulatory gene). This column shows the name of the gene encoding the TF. Gene names are separated by a vertical bar; although most are encoded by a single gene, there are some heterodimers, for instance, IHF (*ihfA ihfB*), HU (*hupA hupB*), etc.
3. bnumber (see Note 7) of the regulatory gene (45).
4. Gene regulated by the TF (regulated gene).
5. bnumber of the regulated gene.

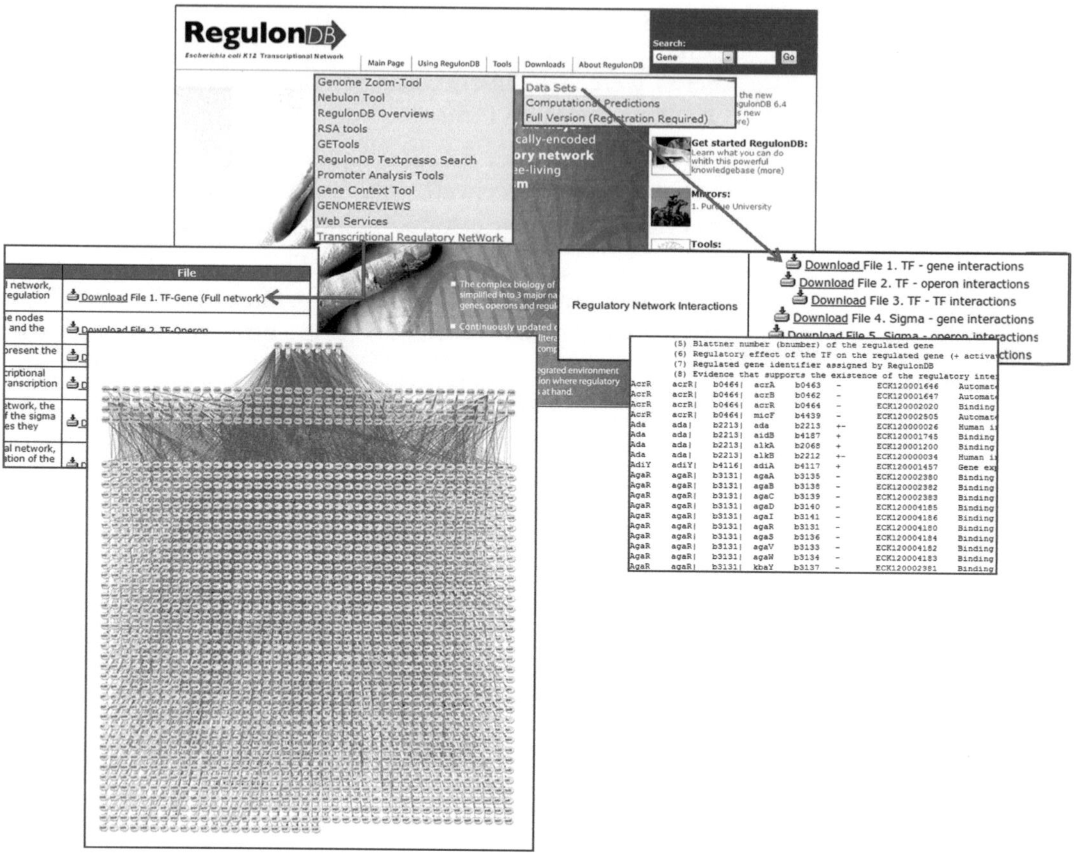

Fig. 1. Downloading of the flat file and image of TF-gene network from RegulonDB website.

6. Regulatory effect of the TF on the regulated gene (+ activator; – repressor; ± dual; ? unknown). The dual function indicates that the TF can act as an activator and repressor; this dual function can be executed either under a single promoter or affecting different promoters.
7. Identifier of regulated gene assigned in RegulonDB.
8. Evidence. This is the evidence that supports the existence of the regulatory interaction.

3.1.3. Downloading the Images

1. At the main menu of RegulonDB, under the menu *Tools*, choose the option *Transcriptional Regulatory NetWork*. It displays a list with imaging options of genetic networks.
2. From the table of options, choose the TF-gene network. The figure legend describes the properties that allow grouping of network elements, as well as the colors of each element.

3.2. Case Study 2: Obtaining Genetic Networks Through Web Services

This section explains how to get the transcription network of TFs and their regulated genes using web services. The protocol describes how to test the *getTFGene* web service with the soapUI tool and how to execute the Perl client available for this service.

3.2.1. Accessing the NetWork Web Service, Its WSDL, and Methods

1. After accessing the RegulonDB website (as described above), go to the *Tools* menu and select the *Web Services* option. A page will be displayed that contains the list of currently available Web services.
2. Go to the row *NetWork Web Service* and click on "view link" in the WSDL column. WSDL (Web Service Description Language) is a standard for describing Web services, with a formal description of the supported methods and their parameters. As you can see, the WSDL document is not very informative for human readers.
3. A readable description (for humans) of the methods and their parameters can be visualized by clicking the link *View* in the *Methods* column.

3.2.2. Testing the Web Service with soapUI

1. Open the *soapUI* tool, go to the *File* menu, and choose *New soapUI Project*.
2. Type the following information:

   Project Name: Genetic NetWorks – RegulonDB

   Initial WSDL/WADL (description languages):

   http://regulondb.ccg.unam.mx/webservices/NetWork.jws?wsdl

   Enable the option "Create sample request for all operations?"

   Click on the OK button.
3. Go to *getTFGene* for the project (at the left side of the tool browser).
4. Click on the expand/collapse button and double click on the *Request 1* label.
5. Click on the green *Submit* button in *Request 1* window to submit your request.
6. The response is shown at the right side of the window.

   Each step of the above procedure is shown in Fig. 2.

3.2.3. Accessing and Executing TF-Gene Network Client

1. From the Web service page, click on the *NetWork service name* link. A list of clients is shown.
2. Go to the row *ClientNetWorkgetTFGene* and click on the *Download* link from the Perl column.
3. The file ClientNetWorkgetTFGene.pl must be downloaded in your computer.

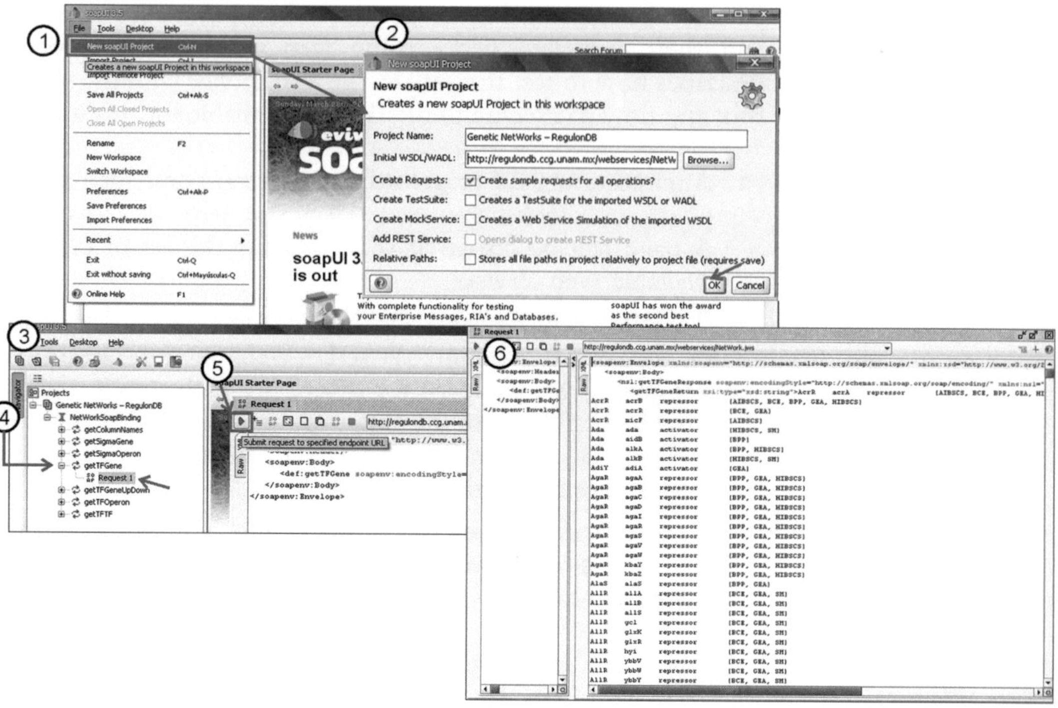

Fig. 2. Testing of the Web service with soapUI.

4. Verify if the file ClientNetWorkgetTFGene.pl has your Perl location path (first line).
5. Open a terminal and execute the following command:

   % perl ClientNetWorkgetTFGene.pl

Figure 3 summarizes the access and results of execution of ClientNetWorkgetTFGene.pl client.

3.3. Case Study 3: Obtaining Genetic Networks Through SQL Queries

For this case study, it is necessary to previously request access to RegulonDB by sending an e-mail to regulondb@ccg.unam.mx

3.3.1. Setting the Connection in Dbvisualizer

1. Open the DbVisualizer database tool.
2. Click on the icon *Create new database connection.*
3. In the window *Use Connection Wizard?*, click on the *Use Wizard* option.
4. Type *RegulonDB_yumkax* as a connection alias and click on the *Next* button.
5. Choose MySQL Database Driver.
6. Type the following information:

   Server: yumkax.ccg.unam.mx

   Port: 3306

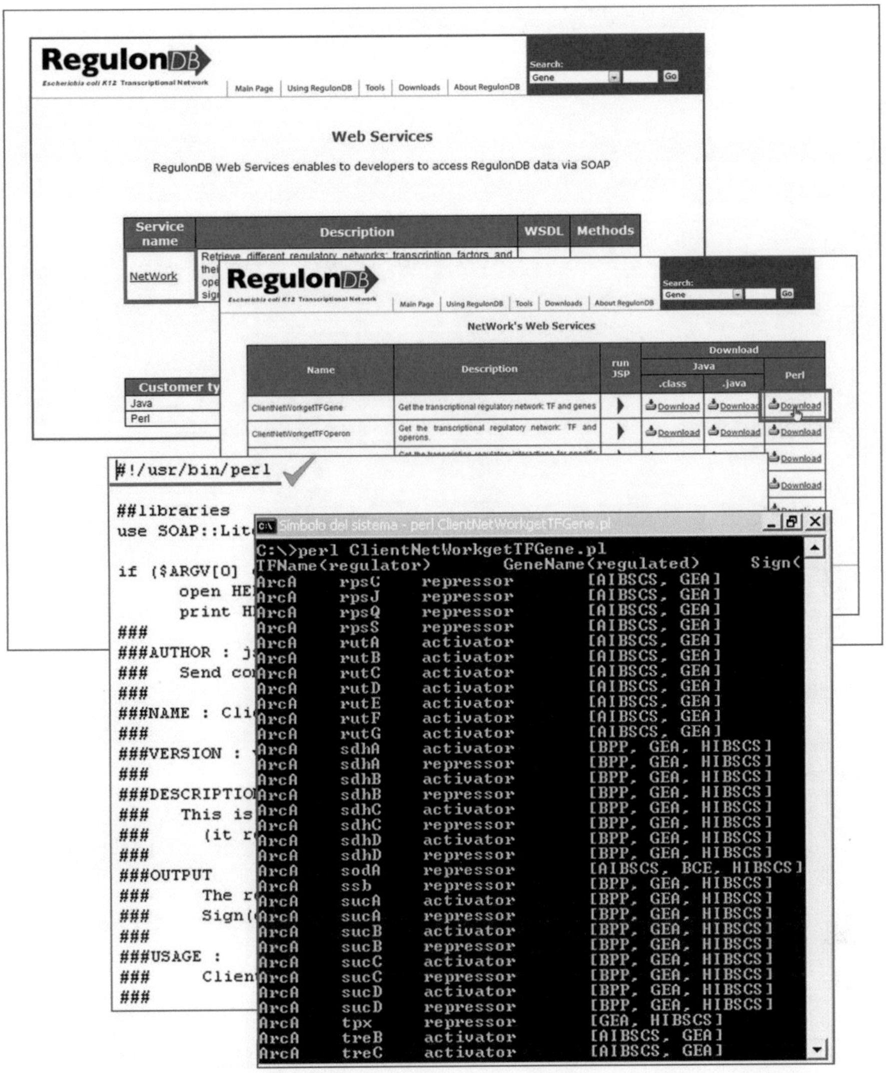

Fig. 3. Accessing and executing of TF-gene network client.

Database: regulondb

Userid: network_guest

Password: ***** (sent by e-mail from regulondb@ccg.unam.mx)

7. Click on the *Test Connection* option. If the connection was successfully established, click on the *Finish* button.

The different windows displayed in the process are shown in Fig. 4. The next section only works if the connection with RegulonDB is established.

3.3.2. Getting Data Using SQL Query from RegulonDB

1. In the Connection Browser (at the left side of the window), click on RegulonDB from the *RegulonDB*_yumkax connection.

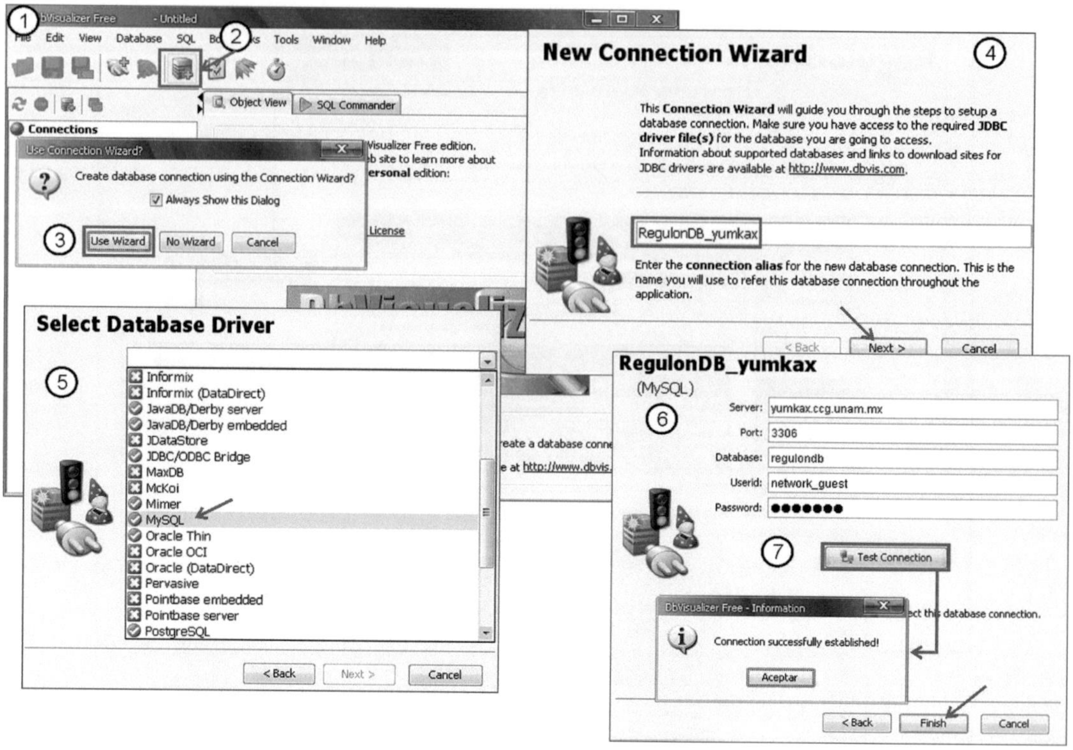

Fig. 4. DbVisualizer configuration.

2. Go to the SQL commander button, at the right side of the window.
3. To extract the genetic network between a TF and its regulated genes, type the following query:

```
SELECT
        regulator_name,
        regulated_name,
        function_interaction
FROM
        genetic_network
WHERE
        regulator_type = 'tf' AND
        regulated_type = 'gene'
ORDER BY regulator_name
```

4. Click on the *Execute* green button.

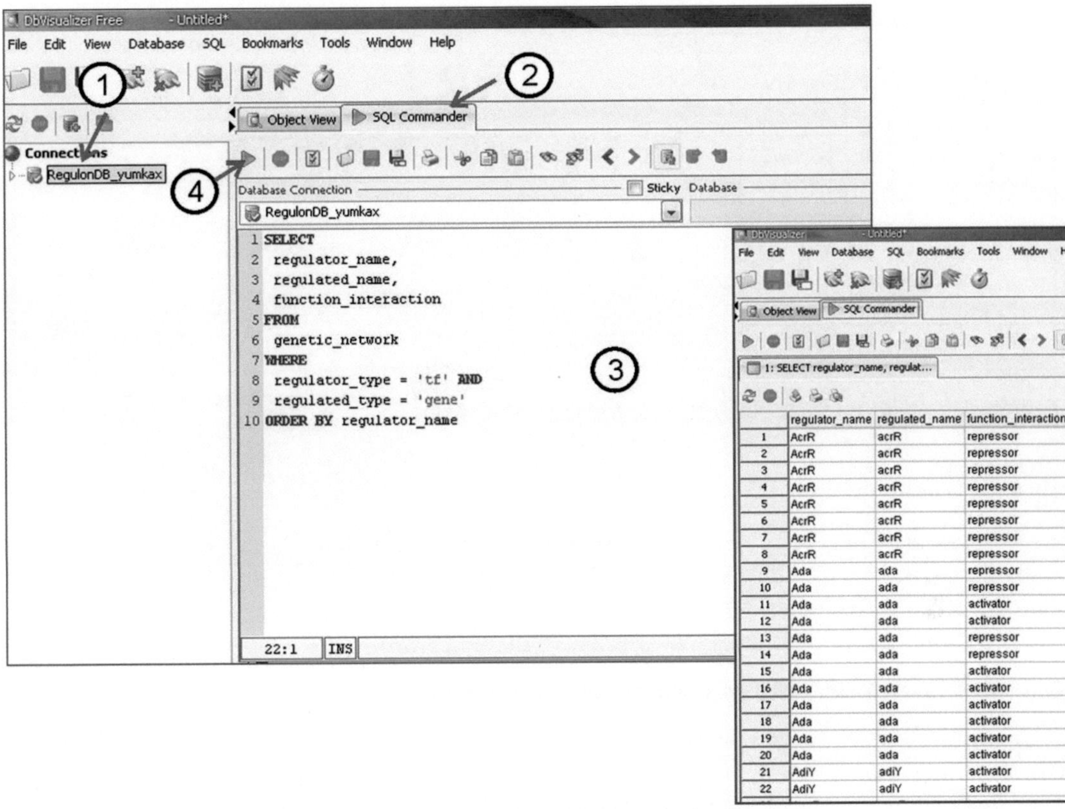

Fig. 5. Getting data using SQL query from RegulonDB.

The results of the query display three columns: the regulator name is the Transcription Factor; the regulated name is the target gene; and the function interaction is the effect of the regulation (see Fig. 5).

All the interactions of the different genetic networks are contained in the *genetic_network* table. The different type of networks can be obtained using the regulator_type and regulated_type fields, for example, in this case the TF-gene network is getting using *tf* and *gene* as values.

3.3.3. Using CytoSQL to Extract and Draw the Genetic Network

1. Open CytoScape and from the plug-in menu go to the CytoSQL option.
2. Click on the *Adjust database connection* button.
3. Type the following parameters:

   Driver: com.mysql.jdbc.Driver

   URL: jdbc:mysql://yumkax.ccg.unam.mx

   Database name: regulondb

   Username: network_guest

   Password: ***** (sent by e-mail from regulondb@ccg.unam.mx)

   Click on the OK button.

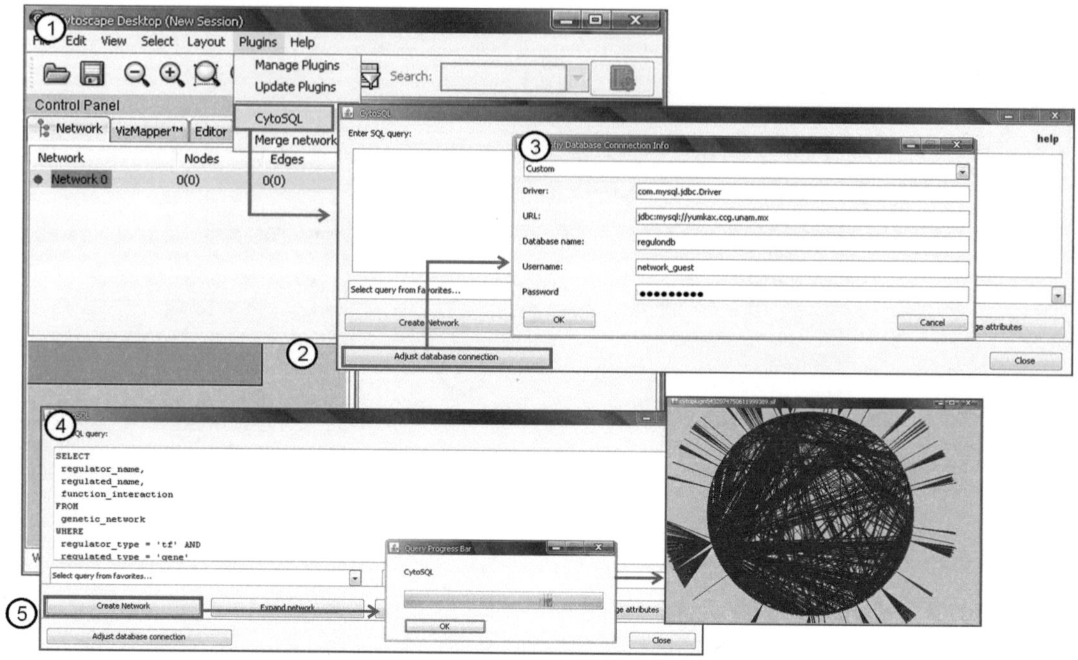

Fig. 6. Extracting and drawing the genetic network by using Cytoscape.

4. Type your query from step 3 in Subheading 3.2 in the query box of the plug-in.
5. Click on the *Create NetWork* button to obtain the data from the RegulonDB database.
6. The *Define attribute type* window must be displayed, it means the data was transferred successfully, after that the network is drawn (see Fig. 6).

4. Notes

1. The edges represent a regulatory effect (positive or negative sign).
2. Historically defined as operators.
3. The database summary is showed in http://regulondb.ccg.unam.mx/html/Database_summary.jsp.
4. The evidences associated to all RegulonDB objects are classified as "strong" or "weak," based on the confidence level of the experiment or prediction that supports objects and their relationships http://regulondb.ccg.unam.mx/evidenceClassification.jsp.

5. Different datasets are available from the Download menu in RegulonDB.
6. RegulonDB is distributed in different database manager systems and the users can install it in their own computers.
7. Bnumber is the locus tag assigned to *E. coli* genes.

Acknowledgments

We acknowledge the technical support of Shirley Alquicira and Víctor del Moral Chávez.

This work was funded by the National Institutes of Health, grant numbers R01 GM071962-05 and by UNAM, PAPIIT grant number IN214905. We acknowledge support from CONACYT PROINNOVA-2010 project number 134817.

References

1. Gama-Castro S, Jimenez-Jacinto V, Peralta-Gil M, Santos-Zavaleta A, Penaloza-Spinola MI, Contreras-Moreira B, Segura-Salazar J, Muniz-Rascado L, Martinez-Flores I, Salgado H, Bonavides-Martinez C, Abreu-Goodger C, Rodriguez-Penagos C, Miranda-Rios J, Morett E, Merino E, Huerta AM, Trevino-Quintanilla L, Collado-Vides J. (2008) RegulonDB (version 6.0): gene regulation model of *Escherichia coli* K-12 beyond transcription, active (experimental) annotated promoters and Textpresso navigation. Nucleic Acids Res, 36: D120–D124.
2. Salgado H, Gama-Castro S, Peralta-Gil M, Diaz-Peredo E, Sanchez-Solano F, Santos-Zavaleta A, Martinez-Flores I, Jimenez-Jacinto V, Bonavides-Martinez C, Segura-Salazar J, Martinez-Antonio A, Collado-Vides J. (2006) RegulonDB (version 5.0): *Escherichia coli* K-12 transcriptional regulatory network, operon organization, and growth conditions. Nucleic Acids Res, 34:D394–D397.
3. Salgado H, Gama-Castro S, Martinez-Antonio A, Diaz-Peredo E, Sanchez-Solano F, Peralta-Gil M, Garcia-Alonso D, Jimenez-Jacinto V, Santos-Zavaleta A, Bonavides-Martinez C, Collado-Vides J. (2004) RegulonDB (version 4.0): transcriptional regulation, operon organization and growth conditions in *Escherichia coli* K-12. Nucleic Acids Res, 32:D303–D306.
4. Salgado H, Santos-Zavaleta A, Gama-Castro S, Millan-Zarate D, Diaz-Peredo E, Sanchez-Solano F, Perez-Rueda E, Bonavides-Martinez C, Collado-Vides J. (2001) RegulonDB (version 3.2): transcriptional regulation and operon organization in *Escherichia coli* K-12. Nucleic Acids Res, 29:72–74.
5. Salgado H, Santos-Zavaleta A, Gama-Castro S, Millan-Zarate D, Blattner FR, Collado-Vides J. (2000) RegulonDB (version 3.0): transcriptional regulation and operon organization in *Escherichia coli* K-12. Nucleic Acids Res, 28:65–67.
6. Salgado H, Santos A, Garza-Ramos U, van Helden J, Diaz E, Collado-Vides J. (1999) RegulonDB (version 2.0): a database on transcriptional regulation in *Escherichia coli*. Nucleic Acids Res, 27:59–60.
7. Huerta AM, Salgado H, Thieffry D, Collado-Vides J. (1998) RegulonDB: a database on transcriptional regulation in *Escherichia coli*. Nucleic Acids Res, 26:55–59.
8. Keseler IM, Bonavides-Martinez C, Collado-Vides J, Gama-Castro S, Gunsalus RP, Johnson DA, Krummenacker M, Nolan LM, Paley S, Paulsen IT, Peralta-Gil M, Santos-Zavaleta A, Shearer AG, Karp PD. (2009) EcoCyc: a comprehensive view of *Escherichia coli* biology. Nucleic Acids Res, 37:D464–D470.
9. Vicente M, Chater KF, De Lorenzo V. (1999) Bacterial transcription factors involved in global regulation. Mol Microbiol, 33:8–17.
10. Müller-Hill B. (2001) Bacterial Transcription Regulation. In *Encyclopedia of Life Sciences* (Article Online Posting Date: March 11,

2004, Cologne, Germany), John Wiley & Sons, Chichester, 1–6.
11. Busby J. (2007) Transcription Activation at Bacterial Promoters. In *Encyclopedia of Life Sciences* (Article Online Posting Date: January 15, 2007), John Wiley & Sons, Chichester, 1–9.
12. Jacob F, Monod J. (1961) Genetic regulatory mechanisms in the synthesis of proteins. J Mol Biol, 3:318–356.
13. Martinez-Antonio A, Salgado H, Gama-Castro S, Gutierrez-Rios RM, Jimenez-Jacinto V, Collado-Vides J. (2003) Environmental conditions and transcriptional regulation in *Escherichia coli*: a physiological integrative approach. Biotechnol Bioeng, 84:743–749.
14. Browning DF, Busby SJ. (2004) The regulation of bacterial transcription initiation. Nat Rev Microbiol, 2:57–65.
15. Geissmann T, Marzi S, Romby P. (2009) The role of mRNA structure in translational control in bacteria. RNA Biol, 6:153–160.
16. Stock JB, Stock AM, Mottonen JM. (1990) Signal transduction in bacteria. Nature, 344:395–400.
17. Edwards JS, Covert M, Palsson B. (2002) Metabolic modelling of microbes: the flux-balance approach. Environ Microbiol, 4:133–140.
18. Ishihama A. (1988) Promoter selectivity of prokaryotic RNA polymerases. Trends Genet, 4:282–286.
19. Hochschild A, Dove SL. (1998) Protein-protein contacts that activate and repress prokaryotic transcription. Cell, 92:597–600.
20. Hengge-Aronis R. (1993) Survival of hunger and stress: the role of rpoS in early stationary phase gene regulation in *E. coli*. Cell, 72:165–168.
21. Hengge-Aronis R, Lange R, Henneberg N, Fischer D. (1993) Osmotic regulation of rpoS-dependent genes in *Escherichia coli*. J Bacteriol, 175:259–265.
22. Ishihama A. (2000) Functional modulation of *Escherichia coli* RNA polymerase. Annu Rev Microbiol, 54:499–518.
23. Huerta AM, Collado-Vides J. (2003) Sigma70 promoters in *Escherichia coli*: specific transcription in dense regions of overlapping promoter-like signals. J Mol Biol, 333:261–278.
24. Mendoza-Vargas A, Olvera L, Olvera M, Grande R, Vega-Alvarado L, Taboada B, Jimenez-Jacinto V, Salgado H, Juarez K, Contreras-Moreira B, Huerta AM, Collado-Vides J, Morett E. (2009) Genome-wide identification of transcription start sites, promoters and transcription factor binding sites in *E. coli*. PLoS One, 4:e7526.
25. Benitez-Bellon E, Moreno-Hagelsieb G, Collado-Vides J. (2002) Evaluation of thresholds for the detection of binding sites for regulatory proteins in *Escherichia coli* K12 DNA. Genome Biol, 3:RESEARCH0013.
26. Murray PR, Rosenthal KS, Pfaller MA. (2006) *Microbiología Médica Genética bacteriana* (Softcover Edition), 5th ed., Elsevier Science Health Science, España, 976.
27. Perez-Rueda E, Collado-Vides J. (2000) The repertoire of DNA-binding transcriptional regulators in *Escherichia coli* K-12. Nucleic Acids Res, 28:1838–1847.
28. Ronen M, Rosenberg R, Shraiman BI, Alon U. (2002) Assigning numbers to the arrows: parameterizing a gene regulation network by using accurate expression kinetics. Proc Natl Acad Sci U S A, 99:10555–10560.
29. Jin VX, Leu YW, Liyanarachchi S, Sun H, Fan M, Nephew KP, Huang TH, Davuluri RV. (2004) Identifying estrogen receptor alpha target genes using integrated computational genomics and chromatin immunoprecipitation microarray. Nucleic Acids Res, 32:6627–6635.
30. Xiong M, Zhao J, Xiong H. (2004) Network-based regulatory pathways analysis. Bioinformatics, 20:2056–2066.
31. Thieffry D, Huerta AM, Perez-Rueda E, Collado-Vides J. (1998) From specific gene regulation to genomic networks: a global analysis of transcriptional regulation in *Escherichia coli*. Bioessays, 20:433–440.
32. Martinez-Antonio A, Janga SC, Thieffry D. (2008) Functional organisation of *Escherichia coli* transcriptional regulatory network. J Mol Biol, 381:238–247.
33. Collado-Vides J, Salgado H, Morett E, Gama-Castro S, Jimenez-Jacinto V, Martinez-Flores I, Medina-Rivera A, Muniz-Rascado L, Peralta-Gil M, Santos-Zavaleta A. (2009) Bioinformatics resources for the study of gene regulation in bacteria. J Bacteriol, 191:23–31.
34. Resendis-Antonio O. (2009) Filling kinetic gaps: dynamic modeling of metabolism where detailed kinetic information is lacking. PLoS One, 4:e4967.
35. Gutierrez-Rios RM, Freyre-Gonzalez JA, Resendis O, Collado-Vides J, Saier M, Gosset G. (2007) Identification of regulatory network topological units coordinating the genome-wide transcriptional response to glucose in *Escherichia coli*. BMC Microbiol, 7:53.
36. Martinez-Nunez MA, Perez-Rueda E, Gutierrez-Rios RM, Merino E. (2009) New insights into the regulatory networks of paralogous genes. Microbiology, 156(Pt 1):14–22.

37. Rodriguez-Caso C, Corominas-Murtra B, Sole RV. (2009) On the basic computational structure of gene regulatory networks. Mol Biosyst, 5:1617–1629.
38. Huttenhower C, Mutungu KT, Indik N, Yang W, Schroeder M, Forman JJ, Troyanskaya OG, Coller HA. (2009) Detailing regulatory networks through large scale data integration. Bioinformatics, 25:3267–3274.
39. Stoebel DM, Hokamp K, Last MS, Dorman CJ. (2009) Compensatory evolution of gene regulation in response to stress by *Escherichia coli* lacking RpoS. PLoS Genet, 5:e1000671.
40. Vazquez CD, Freyre-Gonzalez JA, Gosset G, Loza JA, Gutierrez-Rios RM. (2009) Identification of network topological units coordinating the global expression response to glucose in *Bacillus subtilis* and its comparison to *Escherichia coli*. BMC Microbiol, 9:176.
41. Hervas AB, Canosa I, Little R, Dixon R, Santero E. (2009) NtrC-dependent regulatory network for nitrogen assimilation in *Pseudomonas putida*. J Bacteriol, 191:6123–6135.
42. Shen-Orr SS, Milo R, Mangan S, Alon U. (2002) Network motifs in the transcriptional regulation network of *Escherichia coli*. Nat Genet, 31:64–68.
43. Barabasi AL, Oltvai ZN. (2004) Network biology: understanding the cell's functional organization. Nat Rev Genet, 5:101–113.
44. Sand O, Thomas-Chollier M, van Helden J. (2009) Retrieve-ensembl-seq: user-friendly and large-scale retrieval of single or multi-genome sequences from Ensembl. Bioinformatics, 25:2739–2740.
45. Blattner FR, Plunkett G, 3rd, Bloch CA, Perna NT, Burland V, Riley M, Collado-Vides J, Glasner JD, Rode CK, Mayhew GF, Gregor J, Davis NW, Kirkpatrick HA, Goeden MA, Rose DJ, Mau B, Shao Y. (1997) The complete genome sequence of *Escherichia coli* K-12. Science, 277:1453–1462.

Chapter 11

Browsing Metabolic and Regulatory Networks with BioCyc

Mario Latendresse, Suzanne Paley, and Peter D. Karp

Abstract

The BioCyc database collection at BioCyc.org integrates genome and cellular network information for more than 1,100 organisms. This method chapter describes Web-based tools for browsing metabolic and regulatory networks within BioCyc. These tools allow visualization of complete metabolic and regulatory networks, and allow the user to zoom-in on regions of the network of interest. The user can find objects of interest such as genes and metabolites within the networks, and can selectively examine the connectivity of the network.

The EcoCyc database within the BioCyc collection has been extensively curated. The descriptions within EcoCyc of the *Escherichia coli* metabolic network and regulatory network were derived from thousands of publications. Other BioCyc databases received moderate levels of curation, or no curation at all. Those databases receiving no curation contain metabolic networks that were computationally inferred from the annotated genome sequences of each organism.

Key words: Regulatory network, Metabolic network, Cellular network, Web interface, Highlighting, Regulatory subnetwork, Browsing, Genome database, Metabolic database

1. Introduction

Network visualization tools are among the most powerful known mechanisms for allowing scientists to rapidly comprehend the complexities of biological networks. Here, we explain the usage of a set of Web-based tools for the visualization and interrogation of both metabolic and regulatory networks. These tools are in use both for the BioCyc (1) collection of Pathway/Genome Databases (PGDBs) available from SRI International at BioCyc.org, and for a growing number of other PGDBs available around the Web for organisms such as mouse and yeast, and for many plants and bacteria. A summary of those additional PGDBs is available in ref. 2.

Jacques van Helden et al. (eds.), *Bacterial Molecular Networks: Methods and Protocols*, Methods in Molecular Biology, vol. 804, DOI 10.1007/978-1-61779-361-5_11, © Springer Science+Business Media, LLC 2012

Several considerations shaped the design of our network visualization tools. (1) Because of the large number of databases to which the tools are applied, and because of the fact that multiple visualizations of the same network at different magnification levels are desired, the visualizations must be produced by fast automatic layout algorithms as opposed to by manual positioning of elements within the diagram by the database authors. (2) Because regulatory and metabolic networks are very different biologically, and because biologists have in the past developed different visual conventions for how elements of such networks are drawn, we have chosen to develop different visualizations for these two types of networks. This approach is in contrast to approaches such as Cytoscape that use a single visual network style for all types of networks (3). (3) Because these networks are large, complex, and highly connected, our tools do not initially show all connections within these networks. When all connections are shown, the resulting diagrams tend to be visually incomprehensible. Instead our approach is to allow the user to selectively add connections of interest to the diagram.

All BioCyc PGDBs contain metabolic network information because of the existence of predictive tools for metabolic networks (4). However, no inference tools for regulatory networks are available for PGDBs, therefore regulatory network information must be defined through manual curation. To date, the only curated regulatory network available in BioCyc is that for EcoCyc (5). However, additional regulatory-network curation projects are in progress and may be available soon. In addition, note that although Pathway Tools supports curation of many types of regulatory information, the regulatory network viewer (the Regulatory Overview) described here currently shows only the network of regulatory interactions resulting from regulation of transcription initiation. We plan to extend the Regulatory Overview to include other regulatory relationships in the future. Currently, other types of individual regulatory interactions, including substrate-level regulation of enzyme activity, regulation by small RNAs, and attenuation, are shown on gene pages and transcription unit pages within Pathway Tools based Web sites.

This chapter does not provide a complete description of how to use the BioCyc Web site. We focus here on the metabolic and regulatory network visualization tools, in part because these tools are quite new, and in part because they are among the more complex aspects of the Web site. The BioCyc Web site provides additional documentation on its usage under the Help item in the menu bar, and through Web tutorials (6). The Pathway Tools software that powers the BioCyc Web site and Web sites for other PGDBs was described previously in ref. 4.

Frequent users of BioCyc often install BioCyc PGDBs and the associated Pathway Tools software locally on their own computer

(7) because local usage provides faster response and additional functionality not present in the Web version. A comparison of the Web and desktop modes of Pathway Tools is available in ref. 8.

1.1. Design of the Regulatory Overview

A simple approach to analyze the transcriptional regulatory relationships between genes of a specific organism is to display these relationships using a *network*, that is, a graph where nodes represent genes and arcs represent the relationships from regulators to regulatees. We call the complete regulatory network visualization generated by our tools the *Regulatory Overview.*

One of the main issues in displaying a Regulatory Overview of an organism is the limited size of a computer window. For *Escherichia coli*, around 2,300 genes are involved in the Regulatory Overview because they are either transcriptional regulators or they are regulated by a transcriptional regulator according to EcoCyc. Compact layouts must be used to be able to fit this number of genes on a typical monitor size and still provide readability and structure. In particular, we have maintained the constraint that gene icons should not overlap.

We have used the following approach to make the Regulatory Overview more compact, and to add a new level of biological organization to the diagram. We partitioned the nonregulator genes into groups, such that two genes are in the same group if they are regulated by the same set of genes. Note that genes within a group may respond in different ways to these regulators – some positively and some negatively. Each group of nonregulatory genes is laid out compactly in a leaf shape with a short stem which gives the group a triangular look. Then the groups are spread around on a ellipse as wide and as high as the user's computer window would allow it. The remaining genes, the regulators, are laid out as ellipses too, inside this larger ellipse. We have chosen to use two such inner ellipses. The result can be seen for *E. coli* in Fig. 1. This is the "nested ellipses" layout.

We use a second layout, called "top to bottom," that in some cases is better at representing a regulatory network compared to the nested ellipses. An example, for the complete regulatory network of *E. coli*, is shown in Fig. 2. It is actually more compact than the nested ellipses layout due to the very small size of the icons used for the bottom row representing the nonregulator genes.

Since displaying all, or even most, relationships within the network would greatly clutter the display, our approach is to let the user select the genes for which relationships should be shown. Therefore, when first displayed, the Regulatory Overview does not show any regulatory relationships.

1.1.1. Study Case 1: Displaying a Regulatory Subnetwork

It is likely that an analysis is done on a subset of the genes of an organism and not the complete regulatory network. A natural approach to select this subset is to first display the entire

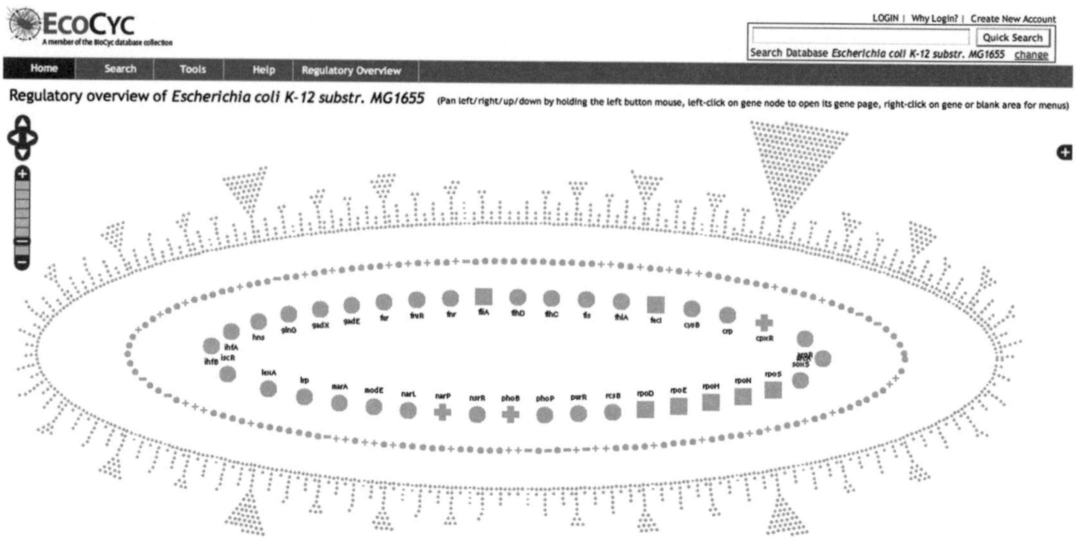

Fig. 1. The entire Regulatory Overview for *E. coli* after selecting the command `Tools→Regulatory Overview`. In this case, zoom level 2 was automatically selected as the best zoom level given the size of the browser window. The names of the genes are displayed for the inner-most ellipse, but not for the other two ellipses as not enough space is available for the names. This display uses the "nested ellipses" layout. The outer most ellipse has all the genes that do not regulate any gene but are regulated by at least one gene. Genes that are regulated by the same set of genes are grouped together, forming a leaf shape. The two inner ellipses contains the regulator genes. The inner-most ellipse has the genes that regulate the largest numbers of genes. They are displayed in ascending alphabetical order starting from the right side. This layout is not based on the locations of the genes on the genome. Each circle, square, minus, or plus icon represents a gene. A square represents a sigma factor, a plus represents a gene for which all its regulators are activators, a minus represents a gene for which all its regulators are inhibitors, and a circle represents a gene that has activators and inhibitors. No regulation relationships are shown in the diagram. Relationship arrows can be added with the searching and highlighting operations.

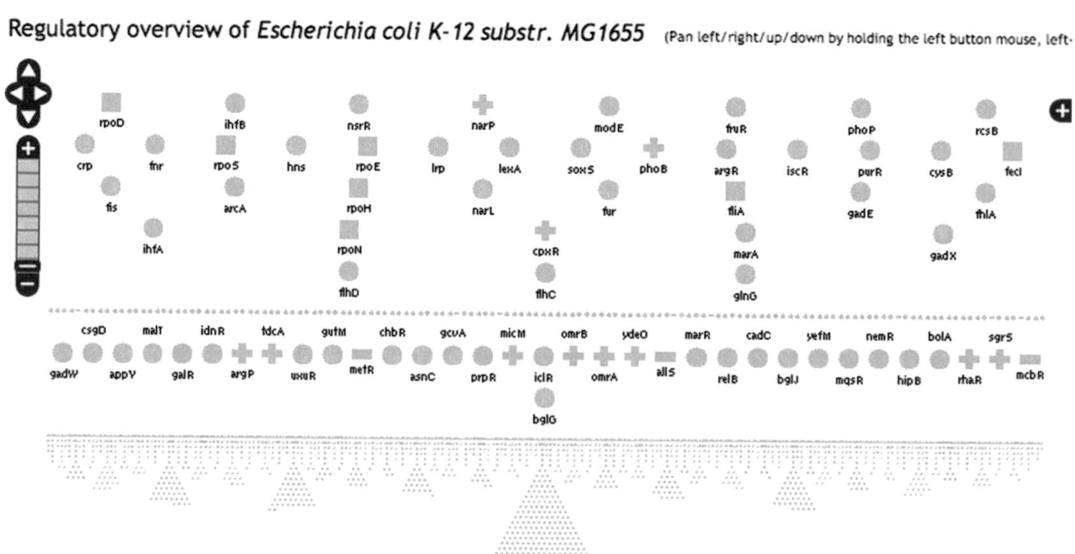

Fig. 2. The entire Regulatory Overview of *E. coli* using the "top to bottom" layout. The gene name appears near an icon if space is available. No regulatory relationships are shown, that is, no arrows are displayed. This is the most compact overview available. The genes on the same row do not directly regulate each other. The bottom row has all the genes that do not regulate any gene but are regulated by at least one gene (this row corresponds to the outer most ellipse of the "nested ellipses" layout). The top row has the regulators that regulate the largest number of genes.

regulatory network then let the user select a subset of genes, using various search and highlighting commands. Then redisplay the regulatory network based only on the selected genes. In this case a *subnetwork* is displayed. Subsection 3.6.1 presents a concrete example to display such a subnetwork. It is based on Subsection 3.5 that introduces the various search and highlighting commands for the Regulatory Overview.

1.2. Design of the Cellular Overview

The Cellular Overview depicts the biochemical machinery of an organism as described in a database. Whereas the Regulatory Overview shows genes and the regulatory relationships between them, the Cellular Overview shows metabolites and the reactions, enzymes, and pathways that interconvert them.

At low zoom level, the shapes of the metabolic icons represent different compound classes. The different shapes used are as follows:

- Triangle: Amino Acids
- Square: Carbohydrates and Derivatives
- Diamond: Proteins and Modified Proteins
- Vertical Ellipse: Purines
- Horizontal Ellipse: Pyrimidines
- T: tRNAs
- Circle: All other compounds
- Filled shape: Phosphorylated compound

At low zoom levels, each icon in the overview (such as the small circles or triangles) represents a single metabolite, and each blue connecting line represents a single bioreaction. Reactions are organized into pathways. Because some metabolites participate in multiple reactions, and some reactions belong to multiple pathways, a single metabolite or reaction may appear in multiple places in the diagram.

At higher zoom levels, the icons are replaced by the corresponding metabolite names, and more detailed information such as enzyme and pathway names appear in the diagram. Figure 3 shows the entire Cellular Overview diagram for *E. coli* at a low zoom level, and Fig. 4 shows a section of the same diagram at a high zoom level. Users can pan the Cellular Overview to see different portions of it, but unlike the Regulatory Overview, the Cellular Overview does not support display of customized subnetworks (although the user can visit the detail page for each individual pathway).

Pathways are grouped together into related clusters, such as all pathways involved in cofactor biosynthesis, or all pathways involved in amino acid degradation. Connections between pathways are not shown. In general, energy metabolism pathways, such as glycolysis and the TCA cycle, appear near the middle of the diagram, with anabolic pathways to the left and catabolic

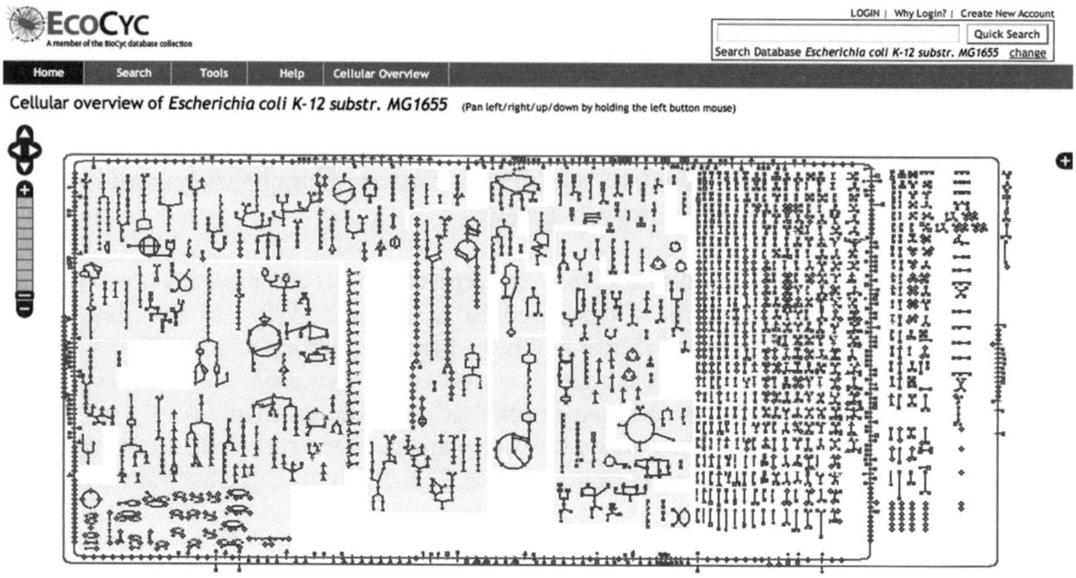

Fig. 3. The cellular overview of *E. coli* right after selecting the command `Tools→Cellular Overview`. The zoom ladder is at level 1, the lowest level, so that we can see all the metabolic pathways. At this zoom level, not enough space is available to display the names of compounds, enzymes, and pathways.

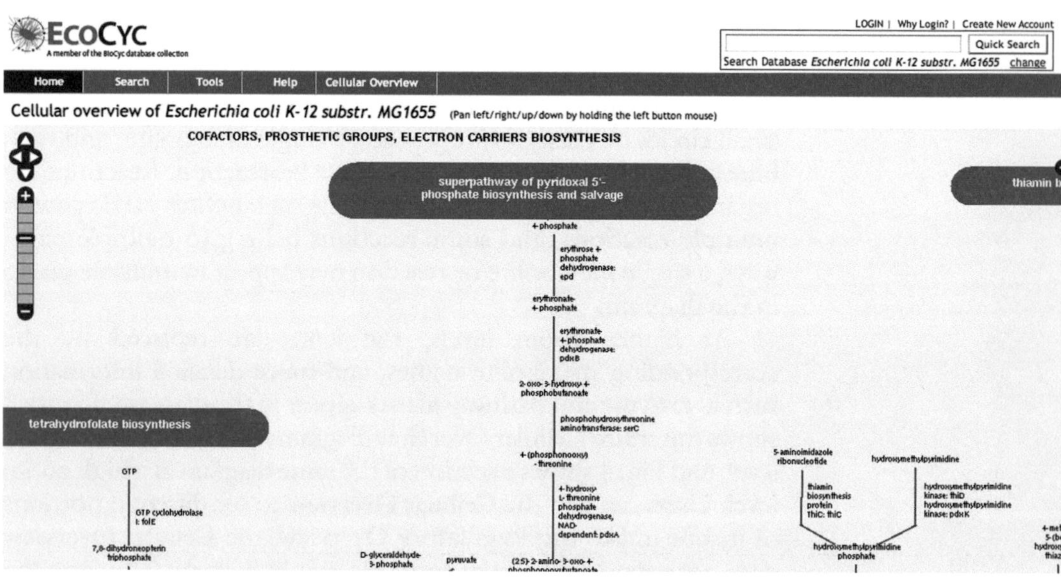

Fig. 4. The Cellular Overview of *E. coli* at zoom level 7. Notice that the blue marker on the zoom ladder, on the left side of the Web page, is at step 7. The compound, enzyme, and pathway names are displayed at this zoom level. Due to the size of the entire overview, only a small portion can be seen at any one time. The user can scroll the overview to the left, right, down, or up, by holding the left mouse button and moving the mouse in the direction the overview should move. This operation is called "panning".

pathways to the right. The existence of anaplerotic pathways prevents rigid classification, however. If present, signal transduction pathways run along the bottom of the diagram. At the far right lies a collection of individual reactions that have not been assigned to any pathway.

Surrounding the diagram is a border representing the cellular membranes of the organism. For organisms that have both a plasma membrane and a periplasmic membrane, such as Gram-negative bacteria, a double border is shown. Transport reactions are shown crossing the appropriate membrane(s), and reactions that take place in the periplasm or extracellular region are drawn between or outside the membranes. In addition, icons for proteins that have not been annotated as enzymes but which have been curated with a location on the membrane or in the periplasmic or extracellular regions will be displayed in the appropriate location.

1.2.1. Study Case 2: Highlighting Metabolic Pathways Based on Genes

The Cellular Overview provides searching and highlighting of reactions, compounds, and pathways. Searching based on gene names is also provided but in this case a mapping is applied from genes to different metabolic objects. This study demonstrates how to search and highlight based on gene names using the basic commands and a file containing gene names. This study is described in Subheading 3.2.

2. Material

2.1. Software

The network visualization tools for BioCyc and other Pathway Tools powered Web sites do not require any special software other than a Web browser such as Firefox, Safari, or Chrome. The network visualization tools are implemented in the commonly available JavaScript language running in such browsers. The Internet Explorer (IE) browser (versions 6–8) may exhibit a slow response due to its slower JavaScript engine, therefore we do not recommend use of IE when using the Regulatory and Cellular Overviews.

2.2. File Formats

For the Cellular Overview, it is possible to highlight reactions, enzymes, and pathways based on a set of gene names listed in a text file. Each line of the file must contain at most one gene name. Any gene name that cannot be recognized within the currently selected organism will be ignored. Subheading 3.2 shows a method to highlight pathways and enzymes using such a file.

3. Methods

The methods we describe share some common operations to both overviews that are covered first. These are called "Procedure." Procedure 1 is the common way to start a method. Procedures 2 and 3 are independent and applied to display a Regulatory Overview or the Cellular Overview. After the procedure descriptions, we present methods to browse, search, highlight, and create a regulatory subnetwork.

3.1. Procedure 1: Getting Started with Network Browsing

1. Access the Web site +biocyc.org+ using a Web browser (see Note 1). We recommend Firefox, Safari, or Chrome. Internet Explorer works too, but it is slower than the other three browsers.
2. Select the organism whose networks you wish to browse using the "change organism database" link located on the right side of the Web page. See Fig. 5 for the location of the change organism database link.
3. Make the main window of your browser as wide and as high as possible. The overviews are best viewed on large computer monitors.

At this point you can either browse the Regulatory Overview (if available for the selected organism) or the Cellular Overview.

3.2. Procedure 2: Displaying a Regulatory Overview

- Apply "Procedure 1" from Subheading 1.
- Select the command `Tools→Regulatory Overview` to display the Regulatory Overview of the currently selected organism. If the selected organism has a Regulatory Overview the command is selectable otherwise it is grayed out and it cannot be selected. Currently, in December 2009, only *E. coli K-12 substr. MG1655* has a Regulatory Overview. When the Regulatory Overview is displayed, it will be as wide and as high as the browser window will allow it.

The common browsing operations described in Subheading 4 are available at this point.

Fig. 5. Near the top-right corner of BioCyc.org Web pages, the link labeled "change organism database" allows the user to select a specific organism database change organism database.

3.3. Procedure 3: Displaying a Cellular Overview

- Apply "Procedure 1" from Subheading 1.
- Select the command `Tools→Cellular Overview` to display the Cellular Overview of the currently selected organism. The lowest zoom level is automatically chosen and a complete Cellular Overview should be displayed (see Fig. 3 for an example using *E. coli*).

The common browsing operations described in Subheading 4 are available at this point.

3.4. Method 1: Browsing an Overview

Once an overview is displayed, the user can navigate through the network by zooming in or out, moving it, and navigating to Web pages describing the objects of the overview. In this subsection, we review the basic common operations to browse the Regulatory or Cellular Overview.

1. Open a Regulatory or Cellular Overview as described in Subheadings 2 or 3.
2. Zoom-in or zoom-out the overview by using the zoom ladder on the left of the overview (see Fig. 6). This can be done in three ways: clicking the minus or plus icon, sliding the blue marker, or clicking a ladder step to go directly to a zoom level. Depending on the zoom level, more details are displayed such

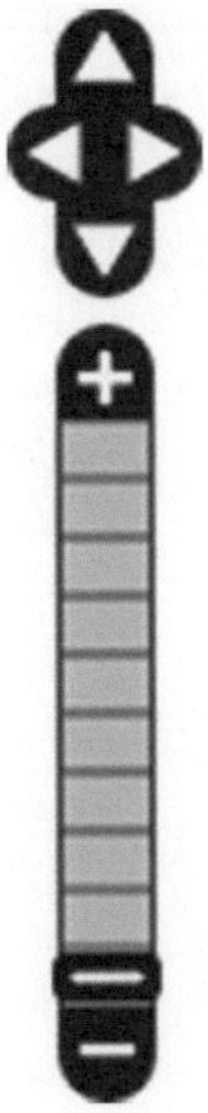

Fig. 6. The zoom ladder. This particular representation shows the zoom ladder at its lowest level. The current zoom level is identified by the blue marker. It is right above the minus icon near the bottom. To change the zoom level, you can either click the plus or minus icon, slide the blue marker along the ladder, or directly click on a ladder step. Above the zoom ladder are four directional buttons to pan right/left/down/up the entire overview. Panning can also be done by holding the left mouse button directly on the overview and moving it in the desired direction.

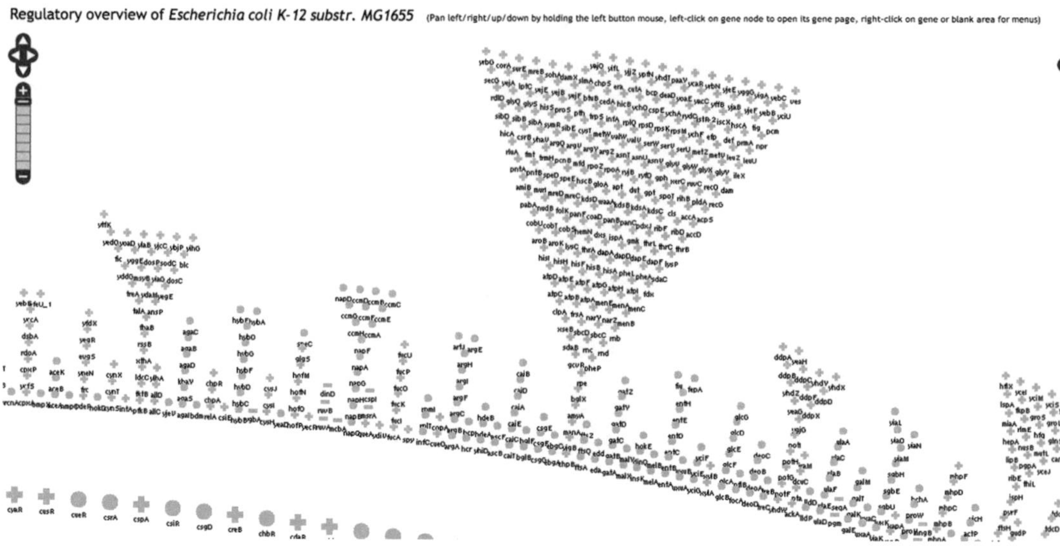

Fig. 7. A zoom-in to the last zoom level of *E. coli* and toward the outer most ellipse.

as gene names, for the Regulatory Overview (see Fig. 7), or enzyme, compound, and pathway names, for the Cellular Overview (see Fig. 4). *Double-left-clicking in a blank area* of the overview does also a zoom-in operation.

3. You can interrogate and navigate the overviews using the following operations.

   (a) *Mousing over an icon* reveals a tooltip window describing succinctly the object represented by the icon.

   (b) *Left-clicking on a icon* such as a gene opens a new browser window (or tab, depending on your set of preferences in your browser) with information about the object represented by the icon. The resulting information page might describe a gene (Regulatory Overview), a pathway, a reaction, a compound, or an enzyme (Cellular Overview).

   (c) *Holding the left-button and moving the mouse left/right/up/down* allows to pan the overview left, right, up, or down. You need to hold down the mouse button while moving the mouse to do the panning. You can also pan the overview by clicking the appropriate arrow above the zoom ladder as shown in Fig. 6.

   (d) The menu bar commands `Cellular Overview→Show Legend` will display a legend explaining the meaning of the icons used in the cellular overview (`Regulatory Overview→Show Legend` performs the analogous operation). These commands are also available by right-clicking in a blank area of the window.

(e) Commands `Cellular Overview→Help` and `Regulatory Overview→Help` will open a new window (depending on your browser setting) containing full documentation of how to use the overviews (also available from the right-click menu).

Next we focus on how to display a regulatory subnetwork using the searching and highlighting commands. The next section reviews the searching and highlighting commands for the Cellular Overview.

3.5. Method 2: Searching and Highlighting Within the Regulatory Overview

This method describes how to find genes of interest within the Regulatory Overview, and how to add regulatory relationships to the diagram.

Searching always entails highlighting since once the genes are found, they are highlighted. Searching might also involve adding relationship arrows between the genes found. This is user selectable for all searching.

Searching (and highlighting) based on a particular gene, entails, in most cases, adding relationship arrows. This type of search is done by right-clicking on a gene icon. Adding the relationships is actually the main intended purpose of the commands selectable from right-clicking on a gene icon.

In the following, we show the typical steps to search, highlight, and add relationship arrows in a Regulatory Overview.

1. Open a Regulatory Overview as described in Subheading 2.
2. Apply a series of search and highlighting commands. For each highlighting command, one color is used to highlight the gene icons and, if any, the relationship arrows but a new color is used for each executed command. Two sets of commands are available, one set by right-clicking on a gene; the second set from the menu bar (or by right-clicking in a blank area of the overview). We first present the list of commands from the menu bar or right-click in a blank area.
   (a) `Regulatory Overview→Highlight Gene By Name or Frame ID` highlights at most one gene. It is essentially a search command since you might not know the location of that gene in the regulatory network. Every object of each database at `BioCyc.org` has a unique identifier, called a "frame id." You can see this frame id in the URL when visiting the Web page describing this object. So, a search based on a frame id is very precise.
   (b) `Regulatory Overview→Highlight Genes By Substring` might highlight several genes. Selecting the command opens a panel by which you can enter a string of characters. Click the button labeled

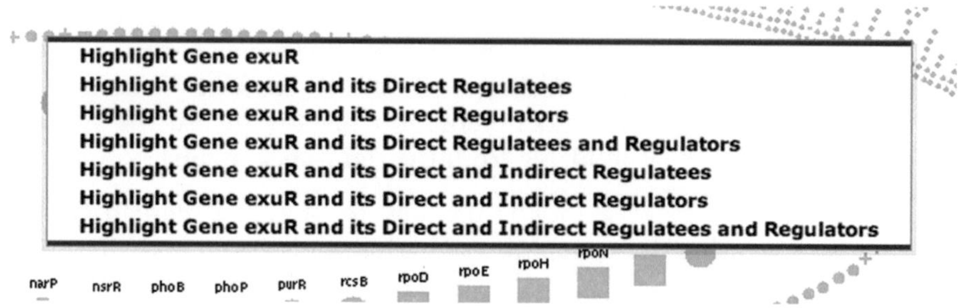

Fig. 8. The right-clicking menu on a gene. This particular menu opened when the mouse was right-clicked on gene `exuR`. Since this gene has regulators and regulatees, all highlighting commands are provided. The user can select one command which would highlight the gene icon and the corresponding regulator and/or regulatee icons as well as displaying arrows to show the direction of regulation between them.

"Highlight" in the panel, the genes highlighted have a name that contains the given string (this is a case-insensitive search). The option `Include Relationships Arrows Between the Genes Found` adds relationship arrows between the highlighted genes.

(c) `Regulatory Overview→Highlight Genes By Gene Ontology Terms` enables you to select one or more Gene Ontology (GO) terms. The genes that produce proteins annotated with the selected GO terms will be highlighted. The option `Include Relationships Arrows Between the Genes Found` adds relationship arrows between the highlighted genes (see Note 2).

Right-clicking on a gene will open a menu of highlighting commands specific to that gene. The menu may contain from one to seven commands. Since some genes do not have any regulators or/and any regulatees, this list of commands may vary from gene to gene. Here are the list of all possible commands available from this menu where *name* will be the gene name (e.g., `trpA`) on which the right-clicking was done (see Fig. 8).

- **Highlight Gene** *name* highlights only the selected gene.
- **Highlight Gene** *name* **and its Direct Regulatees** highlights the selected gene and all its direct regulatees are highlighted and relationship arrows are displayed from the selected gene to its regulatees.
- **Highlight Gene** *name* **and its Direct Regulators** highlights the selected gene and all its direct regulators are highlighted and relationship arrows are displayed from the regulator genes to the selected gene.
- **Highlight Gene** *name* **and its Direct Regulatees and Regulators** combines the two previous commands.

- **Highlight Gene *name* and its Direct and Indirect Regulatees** highlights the selected gene and all its direct regulatees and indirect regulatees are highlighted and relationship arrows are displayed from regulators to regulatees.
- **Highlight Gene *name* and its Direct and Indirect Regulators** highlights the selected gene and all its direct regulators and indirect regulators are highlighted and relationship arrows are displayed from regulators to regulatees.
- **Highlight Gene *name* and its Direct and Indirect Regulatees and Regulators** combines the two previous commands.

Finally, commands `Cellular Overview→Clear All Highlighting` and `Regulatory Overview→Clear All Highlighting` will clear all highlighted items in each overview (also available from the right-click menu). The effects of these commands cannot be undone.

3.6. Method 3: Displaying a Regulatory Subnetwork

The Regulatory Overview enables you to display your own regulatory subnetwork of genes from the regulatory data. You cannot change the regulatory data, but you can select a subset of genes to display. The selection of the genes can be done in many different ways through highlighting and searching operations as described in Subheading 5.

1. Apply a series of searching and highlighting commands as described in Subheading 5.
2. If during the previous highlighting operations you highlighted a set of genes you do not want to see part of your subnetwork, you can deactivate this set by using the switch panel on the right of the Web page (see Fig. 9). You can easily activate or deactivate several sets of highlighted genes by ticking or unticking the square boxes.
3. Once you have the right sets of highlighted genes active, select the command `Regulatory Overview→Redisplay Highlighted Genes Only` (also accessible from right-clicking in a blank area of the overview). The server will generate a new regulatory network for this set of genes. All highlighting will remain active.
4. You can always redisplay the entire overview by selecting the command `Regulatory Overview→Display Complete Regulatory Overview`, reactivate some highlighting or add more highlighted genes, then use the `Redisplay Highlighted Genes Only` command again. In this manner, you can iterate this method until you have a satisfactory subnetwork corresponding to the set of genes to analyze.

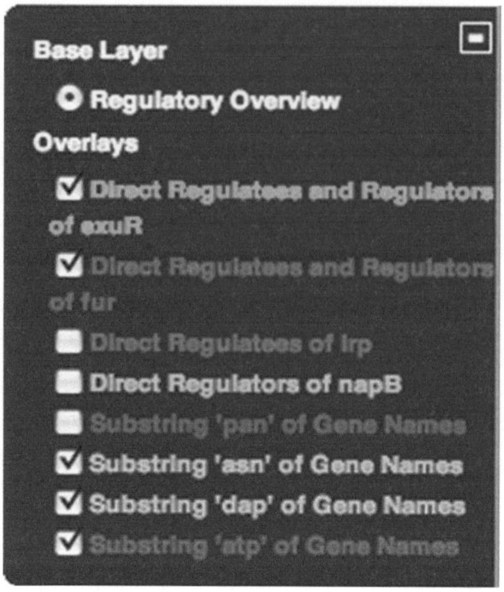

Fig. 9. This is a particular switch panel after highlighting several groups of genes in the Regulatory Overview. Each entry under "Overlays" can be activated or deactivated by ticking the corresponding white square. A tick mark indicates that the overlay is active. When an overlay is not active, all its genes are not highlighted and do not participate in the command "Redisplay Highlighted Genes Only." An inactive overlay is not destroyed and can be activated again by simply reticking the square box. To remove all overlays, use the command `Clear All Highlighting` available under the menu bar `Regulatory Overview` or from right-clicking in a blank area of the overview. Removing all highlighting (i.e., overlays) cannot be undone. To close the panel, click the minus icon (a plus icon will replace the panel), to (re)open it, click the plus icon.

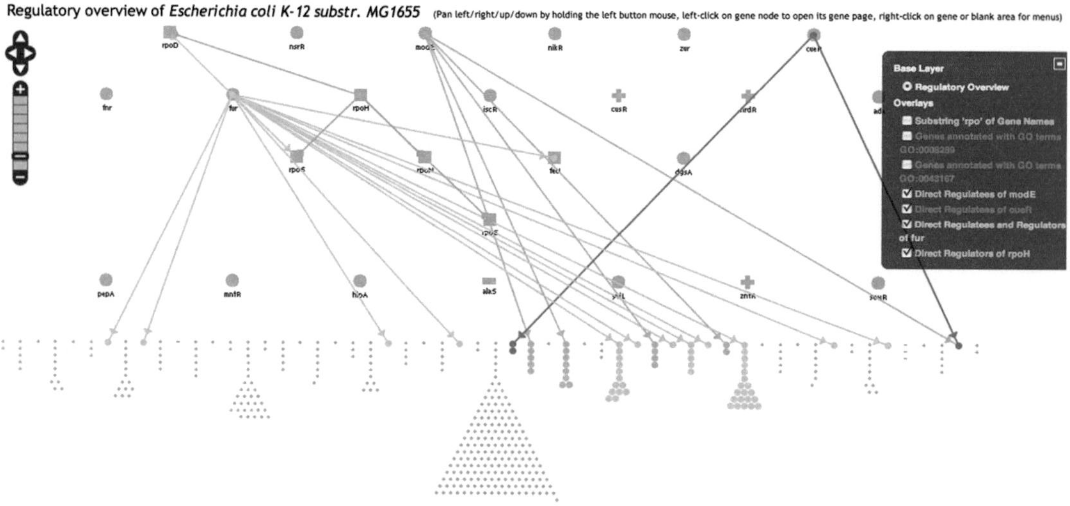

Fig. 10. The regulatory subnetwork resulting from the method of Subheading 3.6.

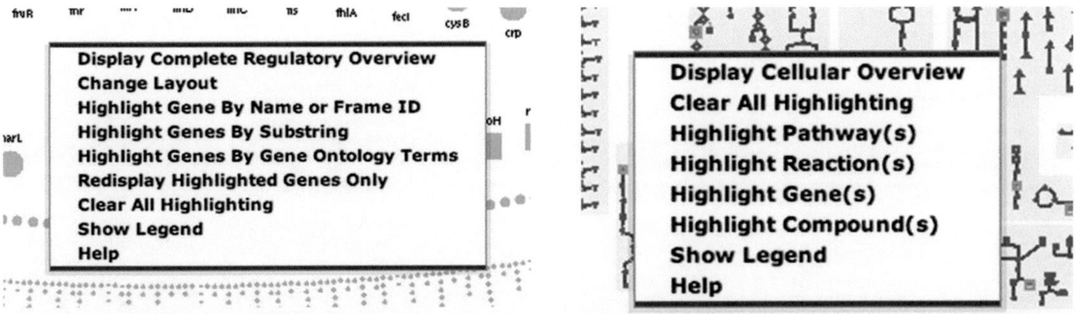

Fig. 11. The general Regulatory Overview menu command on the left and the general Cellular Overview menu command on the right when right-clicking in a blank area of the overview. The commands are the same as the ones provided from the menu bar under "Regulatory Overview" and "Cellular Overview," respectively.

3.6.1. Study Case 1: Displaying a Regulatory Subnetwork

We show a specific example of displaying a regulatory subnetwork based on specific sets of genes for *E. coli*. The result will be the subnetwork shown in Fig. 10.

1. Apply "Procedure 2" from Subheading 2 and select the organism database *E. coli K-12 substr. MG1655* when doing the second step of "Procedure 1."
2. The following three commands are accessible from the menu bar and by right-clicking in a blank area of the overview. You can try right-clicking in a blank area now and you should get the menu as shown on the left of Fig. 11.
3. Apply a substring search using the command `Regulatory Overview→Highlight Genes By Substring` and using the substring "rpo." (This will highlight several sigma factors.)
4. Apply the search command `Regulatory Overview→ Highlight Genes By Gene Ontology Terms` twice, one for GO term GO:0008289 (*lipid binding*) and one for GO term GO:0043167 (*ion binding*). Both of these GO terms are under category *molecular function*, subcategory *binding*. When searching by a GO term, it is usually more precise to select it by ticking its corresponding white square box on its left. You will probably need to expand the GO tree hierarchy to the desired term by clicking the plus icons. Note that clicking a GO term text in the GO tree do not select it, but rather, opens a Web page describing the term (see Fig. 12).
5. Apply the command `Regulatory Overview→Redisplay Highlighted Genes Only`. This will display a subnetwork based on the highlighted genes. At this point you do not have any regulatory relationships shown. The layout used is now "top to bottom" and no longer "nested ellipses."

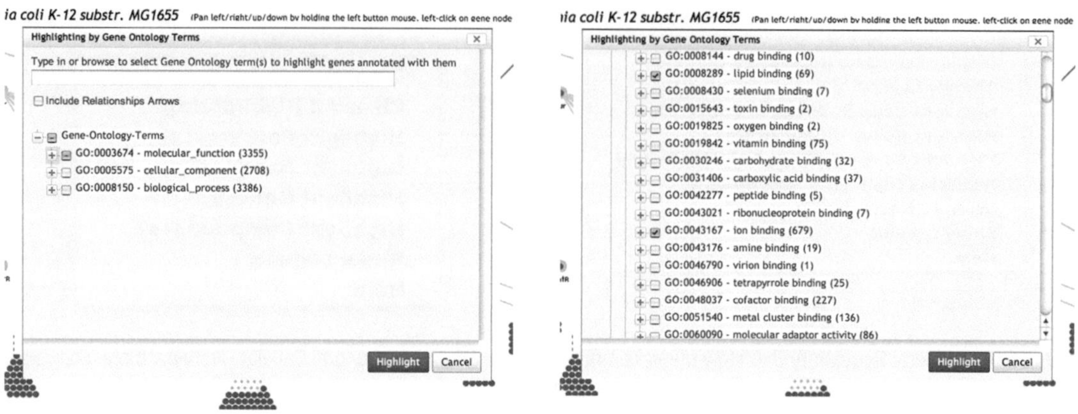

Fig. 12. The ontology dialog for highlighting based on Gene Ontology terms. On the left, the dialog as it opens up from the command Highlight `Genes By Gene Ontology Terms`, on the right, after clicking two plus icons to expand the subcategory *binding* of category *molecular function*.

6. Using the switch panel, on the right of the Web page as shown in Fig. 9, deactivate all the overlays by unticking each white square box. This will remove all highlighting of the gene nodes but without removing the overlays. We will add regulatory relationships using the following commands by right-clicking on specific genes.
7. Right click on genes `modE` and `cueR` and apply the command `Highlight Gene` *name* `and its Direct Regulatees` to each gene. You might need to zoom-in if you do not see the gene names `modE` or `cueR`. Use the zoom ladder to do so, or double-click in a blank area of the overview.
8. Right click on gene `fur` and apply the command `Highlight fur Direct Regulatees and Regulators`.
9. Apply the command `Highlight rpoH Direct Regulators`. You need to right-click on gene `rpoH` to access this command.
10. You should now have the subnetwork as shown in Fig. 10.

3.7. Method 4: Searching and Highlighting in the Cellular Overview

The Cellular Overview includes a variety of commands for searching and highlighting the multiple types of objects that the overview contains, including searching for pathways, reactions, compounds, genes, and proteins. Note that gene searches result in highlighting of reactions and proteins within the Cellular Overview because genes do not have their own graphical icons within this diagram.

As for the Regulatory Overview, the switch panel (see Fig. 9) is used to show the highlighting operations that were done. Each highlighting operation creates a new overlay with its own color. The name of each overlay is based on the operation done. You can activate or deactivate an overlay by ticking or unticking, respectively, its corresponding white square box. Deactivating an overlay simply removes the corresponding highlighting in the overview. It does not remove the overlay since you can reactivate it. An individual overlay cannot be deleted. All overlays can be deleted at once by using the command `Clear All Highlighting` available under the menu bar "`Cellular Overview`" or by righ-clicking in the overview.

3.7.1. Highlighting in the Cellular Overview from Basic Commands

There are four major commands to search and highlight objects in the Cellular Overview. They are accessible from the menu bar and by right-clicking on the Cellular Overview (see right of Fig. 11). These commands are:

- `Cellular Overview→Highlight Pathway(s)`
- `Cellular Overview→Highlight Reaction(s)`
- `Cellular Overview→Highlight Gene(s)`
- `Cellular Overview→Highlight Compound(s)`

They all have the same two subcommands, namely `By Name or Frame ID` and `By Substring`. The command `Highlight Gene(s)`, has a third subcommand, namely `From File`. This special subcommand will be covered in the next subsection.

The highlighting-by-name searches perform exact matching on the name that the user enters, such as the gene name "trpA." The search is not case-sensitive. Every object of each database at `BioCyc.org` has a unique identifier, called a "frame id." This identifier is shown in the URL when displaying the object page.

A `By Substring` search allows searching based on partial object names. It is not case-sensitive. Such a search might result in many objects being highlighted. For example, searching for "degradation" for pathways will usually result in several pathways being highlighted.

The highlighting based on gene names is special. The genes found, either by the subcommands `By Name or Frame ID` or `By Substring`, are mapped to enzymes and reactions. A gene is mapped to its product (according to its database data) and reactions catalyzed by these products. These products and reactions are highlighted. This is so, since no gene icons are shown in the Cellular Overview.

For example, here are the steps to highlight all compounds with the substring "hydro" for the organism *Mycobacterium tuberculosis H37Rv.*

1. Apply "Procedure 3" in Subheading 3 and select the organism database *Mycobacterium tuberculosis H37Rv* when doing the second step of "Procedure 1."
2. By right-clicking in the Cellular Overview the menu as seen on the right of Fig. 11 is shown.

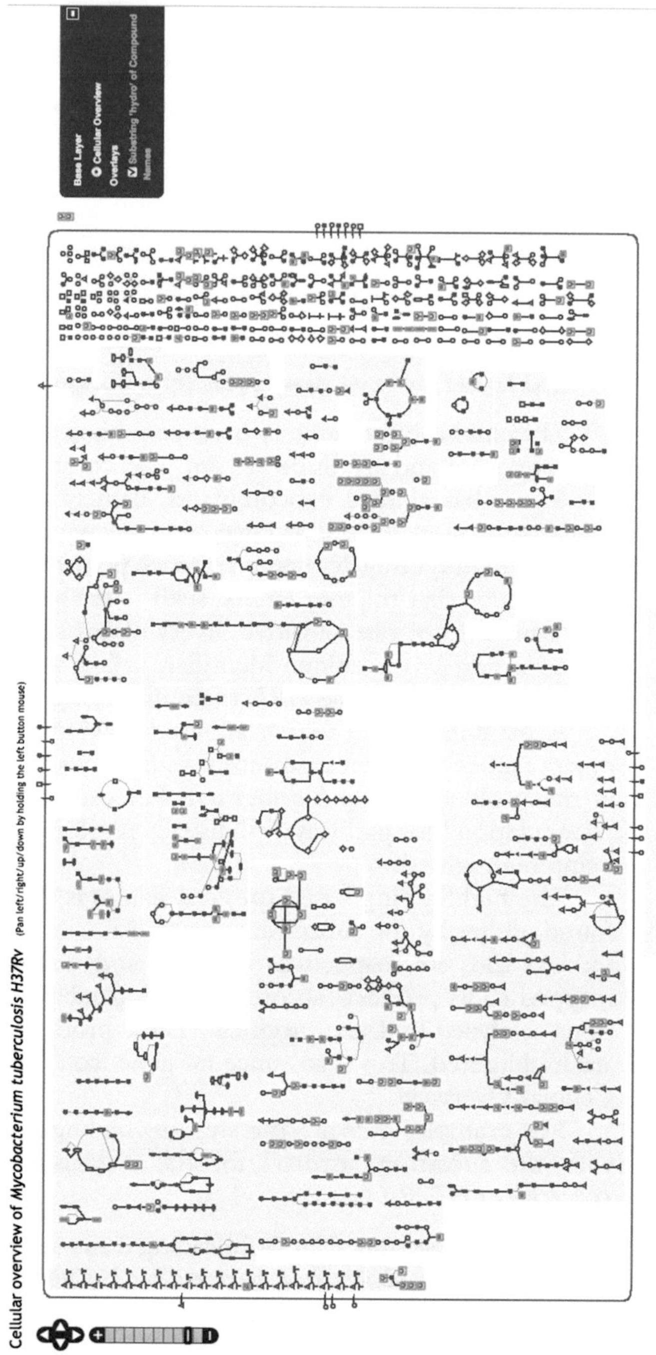

Fig. 13. The cellular overview of *Mycobacterium tuberculosis H37Rv* after highlighting all compounds that have "hydro" in their name.

3. Mouse over the command `Highlight Compound(s)`. A submenu will open with the subcommands `By Name or Frame ID` and `By Substring`. Select the subcommand `By Substring`. A small dialog window opens up where you can enter a substring.
4. Enter the substring "hydro" in the text box of the dialog. Click the button "Highlight." The dialog will disappear and all the compounds found will be highlighted with a specific color. See Fig. 13 for the final result.

3.7.2. Study Case 2: Highlighting Metabolic Pathways Based on Genes

1. Create a file of gene names. This can be done by simply typing the names of the genes (or their database identifiers) in a file using your favorite file editor. There must be one gene name per line. Choose a meaningful file name so that it is easy to remember. You must save the file on your computer for later access (see Note 3).
2. Apply "Procedure 3" from Subheading 1. The organism, selected in the second step of "Procedure 1," must be related to the gene names of your file.
3. Display a Cellular Overview as described in Subheading 3.
4. Select the command `Highlight Genes→From File` by right-clicking in a blank area. Type the name of your file (with a directory path) or use the browse button to find it on your local disk. Click the "Highlight" button. The file is transferred to the server, which extracts the gene names and finds the enzymes and reactions associated with the gene names. Then, the reactions and enzymes found are highlighted.

4. Notes

1. You can access some specific databases by using different URLs. For example, EcoCyc is also available at `ecocyc.org`.
2. If you are displaying a subnetwork, instead of the complete regulatory network, the search for genes apply to the subnetwork. Therefore, it is possible that a message is displayed stating that no genes are found with the selected GO terms although some genes exist in the organism with the selected GO terms.
3. The gene names will be verified by the server once the file is uploaded. If it turns out that a gene name cannot be found, that gene will be ignored, but processing will proceed for the other gene names and highlighting will be done for all gene names that were found. Also note that the file cannot be a

binary file but a text file based on 8-bit ASCII encoding. Most spreadsheet or word processing programs such as Microsoft Word do not, by default, generate text files. You need to explicitly save your file as "text file" when using such programs.

Acknowledgments

This work was supported by grant GM75742 from the National Institutes of Health. The contents of this chapter are solely the responsibility of the authors and do not necessarily represent the official views of the National Institutes of Health.

References

1. Caspi R, Altman T, Dale J, Dreher K, Fulcher C, Gilham F, Kaipa P, Karthikeyan A, Kothari A, Krummenacker M, Latendresse M, Mueller L, Paley S, Popescu L, Pujar A, Shearer A, Zhang P, Karp P. (2010) The Meta-Cyc database of metabolic pathways and enzymes and the BioCyc collection of Pathway/Genome Databases. Nucleic Acids Res, 38. (Database Issue): D473–D479.
2. BioCyc Database Collection. [http://BioCyc.org/otherpgdbs.shtml].
3. Shannon P, Markiel A, Ozier O, Baliga NS, Wang JT, Ramage D, Amin N, Schwikowski B, Ideker T. (2003) Cytoscape: A software environment for integrated models of biomolecular interaction networks. Genome Res, 13 (11):2498–2504.
4. Karp P, Paley S, Krummenacker M, Latendresse M, Dale J, Lee T, Kaipa P, Gilham F, Spaulding A, Popescu L, Altman T, Paulsen I, Keseler I, Caspi R. (2009) Pathway Tools version 13.0: Integrated software for pathway/genome informatics and systems biology. Brief Bioinformn, 11(1):40–79.
5. Keseler I, Bonavides-Martinez C, Collado-Vides J, Gama-Castro S, Gunsalus R, Johnson DA, Krummenacker M, Nolan L, Paley S, Paulsen I, Peralta-Gil M, Santos-Zavaleta A, Shearer A, Karp P. (2009) EcoCyc: A comprehensive view of *E. coli* biology. Nucleic Acids Res, 37: D464–D470.
6. BioCyc Webinars. [http://biocyc.org/webinar.shtml].
7. BioCyc Downloads. [http://biocyc.org/download.shtml].
8. Comparison of BioCyc Desktop Mode and Web Mode. [http://biocyc.org/desktop-vs-web-mode.shtml].

Part II

Topological Analysis of Bacterial Networks

Chapter 12

Algorithms for Systematic Identification of Small Subgraphs

Joseph Geraci, Geoffrey Liu, and Igor Jurisica

Abstract

The ability to analyze large biological networks proves to be a computationally expensive task, but the information one can gain is worth the cost and effort. In cancer research for example, one is able to derive knowledge about putative drug targets by revealing the strengths and weaknesses inherent in a protein–protein interaction (PPI) network. Further, network analyses can be used to optimize high-throughput genetic and proteomic experiments. In addition, the study of biological networks is now an active part of molecular biology. In this chapter, we review techniques for studying biological networks in general but with a focus on PPI networks, including an example of a bacterial PPI network. After a brief introduction, we concentrate on methods based on the analysis of subnetworks, namely, graph motifs and graphlets.

Key words: Protein–protein interaction networks, Network topology, Subgraph, Graph motifs, Graphlets

1. Introduction

1.1. Preamble

A great amount of energy has been put into constructing, analyzing, and modeling biological networks, namely, metabolic networks, transcriptional regulatory networks, and protein–protein interaction (PPI) networks (1). Modeling network topology can significantly increase our understanding of the biology of individual proteins and the networks they form to improve annotation of uncharacterized proteins and to optimize planning of high-throughput experiments and validation studies. It is not merely "yet another model," as has been shown in refs. 2–4: knowing (or assuming) the network model enables one to optimize experiments and maximize detected interactions while minimizing cost. Furthermore, there is a significant relation between the structure of a network, its function, and the function of its components (5).

Jacques van Helden et al. (eds.), *Bacterial Molecular Networks: Methods and Protocols*, Methods in Molecular Biology, vol. 804, DOI 10.1007/978-1-61779-361-5_12, © Springer Science+Business Media, LLC 2012

It must be stressed that modeling for the sake of modeling and artificial theoretical principles are more dangerous than helpful. As practiced in physics, models and theories should be replaced or updated when they fail to model new data. Thus, we must strive to ensure that it is not our biases and hopes (or noise in the data) that we model (6). With this in mind, note that the analysis of network topology has had a long history in biology and other fields, where large networks are commonplace (7–12).

1.2. Network Topology and Structure

An approach has emerged that seeks to characterize networks in terms of the existence and frequency of specific, small *subnetworks*. There are many different algorithms to find these subnetworks, but importantly there are various types of subnetworks that can be identified at different frequencies in biological networks, the simplest of which is a single node. The main focus of this chapter is to discuss methods that attempt to characterize networks in this way. Thus, we briefly review efforts to classify the relative importance of nodes within biological networks, specifically PPIs. First, we give a brief review of some of the elementary concepts used in network theory.

1.3. Review and Definitions

Network theory (also known as graph theory) is a mathematical methodology for studying objects that are formed from nodes and edges. A node is an abstract object, and edges are lines that connect nodes to each other. Usually, one does not care about the actual positions of the nodes and edges but, instead, the relationships between the nodes. At other times, the geometry of the network is important and one is actually interested in some notion of distance between nodes other than the number of edges that separate two nodes. It should also be noted that networks can have different topologies in different tissues, over time, or under different conditions; thus, it is possible to also consider and model the time dynamics of biological networks, since networks and their properties evolve (4, 13–19). Some basic definitions from network theory follow.

Definition 1: *A graph* G is an ordered pair, $G = (V,E)$, where V is a set called the *vertices* or *nodes*, and E is a set consisting of two element subsets of V called edges, i.e., an edge corresponds to an unordered (or ordered) pair of vertices.

Definition 2: *The degree* of a node is the number of edges emanating from it.

Definition 3: *The size* of a graph is the number of nodes it comprises.

Definition 4: *A hub* is a node whose degree is high compared to the average degree of all other nodes in the network.

Definition 5: *A digraph* is a graph, where each edge is labeled with a specific direction, indicating that an edge begins at one node and ends at another.

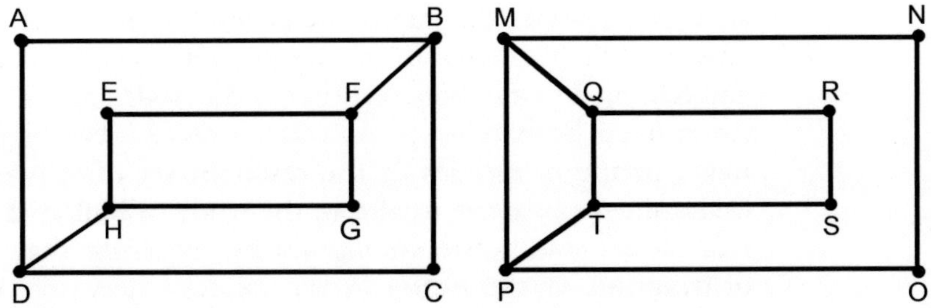

Fig. 1. Two nonisomorphic graphs. An illustration of two networks that are very similar but are not isomorphic.

Definition 6: *A subgraph H* of a given graph *G* is a graph whose edges and nodes also belong to *G*. Thus, every graph is a subgraph of itself.

Definition 7: *Two nodes belonging to the same graph are said to be adjacent* if they are connected by an edge.

Definition 8: *Two graphs G* and *H* are said to be *isomorphic* if there is a way of labeling the nodes of *H* and *G* so that all nodes that are adjacent in *H* are also adjacent in *G*. Otherwise, the graphs are considered nonisomorphic, as shown in Fig. 1.

Definition 9: *A connected component* of a network *G* is a maximal subnetwork *H* of *G* for which any node in *H* can reach (via a pathway of edges) any other node in *H*. However, there may be no pathway to other nodes in *G*. One can imagine an island with no bridge from the mainland – the island and the mainland are two distinct components, since they are each internally connected, but disconnected from each other.

1.4. Random Graphs

Before we review algorithms for subgraph identification, we need to discuss enrichment analysis (20) and random graphs. After specific subgraphs are identified, the next question is whether this finding has any meaning, i.e., whether these subgraphs are more frequent in a given network compared to "a random network." Note that there are several types of random graphs with very different properties. Furthermore, biological networks, such as PPIs, are quite large and complex. It is, thus, desirable to have a way of efficiently comparing networks to determine what subgraphs are enriched in actual biological networks using a statistically significant measure. However, what is deemed significant depends strongly on what particular way one defines a random network. Ultimately, one is interested in a model for biological networks, and random networks are natural candidates. The question remains which particular random model is best suited as a model for biological networks.

Random networks used in biological research are mostly based on a model introduced in refs. 21, 22, and these networks are known as ER random graphs or ER networks, named after the authors. In this model, one fixes the number of nodes n and edges are added with some probability p. Thus, the probability p corresponds to the chance that two nodes become connected by an edge. The set of graphs obtained in this way is denoted by $G(n, p)$. One can envision beginning with no edges and n nodes and adding an edge between two of the nodes one at a time with some fixed probability p. As you consider increasingly large n, many of the properties of the resultant set of networks can be determined (23). Eventually, in the study of biological networks, this model was bolstered slightly by requiring that the degree distribution of the nodes in the random network remains the same as the PPI it is modeling (24). This particular restriction belongs to a class of networks known as Erdos-Renyi Degree Distribution (ER-DD) random graphs (25). In Subheading 3, we briefly mention the purpose of considering this particular instance of ER-DD random networks in that it makes the comparison between PPIs and random networks more stringent.

Several global network properties are useful when comparing individual networks. A network is considered to be *scale-free* if the probability distribution of node degrees, $P(k)$, is given by $P(k) \sim k^{-\gamma}$ (26), i.e., when the probability distribution follows a *power law*. The intuitive consequence of this definition is that in scale-free networks, it is common to have nodes whose degrees are much larger than the average, and those nodes are called *hubs*. However, there are not many hubs in scale-free networks *when compared to the number of nodes of low degree*. In addition, there are more hubs in scale-free networks than in ER networks. It has been noted that biological networks are scale-free (12) but ER networks are clearly not scale-free, as certain nodes in scale-free networks have a preferential status with regards to degree. Thus, one may consider the aforementioned subclass of ER-DD networks, as this is one way of mimicking the degree distribution in PPIs. Another approach is to consider BA random networks (after Barabási and Albert), which are constructed by giving certain nodes that are randomly of higher degree, preferential treatment. A fitting analogy comes from the World Wide Web, where Web pages that are "hit" more often have their popularity accelerated while those Web pages that are "hit" less often, below some threshold, never become really popular. One can see ref. 25 – interactome supplementary – for details on how to construct a scale-free network. These popular Web pages become hubs in the network, and indeed the Internet initially had scale-free properties. It is important to note that it was believed for many years that the Internet had a scale-free topology; however, this view was recently challenged and rejected (19). "As the Internet increases in size and the technologies connected to it change, we must constantly monitor and reevaluate our assumptions to ensure that our conceptual models correctly represent reality" (27). This attitude should also pervade research in the biological sciences.

Much research has been carried out since the original suggestion that PPI networks are scale-free (10, 25, 28–30). An increasing amount of evidence suggests caution when modeling PPI networks as scale-free graphs. For example, current PPI networks are still mostly affected by false negatives. As such, we only have subnetworks available. Studying sampling properties of networks has shown that subnetworks of scale-free networks are not scale-free (30). In turn, this suggests that if the current PPI networks are scale-free (26), this might not be the case for the complete network (31). However, if a subnetwork is not scale-free, then the parent network is not scale-free either. Interestingly, networks derived from brain research by functional magnetic resonance reveal spatiotemporal dynamics of the brain that have a power law degree probability distribution (32). Based on the sampling results (33), the overall brain function network cannot follow such a probability distribution, which creates a nice parallel to functional sampling from PPI networks. Other work also suggests that scale-free networks are more optimal for traffic and signal transduction, compared to random networks (34). However, it has also been shown that geometric random graphs have many optimal characteristics, including almost perfect coverage in telecommunication networks, which suggests that these networks are less fragile (25). Further, it has been shown that in many networks, links between highly connected components are suppressed to reduce the possibility of cross talk (9). It should be mentioned that in many real networks signals can effectively reach local neighborhoods, but in reduced amounts at larger distances (35, 36).

Using spectral clustering,[1] it has been shown that different networks have different spectral densities of the eigenvalues of the connectivity matrix (37). Importantly, ER random networks follow Wigner's semicircular law, scale-free networks follow a triangular distribution, and hierarchical and PPI networks follow a different distribution.

We need to consider that the correctness of the scale-free hypothesis for PPI networks has been challenged (25, 30, 37–39), and the "graphlet" approach outlined in Subheading 4.2 was constructed to address this challenge. Several criticisms exist in the literature, including the fact that the BA random network is defined in an ambiguous way. In addition, the dependency of the initial conditions (of the growth process of the random network) on the final network is not clearly understood (40, 41). Furthermore, most of our information about biological networks, in particular PPIs, is incomplete (42). It is premature to argue that PPI networks are scale-free and that there is no better model until we have a more complete picture of the interactome. More rigor

[1] Given a set of data points A={$x_1, x_2, \ldots, x_n$), one can define a similarity matrix S, where S_{ij} represents similarity between points x_i and x_j. Spectral clustering techniques use the spectrum of the similarity matrix of the data to partition it into groups.

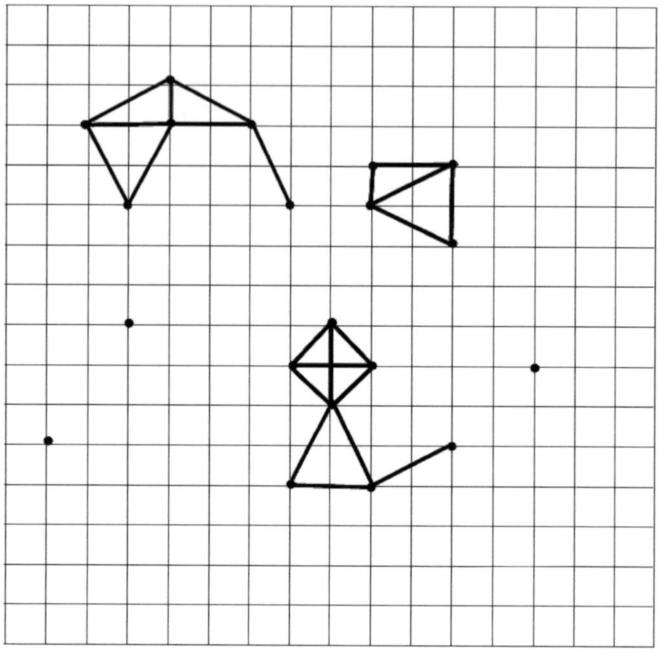

Fig. 2. Geometric graph. An illustration of a network constrained by a metric space.

in this matter should be pursued and alternative models for PPIs considered, especially when they fit the experimental data substantially better than the scale-free model (31). For an overview of the scale-free debate, see ref. 38.

An alternative random network model for PPIs is given by geometric random graphs which are studied in ref. 43 and first considered for biological networks in ref. 25. A geometric graph is constructed as follows: Given a set of points V in a metric space, an edge is drawn between two points in V if the *distance* (according to the predefined metric) is less than some prescribed number r. Thus, V is the node set and the edge set E is determined by the metric space and the placement of points in V within the space (see ref. 25 for details). Now, one obtains a *geometric random graph* when the points in V are distributed in the metric space in an independently and uniformly random way. Figure 2 illustrates a geometric graph. In this case, the metric space is represented by a grid and the distance is defined by estimating horizontal and vertical spacings. Note that in this figure only nodes that are within either two horizontal spacings and one vertical spacing or two vertical spacings and one horizontal spacing are connected by an edge. This metric was chosen arbitrarily to illustrate the concept. One obtains a *random* geometric graph if the points are distributed uniformly and independently of each other.

2. Nodes

The primary notion of importance stems from a simple observation: a node should be considered "important" if it is connected to many other nodes. This basic notion of importance makes intuitive sense biologically. A node that has many edges emanating from it will likely be communicating with many of the other entities represented by the nodes. For example, in a PPI network, a node that shares many partners represents a protein that is involved in many interactions. If one were to disrupt that particular protein's ability to interact, many mechanisms governed by the interactions would be affected. Thus, the *degree* of a node, i.e., the number of edges attached to the node, captures this notion succinctly. Nodes with high degree are often referred to as *hubs* and the removal of a hub can have serious consequences for the topology of a network and also for the underlying biological processes (44). However, it should be noted that the actual relationship between how essential a protein is in a PPI and its degree is not clear and the correlation may be weak (1). Furthermore, it has been proposed that there are two types of hubs – namely, *date* and *party hubs* (45). Party hubs interact with many partners more or less simultaneously, whereas date hubs interact with many other proteins at different times or in different locations. Consequently, party hubs and their interactors are strongly correlated in their expression. This distinction must be taken into consideration because nodes that are date hubs tend to connect portions of the network which are functionally related, and their removal is more likely to disconnect the network, i.e., change the number of *connected components* of the network.

Another natural measure of how important a node is in a biological network deals with how "near" it is to other important nodes. Important nodes should be able to communicate quickly, in some sense, with many other nodes in a network. More specifically, let $\delta(u, v)$ be the minimum number of edges that one would have to traverse to get to node v starting from node u. $\delta(u, v) = \infty$ if there is no set of edges that connect u and v. We have just defined the *distance* (or *geodesic distance*) between two nodes in a graph G. We use this distance function to define three notions of *essentiality*[2] for nodes in a graph.

The first notion is referred to as the *eccentricity* (or *association number*) of a node. Let G be the given network and let $V(G)$ refer to the set of nodes belonging to G. Eccentricity measures how far a node is from the node furthest away from it. It is given by $C_{\in}(u) = \max_{v \in V(G)} \delta(u, v)$.

[2] The essentiality of a node refers to how much of an impact the removal of a given node has on the graph connectivity, and in turn on biological organism survival.

Naturally, we wish to have a function that minimizes the eccentricity, i.e., we wish to retrieve the set of nodes in G whose maximum distance to all other nodes is minimized. This is given by the *centre* of G defined by $C(G) = \{v \in V(G) : C_{\in}(v) = \min_{w \in V(G)} C_{\in}(w)\}$.

The nodes in $C(G)$ seem to be essential because they must carry a lot of traffic and represent some sort of biological "bottleneck" for the network. Thus, a node in a biological network that belongs to the center is likely a key player in the functionality of that network. One can easily alter this definition and allow for a ranking of nodes that qualify as centers by replacing $\min_{w \in V(G)} C_{\in}(w)$ with some desired number n. Thus, the top ranking "center" is the true center and the other $n - 1$ nodes qualify as pseudocenters. It has been suggested that essential proteins tend to belong to biological processes that are densely interconnected (46). It should be noted that there may be a correlation between the number of research papers and hubs. In an early study of 380 yeast genes, Hoffman and Valencia failed to find a correlation between the degree of a protein and its occurrence within Medline (47). However, based on a much larger dataset of 11,334 literature-curated protein interactions, Reguly et al. found that the more a protein is studied, the higher the degree (48). This makes intuitive sense, as proteins that are interesting to the scientific community are more intensely studied, but also proteins with many interacting partners are more likely to co-occur in studies focused on the partners themselves. This must be kept in mind when creating or studying a PPI since a study bias may affect the connectivity of certain proteins.

If one were to sum the distance function over all nodes while holding some node u constant, one obtains the *status*, given by $C_S(u) = \Sigma_{w \in V(G)} \delta(u, w)$. One can use the status to obtain a set of nodes that minimize the average distance to other nodes within G. This is referred to as the *median* of G and is given by $M(G) = \{v \in V(G){:}C_S(v) = \min_{w \in V(G)} C_S(w)\}$.

There are other measures of importance, including the eigenvector centrality measure, which is based on the idea that "important nodes are connected to other important nodes." (See the survey given in ref. 49.)

It is worth mentioning that there are methods which cluster biological networks into hierarchical functional clusters in alternative ways. One noteworthy method, particularly effective for metabolic networks, is given in ref. 50. The method is based on modeling metabolic networks with bipartite graphs by mapping chemical compounds and chemical reactions as two different types of vertices and having edges represent dependencies. The clustering algorithm starts with the less-connected nodes, and progressively incorporates more connected nodes via an iterative process. This provides an alternative approach for understanding metabolic networks, whereby one uses a "growth" process, instead of concentrating on hubs, in order to extract information about the network.

3. Subgraphs

We began by discussing various elementary ways of identifying the important nodes in a biological network. Naturally, we now wish to consider collections of nodes that may be not only statistically significant, but also biologically relevant. A subgraph of a given network G is a graph whose set of nodes is contained in G and whose adjacency relationship (i.e., the edge set belonging to this subset of nodes) is a subset of the adjacency relationship of G. The subgraphs we wish to consider are connected, small, and ideally have biological meaning. An immediate question is whether one can identify certain subgraphs in a given network and if the existence of certain subgraphs can be related to functional aspects, including diseases. For example, in ref. 51, the authors identified subgraphs encompassing specific genes associated with cancer within networks derived from human cellular signaling pathways. We shall review two instances of subgraphs from the scientific literature and present algorithms for their identification within biological networks, namely, *motifs* and *graphlets*. In general, motifs refer to overrepresented, small subgraphs, whereas graphlets include all induced, nonisomorphic subgraphs for a given number of nodes.

In ref. 24, the authors studied diverse networks to identify small subgraphs that occurred more frequently than in randomly generated networks (see Subheading 1.2). These common subgraphs are named motifs and it seems that biochemical and neuronal networks both consist of recurring motifs. These motifs are likely to perform particular information processing tasks (52), and the ability to search networks for the existence of motifs may provide functional insights. Note that the random networks used in their research have the same node characteristics as the real network: the number of incoming and outgoing edges from each node was preserved. This makes the comparison with random networks more stringent and it allows for a model that mimics the node degree distribution of the actual biological network. Furthermore, this criterion was selected to ensure that the subgraphs were characterized as common with a rigorous statistical confidence. However, it can be shown that the results are sensitive to the choice of network randomization (53). Regardless, the search for motifs may in fact lead to a deeper understanding of biological networks and could deepen our understanding of the design principles of evolved networks.

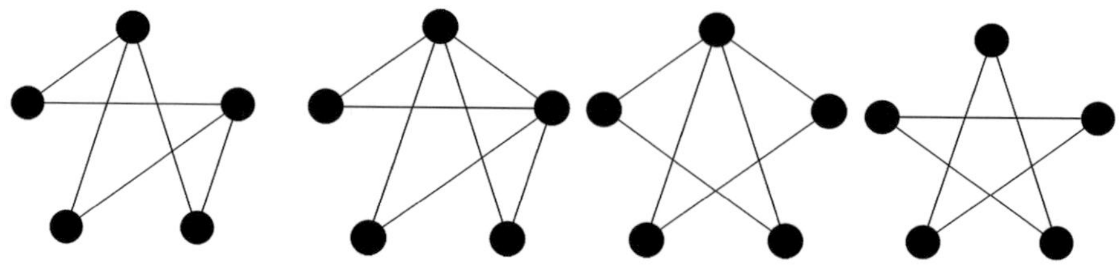

Fig. 3. Five node motifs found in neuronal network from *Caenorhabditis elegans*, ignoring directionality. A neuronal network from *C. elegans* was studied and analyzed for motifs. All of the motifs found were isomorphic to the four networks displayed, ignoring the directionality of the edges.

To overcome these difficulties, in ref. 25, the authors considered the distribution of small subgraphs within a given network, whether or not they appeared more frequently in the given network over random graphs. These are the aforementioned *graphlets*. More precisely, note that an *induced subgraph H* of a given graph *G* is a subgraph of *G* formed from a subset of given nodes belonging to *G*, together with all edges connecting the given nodes. Thus, *H* consists of the given nodes and all the edges in *G* that connect the nodes. A *graphlet* is a small, connected, induced subgraph of a network. The main idea is to study the distribution and biological function of graphlets (e.g., representing sets of proteins) and to characterize networks and their building blocks based on this distribution. In the next section, we outline two algorithms presented in ref. 54 for estimating the graphlet frequency distribution.

Before moving on to algorithms, we provide a brief comparison between motifs and graphlets. The obvious difference gathered by the preceding two paragraphs is that graphlets include all nonisomorphic, small subgraphs up to some predetermined limit in size, i.e., number of nodes. In this chapter, motifs are *distinguished, small subgraphs, in that they appear in random networks significantly less often than they do in actual biological networks*. For example, in ref. 55, they studied a neuronal network from *Caenorhabditis elegans* and determined that the following subgraphs were motifs, as presented in Fig. 3. Note that we ignore the directionality of the edges for simplicity, as most of the current interactions lack this information. However, directionality is an essential feature required to study and model network dynamics.

More specifically, all the five node motifs in this network were isomorphic to these four graphs. Compare this to all five node graphlets (see Fig. 4).

If all of these are considered, it is likely that more information about the network can be extracted and a finer characterization of biological networks is possible. When one searches for graphlets, one counts the existence of any of the induced subgraphs up to some predetermined size. Further, as presented in ref. 54, the frequency of graphlets of actual PPIs are much closer to the random geometric model than to the scale-free random graph

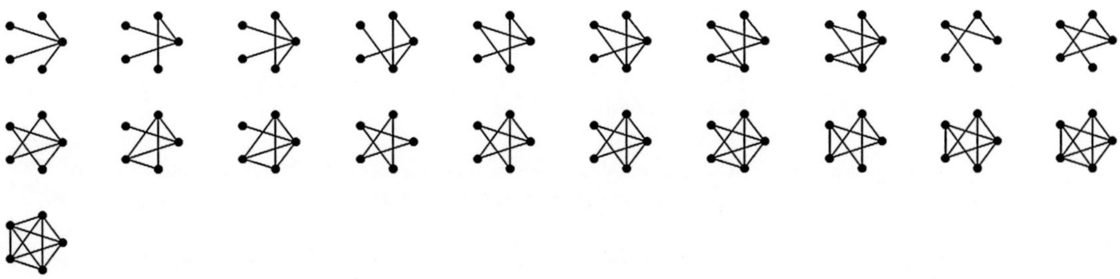

Fig. 4. All five node motifs. All possible undirected networks without loops (edges that start and finish at the same vertex) or multiple edges (duplicate edges between the same two nodes) consisting of five nodes.

model or ER-DD networks. This implies that the random graph model is superior in many ways over the often accepted ER-DD or scale-free networks for PPIs.

Algorithm 1. Identification of Network Motifs in a Directed Graph

Given: A digraph G and a constant k, which determine the number of nodes of each subgraph S_i that one searches for.

Output: A set of motifs of G.

1. As G is given, this is equivalent to having the adjacency or connectivity matrix M_{ij}. $M_{ij} = 1$ if there is an edge between node i and j and $M_{ij} = 0$ otherwise.
2. Scan M for all k × k submatrices corresponding to connected subgraphs consisting of k nodes.
3. Enumerate the subgraphs S_i. This was accomplished in ref. 24 by locating nonzero entries of M and searching through all connected elements $M_{ik} = M_{ki} = M_{jk} = M_{kj} = 1$. One then recursively repeats this process with each of the aforementioned entries of M until one obtains a k node subgraph.

   Now, one has a tally of the S_i in G. In order to determine which of these subgraphs are in fact motifs, we need to compare them with random networks of a prescribed structure.
4. Construct a large number of random networks with the following characteristics.
   (a) Each random network has nodes that correspond to the given network G in that they each shares the same number of incoming and outgoing edges for each corresponding node. It is said that these random networks have the same *degree sequence*.
   (b) Every subgraph of size $k - 1$ appears in the random networks as often as it does in G.
   Note: ER-DD random networks were discussed in Subheading 1.1 and the random networks required to implement this algorithm belong to this class. The details on how to generate this set may be found in ref. 1.
5. Obtain a tally of the subgraphs S_i for each random network.

(continued)

6. One now has a tally of the subgraphs S_i in G which we shall call N^i_G for each subgraph S_i, and a tally of each subgraph in the random networks which we shall denote N^i_R.
7. Decide if each S_i is a motif by testing if
   (a) $N_G^i >> N_R^i$ which, besides meaning that the incidences of S_i are much greater in G than in the random networks, also means that the probability of finding S_i in a random network more often than it appears in G is less than some predetermined amount (say 0.01)
   (b) Each S_i actually appears a minimum of four times in G

   The S_i that satisfy these criteria is output as motifs of G.

Algorithm 2. Enumeration of Subgraphs (Normal + Probabilistic Algorithm)

Given: A network G and a constant k which determine the number of nodes of the subgraphs considered.

Output: A set of k-node subgraphs of G.

The nodes are labeled from 1,. . ., $n=|V|$. The algorithm runs with each node in G as a starting point.

1. Begin at some arbitrary node v.
2. To begin, the first $V_{Sub\text{-}graph}$ is equal to (1).
3. Fill the set of nodes $V_{Extension}$ for v in the following way: If $u \in N(\{v\})$ and the label of u are greater than the label of v (written as $u>v$), then make u a member of $V_{Extension}$.
4. We now have the set $(V_{Sub\text{-}graph}, V_{Extension})$.
5. Remove an arbitrary node w from $V_{Extension}$.
6. Form the set of nodes for which the labels are greater than the label of v and which are members of $N_{excl}(w, V_{Sub\text{-}graph})$ unioned with $V_{Extension}$ and call this new set $V'_{Extension}$. In symbols $V'_{Extension} = \{u \in N_{excl}(w, V_{Sub\text{-}graph}): u > v\} \cup V_{Extension}$.
7. Form $V_{Sub\text{-}graph} = V_{Sub\text{-}graph} \cup \{w\}$.
8. In the nonprobabilistic algorithm, steps 6 and 7 are repeated until one obtains a collection of sets $(V_{Sub\text{-}graph}, V'_{Extension})_i$ (where i is an index of the collection) until for some i $(V_{Sub\text{-}graph}, V'_{Extension})_i$ has $V'_{Extension} = \{\phi\}$. In the probabilistic algorithm, a probability $p_d \in (0,1)$, depending on the loop depth (or the size of $V_{Sub\text{-}graph}$), is introduced. **Thus, with probability p_d, construct a member $(V_{Sub\text{-}graph}, V'_{Extension})_i$.**
9. If the number of nodes in $V_{Sub\text{-}graph}$ is equal to k, then output the graph corresponding to $V_{Sub\text{-}graph}$, and begin at another node v at 1). Otherwise goto 5) and repeat on $V'_{Extension}$ and the updated $V_{Sub\text{-}graph}$ from 7).

4. Algorithms

4.1. Motifs

Let us reiterate the aforementioned distinction between general subgraphs and motifs. Motifs are referred to as subgraphs that have been selected as special in that their appearance in random networks is less likely than occurring in a given biological network. When studying subgraphs, one has to incorporate the notion of graph isomorphism. This ensures that we accurately sample subgraphs that are topologically identical. This means that all topological equivalent motifs are sampled. This in turn ensures that we obtain the correct frequency of a given motif in the given biological network and in all random networks that are generated for the purpose of comparison. Given two graphs G and H, deciding whether a graph H is isomorphic to a subgraph of G is a difficult problem from a complexity theory point of view. This problem is known as the Subgraph Isomorphism problem and it is known to be NP complete (56), i.e., there is no polynomial algorithm that can solve it, unless a much unexpected result in computer science occurs. A natural solution is to limit the size of the motifs that one would search for. This limitation is based on the computing power one has and the algorithm used. In Algorithm 1, we introduce the basic scheme given in ref. 52 and expanded upon in ref. 1 to include PPI networks.

Following the aforementioned work, estimating subgraph frequencies was considered instead of attempting an exact count of all subgraphs. The main problem with exact counts of subgraphs is that the number of subgraphs grows exponentially with increasing k (the number of nodes). Kashtan et al. provided a probabilistic algorithm which estimates the concentration of subgraphs by taking a sample of the given network (57). This work was criticized for having a serious sampling bias, which requires a costly computation to correct. Essentially, the problem occurs because the subgraph sampling procedure tends to favor certain subgraphs and ignore others depending on the topology of the original network. Further, the preprocessing required to correct this bias involves the computation of a function that can scale as much as $O(k^k)$. These problems were addressed in ref. 58 and outlined in Algorithm 2. Software based on the following algorithm is publicly available at http://theinf1.informatik.uni-jena.de/motifs/.

Algorithm 2 outlines a probabilistic way of estimating motif frequencies by first approximately enumerating all subgraphs consisting of k nodes, and then calculating the motif significance, i.e., determining whether the subgraphs are motifs by comparing their frequencies in 1,000 randomly generated networks (having characteristics identical with the given network). The basic idea behind the algorithm is that one starts at an arbitrary node v of the given network G for which the nodes have been labeled $\{1,\ldots,n = -V|\}$ (the labels are applied arbitrarily but fixed throughout the algorithm). The algorithm constructs a set of connected nodes (the desired subgraph) by proceeding as follows: Begin at node v and move to a neighboring node if its label is greater than v's. This node became a member of the subgraph set S and each new member must be a neighbor of the newly added node but not to a node already in the set. This avoids backtracking and gives the algorithm an efficient tree structure. S is constructed in this way until there are k unique nodes within it for each connected subgraph of size k.

We begin with some necessary definitions taken from ref. 12.

Definition 10: *The open neighborhood* of a set of nodes $V' \subset V$, denoted by $N(V')$, is the set of all nodes that are members of $V \backslash V'$ (the set V with all elements of V' removed from V), which are adjacent to at least one node in V'.

Definition 11: *The exclusive neighborhood* of a node v', denoted by $N_{excl}(v, V')$, is the set of all nodes which are neighbors of v but do not belong to $V' \cup N(V')$.

Step 8 in the probabilistic case of Algorithm 2 is responsible for reducing the number of subgraphs that are found as we are not interested in an exact enumeration of all connected subgraphs. This reduces the complexity greatly. Recall that what one is interested in is estimating the subgraph concentration (58); thus, we need a notion of how well we are able to estimate the concentration of a class of subgraphs when running a probabilistic algorithm, which is unable to identify all subgraphs.

Definition 12: *Let* $S^i_k(G)$ be all subgraphs of size k in G which are isomorphic to each other. In this way, the $S^i_k(G)$ partition the set of all size k subgraphs into at most M subsets of isomorphic graphs, where M is the number of isomorphic graphs of size k. The *concentration* $C^i_k(G)$ of the set of subgraphs $S^i_k(G)$ is given by $C^i_k(G) = -S^i_k(G)|\cdot(\Sigma_j \mid S^j_k(G)|)^{-1}$.

We require the notion of an unbiased estimator.

Definition 13: *Let* U be a set of size k subgraphs sampled at random from G by some algorithm A. A mapping $C_k^{\sim i}{:}(U, G) \in [0,1]$ is called an *unbiased estimator* with respect to A if the expected value of $C_k^{\sim i}(U, G)$ equals $C_k^i(G)$.

We now present a theorem proven in ref. 58 that gives an *unbiased estimator* for $C^i_k(G)$ with respect to the Algorithm 2.

Theorem 1: *Given a network* G, an integer k, and $p_d \in (0,1]$ for $d \in [1, k]$, let U be a set of size k subgraphs obtained by executing Algorithm 2 (probabilistic) on G using the probabilities p_d. Then, $C_k^{\sim i}(U, G) = -\{G' \in U{:}G' \in S_k^i(G)\}|/|U|$ is an *unbiased estimator* for $C_k^i(G)$.

There are many issues to consider when deciding how to choose the probabilities p_d, and we refer the interested reader to ref. 58 for details. The author provide a heuristic resulting from their analysis of and experience with running the algorithm on real networks. First, the larger the p_d for small values of d, the more subgraphs that are explored and hence, the more thorough the search. In general, the author recommend that the p_d values decrease for increasing d. The intuition behind this is that a small probability in the beginning of the search can block off certain important sets of subgraphs early on. Thus, this provides a more homogenous search pattern, and avoids certain neighborhoods from being ignored and other neighborhoods from being overexplored.

Recall that in Algorithm 1 the generation of a large number of random graphs was required in order to decide whether an identified subgraph qualified as a motif. Wernicke presents a method to avoid this computationally expensive task (58). Recall that in Algorithm 1 we generate random graphs with the same degree sequence (same number of incoming and outgoing edges from each node) as the given network G and then search for candidate subgraphs in the random graphs. Wernicke makes the observation that this process is akin to randomly choosing subsets of k nodes that induce a connected subgraph. Furthermore, in ref. 59, it was observed that the total number of size k subgraphs within a collection of large graphs with the same degree sequence has a small variance. This means that one can estimate the average concentration $\langle C_k^i(G) \rangle$ by first selecting a subset of k nodes and then finding the ratio of graphs with the same degree sequence which have subgraphs induced by these nodes who are members of $S_k^i(G)$.

Definition 14: Let SEQ(G) be the set of all graphs $\mathcal{g}$ that have the same degree sequence as G.

Thus, we have $\langle C_k^i(G) \rangle \approx \langle C^{\sim i}_k(G) \rangle$, where $\langle C^{\sim i}_k(G) \rangle = (\sum_{\mathcal{g} \in \text{SEQ}(G)} |S^i_k(\mathcal{g})|)/(\sum_{\mathcal{g} \in \text{SEQ}(G)} \sum_i |S^i_k(\mathcal{g})|)$.

In ref. 58, Wernicke provides a thorough discussion on how to compute this equation using Monte Carlo methods and sampling subsets of nodes of size k from the input network. This depends on being able to compute the number $|\{\mathcal{g} \in \text{SEQ}(G)\text{: g}[\{v_1,v_2,\ldots,v_k\}] \in S^i_k(G)|$ given the set of nodes $[\{v_1,v_2,\ldots,v_k\}$. The method to do this is computationally expensive, as it requires the enumeration of k^{k-2} trees (spanning trees between the k nodes) in the undirected case (e.g., PPI networks). However, this is not as bad as it sounds because this value is the same for all subgraph classes, and must thus be computed only once. In addition, this method is still more efficient than having to generate a vast collection of random graphs and to search them for subgraphs.

4.2. Graphlets

It has been observed that the degree distribution of biological networks are *scale-free* meaning that the probability distribution of node degrees, $P(k)$, is given by $P(k) \sim k^{-\gamma}$, with $2<\gamma<3$ (26). This means that the node degrees of biological networks tend to deviate from their mean value greatly. In general, we can use local or global properties. By a local property, we mean a measure that involves the study of a local structure, like the degree of the node or the existence of a particular subgraph. Global properties involve some overarching characterization, for example planarity (the ability to draw the network on a piece of paper with no edges intersecting) or the aforementioned node degree distribution. The previously described motif methodology gives a local characterization of biological networks. Studying the graphlet frequency of a network is again a local characterization of the network, and we present two algorithms that estimate this quantity, i.e., the approximate counts of all 3, 4, and 5 node graphlets within a large network (see Fig. 5 for a list of all undirected subgraphs from size 1 to 6). The computation does not scale well, meaning that as we consider searching for subgraphs consisting of more and more nodes, the time required to complete the task increases sharply. This prevents one from being able to search for subgraphs of size 10 or more. With regards to PPIs, it is understood that many of the connections between proteins are still not yet known and thus global measures result in large errors. However, the local structure of PPI networks may be more preserved, especially when well-studied proteins are considered (54).

Przulj et al. demonstrated that geometric random graphs are a superior (in several ways) model for high-confidence PPI networks (25). One way in which this superiority was established was by comparing the graphlet frequency of actual biological PPIs with scale-free random graphs, ER-DD graphs, and random geometric graphs. The graphlet frequencies were closest between actual PPIs and random geometric graphs.

We present two algorithms given in ref. 54 that estimate the graphlet frequency distribution for large networks. They perform well for PPIs and for other networks that can be modeled by random geometric graphs. It should be noted that the presence or absence of hubs does not affect the algorithms' performance as is the case with the algorithm in ref. 57 and this is important in light of the recent arguments against biological networks being scale-free.

We now give an overview of the two algorithms presented in ref. 54.

Algorithm 3 uses the fact that geometric random graphs have sparser boundaries than their interiors, meaning that the nodes in the middle of the network are of higher degree. It also uses the fact that the interior has a uniform structure as it consists of uniformly distributed points in a bounded space. This means that one can

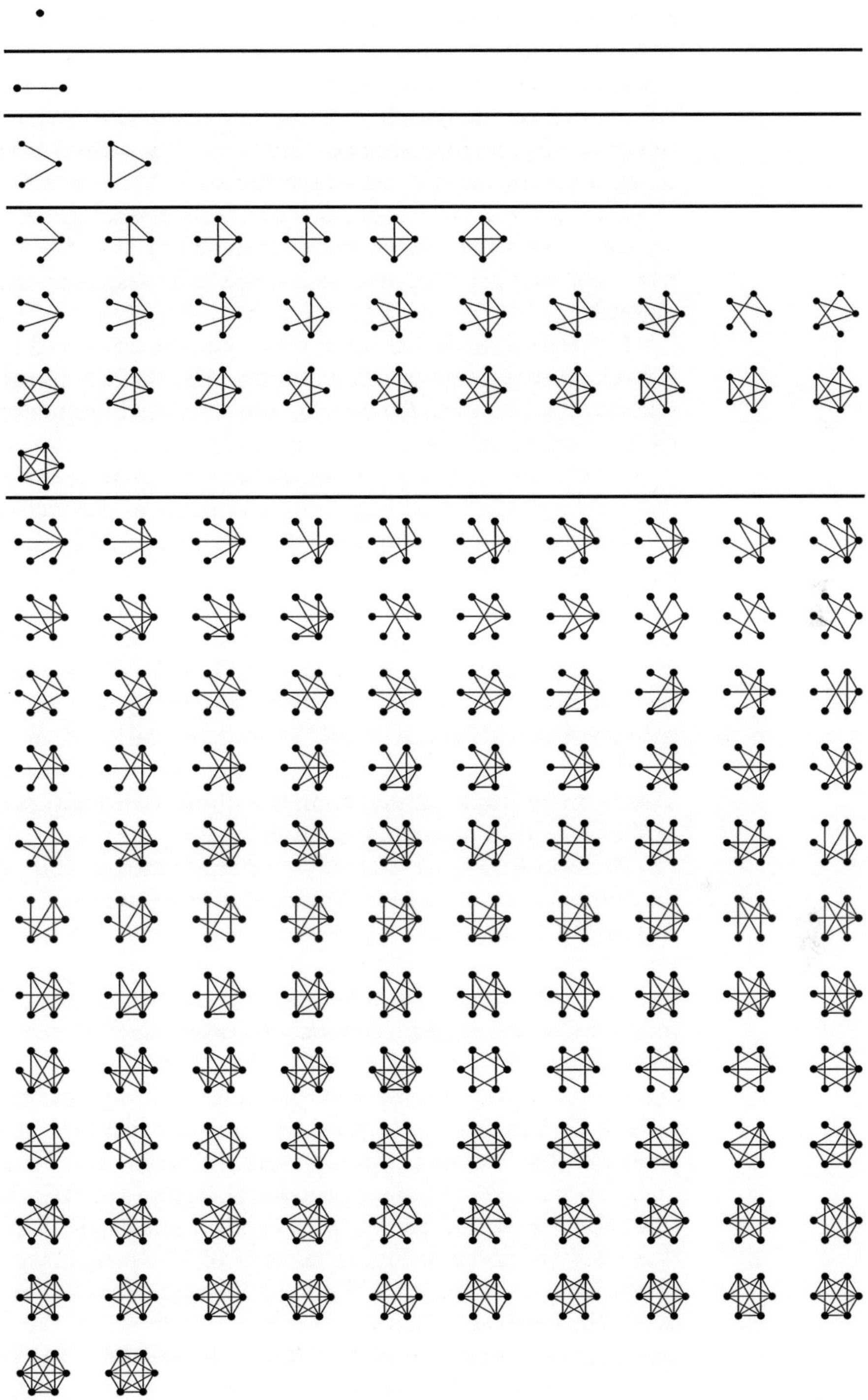

Fig. 5. All graphlets of size 1–6. All possible undirected networks without loops or multiple edges of sizes 1–6.

sample the graphlets on the boundary of the network, which is less computationally intensive than sampling the interior, and due to the uniformity of the interior this boundary sampling results in a calculation of the graphlet frequency distribution that is, after linear scaling, a close approximation to what would be achieved if one were to sample the whole network. The variable x determines which nodes are processed to sample the subgraphs. A small x means that the sampled nodes come from near the fringe and have low degree, thus decreasing the time required to run the algorithm.

Of course, the larger the x, the closer the estimated frequency is to the actual frequency distribution. An excellent feature of this algorithm is that even for $x < 1\%$, the estimated frequency is quite close to the actual one.

In Algorithm 4, any part of the network may be sampled as it is not relegated to the fringe. This algorithm is also well-suited to finding the graphlet frequency distribution of PPIs and random geometric graphs.

What is noteworthy once again is how accurate the procedure is for determining the graphlet frequency distribution for PPIs. In ref. 54, it was observed that as few as $|V|/8$ nodes were required as seeds to the above algorithm in order to get a good approximation to networks with about 1,000 nodes and 2,400 edges. In other words, $|V|/8$ samples per n-node, m-edge graphlet were required. This is much less expensive than the number of samples required in the algorithm presented in ref. 57.

As an example, we created a bacterial network (Fig. 6) solely for the purpose of demonstrating the motif search described in Algorithm 2. We began by creating an interaction network representing the known *Escherichia coli* protein interactome (data from ref. 60) with 2,283 proteins and 7,613 interactions. We then focused on the largest connected component of the network comprising 1,917 proteins and 7,310 interactions. To focus our construction, we singled out penicillin-binding protein 2 and grew it to a depth 1 selection, which means that we selected all proteins that interact directly with it. This new subnetwork included penicillin-binding proteins 3, 5, and 6A. We then took this subnetwork and grew it to a depth 2 selection. The network that resulted, given in Fig. 6, consists of 106 proteins and 1,529 interactions. It illustrates the local interaction network for the penicillin-binding proteins. This was all done on the network visualization software tool NAViGaTOR (ver. 12.1.14) (61).

In order to find the motifs consisting of six nodes in Fig. 6, we created a plug-in for NAViGaTOR, which first translates a given network into a format that is understood by FANMOD (58) (Algorithm 2). The plug-in uses the output from FANMOD to incorporate all the motifs as subgraphs, which may be visualized within the given network using NAViGaTOR (http://ophid.

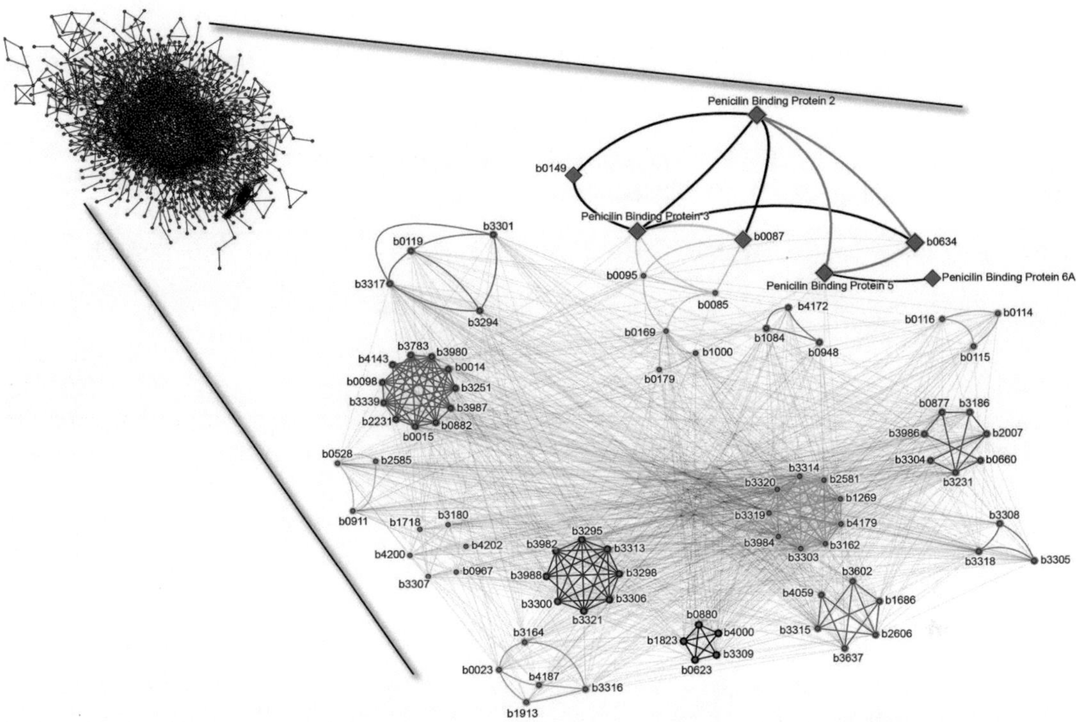

Fig. 6. Identifying dense subgraphs. An *Escherichia coli* protein–protein interaction network comprising 106 proteins and 1,529 interactions representing the local interactome of penicillin-binding proteins 2, 3, 5, and 6A. Data generated from ref. 60. RNSC-predicted complexes in the *E. coli* PPI network from ref. 60 are highlighted. The full network comprises 2,283 proteins and 7,613 interactions, giving it an overall interaction density of 2.9×10^{-3}. Visualization was done in NAViGaTOR (ver. 12.1.14; http://www.ophid.utoronto.ca/navigator) [61]. Color version of Fig. 12.6 as well as NAViGaTOR XML format of the network are provided online: http://www.cs.utoronto.ca/~juris/data/motifs/.

utoronto.ca/navigator). FANMOD also returns a list of nonisomorphic motifs with a significance measure (*Z*-score), and also their frequency within the network, which are presented in Supplemental Table 1 at http://www.cs.utoronto.ca/~juris/data/motifs.

In order to determine the significance of a motif, a comparison is run with the motifs found from 1,000 random graphs with the same node characteristics as the given network. FANMOD computes the significance in terms of a *Z*-score (defined to be the original frequency minus the random frequency divided by the standard deviation). For pedagogical purposes, we chose to illustrate those motifs that were highly significant and had *Z*-scores higher than 200. In addition to the *Z*-score and the frequency of each motif in the given network, the average frequency and standard deviation among all of the randomly generated networks are given.

The last word has not been said with regards to choosing the best model for a particular biological network, but the work on graphlets clearly demonstrates that one should not be too eager to

Table 1
Ten most statistically significant size 6 motifs found in Fig. 6, arranged according to *Z*-scores above 200

ID	Adj.-matrix	Frequency (original) (%)	Mean-freq. (random)	SD (random)	*Z*-Score	*p*-Value
16372449150	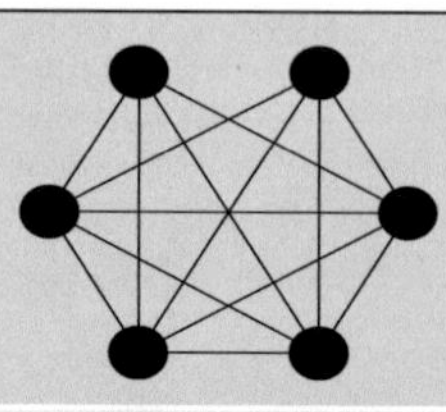	0.04	2.76E−08	5.35E−07	819.07	0
1331541886	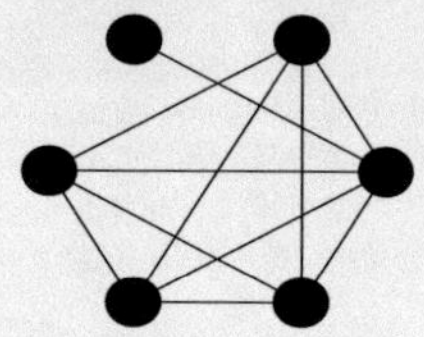	0.19	3.65E−07	5.26E−06	358.33	0
7774125950	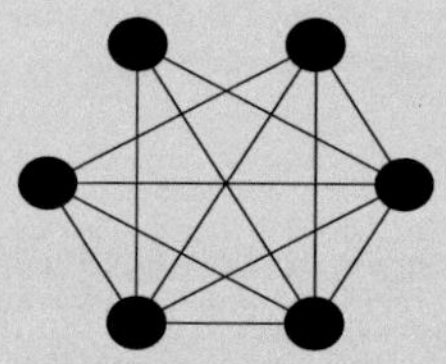	0.13	2.73E−07	3.96E−06	330.25	0
3479027582	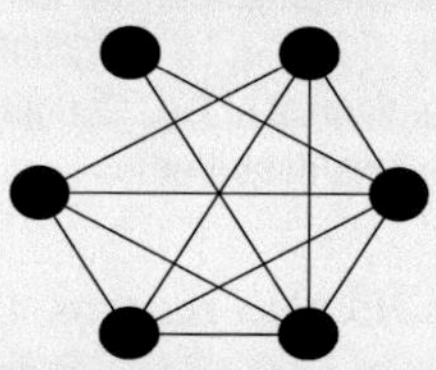	0.15	5.34E−07	7.39E−06	206.35	0
29655678846	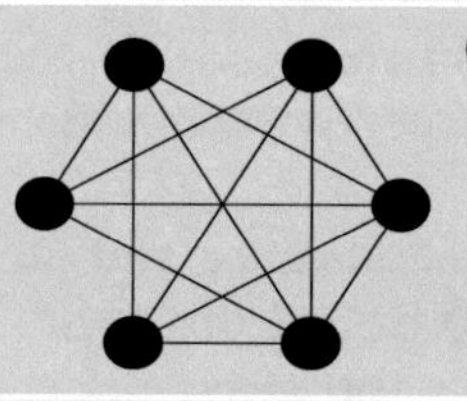	0.06	2.77E−07	2.92E−06	204.83	0
34089189246	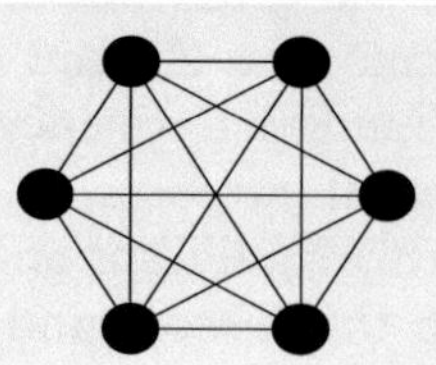	0.01	0%	0	Undefined	0

accept the status quo. A continued exploration of network theory and novel approaches are needed to uncover the methods which will eventually give biologists the tools required to derive real biological facts from the abstract realm of biological networks. We suspect, however, that the vast amount of work done already in other fields of biology, with regards to biological networks, is an asset to bacterial molecular biologists.

Algorithm 3. Time-Limited Node Processing

Input: A network G.

Output: A frequency of each graphlet shown in Fig. 1.

1. Sort the nodes of G by increasing degree and then by decreasing eccentricity, $C_{\in}$. The nodes at the top of this list are of lowest degree and on the boundary of G.
2. Choose a threshold x that determines which top percentage of the list to process. Choose a limiting time T, which determines when the algorithm terminates processing a node.
3. Select a node v from the list. Nodes are chosen in the order they appear in the list.
4. Do an exhaustive search for graphlets that contain v. (Algorithm 1 in the Supplementary Information to ref. 55.) This involves searching for all induced subgraphs of size 3, 4, and 5 involving v.
5. If all the subgraphs are found in time $< T$, then tally the subgraphs, else discard all found subgraphs and ignore v forever.
6. Go to the next node in the list until x% of the list is exhausted.

Algorithm 4. Neighborhood Local Search

Input: A network G and integers n, m, *NUM-MOVES*, and *DIV-FREQth*.

Output: A frequency of each graphlet shown in Fig. 1.

1. Pick a node v in G randomly.
2. Create the set *Neighbors* and place v and all nodes a distance $n - 1$ or less from v into it.
3. Within *Neighbors*, randomly select n connected nodes and call this subset *Subnodes*.
4. Check if the subgraph induced by *Subnodes* in G has m edges. Call this induced subgraph G_s.
5. If $|E(G_s)| = m$, return G_s and stop. Else, search in the neighborhood of G_s for a subgraph with n nodes that is closer to m edges than G_s. This is done as follows: Select a node in the neighborhood of G_s (n-nodes) randomly and a node in G_s and swap them if the number of edges of G_s gets closer to m. Repeat this *NUM-MOVES* times or until a subgraph with m edges is formed.
6. If an n-node, m-edge subgraph is found in less than *NUM-MOVES*, then determine which graphlet in Fig. 1 it is isomorphic to and add to the tally of that particular graphlet.

(continued)

7. While performing step 5, every DIV-FREQth moves and swap a node in G_s with some node in the neighborhood of G_s without concern of an improvement in the number of edges. This is done to help prevent the algorithm from getting stuck in local minima.
8. Repeat the process for some predetermined number of nodes.

5. Concluding Remarks

We have presented a short overview of subnetwork finding algorithms that have become popular in the bioinformatic literature, namely, motif and graphlet search algorithms. Though they are similar in the sense that small subnetworks are being sought, they differ significantly philosophically. The concept of a motif was introduced in order to deal with the extremely expensive task of identifying subnetworks within a network. Motifs are small subnetworks that occur in real networks significantly more often than they do in random networks, and these select subnetworks are used to characterize the real networks in terms of their frequency of appearance. The main criticism of this approach is that the random networks constructed to perform this assessment are scale-free, with little or no attempt to model other characteristics of natural networks. This means that the small set of select subnetworks may exclude important subnetworks. To address this issue, a graphlet approach was introduced which was modeled by geometric random graphs. This has major consequences for subnetwork search algorithms. First off, there is no bias in the subnetworks that one seeks in this setting; all small subnetworks are considered (see Fig. 5). Numerical experiments verified that geometric random graphs model PPIs better than scale-free graphs, in the sense that graphlet frequencies of PPIs follow the distribution of the appearance of graphlets in geometric random networks more closely than they do in scale-free networks. This implies that one has additional structure to work with, meaning that one may assume that PPIs may be modeled by geometric random graphs. This in turn means that when it comes to searching for motifs or graphlets, one can confidently sample the fringe of the PPI and obtain an accurate estimate of the frequency of each subgraph and thereby cut the algorithmic cost significantly. This is due to the fact that geometric random graphs have an interior which have a uniform structure as it consists of uniformly distributed points in a bounded space. Thus, the majority of the information is contained in the fringe.

There is still debate about the best model for PPIs and much work has to be done in order to remedy this ambiguity. For one thing, the proteome is still not understood in its entirety and this causes discrepancies in the topology of the networks constructed, i.e., the connectivity properties. Due to efforts by computer scientists, sophisticated in silico experiments are predicting protein interaction partners and thus, we have reliable information to experiment with. This, of course, cannot take the place of actual biological evidence from the lab.

Future developments in subnetwork search algorithms will also change the way we analyze biological networks. We are still limited by current algorithms and computational power and are currently limited to subnetworks of size 10. As an example, in Fig. 6, we show RNSC algorithm (62, 63) applied to data from ref. 60, and in Supplemental Fig. 1, we show RNSC algorithm applied to data in Fig. 2 from ref. 64. More details on RNSC algorithms are provided online at http://www.cs.utoronto.ca/~juris/data/rnsc.

Improvements in algorithms and our understanding of the global structure of biological networks enable a more systematic and complete analyses. However, a biologically motivated network model is still lacking. It may be the case that some kind of combination between a biologically motivated notion and the aforementioned work on motifs and graphlets provides a superior scheme. One such attempt is given in ref. 65, where the authors presented the duplication–divergence model. The idea is that protein and gene networks evolve through a process of duplicating single genes or proteins and then via some sequence of mutations. This informs a random graph model, where one grows the edge and vertex set as follows: (a) a new node (protein) appears as a duplicate of some randomly chosen node corresponding to the duplication of a gene; (b) this new node takes on all of the connections of the original; and (c) corresponding to the possibility of a mutation, the duplicated node loses some of its edges according to some probability. In this way, a network eventually forms, where genes or proteins that are neighbors are distinct but share common features and functionality. The process outlined in ref. 65 gives networks that are scale-free and whose degree distribution follows a power law. This particular construction suffers certain flaws and a discussion and references may be found in ref. 49; however, this example is meant to highlight the kind of bottom-up modeling that must be considered in order to make progress in this area.

Acknowledgments

This work was supported in part by Genome Canada via the Ontario Genomics Institute, the Canada Foundation for Innovation (grants #12301 and #203383), and CIHR Catalyst grant

(BIO-99745). J.G. is supported by PMH Foundation – Wharton Fund for Head and Neck Cancer Research. G.L. is the Alan Brown Chair in Molecular Genomics; I.J. is the Canada Research Chair in Integrative Computational Biology.

This research was funded in part by the Ontario Ministry of Health and Long-Term Care. The views expressed do not necessarily reflect those of the OMOHLTC.

Authors would like to thank Ali Muhammad for implementing a conversion plug-in for FANMOD algorithm for NAViGaTOR. Supplemental material is provided online at http://www.cs.utoronto.ca/~juris/data/motifs/.

References

1. Wuchty S, Barabasi AL, Ferdig MT. (2006) Stable evolutionary signal in a Yeast protein interaction network. BMC Evol Biol, 6:8.
2. Lappe M, Holm L. (2004) Unraveling protein interaction networks with near-optimal efficiency. Nat Biotechnol, 22:98–103.
3. Schwartz AS, Yu J, Gardenour KR, Finley RL, Jr, Ideker T. (2009) Cost-effective strategies for completing the interactome. Nat Methods, 6:55–61.
4. Wuchty S, Almaas E. (2005) Evolutionary cores of domain co-occurrence networks. BMC Evol Biol, 5:24.
5. Przulj N, Wigle DA, Jurisica I. (2004) Functional topology in a network of protein interactions. Bioinformatics, 20:340–348.
6. Lima-Mendez G, van Helden J. (2009) The powerful law of the power law and other myths in network biology. Mol Biosyst, 5:1482–1493.
7. Yook SH, Oltvai ZN, Barabasi AL. (2004) Functional and topological characterization of protein interaction networks. Proteomics, 4:928–942.
8. Ravasz E, Barabasi AL. (2003) Hierarchical organization in complex networks. Phys Rev E Stat Nonlin Soft Matter Phys, 67:026112.
9. Maslov S, Sneppen K. (2002) Specificity and stability in topology of protein networks. Science, 296:910–913.
10. Yook SH, Jeong H, Barabasi AL. (2002) Modeling the Internet's large-scale topology. Proc Natl Acad Sci U S A, 99:13382–13386.
11. Han JD, Dupuy D, Bertin N, Cusick ME, Vidal M. (2005) Effect of sampling on topology predictions of protein-protein interaction networks. Nat Biotechnol, 23:839–844.
12. Barabasi AL, Albert R. (1999) Emergence of scaling in random networks. Science, 286:509–512.
13. Lieberman E, Hauert C, Nowak MA. (2005) Evolutionary dynamics on graphs. Nature, 433:312–316.
14. Rosvall M, Sneppen K. (2003) Modeling dynamics of information networks. Phys Rev Lett, 91:178701.
15. Zhou H, Lipowsky R. (2005) Dynamic pattern evolution on scale-free networks. Proc Natl Acad Sci U S A, 102:10052–10057.
16. Bhan A, Galas DJ, Dewey TG. (2002) A duplication growth model of gene expression networks. Bioinformatics, 18:1486–1493.
17. Takeuchi F. (2005) Effectiveness of vaccination strategies for infectious diseases according to human contact networks. Jpn J Infect Dis, 58:S16–S17.
18. Stefancic H, Zlatic V. (2005) "Winner takes it all": strongest node rule for evolution of scale-free networks. Phys Rev E Stat Nonlin Soft Matter Phys, 72:036105.
19. Doyle JC, Alderson DL, Li L, Low S, Roughan M, Shalunov S, Tanaka R, Willinger W. (2005) The "robust yet fragile" nature of the Internet. Proc Natl Acad Sci U S A, 102:14497–14502.
20. Schwobbermeyer H, Wunschiers R. (2011) MAVisto: A Tool for Biological Network Motif Analysis. In *Bacterial Molecular Networks.* Edited by Thieffry D, van Helden J, Toussaint A, Springer, New York, 263–280.
21. Erdos P, Renyi A. (1959) On random graphs. I. *Publicationes Mathematicae*, 6:S297.
22. Erdos P, Renyi A. (1960) On the evolution of random graph. *Publ Math Inst Hungar Acad Sci*, 17–61.
23. Bollobas B, Fenner TI, Frieze AM. (1985) An Algorithm for Finding Hamilton Cycles in a Random Graph. In *Proceedings of the 17th Annual ACM Symposium on Theory of Computing*, ACM, Providence, Rhode Island, USA.

24. Milo R, Shen-Orr S, Itzkovitz S, Kashtan N, Chklovskii D, Alon U. (2002) Network motifs: simple building blocks of complex networks. Science, 298:824–827.
25. Przulj N, Corneil DG, Jurisica I. (2004) Modeling interactome: scale-free or geometric? Bioinformatics, 20:3508–3515.
26. Jeong H, Tombor B, Albert R, Oltvai ZN, Barabasi AL. (2000) The large-scale organization of metabolic networks. Nature, 407:651–654.
27. Karagiannis T, Molle M, Faloutsos M. (2004) Long-range dependence: Ten years of Internet traffic modeling. IEEE Internet Computing. Special Issue – Measuring the Internet, 8:57–64.
28. Rzhetsky A, Gomez SM. (2001) Birth of scale-free molecular networks and the number of distinct DNA and protein domains per genome. Bioinformatics, 17:988–996.
29. Lukashin AV, Lukashev ME, Fuchs R. (2003) Topology of gene expression networks as revealed by data mining and modeling. Bioinformatics, 19:1909–1916.
30. Stumpf MP, Wiuf C, May RM. (2005) Subnets of scale-free networks are not scale-free: sampling properties of networks. Proc Natl Acad Sci U S A, 102:4221–4224.
31. Kelly WP, Ingram PJ, Stumpf MPH (2011) The Degree Distribution of Networks: Statistical Model Selection. In *Bacterial Molecular Networks.* Edited by Thieffry D, van Helden J, Toussaint A, Springer, New York, 245–262.
32. Eguiluz VM, Chialvo DR, Cecchi GA, Baliki M, Apkarian AV. (2005) Scale-free brain functional networks. Phys Rev Lett, 94:018102.
33. Stumpf MP, Wiuf C. (2005) Sampling properties of random graphs: the degree distribution. Phys Rev E Stat Nonlin Soft Matter Phys, 72:036118.
34. Toroczkai Z, Bassler KE. (2004) Network dynamics: jamming is limited in scale-free systems. Nature, 428:716.
35. Trusina A, Rosvall M, Sneppen K. (2005) Communication boundaries in networks. Phys Rev Lett, 94:238701.
36. Rosvall M, Gronlund A, Minnhagen P, Sneppen K. (2005) Searchability of networks. Phys Rev E Stat Nonlin Soft Matter Phys, 72:046117.
37. de Aguiar MA, Bar-Yam Y. (2005) Spectral analysis and the dynamic response of complex networks. Phys Rev E Stat Nonlin Soft Matter Phys, 71:016106.
38. Keller EF. (2005) Revisiting "scale-free" networks. BioEssays, 27:1060–1068.
39. Baiesi M, Manna SS. (2003) Scale-free networks from a Hamiltonian dynamics. Phys Rev E Stat Nonlin Soft Matter Phys, 68:047103.
40. Bollobas B, Riordan O, Spencer J, Tusnady G. (2001) The degree-sequence of a scale-free random graph process. Random Struct Algorithms, 18:279–290.
41. Bollobas B, Riordan O. (2002) Mathematical Results on Scale-Free Graphs. In *Handbook of Graphs and Networks.* Edited by Bornholdt S, Schuster HG, Wiley, Berlin, 1–34.
42. von Mering C, Krause R, Snel B, Cornell M, Oliver SG, Fields S, Bork P. (2002) Comparative assessment of large-scale data sets of protein-protein interactions. Nature, 417:399–403.
43. Penrose M. (2003) *Random Geometric Graphs.* Oxford University Press, Oxford.
44. Uetz P, Giot L, Cagney G, Mansfield TA, Judson RS, Knight JR, Lockshon D, Narayan V, Srinivasan M, Pochart P, Qureshi-Emili A, Li Y, Godwin B, Conover D, Kalbfleisch T, Vijayadamodar G, Yang M, Johnston M, Fields S, Rothberg JM. (2000) A comprehensive analysis of protein-protein interactions in *Saccharomyces cerevisiae.* Nature, 403: 623–627.
45. Han JD, Bertin N, Hao T, Goldberg DS, Berriz GF, Zhang LV, Dupuy D, Walhout AJ, Cusick ME, Roth FP, Vidal M. (2004) Evidence for dynamically organized modularity in the yeast protein-protein interaction network. Nature, 430:88–93.
46. Zotenko E, Mestre J, O'Leary DP, Przytycka TM. (2008) Why do hubs in the yeast protein interaction network tend to be essential: reexamining the connection between the network topology and essentiality. PLoS Comput Biol, 4:e1000140.
47. Hoffmann R, Valencia A. (2003) Life cycles of successful genes. Trends Genet, 19:79–81.
48. Reguly T, Breitkreutz A, Boucher L, Breitkreutz BJ, Hon GC, Myers CL, Parsons A, Friesen H, Oughtred R, Tong A, Stark C, Ho Y, Botstein D, Andrews B, Boone C, Troyanskya OG, Ideker T, Dolinski K, Batada NN, Tyers M. (2006) Comprehensive curation and analysis of global interaction networks in *Saccharomyces cerevisiae.* J Biol (Online), 5:11.
49. Mason O, Verwoerd M. (2007) Graph theory and networks in Biology. IET Syst Biol, 1:89–119.
50. Gagneur J, Jackson DB, Casari G. (2003) Hierarchical analysis of dependency in metabolic networks. Bioinformatics, 19: 1027–1034.
51. Awan A, Bari H, Yan F, Moksong S, Yang S, Chowdhury S, Cui Q, Yu Z, Purisima EO, Wang E. (2007) Regulatory network motifs and hotspots of cancer genes in a mammalian cellular signalling network. IET Syst Biol, 1:292–297.

52. Shen-Orr SS, Milo R, Mangan S, Alon U. (2002) Network motifs in the transcriptional regulation network of *Escherichia coli.* Nat Genet, 31:64–68.

53. Artzy-Randrup Y, Fleishman SJ, Ben-Tal N, Stone L. (2004) Comment on "Network motifs: simple building blocks of complex networks" and "Superfamilies of evolved and designed networks". Science, 305:1107; author reply 1107.

54. Przulj N, Corneil DG, Jurisica I. (2006) Efficient estimation of graphlet frequency distributions in protein-protein interaction networks. Bioinformatics, 22:974–980.

55. Kashtan N, Itzkovitz S, Milo R, Alon U. (2004) Topological generalizations of network motifs. Phys Rev E Stat Nonlin Soft Matter Phys, 70:031909.

56. Garey M, Johnson DS. (1979) *Computers and Intractability: A Guide to the Theory of NP-Completeness.* W.H. Freeman and Company, San Francisco.

57. Kashtan N, Itzkovitz S, Milo R, Alon U. (2004) Efficient sampling algorithm for estimating subgraph concentrations and detecting network motifs. Bioinformatics, 20:1746–1758.

58. Wernicke S. (2006) Efficient detection of network motifs. IEEE/ACM Trans Comput Biol Bioinform, 3:347–350.

59. Milo R, Itzkovitz S, Kashtan N, Levitt R, Shen-Orr S, Ayzenshtat I, Sheffer M, Alon U. (2004) Superfamilies of evolved and designed networks. Science, 303:1538–1542.

60. Peregrin-Alvarez JM, Xiong X, Su C, Parkinson J. (2009) The modular organization of protein interactions in *Escherichia coli.* PLoS Comput Biol, 5:e1000523.

61. Brown KR, Otasek D, Ali M, McGuffin MJ, Xie W, Devani B, Toch IL, Jurisica I. (2009) NAViGaTOR: network analysis, visualization and graphing toronto. Bioinformatics, 25:3327–3329.

62. King AD, Przulj N, Jurisica I. (2004) Protein complex prediction via cost-based clustering. Bioinformatics, 20:3013–3020.

63. King AD, Przulj N, Jurisica I. (2011) Protein Complex Prediction with RNSC . In *Bacterial Molecular Networks.* Edited by Thieffry D, van Helden J, Toussaint A, Springer, New York, 297–312.

64. Savas S, Geraci J, Jurisica I, Liu G. (2009) A comprehensive catalogue of functional genetic variations in the EGFR pathway: protein-protein interaction analysis reveals novel genes and polymorphisms important for cancer research. Int J Cancer, 125: 1257–1265.

65. Vazquez A, Flammini A, Maritan A, Vespignani A. (2003) Global protein function prediction from protein-protein interaction networks. Nat Biotechnol, 21:697–700.

Chapter 13

The Degree Distribution of Networks: Statistical Model Selection

William P. Kelly, Piers J. Ingram, and Michael P.H. Stumpf

Abstract

The degree distribution has been viewed as an important characteristic of network data. Many biological networks have been labelled scale-free as their degree distribution can be *approximately* described by a power-law probability distribution.

This chapter presents a formal statistical model selection procedure that can determine which functional form, from a collection of specified models, best describes the degree distribution of network data. The degree distribution found for empirical data is viewed as belonging to a class of probability models and the model which best describes the data is determined in a maximum likelihood framework.

In conclusion, it is important to note that these statistical tests do not confirm the true underlying distribution of the observed data, but instead show which models from a chosen set best describe the data. In reality, these approaches should be viewed as providing evidence for which probability models do not adequately (or optimally) describe the data, and give an indication of the underlying sampling and true interaction properties of the system considered.

Key words: PIN, Protein, Networks, Likelihood, Akaike information criterion, Power-law, Scale-free, Poisson, Model selection

1. Introduction

Networks have been a prominent topic in systems biology for over a decade now (1–4) and beyond. The topological structures that connect proteins, metabolites, and other molecules have been recognised as underlying many dynamic and evolutionary processes of biological systems (5–7), and rapid technological progress in molecular biology and genetics (8, 9) has provided us with vast amounts of data on the possible inter-relationships between different molecules, in particular the rich set of interactions among the proteins in a species (10, 11).

The degree distribution of a network has been a popular summary statistic in the biological arena (12, 13), even though

Jacques van Helden et al. (eds.), *Bacterial Molecular Networks: Methods and Protocols*, Methods in Molecular Biology, vol. 804, DOI 10.1007/978-1-61779-361-5_13, © Springer Science+Business Media, LLC 2012

it will generally not accurately summarise the biologically relevant or important properties of the system (14); this observation has also been made in the context of research into the structural determinants and characteristics of the internet (15, 16).

While previously mere visual inspection of the shape of the degree sequence was used in order to assess its properties, we really require thorough statistical analysis in order to assess whether a network adheres to a particular probability model; or rather which model best fits the observed data. The formal procedures described here are tests to determine the most likely distribution from a clearly (and a priori) defined set of candidate models. Accordingly, as shown by the title of (17) (*Some protein interaction data do not exhibit power law statistics*), these tests are better suited for initial analysis with a view of excluding possible distributions, or to provide comparisons, rather than for determining the nature of a particular model, see Note 1.

This chapter is concerned only with methods for studying the degree distribution of a network (as previously described in refs. 18–20). For a given probability model, the parameters which describe the degree data best are found and then the models are compared in order to assign which one best describes the empirical observations. Model selection in this sense is always contingent on the available data and the set of candidate models considered. A likelihood approach is used to assess to what extent data are described by a given probability model. Section 1 introduces the concepts which are employed later in Section 2. Case study data, and where it can be downloaded, are outlined in Section 3.

1.1. Notation

To motivate the methods, suppose that a collection of data, D, has been found which can be viewed as a simple, undirected network. The network, $N \sim (\nu, \varepsilon)$, has nodes, $\nu = \{\nu_1, \nu_2, \ldots, \nu_n\}$ and edges, $\varepsilon = \{e_1, e_2, \ldots, e_m\}$, between distinct pairs of nodes from V. The numbers of nodes with given degree, $k = \{k_0, k_1, k_2, \ldots, k_{n-1}\}$, in the network are considered as independent observations (where their sum is an even number). For the interaction datasets considered here, which are a collection of interaction reports, we only have data on proteins with at least 1 interaction, so set $k_0 = 0$. For a degree distribution model, M, the probability of the degrees, K, and model parameters, θ, $P(k; \theta)$ is considered. The likelihood, $L(\theta)$, of the parameters (for the model) is considered and then maximum likelihood estimation (MLE) is used to find the parameters, $\hat{\theta}$, which best characterise (in a maximum likelihood sense) the empirical data.

1.2. Maximum Likelihood Inference

To construct a likelihood function for the degree distribution, a candidate model for the degree of each node is proposed. In order to allow easier likelihood calculations the simplifying assumption is made that the node degrees are regarded as independent, and a discrete probability distribution (see Note 2), $P(k; \theta)$, is defined such that

$$\sum_{i=1}^{n-1} P(i; \theta) = 1, \tag{1}$$

where n is the number of nodes in the network (see Note 3), and $P(I; \theta) >$ for all $i \in [1, n\text{–}1]$. While not strictly realistic, this simplifying assumption is made to enable tractable likelihood analyses on large complex datasets.

The composite likelihood of the model, M, given the observed positive degrees, $k = \{k_1, k_2, \ldots, k_{n-1}\}$ and the candidate degree distribution shown in Eq. 1, is defined by,

$$\begin{aligned} L(\theta) &= \mathbb{P}(K; \theta) \\ &= \prod_{i=1}^{n-1} \mathbb{P}(k_i; \theta). \end{aligned} \tag{2}$$

This is true as independence of each node is assumed and thus the product of the probabilities for each node makes up the complete composite likelihood for the observed network data. The log-likelihood is used throughout this chapter, and is defined as,

$$\log(L(\theta)) = \sum_{i=1}^{n-1} \log(\mathbb{P}(k_i; \theta)). \tag{3}$$

The maximum likelihood estimate (MLE), $\hat{\theta}$, of the parameter set θ maximises the value of Eq. 3 for the given model. This relates to the parameter values for which the observed data is more probable than for any other possible parameter value. The likelihood values are then used in order to choose between the degree distribution models considered.

1.3. Model Selection

The main aim of this analysis is to evaluate which model (of our candidate set $\{M_1, \ldots, M_j\}$) best describes the empirical degree data. This could be achieved by simply using the likelihoods found for the MLE parameters for each model. However, this does not take account of the number of parameters (the complexity) of each model, and thus is a poor means of inferring the best model. The Akaike- information criterion (AIC) is used here to choose between the different models; it is defined as,

$$\text{AIC} = 2\left(-\log\left(L\left(\hat{\theta}\right)\right) + d\right), \tag{4}$$

where d is the dimension of the parameter vector required to define the model and $\hat{\theta}$ is the MLE for the model parameters. The AIC biases against overly complicated models so that a model with more parameters is only accepted as superior if it captures sufficiently more information about the observed data than a simpler one.

Thus, the AIC aims to discover the model which best explains the observed data with a minimum number of parameters. Assuming the same number of parameters, this is the distribution which best fits the distribution of observed data based on the Kullback Leibler-divergence (21).

The Bayesian information criterion (BIC) can also be used to assess the possible models, an alternative measure that takes account of the number of samples (see Note 4) and, unlike the AIC, it is unbiased as the amount of data approaches the limit of infinity.

For each of the models, M_i, the MLE ($\hat{\theta}_i$) parameters and the AIC (A_i) can now be found. Comparisons between the different models are all made by comparing the relative differences,

$$\Delta_i = A_i - \min_j (A_j) \tag{5}$$

where i refers to the ith model. Equation 5 defines the relative differences between the AIC for each model. This is now used to define the likelihood for each model i conditional on the data and the model set (22):

$$L(M_i|D, M_1, \ldots, M_J) = exp\left(-\frac{\Delta_i}{2}\right). \tag{6}$$

These likelihoods are normalised to produce Akaike weights, w_i, for each model ($i \in \{1,2\ldots, J\}$),

$$w_i = \frac{exp(-\frac{\Delta_i}{2})}{\sum_{j=1}^{J} exp\left(-\frac{\Delta_j}{2}\right)}. \tag{7}$$

The Akaike weight, w_i, can be interpreted as our belief (or the relative probability) that model i is the best representation of the observed data (from the J different candidate models). The support is found using the ratio of these weights and the analysis must be repeated if the set of models is altered.

1.4. Study Case 1: Bacterial Interactome Data

A variety of protein interaction datasets are available for bacterial organisms (see Chapter 2 by Bouveret and Brun, for an overview of bacterial interaction data). For the case study considered in this chapter, high-throughput yeast two-hybrid data from *Campylobacter jejuni* are used (23). *C. jejuni* is a foodborne pathogen which is a major cause of gastroenteritis. The data consist of 11,687 distinct interactions from approximately 80% of the predicted proteome, and these are used as an approximation for the complete interactome in the following case study.

1.5. Study Case 2: Yeast Interactome Data

Yeast protein interaction data provide a second case study of the methods described above. *Saccharomyces cerevisiae* has approximately 6,000 different proteins (24), and over 100,000 (redundant) reports of putative interactions (25). Two study datasets are used: the first comes from an early yeast two-hybrid experiment on budding yeast (8) and the second from a more recent interactome study (26) to complement the analysis performed on the *C. jejuni* interactome.

The methods described above are implemented on some protein interaction data using an extension called netZ for R. As well as the two case study datasets there is also a further selection of experimental interaction datasets from published sources included with the package netZ, which is now introduced.

2. Material

2.1. Software

The statistical analyses described in this chapter can be completed using tools contributed to the statistical computing program R which are available from `http://cran.R-project.org` (see "Network analysis with R" Chapter 19 by Le Meur and Gentleman this volume). There is also a dedicated R extension which deals with the task of evaluating the likelihood of particular standard probability models for network data, `netZ`. This extension is used throughout the section and the output which can be found for any novel network data. `netZ` can be downloaded from

`http://www.imperial.ac.uk/theoreticalsystemsbiology/data-software`.

The examples which follow assume that the user has a version of R (at least version 2.7), `netZ`, and the `gsl` package installed.

2.2. File Formats

The R extension `netZ` requires the input to be provided as an edge list. Comma separated files are the simplest means of inputting network data as shown in Fig. 1. The only requirement for the majority of the functions is that the `bait` and `prey` for each found interaction can be provided as distinct vectors, for instance as `edges = c (bait, prey)`, where `bait[i]` and `prey[i]` have been seen to interact in the system of interest. Any list of the interactions can be used with any

```
Cj0442,Cj0180
Cj0335,Cj0180
Cj0326,Cj0715
```

Fig. 1. An example section of an input file, taken from the file `cjejuni.rda` stored within the `netZ` package. This shows the expected format for input network files, although the protein identifications can be any alphanumeric string.

type of alphanumeric identification (e.g. SGD tags) for each node. Self-interactions and repeated data can also be included although they will be ignored when analysing the possible degree distributions for the network data.

2.3. Sample Files

`netZ` contains a collection of sample data files, stored in the folder `netZ/data/`. These can also be accessed through the examples in the R extension using the `data()` command to load each of the network edge sets. The sample data files form edge lists, for protein interactions, reported in a selection of different articles, as well as the case study data. The *C. jejuni* data used is stored in the file `cjejuni.rda` and there are also a collection of data files from *S. cerevisiae* experiments that are named via their PUBMED identification numbers (`http://www.ncbi.nlm.nih.gov/pubmed/`) from BioGRID (27).

3. Methods

For the purpose of illustrating the techniques, the considered data are *assumed* to be representations of the complete network of interest (see Note 5).

The following describes how to use net Z to analyse the degree distribution of any network set using the case study data from *C. jejuni* (23) and *S. cerevisiae* (8).

1. If gsl has not been installed within R then run install. `packages(pkgs = "gsl")` before installing netZ.
2. Download and install netZ using the instructions at: `http://www.imperial.ac.uk/theoreticalsystemsbiology/data-software`
3. Start R and load `netZ`:

   ```
   library(netZ)
   ```
4. Load the yeast two-hybrid dataset by Parrish et al. (23), which can also be found in the package at `/netZ/data/cjejuni.rda`, and assign the variable `Y2H`:

   ```
   data(cjejuni) Y2H <- cjejuni
   ```
5. Print out first 5 items of this edge list to see the format `netZ` requires for the further analysis.

   ```
   Y2H[1:5,]
   ```
6. Store the degree sequence, and other sampling information for the dataset, noting that in this case there are no redundant data in the dataset:

   ```
   Y2Hdeg <- degseq(Y2H)
   ```
7. `Y2Hdeg` stores degree sequence information for the interaction data. This includes three different lists: `degree`, `indegree`, and

outdegree. This enables the likelihood analysis to be applied to either directed or undirected networks. In the case of directed networks, the input (bait, prey) data are assumed to be directional, where the in-degree of each protein reflects the number of occurrences in the first vector (bait), while the out-degree reflects the number of times each protein occurs in the second vector (prey). The undirected degree sequence is used to find the likelihoods and Akaike weights now. It can be easily accessed using:

```
Y2Hdeg$
# The number of proteins with one edge:
Y2Hdeg$degree[1]
```

8. Plot the degree distribution Y2Hdeg$degree using both frequency-degree and rank-degree plots on a log-log scale. Both plots provide an illustration of the best fit parameters for the tested models. These plots show some of the properties of the dataset, which features a large proportion of proteins with low degree, and a small proportion of hub proteins; by themselves, however, they do not allow us to decide which probability model best describes the degree distribution:

```
plot.freqrank(Y2Hdeg$degree,freq = TRUE)
```

9. Calculate the likelihoods for the standard degree distribution models as shown in Table 1. These are powerlaw, poisson, exponential, lognormal, and stretchedexp (stretched exponential).

```
Y2Hlike <- mmi.likelihoods(Y2H)
```

10. The Y2Hlike variable contains information for the best fit parameters and likelihoods for the five tested degree distribution models.

```
#The complete variable:
Y2Hlike
#The different outputs stored in the likelihood
   output: names(Y2Hlike)
```

dseq and sampled provide information about the underlying dataset. These detail the number of nodes in the network and the degree sequence being analysed. The remaining five attributes (poisson, exponential, powerlaw, lognormal and stretchedexp) provide the maximum likelihood and associated parameters for each degree distribution model, for instance for the log-normal distribution:

```
# Parameters:
Y2Hlike$lognormal$var
#Maximum likelihood (for the parameter values
   above): Y2Hlike$lognormal$lik
```

Table 1
Degree distributions. The degree distribution models are discretised forms of the $\mathbb{P}(d(v_i) = k;)$ shown here. In netZ, the param vector denotes the shown here for poisson, exponential and power law models. For the log-normal and stretched exponential models, where two parameters are used to define the distribution, `param` $= (\sigma, \mu)$

Distribution	$\mathbb{P}(d(v_i) = k; \theta)$
Poisson	$\frac{\theta^k}{k!} exp(-\theta) : k \geq 0$
Exponential	$\theta exp(-\theta k) : k \geq 0$
Power law	$\frac{k^{-\theta}}{\zeta(\theta)} : k > 0$
Log-normal	$\frac{1}{\sqrt{2\pi k\sigma}} exp\left(-\frac{(\log(k)-\mu)^2}{2\sigma^2}\right) : k \geq 0$
Stretched exponential	$\frac{\sigma}{\mu}\left(\frac{k}{\mu}\right)^{\sigma-1} exp\left(-\left(\frac{k}{\mu}\right)^{\sigma}\right) : k > 0$

11. MLEs for the parameters can now be used to show fits to the degree sequence data. The rank-degree and frequency-degree plots are now reproduced with the best fit lines from the different models and are shown in Fig. 5:

```
par(mfcol = c(1,2))
plot.modelfreq(Y2H, Y2Hlike, title = "")
plot.modelrank(Y2H, Y2Hlike, title = "")
par(mfcol = c(1,1))
```

12. The likelihoods and MLE for the parameters for each model are now used to determine the Akaike weights for the considered models. The following function displays the results (as shown in Fig. 6) of both the likelihood analysis and the Akaike weights for the considered models:

```
plot.modelsummary(Y2Hlike)
```

13. If the Akaike weights or the probability density of the Akaike weighted degree distribution are required explicitly, they can be found using:

```
Y2Hakaike <- mmi.akaike(Y2Hlike,mmi = TRUE)
print(Y2Hakaike)
```

14. The probability density functions for each of the discrete models tested at their MLE parameter values, can also be viewed in order to compare them to the Y2H data and each other. The degree sequences could also be shown on a log axis by passing the plot argument `log = x` to the same function:

```
#Degree sequences for the fitted models and input
   data:          plot.degreesequences(Y2Hlike,
   Y2Hakaike,Y2Hdeg)
```

```
Now the same plot with a log axis: plot.degreese-
   quences(Y2Hlike,Y2Hakaike,Y2Hdeg, log = x)
```

The above example can also be completed using a demo stored in `/netZ/demo/` or by running demo(Y2H). This analysis can be completed on other datasets stored in the R extension, including the early yeast two-hybrid data from *S. cerevisiae* (8). This can be performed using demo(ITO) or by replacing the commands in step 4 above with:

```
data(pm10688190)
Y2H <- pm10688190
```

3.1. Application of the Method to C. jejuni Interaction Network Data

The *C. jejuni* data (23) contains information about 1,091 different proteins. The complete set has 3,209 protein interactions. Figure 2 shows two different degree sequence plots for this dataset. Figure 5 shows the degree sequence for this dataset along with the best fits for each of the assessed degree distributions (which have been truncated to reflect the number of possible proteins and discretised to form proper densities over the possible degrees – from 1 to 1,090 – for each protein). A frequency-degree plot, shown on the left, does not allow us to determine which model best characterises the data, although it may give the impression that various probability distributions fit the data well (as seen in Fig. 5). The right hand plot shows the same information for a rank-degree plot. In this plot, each node is assigned a rank based on its place in an ordered list of the degrees of all nodes. The rank-degree plot highlights how well each distribution fits the highly connected nodes (the hubs) better than the frequency-degree plot, showing how the best fit power law parameters predict a higher number of hubs than are apparent in the real data.

Figure 3 shows the likelihood results for the standard models implemented in netZ, and the Akaike weights (as determined by AIC) for each of the models. Of the five candidate models, the log normal distribution fits best, with an Akaike weight of close to 1. The stretched exponential model (which tends to the log normal for appropriate choice of parameters) attracts the second highest weight, although all the other models have insignificant weights

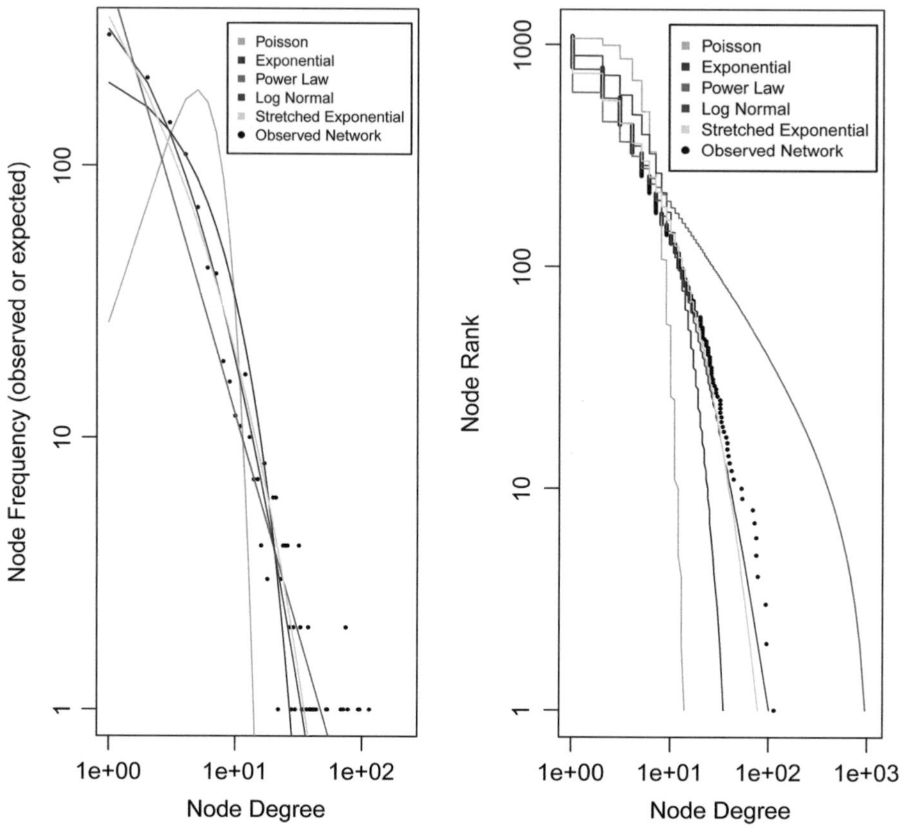

Fig. 2. The frequency-degree and rank-degree plots for the *C. jejuni* interaction data are shown overlaid with the best fit lines for each of the tested degree distribution probability models.

attached to them. Accordingly, it can be said that of the assessed distributions, the *C. jejuni* data is best represented as belonging to a truncated log normal probability density with parameter values 0.84 and 1.20 (found from `Y2Hlike$lognormal` in the example). Use of the BIC rather than the AIC has little impact on the results (see Note 4 for implementation instructions).

Finally, Fig. 4 shows the probability densities for the MLE parameters of each of the distributions (and the final Akaike-weighted distribution). For the data, the best fit power law distribution in general results in a higher number of singleton proteins (with degree 1) and too great a number of proteins with high degrees (greater than 5) than are seen in the data. The random and exponential distributions, using MLE parameters, underestimate the number of proteins with degree 1 while also overestimating the proportion of proteins with intermediate degrees. This shows the contrast between the so-called normal and fat-tailed distributions. However, the log normal and stretched exponential distributions produce better overall fits to the observed data.

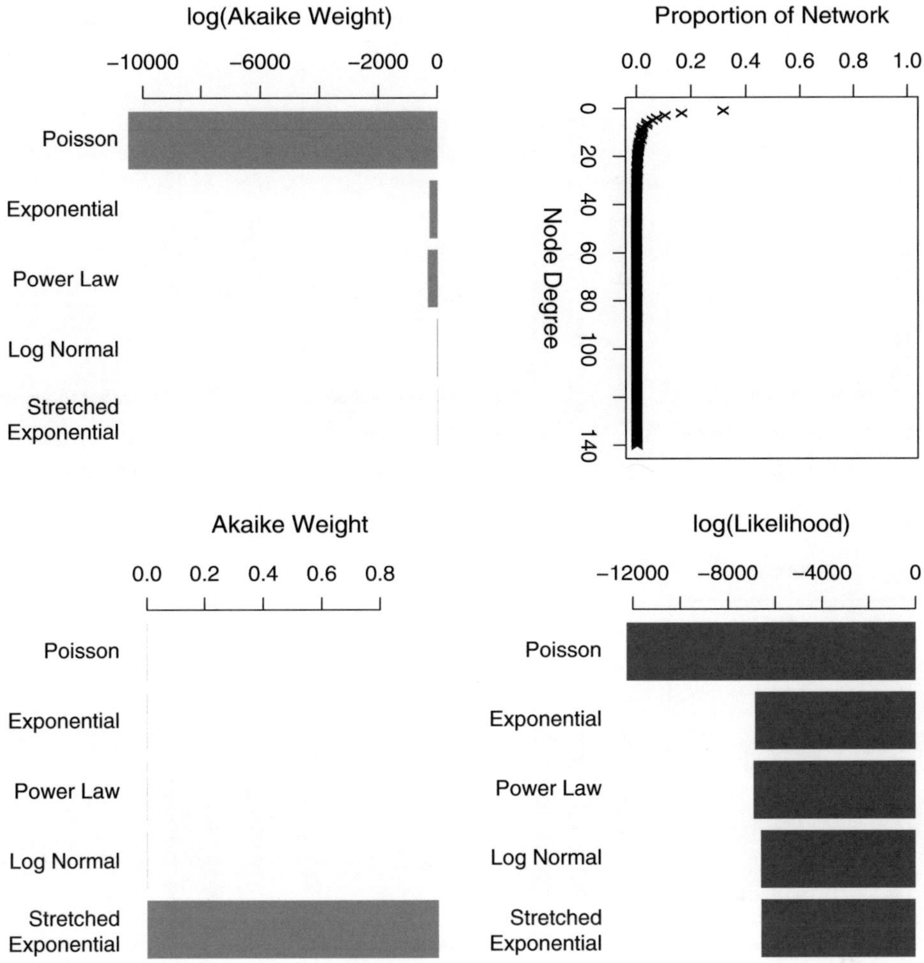

Fig. 3. The degree sequence for *C. jejuni* interaction data is shown alongside the likelihoods calculated for each of the standard models and the Akaike weights which determine the model selected to be represent the data.

3.2. Application of the Method to S. cerevisiae *Data*

The *S. cerevisiae* network dataset (8) contains 735 different interactions among 766 distinct proteins. These interactions were reported in an early yeast two-hybrid experiment. For the degree sequence analysis, there are 734 different interactions (as a single self-interaction is excluded) (Fig. 5).

Figure 6 shows the likelihood results for the *S. cerevisiae* data. Once again the log normal distribution fits best (which is found to be true for most yeast two-hybrid datasets of reasonable size), with an Akaike weight of more than 0.8. The stretched exponential model attracts the second highest weight. Unlike in the first example, the weight attached to this second distribution is far higher for this dataset, which is a reflection of the more small-scale nature of this second interaction dataset. Overall, the yeast data is best

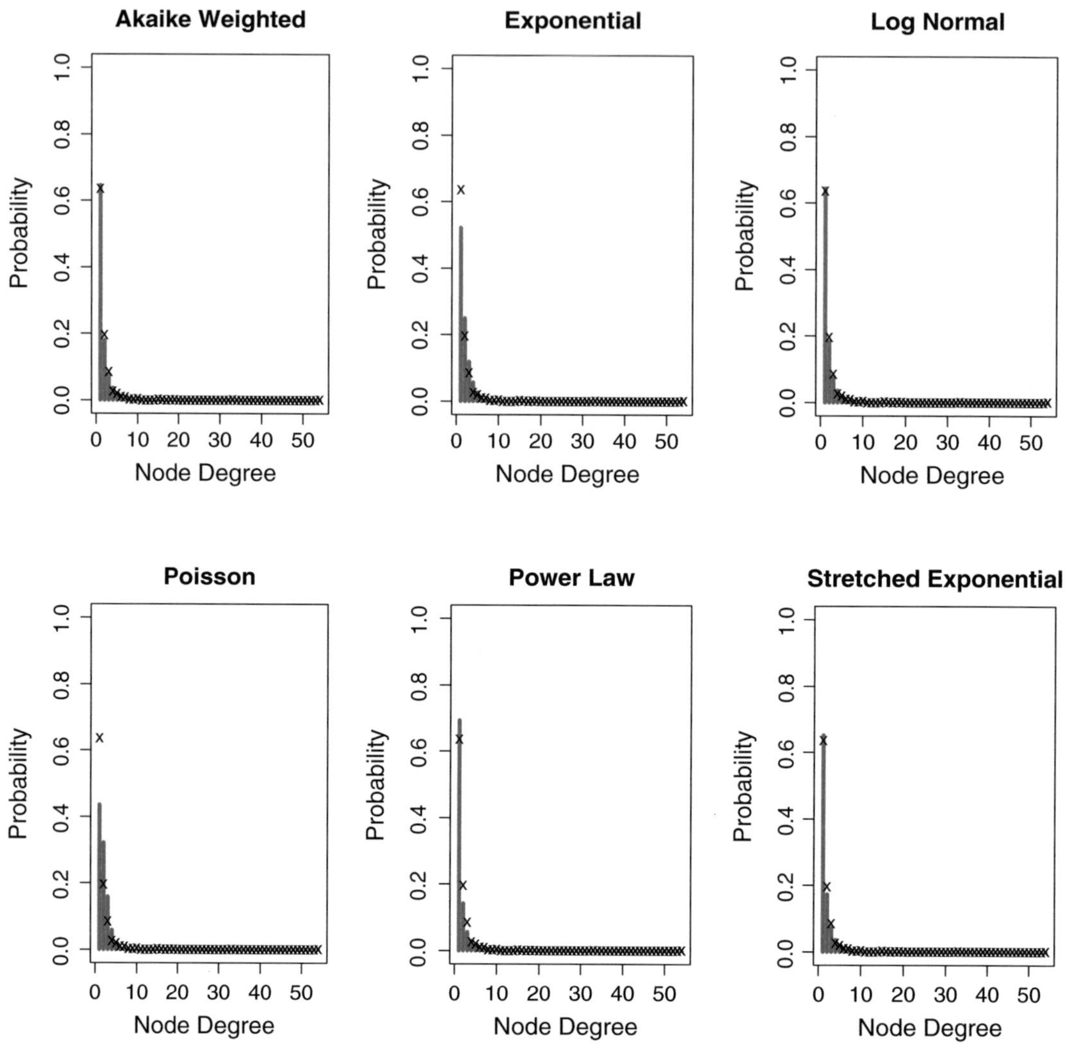

Fig. 4. The probability densities for the MLE parameters of each of the standard models are shown, alongside the Akaike weighted degree distribution. Overlaid on each plot is the actual observed degree data found for the *C. jejuni* interaction data.

represented as belonging to a truncated log normal probability density with parameter values −0.82 and 1.19.

Finally, a second yeast dataset from a more recent large proteome study is shown in Fig. 7. These data are best represented by the stretched exponential distribution. The Akaike weight for this network gives this distribution a weight of close to 1 (access after running `demo(KROGAN)` by using `KROGANakaike $stretchedexp`). Accordingly, of the five models, we may say that the data are best approximated using the stretched exponential distribution.

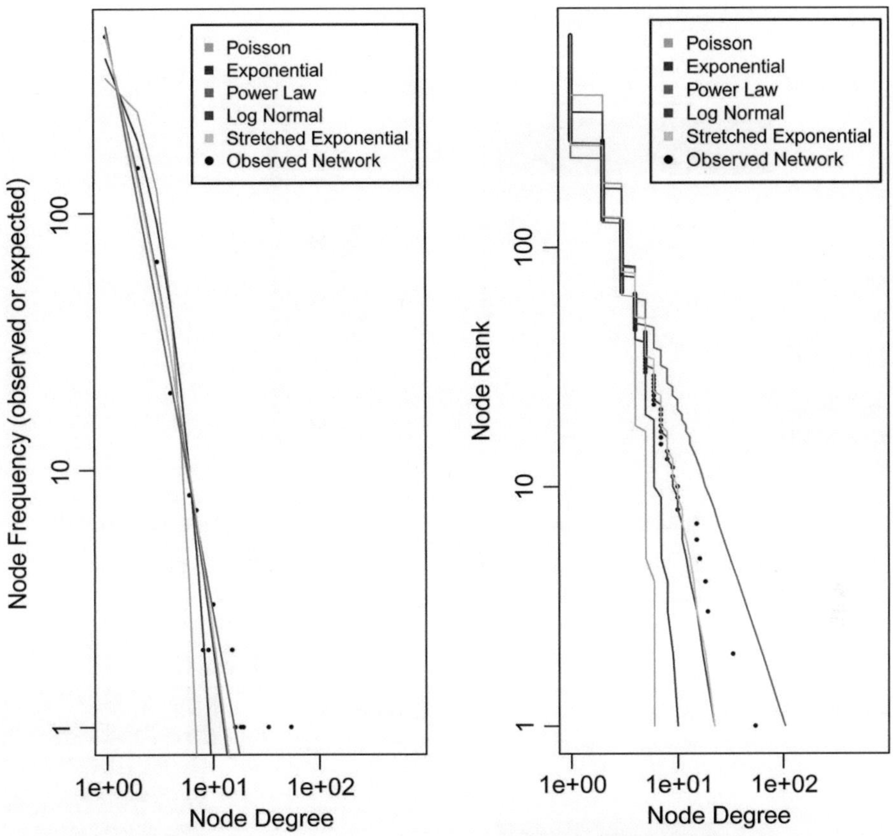

Fig. 5. The frequency-degree and rank-degree plots for the *S. cerevisiae* yeast two-hybrid interaction data are shown overlaid with the best fit lines for each of the tested degree distribution probability models.

This is a perspective on the best fit for a given selection of models, rather than an assessment of how well the actual models fit the data. Applying a goodness-of-fit procedure on top of model selection would be double-dipping into the data and is not justifiable. While it is in principle possible to use goodness-of-fit measures for basic model selection, this framework is really only suitable to reject unrealistic models. Ranking competing models in view of such approaches is problematic and should always be avoided in favour of formal model selection procedures.

4. Notes

1. Frequency-degree logarithmic plots have often been used in the literature for a rudimentary analysis of the data and to detect power-law properties. However, the observation that power-law distributed data appears as a straight line on these

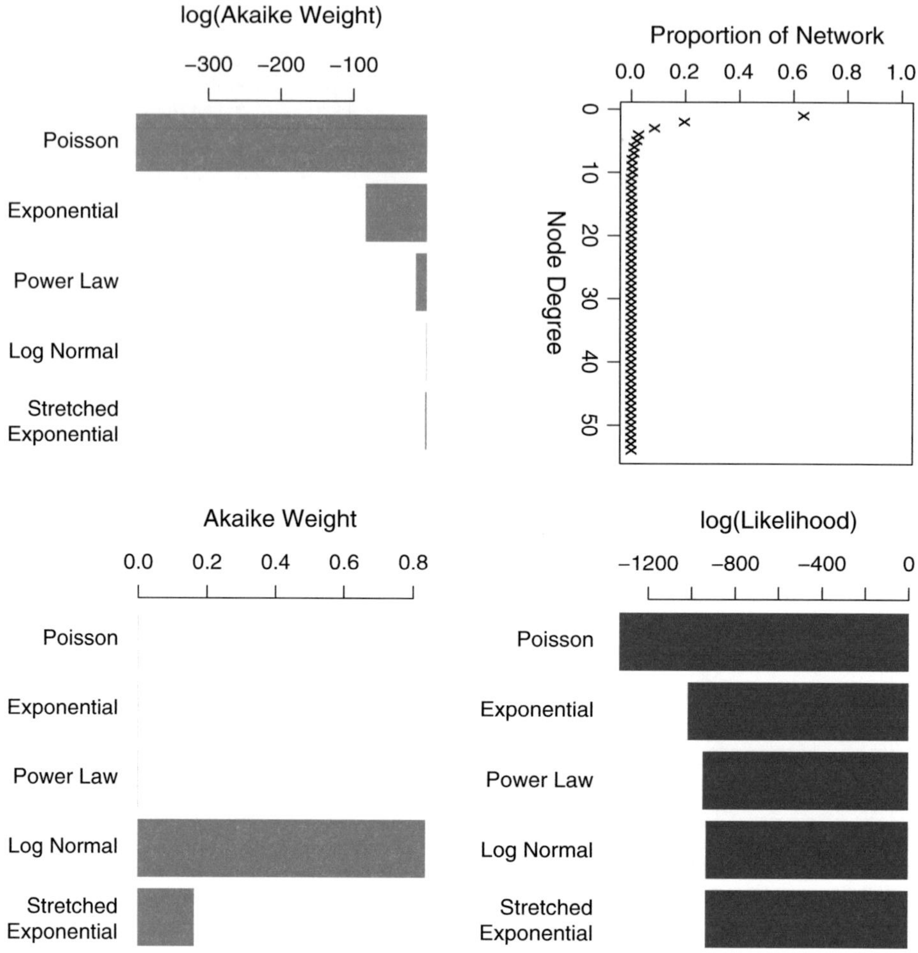

Fig. 6. The degree sequence for *S. cerevisiae* yeast two-hybrid interaction data is shown alongside the likelihoods calculated for each of the standard models and the Akaike weights which determine the model selected to represent the data

plots could be misleading as the same property can be exhibited for exponentially distributed data (as technically described in ref. 17). To combat this possible misinterpretation, these diagnostic plots should always show the node degree against the node's rank (so for a node list $\{v_1,\ldots,v_n\}$ where $d_i \geq d_{i+1}$ the rank of node v_j is j). These plots should also follow a straight line if the data from drawn from a power-law probability distribution, but not for exponentially distributed data.

2. The degree distributions used in `netZ` are truncated, discrete probability distributions found from a selection of continuous probability density functions. The functions used are: Poisson, exponential, power-law, log-normal, and stretched exponential. The first three distributions have one parameter, while the

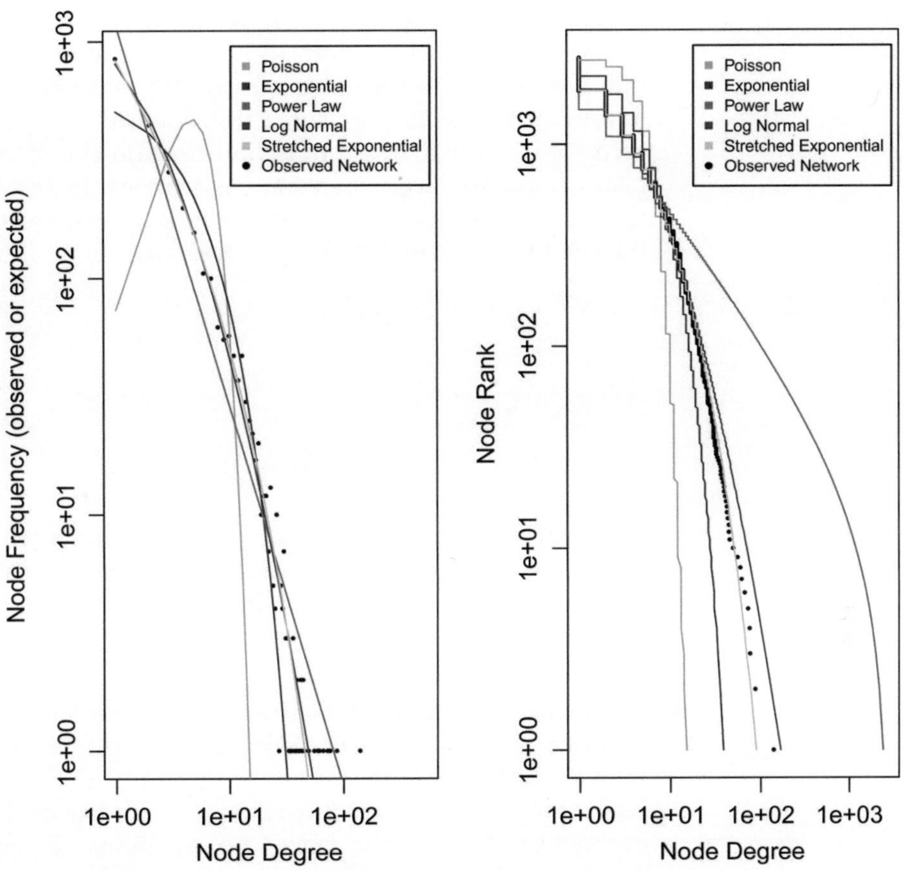

Fig. 7. The frequency-degree and rank-degree plots for the *S. cerevisiae* large-scale Krogan et al. data are shown overlaid with the best fit lines for each of the tested degree distribution probability models.

log-normal and stretched exponential distributions have two parameters (the functions can be found in the help file in `netZ` for `distributions`). In order to produce a degree distribution for the network nodes, these functions are both truncated (so that they can only take values between 0 and $n - 1$ when there are n nodes in the considered network) and discretised (as the degree must be an integer). This is completed by taking any given distribution defined on a superset of the integers, $P_R(K,\theta)$ and then defining the degree probability distribution for the node as,

$$\mathbb{P}(K;\theta) = \frac{\mathbb{P}_R(K;\theta)}{\sum_{i=0}^{n-1} \mathbb{P}_R(i;\theta)}, \tag{8}$$

for any integer $K \in [0,n - 1]$.

3. In general, protein interaction data are a collection of positive reports about the existence of binding between pairs of proteins. Accordingly, it may not be possible for an observed

protein to have degree zero. Indeed, the data only provide a view regarding those sampled proteins which have at least one interaction. When using subsampled network data, it is important to define the degree distribution in terms of the seen data, rather than assuming information about the complete proteome. So, if there are N proteins in the studied proteome but only n proteins found in the interaction dataset, then our probability distribution model is,

$$\sum_{i=1}^{n-1} \mathbb{P}(i;\theta) = 1, \tag{9}$$

rather than,

$$\sum_{i=0}^{N-1} \mathbb{P}(i;\theta) = 1. \tag{10}$$

This distinction is also important when assuming subsampling as it is important to define the complete network in terms of how many proteins exhibit at least one interaction, rather than just the number of different proteins found in the organism's proteome (see Note 5).

4. As an alternative to the AIC, the BIC can be used to find the Bayesian equivalent to the Akaike weights, the so-called Schwarz-weights (28). This takes account of the number of nodes found in the data as well as the number of different parameters. For `Y2H` it makes no significant difference to the output, although for very large networks it may change the results. The BIC is defined as

$$\mathrm{BIC} = -2\log\left(L\left(\hat{\theta}\right)\right) + d\log(n), \tag{11}$$

where n is the number of nodes found in the network. This is a slight alteration to the AIC which is found in Eq. 4. To use the BIC to find weights in either `plot.modelsummary` or `mmi.akaike` add the code `IC = "BIC"`. For the `COMP` dataset use: `plot.modelsummary(COMPlike,IC = "BIC")`

5. In reality, the interaction data considered are in general subsamples from a complete interaction network. Therefore, the methods are being used to determine the most likely degree distribution for only the provided data, rather than the complete network. The results can also be viewed as providing information about the different experimental techniques which have produced the data. Alternatively, subsampled network data can also be used and the complete degree sequences assessed using `netZ` (see ref. 20). A binomial model for the sampling is assumed, and this will alter the MLE parameters found from the calculations for a given dataset, dependent on the size of the

complete network. If the complete network is thought to have n nodes with at least one interaction (and the observed data has less than n nodes), then the following is used to find the likelihoods for a dataset `net`:

```
sublikelihoods <- mmi.likelihoods(net,nodes = n)
```

Similarly, for the Akaike function, the same alteration is made to the code described for Y2H data:

```
subakaike <- mmi.akaike(net,nodes = n)
```

Acknowledgments

This work has been funded by the Wellcome Trust and the BBSRC. MPHS is a Royal Society Wolfson Research Merit Award holder.

References

1. Sanchez C, Lachaize C, Janody F, Bellon B, Röder L, Euzenat J, Rechenmann F, Jacq B. (1999) Grasping at molecular interactions and genetic networks in drosophila melanogaster using FlyNets, an internet database. Nucleic Acids Res, 27(1):89–94.
2. Xenarios I, Rice D, Salwinski L, Baron M, Marcotte EM, Eisenberg D. (2000) Dip: the database of interacting proteins. Nucleic Acids Res, 28(1):289–291.
3. Legrain P, Wojcik J, Gauthier JM. (2001) Protein–protein interaction maps: a lead towards cellular functions. Trends Genet, 17(6):346–352.
4. Lehner B, Fraser AG. (2004) A first-draft human protein-interaction map. Genome Biol, 5(9):R63.
5. Zhang J. (2003) Evolution by gene duplication: an update. Trends Ecol Evol, 18(6): 292–298.
6. Qin H, Lu H, Wu W, Li W-H. (2003) Evolution of the yeast protein interaction network. Proc Natl Acad Sci U S A, 100(22): 12820–12824.
7. Hakes L, Pinney JW, Robertson DL, Lovell SC. (2008) Protein-protein interaction networks and biology – what's the connection? Nat Biotechnol, 26(1):69–72.
8. Ito T, Chiba T, Ozawa R, Yoshida M, Hattori M, Sakaki Y. (2001) A comprehensive two-hybrid analysis to explore the yeast protein interactome. Proc Natl Acad Sci, 98 (8):4569–4574.
9. Giot L, Bader JS, Brouwer C, Chaudhuri A, Kuang B, Li Y, Hao Y, Ooi C, Godwin B, Vitols E, Vijayadamodar G, Pochart P, Machineni H, Welsh M, Kong Y, Zerhusen B, Malcolm R, Varrone Z, Collis A, Minto M, Burgess S, McDaniel L, Stimpson E, Spriggs F, Williams J, Neurath K, Ioime N, Agee M, Voss E, Furtak K, Renzulli R, Aanensen N, Carrolla S, Bickelhaupt E, Lazovatsky Y, DaSilva A, Zhong J, Stanyon CA, Finley R, White K, Braverman M, Jarvie T, Gold S, Leach M, Knight JR, Shimkets R, McKenna M, Chant J, Rothberg J. (2003) A protein interaction map of drosophila melanogaster. Science, 302(5651):1727–1736.
10. Tarassov K, Messier V, Landry CR, Radinovic S, Serna Molina MM, Shames I, Malitskaya Y, Vogel J, Bussey H, Michnick SW. (2008) An in vivo map of the yeast protein interactome. Science, 320(5882):1465–1470.
11. Kelly WP, Stumpf MPH. (2008) Protein-protein interactions: from global to local analyses. Curr Opin Biotechnol, 19:396–403.
12. Jeong H, Tombor B, Albert R, Oltvai ZN, Barabási AL. (2000) The large-scale organization of metabolic networks. Nature, 407 (6804):651–654.
13. Reguly T, Breitkreutz A, Boucher L, Breitkreutz B-J, Hon G, Myers C, Parsons AB,

Friesen H, Oughtred R, Tong A, Stark C, Ho Y, Botstein D, Andrews BJ, Boone C, Troyanskya O, Ideker T, Dolinski K, Batada NN, Tyers M. (2006) Comprehensive curation and analysis of global interaction networks in *Saccharomyces cerevisiae.* J Biol, 5(4):11.

14. Kelly WP. (2009) On the analysis of protein interaction networks. PhD thesis.
15. Doyle J, Alderson D, Li L, Low L, Roughan M, Shalunov S, Tanaka R, Willinger W. (2005) The "robust yetfragile" nature of the internet. Proc Natl Acad Sci, 102(41):14497–14502.
16. Willinger W, Alderson D, Doyle J. (2009) Mathematics and the internet: a source of enormous confusion and great potential. Not Am Math Soc, 56(5):586–599.
17. Tanaka R, Yi T-M, Doyle J. (2005) Some protein interaction data do not exhibit power law statistics. FEBS Lett, 579(23):5140–5144.
18. Stumpf MPH, Ingram P. (2005) Probability models for degree distributions of protein interaction networks. EPL (Europhys Lett), 71(1):152–158.
19. Stumpf MP, Ingram P, Nouvel I, Wiuf C. (2005) Statistical model selection methods applied to biological networks. Lect Notes Comput Sci, 65–77.
20. Stumpf MPH, Thorne T. (2006) Multi-model inference of network properties from incomplete data. J Integr Bioinform, 3(2):32.
21. Burnham K, Anderson DR. (1998) *Model Selection and Inference: A Practical Information-Theoretic Approach.* Springer, New York.
22. Akaike H. (1983) Information measures and model selection. Bull Inst Int Statist, 50(1):277–290.
23. Parrish JR, Yu J, Liu G, Hines JA, Chan JE, Mangiola BA, Zhang H, Pacifico S, Fotouhi F, DiRita VJ, Ideker T, Andrews P, Finley RL. (2007) A proteome-wide protein interaction map for *Campylobacter jejuni.* Genome Biol, 8(7):R130.
24. Hirschman JE, Balakrishnan R, Christie KR, Costanzo MC, Dwight SS, Engel SR, Fisk DG, Hong EL, Livstone MS, Nash R, Park J, Oughtred R, Skrzypek M, Starr B, Theesfeld CL, Williams J, Andrada R, Binkley G, Dong Q, Lane C, Miyasato S, Sethuraman A, Schroeder M, Thanawala MK, Weng S, Dolinski K, Botstein D, Cherry JM. (2006) Genome snapshot: a new resource at the *Saccharomyces* genome database (sgd) presenting an overview of the *Saccharomyces cerevisiae* genome. Nucleic Acids Res, 34(Database issue):D442–D445.
25. Stark C, Breitkreutz B-J, Reguly T, Boucher L, Breitkreutz A, Tyers M. (2006) Biogrid: a general repository for interaction datasets. Nucleic Acids Res, 34(Database issue): D535–D5396.
26. Krogan NJ, Cagney G, Yu H, Zhong G, Guo X, Ignatchenko A, Li J, Pu S, Datta N, Tikuisis AP, Punna T, Peregrín-Alvarez JM, Shales M, Zhang X, Davey M, Robinson MD, Paccanaro A, Bray JE, Sheung A, Beattie BK, Richards DP, Canadien V, Lalev A, Mena F, Wong P, Starostine A, Canete MM, Vlasblom J, Wu S, Orsi C, Collins SR, Chandran S, Haw R, Rilstone JJ, Gandi K, Thompson NJ, Musso G, St Onge P, Ghanny S, Lam MHY, Butland G, Altaf-Ul AM, Kanaya S, Shilatifard A, O'Shea E, Weissman JS, Ingles CJ, Hughes TR, Parkinson J, Gerstein M, Wodak SJ, Emili A, Greenblatt JF. (2006) Global landscape of protein complexes in the yeast *Saccharomyces cerevisiae.* Nature, 440(7084):637–643.
27. Breitkreutz B-J, Stark C, Reguly T, Boucher L, Breitkreutz A, Livstone M, Oughtred R, Lackner DH, Bähler J, Wood V, Dolinski K, Tyers M. (2008) The biogrid interaction database: 2008 update. Nucleic Acids Res, 36(Database issue):D637–D640.
28. Schwarz G. (1978) Estimating the dimension of a model. Ann Stat, 6(2):461–464.

Chapter 14

MAVisto: A Tool for Biological Network Motif Analysis

Henning Schwöbbermeyer and Röbbe Wünschiers

Abstract

Data from high-throughput experimental methods are currently being used to construct complex biological networks. These include regulatory gene networks, regulatory protein–DNA networks, protein–protein interaction networks, or metabolic networks. Independent of its type, every network can be characterized by a number of parameters such as number of nodes, number of edges connecting nodes, direction and weight of edges, in- and out-degree of nodes, etc. One can draw an analogy of such rather simple network parameters to the primary sequence of proteins or nucleic acids. More insight can be gained by an analysis of the secondary and tertiary structure of biomolecules, which often contain motifs. The same holds for biological networks. The occurrence and frequency of certain motifs or pattern characterize the topology and often the functional space of a network. Here, we describe the utilization of the free software MAVisto, which was designed to mine networks for typical motifs by combining a flexible motif search algorithm with interactive exploration methods and sophisticated visualization techniques.

Key words: Gene-regulatory network, Graph drawing, Information visualization, Network analysis, Network motifs

1. Introduction

Biological networks capture processes in living beings. High-level networks describe processes and interactions in ecosystems. At the other end of the scale are molecular networks describing the interaction between molecules. Thus, a biological network may represent multiple interactions between its components. Graphs, consisting of nodes connected by edges, are the mathematical way to describe a network. Prominent examples are

- Disease–Gene Networks: A graph with two sets of nodes. One set represents all known diseases of a given organism while the other set represents all known genes of this organism. Diseases and genes are connected by an edge if the gene is involved in the disease.

Jacques van Helden et al. (eds.), *Bacterial Molecular Networks: Methods and Protocols*, Methods in Molecular Biology, vol. 804, DOI 10.1007/978-1-61779-361-5_14, © Springer Science+Business Media, LLC 2012

- Metabolic Networks: Here, nodes represent metabolites while edges represent enzyme catalysed biochemical conversions of metabolites. Depending on the energetics of biochemical reactions, edges might be directed.
- Protein–Protein Interaction Networks: Edges represent the interaction of proteins, e.g. in signalling cascades or multi-subunit enzyme complexes, which in turn are represented by a set of nodes.
- Co-Expression Networks: These networks describe the similarity of expression patterns of different genes. Genes, represented by nodes, are connected by edges if they show similar expression patterns, i.e. they are equally expressed under equal conditions.
- Gene-Regulatory Networks: Also called transcriptional networks. They consist of two sets of nodes, one set representing transcription factors and the other set representing genes. Edges represent binding of transcription factors to the gene's promoter or enhancer region.

Pregenomic approaches led to the probably best known biological network: cell metabolism. In the early 1940s, biochemist Donald Nicholson was among the first to put together individual biochemical reactions described by Embden, Meyerhof, Parnass, Cori, and others to what is well known as glycolysis (1). He was probably stimulated by Henry C. Beck's novel schematic map of the London Underground from 1933 (2). By the 1950s dozens of pathways have been described. Yet, it took another decade until the first metabolic pathway chart, drawn by Donald Nicholson, was printed in 1960. Independently, Gerhard Michal developed his metabolic pathway map (which is now available from Roche Applied Science) and later his book (3, 4). Nowadays, databases like MetaCyc (5) and KEGG (6, 7) collect and visualize metabolic pathway data. From the metabolic network shown in Fig. 1 it becomes obvious that some kind of network analysis is necessary to extract important network attributes.

During the past years, biological network inference has been greatly improved by high-throughput application of modern *omics technologies such as genome wide gene-expression analysis or protein–protein interaction detection. This led to the expansion of the biological network concept to co-expression and gene-regulatory networks as summarized above. Moreover, the wealth of data and its integration in the context of systems biology led to expansion of graph theory for biological network analysis. The benefit of having a network graph is twofold. First, the biological network becomes accessible to descriptive analysis. For example, network attributes such as the degree distribution [number of incoming (in-degrees) and out-going edges (out-degree)], hubs (nodes with high in- and out-degree), or connectivity

Fig. 1. Metabolic network derived from 5,706 enzymatic reactions stored in the Kyoto Encyclopedia of Genes and Genomes (KEGG). Metabolites are represented by *nodes* and enzymatic reactions by *edges*.

(edge-to-node ratio) characterize a network and point to important network components. Second, the data can be used for simulation and prediction, both important aspects for, e.g., the rising field of synthetic biology (8).

1.1. Network Motifs

One characteristic property of networks (or graphs) are network motifs. These are usually directed subgraph, consisting of a few nodes or edges. Such motifs are embedded in a larger directed graph. For molecular biological systems nodes may be sets of metabolites, proteins (e.g. enzymes or transcription factors), genes or transcripts, and so forth, while edges are used to model particular (e.g. functional) relations between these nodes. Some network motifs may perform specific regulatory functions (9). However, it can generally not presumed that motifs necessarily fulfil a function. Employed as a topological analysis technique, enumeration of all motifs consisting of a given number of nodes oredges summarizes the local connectivity patterns that compose a complex network (10, 11).

MAVisto (Motif Analysis and Visualization tool) is a free software for the discovery of motifs in any given network (12). It implements an adaptable motif search algorithm (13) and provides a dynamic result view. Furthermore, networks can be created and edited. Network visualization is optimized to create human readable views. This is performed by an advanced force-directed layout algorithm (14) that tries to preserve the motif layout. The motif search algorithm can be set to discover all motifs with a given number of nodes or edges. Three types of motif frequencies and their statistic relevance are returned.

1.2. Motif Frequency

The frequency of a particular motif in a network describes its number of occurrences. However, three different cases can be discriminated.

- Frequency concept F1: All occurrences of a motif are counted, even if nodes and edges are shared between motifs.
- Frequency concept F2: Only nodes but no edges may be shared between motifs.
- Frequency concept F3: Neither nodes nor edges may be shared between motifs, i.e. motifs may not overlap.

The first case captures the full potential of a motif by counting all occurrences of this motif. The covering of nodes by motif occurrences is the basis for a new centrality measure, the motif-based centrality (15).

1.3. Motif Frequency Significance

In statistics, a result is considered significant if the probability that this result has occurred by chance is low. Testing for statistical significance requires the formulation of a *null hypothesis* based on the distribution of these results. If the probability of a given result falls below a predefined threshold the null hypothesis is rejected, and the result is considered as statistically significant. For the analysis of network motifs, a commonly used null hypothesis is based on the frequency distributions of motifs in random networks. It is assumed that the structure of random networks is generated by a process free of any type of selection influencing the distribution of motifs. Thus, statistical significant motifs may occur due to functional constraints or design principles that have shaped the network structure (16).

After motif discovery, MAVisto can calculate p-values and Z-scores based on a randomization method that preserves the degree of all nodes. A widely used algorithm for network randomization arbitrarily reconnects the edges (17, 18). The algorithm rewires two edges $x = \{a, b\}$ and $y = \{c, d\}$ such that a becomes connected to $d(x' = \{a, d\})$ and c becomes connected to $b(y' = \{a, d\})$ without creating duplicated edges. This reconnection step is repeated many times to guarantee that the newly created network is a properly randomized version of the original network.

1.3.1. Z-Score

In order to judge if the frequency of a certain motif in a network is under- or overrepresented, comparisons with random networks are performed. The Z-score has been proposed as a metric of statistical under- or overrepresentation (16). It is defined as the frequency difference of a motif $F(m)$ in a network to its mean frequency $\overline{F_r(m)}$ in a sufficiently large set of randomized networks, divided by the standard deviation $\sigma_r(m)$ of the frequency values of the randomized networks (16) (Eq. 1). A positive Z-score indicates that a motif is overrepresented in the real network, whereas a negative score indicates underrepresentation in the real network compared to random graphs.

$$Z(m) = \frac{F(m) - \overline{F_r(m)}}{\sigma_r(m)}. \tag{1}$$

1.3.2. p-Value

The p-value represents the probability that a motif appears an equal or greater number of times in randomized networks than in the analysed network. Therefore, a very low p-value means that it is unlikely to find a motif by chance with the given or a higher frequency. Motifs are considered as statistically significant if the p-value is less than 0.01. For calculation of the p-value see Eq. 2: F_1 is the given motif frequency, N is the number of random networks considered, and F_r is the motif frequency in a random network. The Kronecker delta $\delta_{c(n)}$ equals 1 if the condition $c(n)$ holds, otherwise it equals 0. For meaningful statistics at least 1,000 random networks should be included.

$$P(m) = \frac{1}{N} \sum_{n=1}^{N} \delta_{c(n)} \qquad c(n) : F_{r_n}(m) \geq F_1(m). \tag{2}$$

1.4. Study Case: Analysis of a Transcriptional-Regulatory Network of Escherichia coli Gene-Regulatory Network of Yeast

To demonstrate the capacity of MAVisto to find motifs in biological networks, we chose to concentrate on a well analysed transcriptional-regulatory network of the prokaryote *Escherichia coli* (11). Martínez-Antonio et al. determined how about one-third of the transcriptional regulators and sigma factors encoded in *E. coli*. Interact with each other. Just as maps of metabolic networks describe the potential pathways that may be used by a cell to accomplish metabolic processes, this network of regulator interactions describes potential pathways *E. coli* cells can use to regulate global gene-expression programs.

In the study, 115 transcription and seven sigma factors, i.e. around one-third of the total predicted transcription factors in *E. coli* were analysed by Martínez-Antonio et al. Data were extracted from RegulonDB (19).

2. Material

2.1. Software

MAVisto is platform independent and available free of charge as a Java webstart application at http://mavisto.ipk-gatersleben.de. Version 2.7.0 was used for all work presented in this chapter. MAVisto is written in Java and is based on Gravisto (http://gravisto.fim.uni-passau.de), an editor for graphs and a toolbox for implementing graph algorithms (20, 21). Prerequisite to launch the application is the installation of the Java runtime environment (JRE), which can be downloaded for all common computer operating systems at http://www.java.com/getjava. Users can either start the application from the web page or from the terminal (shell) with the command

```
javaws http://mavisto.ipk-gatersleben.de/mavisto.
jnlp
```

If you download MAVisto's executable file *mavisto.jnlp* to your local hard drive, you may launch MAVisto under MacOSX and Linux without web access with

```
javaws /Users/All/MethodsMolBiol/mavisto.jnlp
```

where */Users/All/MethodsMolBiol/* has to be substituted by your path to *mavisto.jnlp*. Under Windows the command may look like

```
javaws C:\\MethodsMolBiol\mavisto.jnlp
```

Figure 2 shows a screen shot of MAVisto after successful launch.

2.2. File Formats

MAVisto supports the Pajek (*.net*) and GML (*.gml*) graph file formats for loading and saving. Additionally, a simple text (*.txt*) file format is supported. A beginners guide to biological data processing and file reformatting can be found in ref. 22.

2.2.1. Raw Text Format

The simplest format to describe a network is the two-column raw text file format. Each line contains two node-identifiers specifying an edge and optionally a third string specifying the edge label. Such a network can easily be extracted from, e.g. an Excel sheet. Nodes but no edges can be annotated (Fig. 3).

Fig. 2. How MAVisto should look like after successful launching.

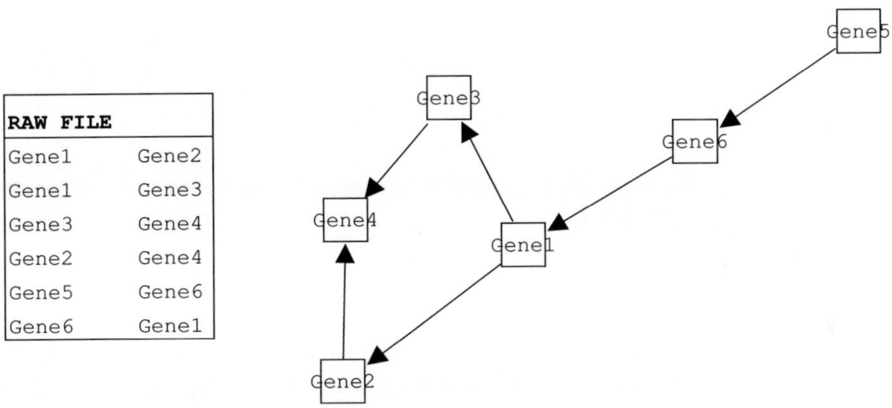

Fig. 3. A regulatory cascade loaded from a simple text file.

2.2.2. Pajek Format

The Pajek (Slovene word for spider) format has been developed in parallel to the Pajek software, which serves for large network analysis (23). The PajekWiki (http://pajek.imfm.si/doku.php) links to some tools that allow the conversion of text and Excel files to Pajek format. Figures 4 and 5 give you a taste how Pajek files look like.

2.3. Graph Format Language

Graph Format Language (GML) is a rich graph format language supported by many other network visualization packages. In contrast to the Pajek format, GML files can store graph-layout information in addition to the network itself. The GML file format specification is available at http://www.infosun.fmi.uni-passau.de/Graphlet/GML/. Figure 6 gives an example how a network is described in GML.

```
PAJEK FILE
*vertices 3
1 Gene-1
2 Gene-2
3 Gene-3
*arcslist
1 2 3
2 3
```

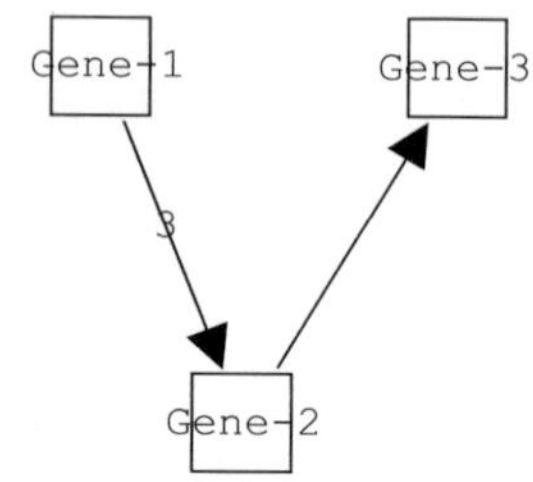

Fig. 4. *Directed graph* with *three nodes* and *two edges*, one of which has weight three, in simple Pajek format.

```
PAJEK FILE
*vertices 3

2 Gene-2
3 Gene-3
*edges
1 2 0.2
1 3 0.9
2 3 1.0
```

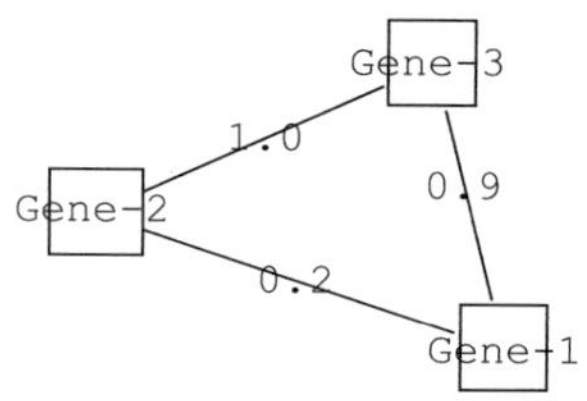

Fig. 5. *Unidirected graph* with *three nodes, three edges*, and *edge weights*, in extended Pajek format.

2.4. Sample File

Example biological networks, such as the one analysed in this chapter, are always artificial separations from an unexplainable complex whole. Even transcriptome- or proteome-wide measurements will fail to capture all interactions occurring in cells, tissue, or organisms. Still, analysis of such isolated subnetworks profoundly influences contemporary biology and led to the field of systems biology and facilitate synthetic biology. Here, the understanding of functional network units is crucial in order to implement new biological function, e.g. metabolism, in artificial cells that grow and divide. All software demonstrations shown below were performed with the core components of a gene-regulatory network of *E. coli* as described in Subheading 1.4. Only the largest, loop-free connected components, composed of 122 nodes and 355 edges, were extracted and converted into the Pajek-format. This ready-to-use file is available for download at http://www.hs-mittweida.de/wuenschi. The original data set has been published by Martínez-Antonio et al. (11) and was extracted from http://regulondb.ccg.unam.mx.

```
GML FILE
graph [
    comment "This is a sample graph"
    directed 1
    label "MyGraph"
    node [
        id 1
        label "ONE"
    ]
    node [
        id 2
        label "TWO"
    ]
    node [
        id 3
        label "THREE"
    ]
    edge [
        source 1
        target 2
        label "1->2"
    ]
    edge [
        source 2
        target 3
        label "2->3"
    ]
    edge [
        source 3
        target 1
        label "3->1"
    ]
]
```

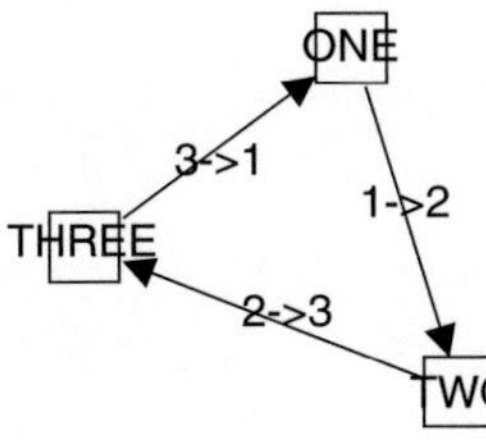

Fig. 6. A feedback loop loaded from a GML file.

3. Methods

For the following sections, we assume that you installed Java, downloaded the example network file (Subheading 2.4) and launched MAVisto (Subheading 2.1). In order to get you as fast as possible to a level that allows you to analyse your own data, we combine both the description of MAVisto and its utilization.

We focus on the identification of three-node feed-forward loops (see network in Fig. 11) in the gene-regulatory network of yeast. Three-node feed-forward loop motifs contain a regulator that controls a second regulator and have the additional feature that both regulators bind a common target gene. The feed-forward loop architecture has been highly favoured during the evolution of transcriptional regulatory networks in *E. coli* and yeast (10). Rather than diving into biology, let us dive into the software – Have a lot of fun.

Fig. 7. Menu bar. The *icons* represent from *left* to *right*: *new* and *open* network, *save* and *save as* network, *export* network view to PDF, *cut*, *copy*, and *paste* network elements, *undo* and *redo* last action, *toggle graph state* between directed or undirected, *select edge color*, *zoom out*, *in*, to 100% and *fit to view*.

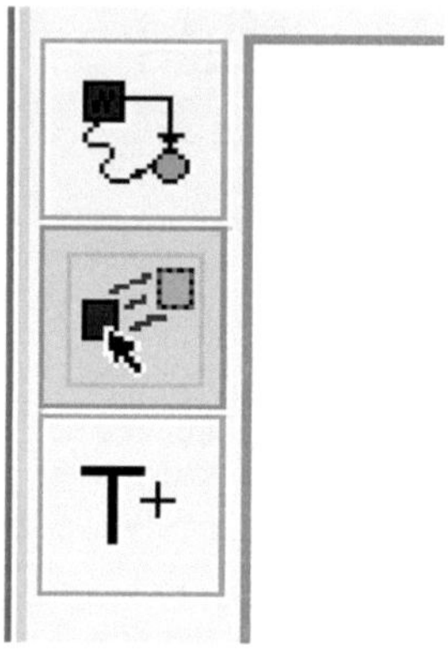

Fig. 8. Editor bar. The *icons* represent from *top* to *bottom*: *edit* network, *move selected nodes*, and *edit node or edge label*.

3.1. Network Import and Editing

Once MAVisto has been launched, the network to be analysed may be created using the editor functionality of MAVisto. However, most user may want to load the network from a properly formatted file. Select *New* or *Open* from the file menu or from the menu bar, respectively (Fig. 7).

The editor functionality allows both to create and manipulate a network (Fig. 8). In the editing mode new nodes can be inserted by clicking on a free area in the network pane. Edges between nodes can be inserted by selecting a node as source, holding the left mouse button and subsequently selecting a second node as target. In the move mode the position of selected nodes can be changed and selected nodes and edges can be removed from the network by pressing the delete key. In the node label editing mode, the label of a selected node can be altered.

3.2. Network Layout

The layout is based on a force-directed method that calculates a "force" vector for each node (24). For nodes connected by an edge an attractive force is computed, for all other nodes not connected to each other repulsive forces are computed. The force vector is the sum of attractive and repulsive forces and each node is moved according to the corresponding force. This cycle is repeated until a threshold is reached.

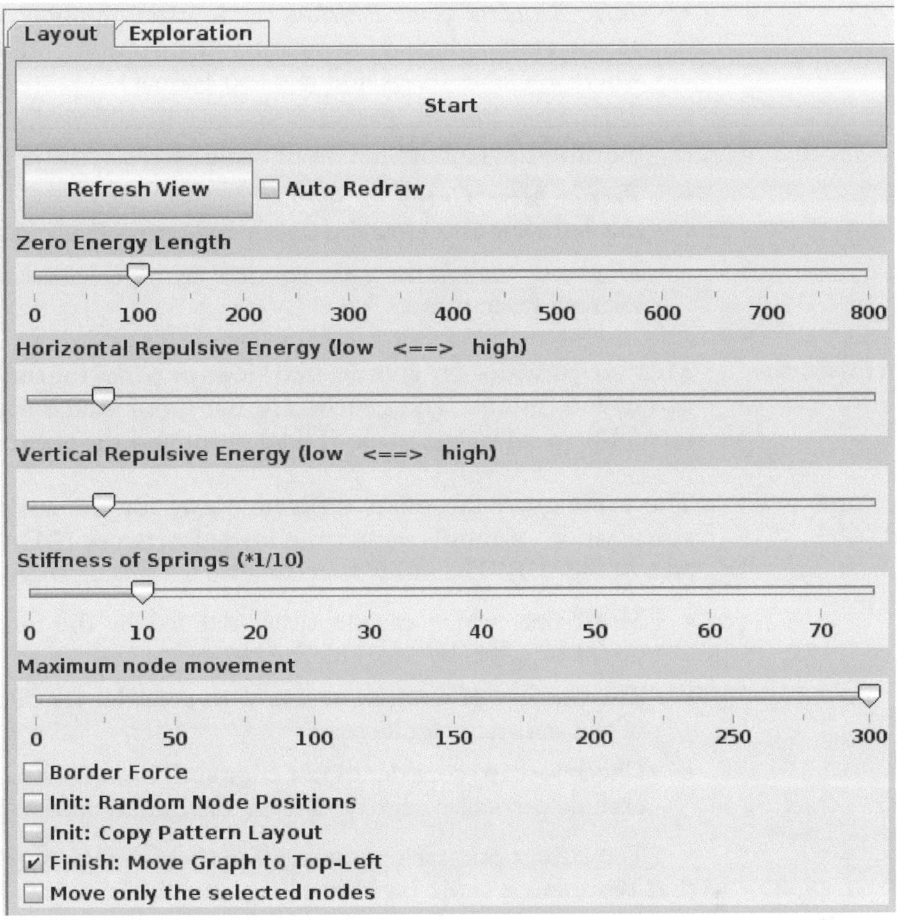

Fig. 9. Layout tab. The layout tab can be found in the *right panel* of the application.

The layout tab of MAVisto comprises a *Start* button for starting the layout algorithm (Fig. 9). A *Refresh View* button is used to redraw the network once the layout algorithm is running. Alternatively, the *Auto Redraw* checkbox can be activated to redraw the network after each cycle. Sliders are available to specify:

- The edge length that is set to zero energy.
- The horizontal repulsive energy for nodes.
- The vertical repulsive energy for nodes.
- The stiffness of springs that affects the edge length.
- The value for the maximum node movement which continuously decreases until the layout reaches a fixed state.

  The checkboxes allow to specify the following options:

- *Border Force* is used to prevent the network from moving outside the top-left view area.

- *Init*: *Random Node Positions* randomizes all node positions at the start, which is mandatory if no layout information where available (Pajek and raw text format).
- *Init*: *Copy Pattern Layout* copies the layout of the selected motif to the highlighted motif occurrences in the network.
- *Finish*: *Move Graph to Top-Left* moves the network to the top-left view area after the layout has been calculated.
- *Move only the selected nodes* restricts layout calculation to the selected node subset.

3.3. Motif Exploration

MAVisto provides five coordinated views or panes for the analysis of network motifs. These views are the motif search and result table, the motif view, the motif fingerprint and the network view.

3.3.1. Motif Search and Result Table

The exploration tab offers different views for the analysis and visualization of motifs within the loaded network (24) (see top area of Fig. 10). The motif search allows for specification of

- Motif size, which can be either defined by the number of nodes or edges by toggling the button.
- The number of random networks to consider for calculation of the statistical significance.
- Checkboxes to consider or not edge and/or node labels and to exclude particular edge types from the randomization process.

The latter option is reasonable if particular edge type(s) exist that represent a static backbone (e.g. metabolic reactions), and

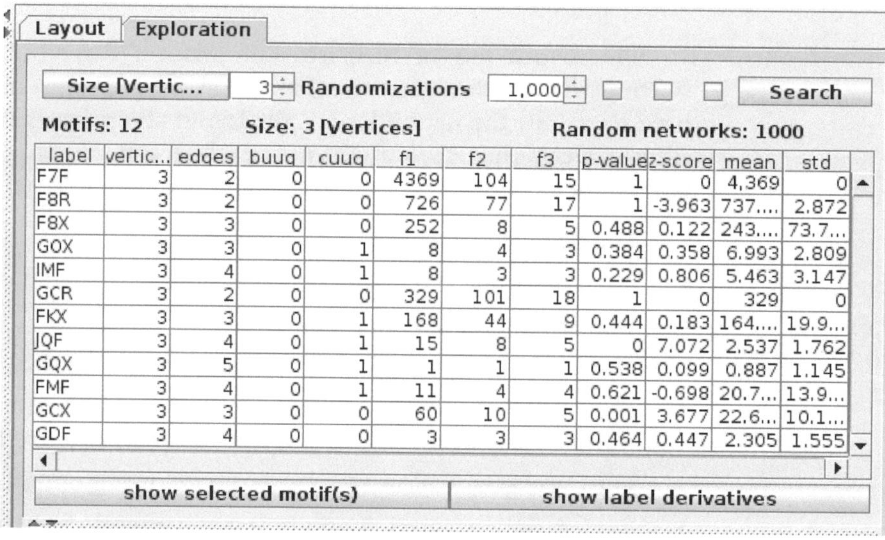

label	vertic...	edges	buug	cuug	f1	f2	f3	p-value	z-score	mean	std
F7F	3	2	0	0	4369	104	15	1	0	4,369	0
F8R	3	2	0	0	726	77	17	1	-3.963	737....	2.872
F8X	3	3	0	0	252	8	5	0.488	0.122	243....	73.7...
GOX	3	3	0	1	8	4	3	0.384	0.358	6.993	2.809
IMF	3	4	0	1	8	3	3	0.229	0.806	5.463	3.147
GCR	3	2	0	0	329	101	18	1	0	329	0
FKX	3	3	0	1	168	44	9	0.444	0.183	164....	19.9...
JQF	3	4	0	1	15	8	5	0	7.072	2.537	1.762
GQX	3	5	0	1	1	1	1	0.538	0.099	0.887	1.145
FMF	3	4	0	1	11	4	4	0.621	-0.698	20.7...	13.9...
GCX	3	3	0	0	60	10	5	0.001	3.677	22.6...	10.1...
GDF	3	4	0	0	3	3	3	0.464	0.447	2.305	1.555

Fig. 10. Exploration tab. The exploration tab can be found in the *right panel* of the application. It contains the motif search menu and results.

only other edge types should be analysed for their occurrence within motifs (e.g. regulatory interactions).

Subsequent to a motif search triggered by hitting the *Search* button, the result table shows information about all discovered motifs. The table presents for each motif

- A label used for motif identification.
- The size as number of nodes and number of edges.
- Information on the topology by giving by the number of branches (buug, node with a degree of three or higher).
- Cycles (cuug, uses the underlying undirected graph).
- The frequency values for the three different frequency concepts F1–F3 (see Subheading 1.2).
- The *Z*-score (see Subheading 1.3.1).
- The *p*-value (see Subheading 1.3.2).
- The mean value and standard deviation of motif frequency in randomized versions of the network.

The table allows for sorting by all criteria and selection of motifs to be displayed in the motif view. Selected table rows can copied into the clipboard and subsequently pasted into any suitable application.

3.3.2. Motif View

The motif view provides a visual representation of the displayed motif(s) (Fig. 11). Furthermore, it is used to control the highlighting of motif occurrences in the network view. The motif view offers the following options:

- Refreshing layout of the selected motif.
- Removing active motif from the motif view pane.

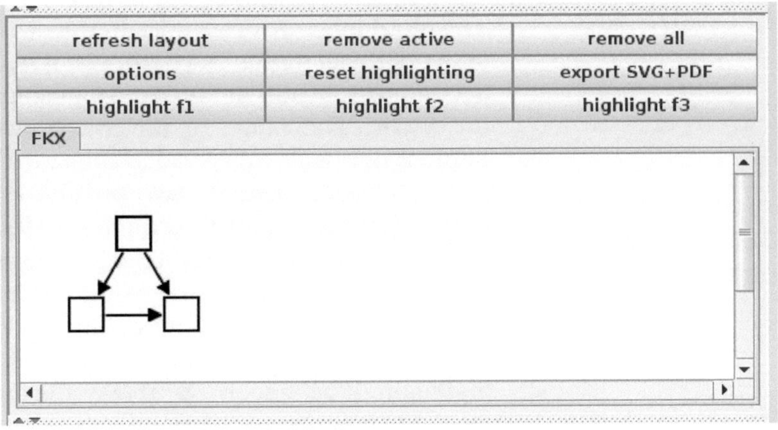

Fig. 11. Display of a three-node feed-forward loop in the motif view.

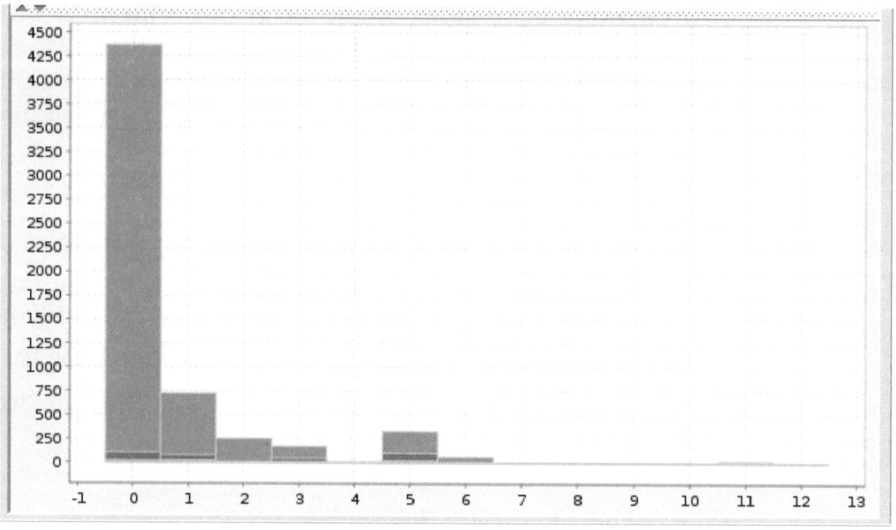

Fig. 12. The Motif Fingerprint. *Bar chart* showing the frequency (*y*-axis) of all 13 possible directed network motifs composed of three nodes (*x*-axis) in the analysed network. Frequency concepts F1–F3 (Subheading 1.2) are represented by (from *top* to *bottom*) *light gray*, *dark gray*, and *gray*, respectively. *Bar 5* is currently active, as highlighted by a *black arrow* above the bar.

- Removing all displayed motifs from the motif view pane.
- Options for motif highlighting (colour or grey; marking of non-overlapping elements of different motif occurrences individual or uniform, selection of colour for network elements not covered by motif matches).
- Resetting the highlighting in the motif view.
- Exporting the displayed motif to PDF (portable file format) and SVG (scalable vector graphics).
- Highlighting the motif occurrences within the network view by frequency concepts F1–F3 (Subheading 1.2).

3.3.3. Motif Fingerprint

The motif fingerprint shows a frequency spectrum of the analysed network for all motifs of a particular size, including motifs with no occurrences. Motifs are ordered alphanumerically by their label at the *x*-axis. The range of the *x*-axis gives an estimation of the number of possible directed graphs for a given number of nodes. For two nodes there are two possible directed graphs, for three nodes there are 13, for four there are 199, for five there are 9,364, and for six there are 1,530,843. The values for the frequency concepts F1–F3 are plotted using red, blue, and green colour as overlaid columns. A yellow column is used as place holder for absent motifs. The currently displayed motif in the motif view is highlighted by an arrow above the corresponding frequency

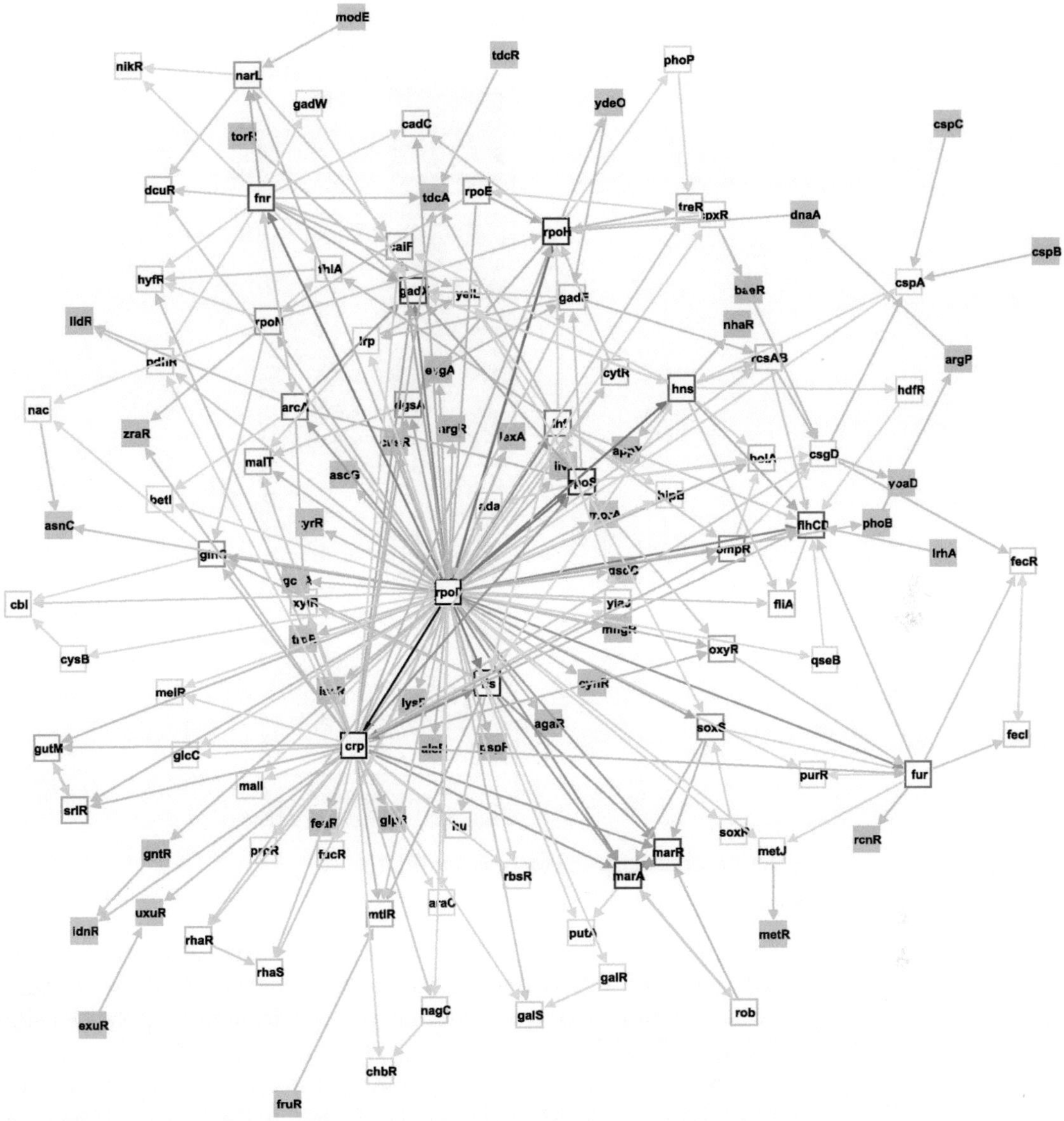

Fig. 13. Network view with *highlighted* occurrences of the three-node feed-forward loop motif using frequency concept F1. *Colours* (*grayscale*) indicate the frequency by which an individual network element is covered by different motif occurrences. The corresponding colour map is displayed in Fig. 14.14. Nodes and edges not covered by any motif occurrences are *coloured* in *grey*.

column. In turn, selecting a frequency column leads to the selection of the corresponding motif within both the motif view and motif table (Fig. 12).

3.3.4. Network View

Apart from network editing, the network view is used to display motif occurrences within the analysed network (Figs. 13 and 14). Depending on the chosen frequency concept

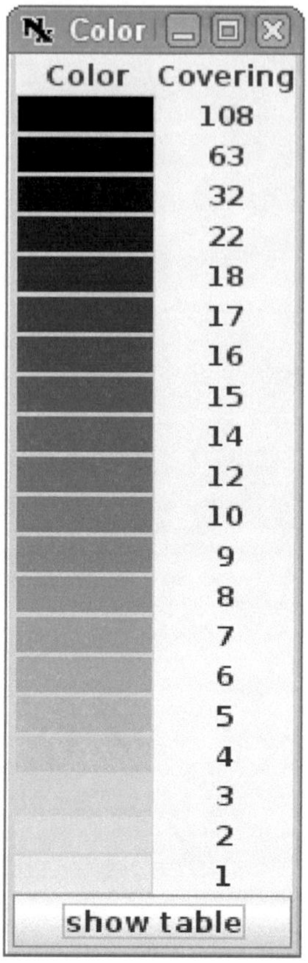

Fig. 14. Colour map window related to Fig. 13.

(Subheading 1.2) nodes and edges may be part of several motifs. The number of participations in different networks is colour-coded. A new window, containing the colour map, provides the legend (Fig. 14).

Frequency concept F3, where neither nodes nor edges are shared between network motifs, allows for a motif preserving layout (Fig. 14). This means, the layout of the selected motif in the motif view pane is copied to all highlighted occurrences within the network, by using the corresponding option within the layout tab.

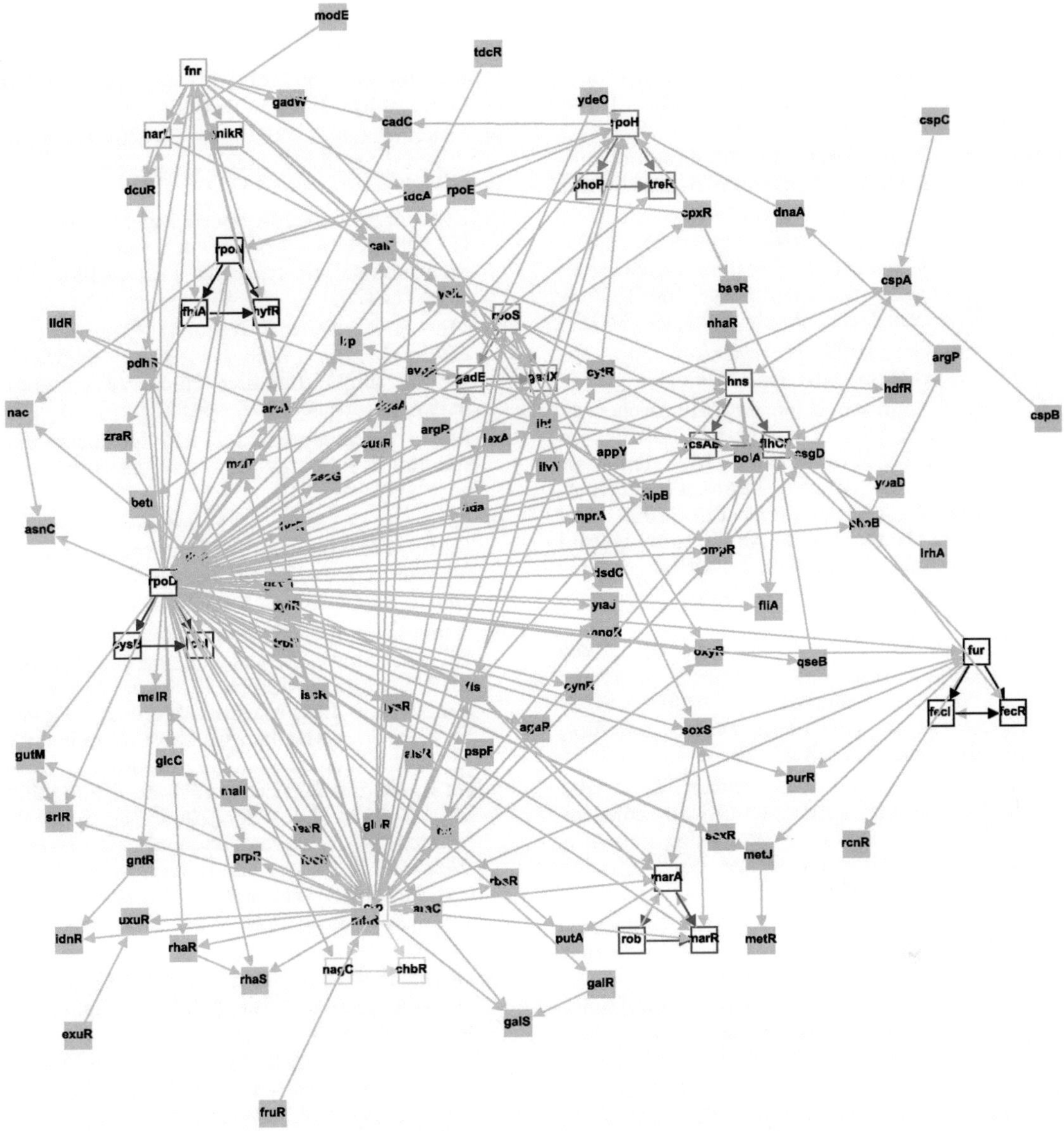

Fig. 15. Network view with *highlighted* occurrences of the three-node feed-forward loop motif and application of the motif preserving layout using frequency concept F3.

Acknowledgments

The authors like to thank Professor Dr. Falk Schreiber, head of the Bioinformatics group at the Leibniz Institute of Plant Genetics and Crop Plant Research (IPK) in Gatersleben/Germany, and his team for support and hosting of MAVisto.

References

1. Cori C, Cori G. (1929) Glycogen formation in the liver from D- and L-lactic acid. J Biol Chem, 81:389–403.
2. Billson C. A history of the London tube maps 2010. [http://homepage.ntlworld.com/clivebillson/tube/tube.html].
3. Appel R, Bairoch A, Hochstrasser D. (1994) A new generation of information retrieval tools for biologists: the example of the ExPASy WWW server. Trends Biochem Sci, 19:258260.
4. Michal G. (1999) *Biochemical Pathways*. Wiley-Spektrum, New York/Heidelberg.
5. Caspi R, Altman T, Dale JM, Dreher K, Fulcher CA, Gilham F, Kaipa P, Karthikeyan AS, Kothari A, Krummenacker M, Latendresse M, Mueller LA, Paley S, Popescu L, Pujar A, Shearer AG, Zhang P, Karp PD. (2008) The MetaCyc Database of metabolic pathways and enzymes and the BioCyc collection of pathway/genome databases. Nucleic Acids Res, 36:D623–D631.
6. Kanehisa M, Goto S, Hattori M, Aoki-Kinoshita K, Itoh M, Kawashima S, Katayama T, Araki M, Hirakawa M (2000) KEGG: Kyoto encyclopedia of genes and genomes. Nucleic Acids Res, 28:27–30.
7. Kanehisa M, Goto S, Furumichi M, Tanabe M, Hirakawa M. (2010) KEGG for representation and analysis of molecular networks involving diseases and drugs. Nucleic Acids Res, 38:D355–D360.
8. Friedland A, Lu T, Wang X, Shi D, Church G, Collins J. (2009) Synthetic gene networks that count. Science, 324:1199–1202.
9. Mangan S, Alon U. (2003) Structure and function of the feed-forward loop network motif. Proc Natl Acad Sci U S A, 100:1198011985.
10. Lee T, Rinaldi N, Robert F, Odom D, Bar-Joseph Z, Gerber G, Hannett N, Harbison C, Thompson C, Simon I, Zeitlinger J, Jennings E, Murray H, Gordon D, Ren B, Wyrick J, Tagne J, Volkert T, Fraenkel E, Gifford D, Young R. (2002) Transcriptional regulatory networks in *Saccharomyces cerevisiae*. Science, 298:799–804.
11. Martínez-Antonio A, Janga SC, Thieffry D. (2008) Functional organisation of *Escherichia coli* transcriptional regulatory network. J Mol Biol, 381:238–247.
12. Schreiber F, Schwöbbermeyer H. (2005) MAVisto: a tool for the exploration of network motifs. Bioinformatics, 21:3572–3574.
13. Schreiber F, Schwöbbermeyer H. (2005) Frequency concepts and pattern detection for the analysis of motifs in networks. Trans Comput Syst Biol, 3:89–104.
14. Fruchterman T, Reingold E. (2000) Graph drawing by force-directed placement. Software Pract Ex, 30:1303–1324.
15. Koschützki D, Schwöbbermeyer H, Schreiber F. (2007) Ranking of network elements based on functional substructures. J Theor Biol, 248 (3):471–479.
16. Milo R, Shen-Orr S, Itzkovitz S, Kashtan N, Chklovskii D, Alon U. (2002) Network motifs: simple building blocks of complex networks. Science, 298:824827.
17. Maslov S, Sneppen K. (2002) Specificity and stability in topology of protein networks. Science, 296:910–913.
18. Shen-Orr S, Milo R, Mangan S, Alon U. (2002) Network motifs in the transcriptional regulation network of *Escherichia coli*. Nat Genet, 31:64–68.
19. Gama-Castro S, Jiménez-Jacinto V, Peralta-Gil M, Santos-Zavaleta A, naloza Spinola MIP, Contreras-Moreira B, Segura-Salazar J, niz Rascado LM, Martínez-Flores I, Salgado H, Bonavides-Martínez C, Abreu-Goodger C, Rodríguez-Penagos C, Miranda-Ríos J, Morett E, Merino E, Huerta AM, no Quintanilla LT, Collado-Vides J. (2008) RegulonDB (version 6.0): gene regulation model of *Escherichia coli* K-12 beyond transcription, active (experimental) annotated promoters and Textpresso navigation. Nucl Acid Res, 36: D120D124.
20. Holleis P, Zimmermann T, Gmach D. (2005) Drawing graphs within graphs. J Graph Alg Appl, 9:7–18.
21. Klukas C, Koschützki D, Schreiber F. (2005) Graph pattern analysis with PatternGravisto. J Graph Alg Appl, 9:19–29.
22. Wünschiers R. (2004) *Computational Biology: Unix, Linux, Data Processing and Programming*. Springer, New York.
23. Batagelj V, Mrvar A. (1998) Pajek – program for large network analysis. Connections, 21:47–57.
24. Klukas C, Schreiber F, Schw¨obbermeyer H. (2006) Coordinated Perspectives and Enhanced Force-Directed Layout for the Analysis of Network Motifs. In *Asia Pacific Symposium on Information Visualisation (APVIS2006), Volume 60 of Conferences in Research and Practice in Information Technology (CRPIT)*. Edited by Misue K, Sugiyama K, Tanaka J, ACS, Tokyo, 39–48.

Chapter 15

Using MCL to Extract Clusters from Networks

Stijn van Dongen and Cei Abreu-Goodger

Abstract

MCL is a general purpose cluster algorithm for both weighted and unweighted networks. The algorithm utilises network topology as well as edge weights, is highly scalable and has been applied in a wide variety of bioinformatic methods. In this chapter, we give protocols and case studies for clustering of networks derived from, respectively, protein sequence similarities and gene expression profile correlations.

Key words: Network clustering, Cluster analysis, Protein sequence similarity, Gene expression profiles

1. Introduction

Networks, also known as graphs, are composite structures in which *nodes* are connected by weighted links, called *edges*. An edge always connects two nodes, and a node may be connected to multiple nodes, called its *neighbours*, by as many edges. The nodes usually represent members from a well-defined class of objects, such as genes, proteins, or experimental conditions. An edge may have associated with it a weight that confers a degree of similarity or dissimilarity between the objects connected by the edge. In network analysis in bioinformatics, edge weights predominantly play the role of similarities, and such is the case here.

The MCL algorithm, or MCL in short, stands for Markov CLuster algorithm. It computes stochastic flow through a network by alternating dissipation and reinforcement steps (1–3). This defines a simple algebraic bootstrapping process in which cluster structure is found as the result. The number of clusters cannot and need not be specified, but a single parameter called *inflation* controls the reinforcement step, so that clusterings at different levels of granularity can be found. This is the only parameter that will be varied.

Jacques van Helden et al. (eds.), *Bacterial Molecular Networks: Methods and Protocols*, Methods in Molecular Biology, vol. 804, DOI 10.1007/978-1-61779-361-5_15, © Springer Science+Business Media, LLC 2012

MCL was first published in refs. 1, 2 and then applied to protein sequence networks (4, 5). It has subsequently been used, among others, for detecting orthologous groups (6), predicting protein complexes from protein interaction networks (7, 8), clustering of genes according to their expression profiles (9–11), and clustering of scientific documents (12). Ensembl families (13) are computed using the same general approach as proposed in ref. 4, and this is detailed in Protocol 1. Protocol 2 provides a recipe for clustering genes according to expression profiles. Both protocols are accompanied by a case study.

The protocols described here are relatively simple and exclusively use command line tools, namely (the program) **mcl** itself, sibling applications that are shipped and installed with **mcl**, and standard UNIX commands. This manner of processing offers many benefits, such as off-line processing, flexible coupling and decoupling of components, adaptability and scalability. The software can be downloaded from http://micans.org/mcl, where documentation and installation guidelines are provided as well. Assistance can be sought and bug reports submitted on a mailing list (http://micans.org/mcl/questions). On Windows **mcl** can run under the freely available Cygwin environment.

No real alternatives exist for Protocol 1, but a more specialised version is available as OrthoMCL (14). Regarding Protocol 2, BioLayout Express3D is a comprehensive tool for analyzing expression data (10, 15), integrating large-scale data analysis and network visualisation with the ability to use **mcl**. A recently published approach using MCL is DBF-MCL (11). Finally, MCL is also integrated in the web services (http://rsat.ulb.ac.be/rsat/index_neat.html) provided through Network Analysis Tools or NeAT (Chapter 18 and ref. 16). Another method for clustering networks is Restricted Neighbourhood Search Clustering (see Chapter 16 and ref. 17). It is a fast and general purpose method, currently utilising absence/presence of edges only without regarding weights.

Protocol 1 describes the analysis and prediction of family relationships between proteins based on their sequence similarity. It starts with a set of files resulting from running **blastall -p blastp** on FASTA files encoding protein sequences, and details the construction, clustering, and analysis of the protein sequence similarity network. The supplementary material gives examples for obtaining FASTA files and running **blastall**.

In Protocol 2 the aim is to group together and analyse genes that show correlated or possibly anticorrelated expression across different conditions. Starting with a table of array expression values for a set of genes measured across different conditions, it describes the necessary steps to create, cluster, and analyse a gene expression profile network.

1.1. Constructing Appropriate Networks

The definition of a network involves several choices. Important among these are the selections of nodes and edges and the similarities assigned to edges. Generally, choices should be based on the appropriateness for the analysis and the methods one intends to apply. We illustrate this in the particular case of MCL and the protocols introduced above.

MCL is affected by differentiation among the edge weights. Often a network has edges between distant objects, in addition to edges between closer objects. Equally often the former are more numerous than the latter, and can play a vital role as third party support in pulling the objects into meaningful groups, i.e. clusters. For what follows it is convenient to use self-identities as a measure for maximally close nodes. As a rule of thumb in the context of MCL, distant nodes, if present, should have similarities between them that are between 1 and 2 orders of magnitude smaller than self-identities. If distant similarities otherwise get bigger or smaller they exert too much or too little pull, respectively.

The raw data in Protocol 1 are *E*-values for BLAST scores between two sequences. A naive transformation would take the inverse of these *E*-values to arrive at a similarity. *E*-values of 1e-160 and 1e-20 are thus mapped to similarities 1e160 and 1e20, with 140 orders of magnitude between them, and the second similarity might as well be zero. The normal, widely used transformation step uses a log-transformation followed by a sign reversal, leading to similarities 160 and 20, with approximately 1 order of magnitude between them.

The raw data in Protocol 2 are (absolute values of) Pearson correlations between different probes on the array across series of experiments. A cutoff is chosen in order to keep the data manageable and meaningful. In our case study 1, we use a cutoff of 0.6. If left at that, similarities vary in the interval [0.6–1]. However, this falls short if we desire to attach much greater importance to correlation values close to one compared to values close to 0.6. We note that correlations in the lower ranges are much more common, yet they are in the same order of magnitude as high correlations and have a comparable impact on MCL. The approach chosen is to shift the interval [0.6–1] to the interval [0–0.4], which results in a gradual fading out of the similarities between more distant probes.

A few concluding remarks are in order. Often a threshold is chosen so that similarities below it are excluded. Examples are a transformed *E*-value cutoff and the correlation cutoff. Ideally the data are transformed in such a manner that the network changes gradually as the cutoff is varied. The transformations given here satisfy this property. Second, other choices can be made in order to tune the network for a given task. A good example is the prediction of orthology and paralogy from simple sequence similarity networks such as performed by OrthoMCL (6). In this case one must be careful to choose species that evenly sample the species subtree in which one is interested. Additionally, OrthoMCL submits the input network to a reduction step in which only putative orthologs and paralogs are considered by using best reciprocal hit derived criteria, followed by an augmentation step adding so-called co-orthologs. Finally, the network construction step should be seen as separate from the clustering method itself. Sometimes these are lumped together, complicating the comparison between different clustering methods.

1.2. Case Study 1: Clustering Based on Sequence Similarity

Sequence similarity is widely used to transfer functional annotation between genes. This process leads to errors, given the entangled nature of gene evolution. Duplications give rise to paralogs, which can then functionally diverge while maintaining high levels of sequence similarity. Many proteins are composed of more than one domain, each with different functional properties and even different evolutionary histories. Complex phylogenetic analysis can sometimes be required to identify the correct evolutionary relationships between genes. A simple but powerful approach involves clustering a network based on sequence similarity. If there is detectably more sequence similarity between orthologs than between paralogs, a good clustering algorithm should be able to separate these two groups.

Orthologs may be difficult to find, particularly if recent duplications have occurred, or if the network contains sequences from more than one group of closely related organisms. In these cases the cluster structure will reflect this, some clusters will mostly contain sequences from closely related organisms and others will contain paralogs. This structure can still be useful for understanding the phylogenetic relationship between organisms, but shall not be further discussed in this chapter.

The first protocol focuses on clustering a protein sequence similarity network. We use the proteomes of 28 fully sequenced genomes of the order Rickettsiales. These intracellular alphaproteobacteria have complex life cycles, involving arthropod and mammalian hosts, and their highly streamlined genomes have become the paradigm of reductive evolution (for a review, see ref. 18). They also represent the closest extant relatives of the ancestor to mitochondria. Beyond having small genomes, Rickettsiales can still be quite different, reflecting several hundred million years of evolution. In this case study, we show examples of how orthologs are separated from paralogs with MCL.

1.3. Case Study 2: Clustering Based on Expression Profiles

The use of microarrays for analysing gene expression is widespread. After measuring the expression of thousands of genes over a variety of conditions, clustering the results can help to visualise the data and gain further functional insights (19). The cluster structure can then be used to predict functional relationships amongst each set of co-expressed genes, assuming that co-expression implies co-regulation.

In bacteria, many genes are organised in operons, which are transcribed in a coordinated manner. Co-regulation of multiple operons can be caused by transcription factors, leading to the formation of so-called regulons. In optimised bacterial genomes, genes that belong to the same functional pathway (e.g. the biosynthesis of an amino acid) are expected to be co-regulated: when the amino acid is present, they can all be turned off and when the amino acid is absent, they can all be turned on. The same logic can be applied to protein complexes, where the level of the protein subunits should be regulated in a coordinated manner. If these kinds of functional relationships demand co-expression, co-expression should then be able to be used to predict them.

Expression data can be noisy: two genes can show apparent co-expression for technical reasons such as probe cross-hybridisation. Given the scale of microarray experiments, where the number of genes is usually much higher than the number of conditions, some level of co-expression can be expected simply by chance. Clustering can help overcome some of these problems, by finding groups of consistently co-expressed genes across multiple conditions. It is worthwhile noting that more elaborate filtering steps can enhance the results. Conserved co-expression, either using different species or comparing paralogous pairs of genes in a single species, has been shown to improve functional prediction derived from co-expression analysis (20).

For this case study, we use Affymetrix *Escherichia coli* expression data from the Many Microbe Microarray Database (21). The dataset comprises 466 different experimental conditions, enabling us to extract clusters of diverse functions. The advantage of working with *E. coli* is the wealth of available annotation, both at the level of gene function and regulation (22, 23). We rely on these annotations to exemplify how different levels of clustering can yield important functional insights.

2. Material

2.1. Software

The software required consists of programs provided in the MCL distribution, available from http://micans.org/mcl/. A version at least as recent as 09–320 should be used. Here, the version number is of the form YY–DDD, encoding the year and the day-in-year. The programs used are **mcl**, **mcxload**, **mcxdump**, **mcxarray**, **clm**, and **mcx**. These are installed by default. The **clm** program implements various analysis modes for clusterings and networks, and is invoked as **clm** *mode*. The **mcx** program similarly supplies analysis modes for networks. An example is **mcx query**, used to gauge graph properties as a graph is subjected to increased levels of edge weight thresholding. Additionally, the script **clxdo** is found in the scripts/ subdirectory of the unpacked MCL source, and the standard UNIX commands **cut** and **sort** are used.

2.2. File Formats

The **mcl** program allows two types of input. The first type is called ABC-format, useful for simple protocols, and triggered by the --abc option to **mcl**. Input is read line by line, each line encoding an edge as two labels separated by whitespace, optionally followed by a numeric similarity. The clustering is output in another line-based format, where a single line encodes a clustering as a list of white-space separated labels.

For more complex protocols such as described here it is useful to have a more abstract representation of a network. This format is obtained by feeding the ABC-format to **mcxload**. It outputs a network in *native mcl format* as well as a *dictionary file* allowing the mapping of the native format onto the original label format. By default **mcl** writes output clusterings in native format as well. The program **clm** has various analysis modes such as **order**, **close**, and **dist**, all of which understand this format. The program **mcxdump** maps results back to the original labels using a dictionary file, which has a simple two-column format consisting of an **mcl** identifier and the label it refers to. **mcxdump** can output ABC-format, line-based cluster dumps, as well as structured tables.

Protocol 1 uses as a starting point files resulting from running **blastall -p blastp** with parameter **–m 8** on FASTA files encoding protein sequences. The **–m 8** options causes **blastall** to output a concise tabular output. The information that is relevant to **mcl** is present in columns 1, 2, and 11, namely the identifiers of the two sequences being compared and the *E*-value.

Protocol 2 uses a table of normalised intensity values from Affymetrix microarrays. This is the standard way of representing processed microarray results for single-channel experiments.

2.3. Sample Files

Case study 1 uses FASTA files downloaded (http://www.uniprot.org/) from Uniprot (24) for 28 genomes of the order Rickettsiales, containing sequences for a total of 31,374 proteins. Case study 2 uses *E. coli* expression data downloaded (http://m3d.bu.edu/norm/) from the Many Microbe Microarrays Database (21). It contains expression values for 4,297 genes measured across 466 conditions.

3. Methods

Both protocols generate clusterings at different levels of granularity, and we consider it useful to analyse the data in this wider context. A rigid strategy for choosing a single clustering is not given. Criteria for such strategies differ from case to case and depend on the research objectives. Clusterings at different levels of granularity may summarise the data equally well, albeit from different points of view. A fine-grained clustering implies a conservative view of family relationships, whereas a coarse-grained clustering implies a more permissive view. MCL can generate meaningful clusters at very different levels of granularity, as demonstrated in case study 2.

Three simple means to compare different clusterings are supplied, namely by granularity, by distance, and by edge weight capture, but it is suggested to use these as measurement tools rather than decision criteria.

For simplicity, both protocols put all generated files in the same directory. A more organised way of working would use subdirectories such as fasta/, blast/, and clusterings/, but we leave these improvements to the practitioner. A general guideline for the commands given below is that after the program name any word that starts with a hyphen is an *option* to the program. If it starts with a single hyphen the option takes an argument, which is the next word. If it starts with two hyphens the option takes no argument and sets a modality.

The first protocol is elaborate in that it shows many ways to assess clusterings. The majority of these are not repeated in the second protocol but are nevertheless applicable there as well.

3.1. Protocol 1: Clustering Protein Sequence Similarity Networks

The first command to run requires that **blastall -p blastp** results were obtained using the **-m 8** parameter. The blastall output files are assumed to reside in the current directory and have names with suffix blast. The supplementary material has examples of running **blastall**. Below the ↩ symbol indicates that the example continues on the following line and that <**ENTER**> or new-line should not be pressed.

1. The following command reads the BLAST results and creates a file in ABC-format.

   ```
   cut -f 1,2,11 *blast > seq.abc
   ```

2. The ABC-format file seq.abc is read by **mcxload**, which creates a network file called seq.mci and stores the label information in a separate file seq.dict.

   ```
   mcxload -abc seq.abc -write-tab seq.dict -o seq.mci↩
         --stream-mirror --stream-neg-log10 -stream-tf 'ceil(200)'
   ```

The first part to the command, on the first (continued) line, represents the standard way of loading networks in ABC-format. The --stream-mirror option ensures that an undirected network is obtained, which is preferable when using **mcl**. The last continued line is *specific to this protocol.* It specifies a negative log-10 transformation and additionally edge weights are capped to a maximum value of 200. The –tf infix in the –stream-tf option stands for "transformation," and several other programs in the MCL suite accept such an option.

3. Now run **mcl** with different inflation parameters. The first argument to **mcl** is always the input file. By default it is assumed to be a network file.

```
mcl seq.mci -I 1.4
mcl seq.mci -I 2
mcl seq.mci -I 6
```

The inflation values here represent a good first set of parameters. Lower and higher inflation values yield respectively coarser and finer clusterings. File names out.seq.mci.I14, out.seq.mci.I20, and out.seq.mci.I60 are automatically created by **mcl**. The option –o can be used to override this.

4. The output can be transformed to label output as follows:

```
mcxdump -icl out.seq.mci.I14 -o dump.seq.mci.I14 ↩
    -tabr seq.dict
```

This generates a format with all labels for a cluster put together on a single line separated by tab characters.

5. Different clusterings output by **mcl** are not guaranteed to be strictly nested, although usually they are *almost fully* nesting, as seen further below. A strictly nesting set of clusterings is created as follows:

```
clm order -prefix P out.seq.mci.I{14,20,60}
```

This generates from the list of clusterings specified a new set of clusterings P1, P2, and P3, each of which is a strict super-clustering of all the following.

6. It is possible to encode the nested clusterings as a tree, in this case of depth just three.

```
clm order -o seq.mcltree out.seq.mci.I{14,20,60}
mcxdump -imx-tree seq.mcltree -tab seq.dict↩
    --newick -o tree.nwk
```

The file seq.mcltree is a concatenation of networks in **mcl**'s native format. It is transformed to Newick format by **mcxdump**. The branch lengths simply indicate the level at which nodes and clusters join, and should not be given any further meaning.

7. Small clusters are to some extent explained by the input graph already consisting of different components. Therefore, it is useful to obtain the baseline coarse clustering consisting of the natural maximum components of the input graph (also known as *connected components*). The program **clm close** does just this.

```
clm close -imx seq.mci --write-cc -o seq.base
```

8. The granularity of a clustering is gauged from the list of cluster sizes. Smaller clusters are sometimes not so interesting, so the second command below specifies a minimum cluster size. We include the previously obtained connected components in this analysis.

```
clxdo granularity seq.base out.seq.mci.I{14,20,60}
clxdo granularity_gq 5 seq.base out.seq.mci.I{14,20,60}
```

9. It is very useful to assess how close two clusterings are. To this end, issue the following:

```
clm dist --chain out.seq.mci.I{14,20,60}
```

The --chain option causes only consecutive clusterings to be compared, whereas the default behaviour is to do all pairwise comparisons. The first number reported is the distance between two clusterings, which may be used to gauge quantitative differences between them. It should be interpreted as the number of nodes required to change location in one clustering in order to obtain the other (25).

In our example we obtain distances 2,977 and 3,949 respectively, translating to node fractions of approximately 9% and 13%. This indicates that the three clusterings are genuinely very different in terms of granularity. In contrast, the clustering obtained with inflation 7 (using –I 7) has a distance of 364 to the result at inflation 6. This translates to a fraction of approximately 1% of nodes that needs relocation between the two clusters, indicating that in this example the clusterings obtained at high inflation levels are very stable. This likely means that the clusters represent tight sets of highly conserved sequences that do not easily split further.

The *dist* mode to **clm** can also be used to judge how far the projections computed by **clm order** are from the original MCL clusterings.

```
clm dist P1 out.seq.mci.I14
clm dist P2 out.seq.mci.I20
clm dist P3 out.seq.mci.I60
```

The answers obtained in our case study are {0, 32, 102}, meaning, for example, that only 102 nodes need relocation between P3 and out.seq.mci.I60. To recap, the projected clusterings are strictly nesting, and additionally very close to the MCL clusterings from which they were derived. This implies that the latter are *almost* fully nesting, and hence represent compatible results at different levels of granularity. To obtain a strictly nested set of clusterings the program **clm order** can be used.

10. Finally, **clm** provides an **info** mode that gives a simple numerical criterion for a clustering. It rewards clusterings both for being granular and for capturing many edges in the input graph. Its criterion lies in the range [0–1] and achieves 1 only for the natural clustering of a graph that falls apart into node sets (components) such that no edges between different components exist and edges are present for all node pairs that fall within the same component. In addition, it is affected by differentiation among the edge weight. It is not intended as an optimisation criterion, but can be used to detect trends and optionally to find bad apples (clusterings).

```
clm info seq.mci out.seq.mci.I{14,20,60}
```

In our example the results are {0.76, 0.82, 0.86}, compared with an efficiency of 0.47 for the baseline clustering of maximal components in the input graph. This suggests satisfactory consistency and performance. As a general guideline, as graphs include more distant relationships (lower similarities) and as inflation is lowered (leading to coarser clusterings), this criterion will degrade. This does not mean those results are invalid. However, a sharp drop in this criterion could be reason to distrust the pertinent clustering.

3.2. Protocol 2: Clustering Gene Expression Profiles

The protocol uses as example a tabular file ecoli.exprs containing expression values for genes across a series of experiments. The program **mcxarray** is used to create a network file and a dictionary file.

1.
```
mcxarray -data ecoli.exprs -co 0.2 -skipr 1 -skipc 1 -tf 'abs()' ↩
-o ecoli20.mci -write-tab ecoli.dict
```

In this step, we create a test network for the purpose of finding a reasonable correlation cutoff. Hence, a very low correlation cutoff of 0.20 is used. The –tf 'abs()' instruction reverses the sign of negative values, implying that we consider correlations and anticorrelations alike. The first line, containing column names, is skipped according to –skipr 1. The –skipc 1 instruction causes the first column in the tabular input file to be skipped, and numerical input will be expected in all other columns. The program is asked to write labels for the rows with –write-tab ecoli.dict. and it deduces that these labels must be found in the first (skipped) column.

2. mcx query -imx ecoli20.mci - -vary-correlation
The correlation cutoff is varied from 0.20 to 0.95 in increments of 0.05, and at each cutoff the graph resulting from removing lower correlations is analysed. Reported attributes of the resulting graph include the size of its largest component, the number of singletons (isolated nodes), and node degree and edge weight statistics (mean, median, and inter-quartile range). In this example, we are prepared to choose a relatively low threshold as the genes were measured across many experiments. We wish to have a small number of singletons (in the initial graph) while keeping the average node degree in check. In this example, we choose a cutoff of 0.60, at which 7% of the nodes are singletons (not connected to any other node), and at which the node degree mean and median are 98 and 36, respectively.
3. mcx alter -imx ecoli20.mci -tf 'gq(0.6), add(−0.6)' -o ecoli60.mci
After choosing a correlation cutoff in the previous step, we create a more selective network by applying this cutoff, as instructed by the gq(0.6) part of the argument to –tf. The correlations are shifted to the interval [0–0.4] by subtracting 0.6 from all the values, increasing the contrast between lower and higher correlations, as required for MCL. It is suggested to set the shift value to the cutoff value.
4. We apply **mcl** with the same set of inflation values as before. After this, the programs from the first protocol can be used as before.

```
mcl ecoli60.mci -I 1.4
mcl ecoli60.mci -I 2
mcl ecoli60.mci -I 6
```

3.3. Application of the Method to Study Case 1: Clustering Based on Sequence Similarity

In Fig. 1 we have plotted the cumulative size distribution of all clusters obtained at different inflation values. For a given cluster size *N* represented along the *X*-axis, the *Y*-axis represents the fraction of all genes contained in a cluster of size at most *N*. Notice that at any inflation value there is a large number of clusters of size 28. These represent well-defined groups of orthologs, containing one protein from each species. In order to visualise what is happening during the clustering process, we highlight three examples and track them across different inflation values (protein annotation was obtained from Uniprot (24)).

RNA polymerase σ^{70} (RpoD) is required for the transcription initiation of housekeeping genes. Rickettsiales also encode a heat-shock sigma factor σ^{32}. The connected components of the network, as well as low inflation clusterings, group all the sigma factors together. At inflation 2 and above, they correctly split into two clusters. The σ^{70} cluster contains 30 proteins, one from 27 of the species and three from *Rickettsia bellii* OSU 85–389 whose σ^{70} is

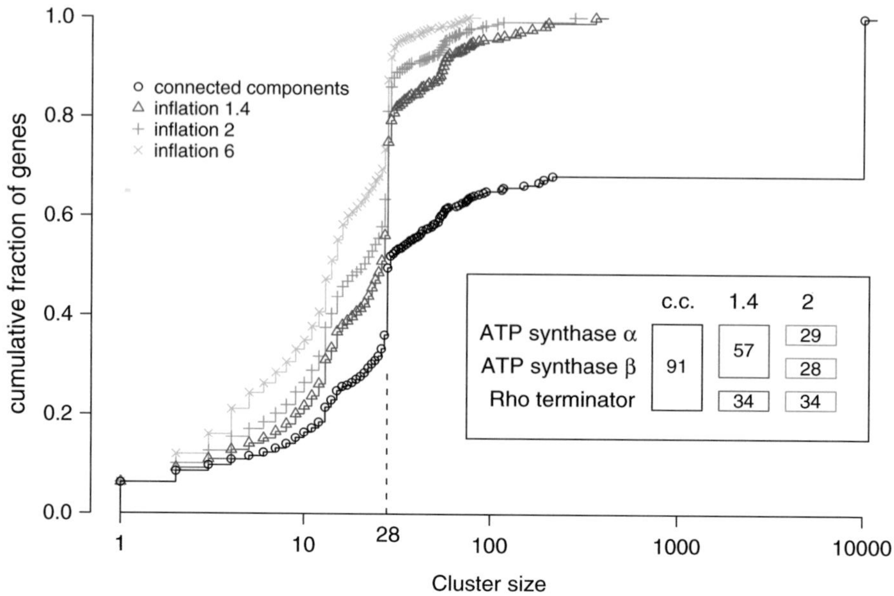

Fig. 1. Cluster size distributions of the Uniprot sequence similarity network.

annotated as being split into three separate proteins. The σ^{32} cluster contains the remaining 27 sigma factors, with *Rickettsia bellii* OSU 85–389 lacking a σ^{32} in its annotated proteome.

Bacterial ATP synthase is a multimeric enzyme comprised of eight subunits. The α and β subunits are homologous, sharing a highly conserved nucleotide-binding domain which is also present in the transcription termination factor Rho. Initially, 91 of these proteins are connected, but different levels of inflation can split them apart. As seen in the inset of Fig. 1, low inflation generates two clusters, one with 57 ATP synthase subunits and one with 34 Rho terminators (several *Anaplasma* and *Ehrlichia* encode two copies of Rho). At inflation values of 2 and above, the ATP synthases correctly split into two clusters. The first one contains 29 α subunits, with *Rickettsia bellii* RML369-C encoding a split protein. The second cluster contains the 28 ATP synthase subunit β proteins.

ATP-binding cassette (ABC) transporters represent one of the largest protein families in bacteria. In the unclustered network, ABC transporters lie in the single largest connected component of 10,012 proteins. Many other families of proteins are also included in this component, for instance DNA polymerase I. This protein in *R. felis* contains a *Rickettsia* palindromic element (26), giving rise to low-score BLAST hits to many unrelated protein families. These low-weight connections are amongst the first to be split by MCL, leading to the separation of DNA polymerase I and other orthologous groups at the lowest levels of inflation. Increasing inflation leads to further splitting of the largest cluster, with high

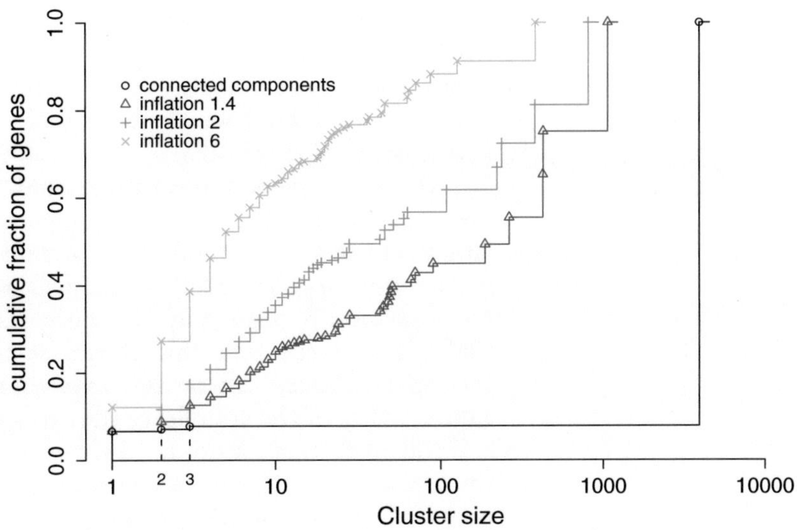

Fig. 2. Cluster size distributions of the *Escherichia coli* gene expression network.

values leading to many ABC transporters being clustered with consistent functional annotation, such as the Zinc import ATP-binding protein ZnuC, the Ribonucleotide ABC transporter, and the Glutamine ABC transporter. At inflation of six the largest remaining cluster contains exclusively ADP/ATP carrier proteins, a family of proteins required by Rickettsiales to obtain ATP from the host cell in exchange for ADP.

3.4. Application of the Method to Study Case 2: Clustering Based on Expression Profiles

The correlation of gene expression values yields a very different network, as seen in Fig. 2. Due to the low correlation cutoff (0.6), almost all the genes are initially part of a single connected component. Increasing inflation continuously breaks the network down, with a tendency at the end towards clusters with two or three genes, mostly representing operons. Unless cited, the information discussed in this section was obtained from EcoCyc (23).

One of the most consistent clusters comprises a set of genes related to motility. This cluster includes the flagellum operons: *flgAMN*, *flgBCDEFGHIJ*, *flgKL*, *flhBAE*, *flhDC*, *fliAZY*, *fliC*, *fliDST*, *fliE*, *fliFGHIJK*, *fliLMNOPQR*, and *motAB*; chemotaxis genes: *cheAW*, *tar-tap-cheRBYZ*, *trg*, and *tsr*; the aerotaxis gene *aer* and several others. This suggests that other members of this cluster are likely to be functionally connected to motility, giving us novel predictions for genes with no known function such as *ymdA*, *ynjH*, *yjcZ*, and *yecR*.

Another interesting cluster contains genes related to the SOS response, which is induced by DNA damage (27). At lower inflation values it also contains many prophage genes. At high inflation levels, all of the prophage genes are split off, leaving a core set of genes all of which are involved in the SOS response and DNA repair: *lexA-dinF*, *dinB*, *dinD*, *dinG*, *dinI*, *recAX*, *recN*, *sulA*, *symE*, *umuDC*, *uvrB*, *ydjM*, and *yebG*. The set of prophage genes which were initially in the cluster could also merit further analysis, since silent bacteriophage genes are known to be induced by the SOS response (27).

Many groups of genes with related functions are less tightly co-regulated than the previous examples. For instance, one of the largest clusters at low inflation contains the genes for all three nitrate reductases encoded by *E. coli* as well as many other reductase enzymes. By using any GO enrichment tool such as DAVID (28) we can see that this cluster is highly enriched for the Biological Process *anaerobic respiration*. At higher inflation values, many of the complexes start to split off into new clusters, and the enrichment score for *anaerobic respiration* falls. Some of the first genes to form a new cluster are *ccmBCDEFGH*, components of the CcmCDE protoheme IX reservoir complex and the CcmEFGH holocytochrome c synthase. At higher inflation these complexes separate into two distinct clusters. Meanwhile the nitrate reductases are also split into different clusters, containing the structural genes for each complex as well as some of its accessory proteins. An interesting point is that a single operon consists of *napFDAGHBC-ccmABCDEFGH*, with alternative promoters allowing distinct transcription of *napFDAGHBC* and *ccmABCDEFGH*. Nevertheless, the clustering process is able to separate all these genes into three groups that are consistent with the complexes they are known to form.

4. Conclusions

With these study cases we have illustrated the power of MCL to cluster networks of different complexity, even without any prior preprocessing. We have shown the relevance of varying the inflation parameter to obtain different clusterings, all of which can be interesting and biologically relevant. We remind the practitioner that different problems require different considerations, and preprocessing the underlying network may be desirable to accommodate focused research questions.

References

1. van Dongen S. (2000) A cluster algorithm for graphs. Tech. rep., National Research Institute for Mathematics and Computer Science in the Netherlands, Amsterdam.
2. van Dongen S. (2000) Graph clustering by flow simulation. PhD thesis, University of Utrecht.
3. van Dongen S. (2008) Graph clustering via a discrete uncoupling process. SIAM J Matrix Anal Appl, 30:121–141.
4. Enright A, van Dongen S, Ouzounis C. (2002) An efficient algorithm for the large-scale detection of protein families. Nucleic Acids Res, 7:1575–1584.

5. Enright AJ, Kunin V, Ouzounis CA. (2003) Protein families and TRIBES in genome sequence space. Nucleic Acids Res, 31:4632–4638.
6. Li L, Stoeckert C, Roos D, OrthoMCL. (2003) Identification of ortholog groups for eukaryotic genomes. Genome Res, 13:2178–2189.
7. Pereira-Leal JB, Enright AJ, Ouzounis CA. (2004) Detection of functional modules from protein interaction networks. Proteins, 54:49–57.
8. Brohée S, van Helden J. (2006) Evaluation of clustering algorithms for protein-protein interaction networks. BMC Bioinformatics, 7:488.
9. Samuel Lattimore B, van Dongen S, Crabbe MJ. (2005) GeneMCL in microarray analysis. Comput Biol Chem, 29:354–359.
10. Freeman TC, et al. (2007) Construction, visualisation, and clustering of transcription networks from microarray expression data. PLoS Comput Biol, 3:2032–2042.
11. Lopez F, et al. (2008) TranscriptomeBrowser: a powerful and flexible toolbox to explore productively the transcriptional landscape of the Gene Expression Omnibus database. PLoS ONE, 3:e4001.
12. Theodosiou T, et al. (2008) PuReD-MCL: a graph-based PubMed document clustering methodology. Bioinformatics, 24:1935–1941.
13. Hubbard TJ, et al. (2009) Ensembl. Nucleic Acids Res, 37:D690–697.
14. Chen F, et al. (2007) Assessing performance of orthology detection strategies applied to eukaryotic genomes. PLoS ONE, 2:e383.
15. Theocharidis A, et al. (2009) Network visualization and analysis of gene expression data using BioLayout Express(3D). Nat Protoc, 4:1535–1550.
16. Brohee S, Faust K, Lima-Mendez G, Sand O, Janky R, Vanderstocken G, Deville Y, van Helden J. (2008) NeAT: a toolbox for the analysis of biological networks, clusters, classes and pathways. Nucleic Acids Res, 36: W444–W451.
17. King AD, Przulj N, Jurisica I. (2004) Protein complex prediction via costbased clustering. Bioinformatics, 20:3013–3020.
18. Darby AC, et al. (2007) Intracellular pathogens go extreme: genome evolution in the Rickettsiales. Trends Genet, 23:511–520.
19. d'Haeseleer P. (2005) How does gene expression clustering work? Nat Biotechnol, 23:1499–1501.
20. van Noort V, Snel B, Huynen MA. (2003) Predicting gene function by conserved co-expression. Trends Genet, 19:238–242.
21. Faith JJ, et al. (2008) Many Microbe Microarrays Database: uniformly normalized Affymetrix compendia with structured experimental metadata. Nucleic Acids Res, 36:D866–870.
22. Gama-Castro S, et al. (2008) RegulonDB (version 6.0): gene regulation model of *Escherichia coli* K-12 beyond transcription, active (experimental) annotated promoters and Textpresso navigation. Nucleic Acids Res, 36:D120–124.
23. Keseler IM, et al. (2009) EcoCyc: a comprehensive view of *Escherichia coli* biology. Nucleic Acids Res, 37:D464–470.
24. Bairoch A, et al. (2009) The universal protein resource (UniProt) 2009. Nucleic Acids Res, 37:D169–D174.
25. van Dongen S. (2000) Performance criteria for graph clustering and Markov cluster experiments. Tech. rep., National Research Institute for Mathematics and Computer Science in the Netherlands, Amsterdam. [http://www.cwi.nl/static/publications/reports/INS-2000.html].
26. Ogata H, Audic S, Barbe V, Artiguenave F, Fournier PE, Raoult D, Claverie JM. (2000) Selfish DNA in protein-coding genes of Rickettsia. Science, 290:347–350.
27. Neidhardt FC, Curtiss R. (1996) *Escherichia Coli and Salmonella: Cellular and Molecular Biology.* 2nd ed. ASM Press, Washington. [Walker GC. The SOS response of *Escherichia coli.* 1400–1416].
28. Huang DW, Sherman BT, Lempicki RA. (2009) Systematic and integrative analysis of large gene lists using DAVID bioinformatics resources. Nat Protoc, 4:44–57.

Chapter 16

Protein Complex Prediction with RNSC

Andrew D. King, Nataša Pržulj, and Igor Jurisica

Abstract

Graph theory analysis of biological networks, such as protein–protein interactions (PPIs), gene regulatory, metabolic, etc., has identified a strong relationship between topology of these networks and the underlying cellular function and biological processes (Sharan et al. Mol Syst Biol 3:88, 2007). We focus on PPI networks, in which nodes correspond to proteins and edges represent interactions among the proteins. The size of these networks is ever growing, and thus efficient identification of various network motifs and dense sub-networks has become necessary. Predicting highly connected sub-graphs in a PPI network is important to biologists as it may help to identify biologically meaningful protein complexes, and with further integrative analysis may lead to identifying dynamic assembly of individual subunits in these complexes.

In this chapter, we describe one method for predicting protein complexes in two steps. The first step is to partition the nodes of a PPI network (i.e. proteins) into highly connected groups or *clusters* using the Restricted Neighbourhood Search Clustering algorithm. This provides a set of clusters that represent candidate complexes.

The second step of the method is to filter the candidate complexes based on three criteria: minimum cluster size, minimum interaction density, and minimum *functional homogeneity*, which reflects the extent to which the proteins of the candidate cluster operate in the same functional group. Candidate complexes passing all three criteria are then put forward as predicted protein complexes. The effectiveness of this method is investigated in the previous studies (King et al. Bioinformatics 20:3013–3020, 2004; Brohee and van Helden BMC Bioinformatics 7:488, 2006; and Moschopoulos et al. BMC Bioinformatics 10 (Suppl 6):S11, 2009).

Key words: Protein–protein interaction network, Protein complex, Graph theory, Graph clustering, Local search, RNSC, Functional homogeneity

1. Introduction

Determining the ways in which proteins work together to achieve a given function is a primary focus of molecular biologists in the post-genomic era. Thus, it is important to identify these working groups, which are known as *protein complexes* or *multi-protein complexes*. In protein–protein interaction (PPI) networks, nodes correspond to proteins and edges correspond to interactions between the proteins.

Jacques van Helden et al. (eds.), *Bacterial Molecular Networks: Methods and Protocols*, Methods in Molecular Biology, vol. 804, DOI 10.1007/978-1-61779-361-5_16,

Thus, it is expected that cellular functions carried out by these groups of interacting proteins are reflected in the structure (or topology) of PPI networks (also called *graphs*) (1–3). Since wet-lab experiments for discovering PPIs are not only noisy (4) but also costly and time consuming relative to the wealth of biological data to be analyzed, it is useful to direct such experiments by predicting protein complexes through computational analyses of existing PPI datasets. One such method involves observing characteristics of known protein complexes and putting forth predicted candidate complexes based on how strongly they display these same characteristics. Relaxing the selection criteria of candidate complexes leads to a broader, less-exclusive set of predicted complexes – as a result, the method is robust in the face of imperfect, high-throughput PPI datasets.

This chapter describes such a method that identifies dense network sub-graphs, *or clusters*, and predicts them as protein complexes based on how well they satisfy certain characteristics of true protein complexes (see below). The method was used to predict complexes based on PPI datasets for yeast *Saccharomyces cerevisiae* (5), fly *Drosophila melanogaster* (6), worm *Caenorhabditis elegans* (7), and bacterium *Escherichia coli*.

The method focuses on three specific characteristics of protein complexes: *density*, *size*, and *functional homogeneity*. *Density* refers to the number of interactions between the proteins of a given complex relative to the maximum possible number, i.e. $m/[n(n-1)/2]$ for a complex with n proteins and m interactions. Naturally, protein complexes are made up of proteins that interact extensively with one another, so we expect a complex to have a high interaction density. Protein complexes are, thus, believed to correspond to dense sub-graphs in a PPI network (8). While protein complexes can be made up of as few as two proteins, small complexes cannot be distinguished from other interacting proteins purely on the basis of network topology by examining data from PPI networks. Furthermore, larger predicted protein complexes are more likely to point us in the direction of a significant group of proteins working together. It should also be noted that protein complexes change dynamically (9), and may also form hierarchical structures of larger complexes being composed of smaller subunits. Thus, we restrict our view to predicted complexes surpassing a minimum size that depends on the PPI network in question. The third characteristic we focus on is functional homogeneity, which expresses the extent to which the proteins in a predicted complex work in the same functional group. Specifically, we consider a predicted complex's functional homogeneity P-value for a given functional group containing a disproportionate number of the complex's proteins. The P-value is the probability that a given protein set would contain so many proteins in a given functional group if the network's functional assignments were permuted at random (see ref. 2 for the formula).

Given a PPI network, the Restricted Neighbourhood Search Clustering (RNSC) algorithm first finds a set of dense sub-graphs or clusters. These clusters serve as the initial predicted complexes. Predicted complexes that are too small, too sparse, or with insufficient functional homogeneity are removed in the next step. The remaining clusters are the putative complexes.

Filtering clusters based on density, size, and homogeneity is straightforward. The main task is to partition the PPI network into dense clusters. This *graph clustering problem* has been studied extensively (see the survey (10)), and there are many approaches with their own advantages and disadvantages. The particular tool we use for clustering PPI networks is *RNSC*, a clustering algorithm developed by the first author and described in detail in refs. 2, 11, 12. RNSC is scalable and efficient, particularly for finding dense sub-graphs in globally sparse networks, such as PPI networks. Unlike clustering algorithms such as Markov clustering algorithm (MCL) (13), RNSC is stochastic and does not give the same output every time it is run. Thus, repeated applications of RNSC may offer a richer set of predicted complexes than a single run. Specifically, multiple runs of the method result in multiple predicted complex sets; protein groups that appear together in a predicted complex consistently over multiple runs are particularly compelling as candidate complexes. Advantages and disadvantages of RNSC and MCL have been explored in refs. 4, 14, and RNSC appears to be particularly effective when paired with simple refinement techniques (4). We now give an overview of the RNSC algorithm, and then discuss technical specifications and methods in Subheadings 2 through 4. Memory and time requirements of the algorithm, which are likely to be a concern only with very large networks, are discussed in Subheading 4.

1.1. An Overview of the RNSC Algorithm

RNSC is a cost-based clustering algorithm. This means that it uses a *cost function* to measure how bad (i.e. high cost) a given clustering of a network is. Given an initial clustering, RNSC makes small modifications that reduce the cost. A modification or *move* consists of assigning one node from its original cluster to a different cluster. This process is repeated until RNSC can no longer find a way to reduce the cost of the current clustering. The best (i.e. lowest cost) clustering found is then given as the output clustering.

Thus, RNSC is what is known as a *local search* algorithm because when it is given a clustering, RNSC only looks at a new clustering that it can reach by moving a single node. Intuitively, one may think of RNSC as an agent with very limited vision trying to reach a point at low altitude by taking one step at a time – the position represents the current clustering while the altitude represents the cost of the clustering.

In the simplest form of RNSC, a clustering is output as soon as the current clustering cannot be improved. However, a local optimum is not necessarily a good solution as it may not be even close to the global optimum (e.g. consider the agent who is searching for low ground, but who has the bad luck to stop in a pothole). For this reason, RNSC employs techniques to avoid stopping at such solutions.

RNSC has a *stopping tolerance*, which means that instead of stopping when the current clustering cannot be improved, RNSC stops when no improvement has been made in a fixed number of moves. To avoid repeatedly visiting the same solution, RNSC uses a *tabu list* of the most recently moved nodes. Nodes on the tabu list cannot be moved until a fixed number of other nodes have been moved. RNSC also employs two approaches to *diversification*, which means that every so often, a little bit of the current clustering is shuffled randomly. *Destructive diversification* involves destroying a randomly chosen cluster while *shuffling diversification* moves some randomly chosen nodes into randomly chosen clusters. While the stopping tolerance and tabu list help RNSC "climb" out of a local optimum, diversification has the effect of "ejecting" RNSC from a local optimum. If the optimum is truly a good solution, RNSC will likely return to it by making normal moves.

There are a few more complications. First, RNSC actually uses two cost functions: a *naïve cost function* and a *scaled cost function*. The naïve cost function is simpler and easier to compute, but the scaled cost function better expresses the quality of a clustering. So RNSC first searches for a good clustering using the naïve cost function. This can be done very quickly. RNSC then switches over to the scaled cost function and, using the naïve clustering as a fairly good starting point, seeks to minimize the scaled cost function, giving an even better final solution faster than it would if it used only the scaled cost function.

Second, RNSC does not choose moves in a deterministic way. Rather, it considers a set of moves that are *near optimal*, meaning that they decrease the cost of the current clustering by almost the maximum amount possible, and then chooses one such near-optimal move at random. Thus, RNSC may not always give the same output from a fixed input. To exploit this fact, RNSC performs a given number of *experiments* – an experiment consists of taking an initial clustering and searching for a good clustering. Thus, RNSC may run several times and take the best overall solution as its final output.

For a detailed description of RNSC, see ref. 10 and http://www.cs.utoronto.ca/~juris/data/rnsc/ (see also Note 1 for a cursory discussion of scalability and resource requirements). We now briefly review three practical applications of the RNSC algorithm.

1.2. Study Case 1: Yeast Complexes in a PPI Network

In ref. 2, we applied our method to the yeast *S. cerevisiae* interactome provided by ref. 3, consisting of the top 11,000 high-, medium-, and low-confidence interactions (henceforth denoted by Υ_{11k}). Using one specific run of RNSC, we predicted 45 protein complexes using size, density, and homogeneity *P*-value cut-offs of 5, 0.7, and 10^{-3}, respectively.

Thirty of the predicted complexes overlapped with known protein complexes; the rest may correspond to new complexes and warrant investigation. For example, a predicted complex with the density of 0.96 and *P*-value of 3.9×10^{-8}, containing the following eight proteins: COP1, RET2, RET3, SEC21, SEC26, SEC27, SEC28, and GPT2, closely matched the known coat protein complex COPI; COPI complex contained exactly the same set of proteins as the predicted complex, with the exception of GPT2 protein; instead, it contained ARF1 protein (3).

Note that some of the clusters identified by RNSC could have overlapped with the known protein complexes, but could have been missed by our approach due to the size, density, and functional homogeneity restrictions that we imposed on the identified clusters. For example, RNSC gave a cluster of two proteins, COX4 and COX6. Although this cluster matched a known protein complex exactly (3), our method did not predict it as a complex, since the cluster with only two proteins was too small to pass the size restriction.

As an example, Fig. 1 shows eight dense putative complexes in the yeast network, as visualized using NAViGaTOR ver. 2.1.14 (http://ophid.utoronto.ca/navigator) (17). RNSC used yeast protein interaction data from the I2D database ver. 1.71 (http://ophid.utoronto.ca/i2d) (15, 16). This network comprised 6,027 proteins and 120,974 interactions. RNSC identified 1,276 multi-protein complexes, of which 702 had density at least 0.8, 93 had size at least 9, and 36 satisfied both criteria.

1.3. Study Case 2: Worm Complexes in *Caenorhabditis elegans* PPI Network

In ref. 2, we also applied our method to the entire worm *C. elegans* interactome provided by ref. 5. We predicted 32 protein complexes using size and density cut-offs of 4 and 0.67, respectively; as no functional annotation data was available for the worm interactome, we did not filter the clusters with respect to functional homogeneity.

A promising prediction was the cluster containing the following eight proteins: F28F8.3, F32A5.7, F40F8.9, T10G3.6, Y39G8C.1, Y62E10A.12, Y71G12B.14, and ZK593.7. Due to its density of 1.00, meaning that every possible interaction was present amongst the eight proteins, it seemed extremely plausible that these proteins were working together as part of a true biological complex (provided the PPIs themselves were true, and not just inferred using a matrix-versus-spoke model).

1.4. Study Case 3: Predicted Complexes in *Escherichia coli* PPI Network

RNSC-predicted complexes in the *E. coli* PPI network from ref. 18 are highlighted in Fig. 2. The full network comprises 2,283 proteins and 7,613 interactions, giving it an overall interaction density of 2.9×10^{-3}. Five large and dense putative complexes have been identified: Cluster #0 with 16 nodes, density 0.675000; Cluster #1 with 15 nodes, density 0.590476; Cluster #2 with 15 nodes, density 0.780952; Cluster #3 with 15 nodes, density 0.733333;

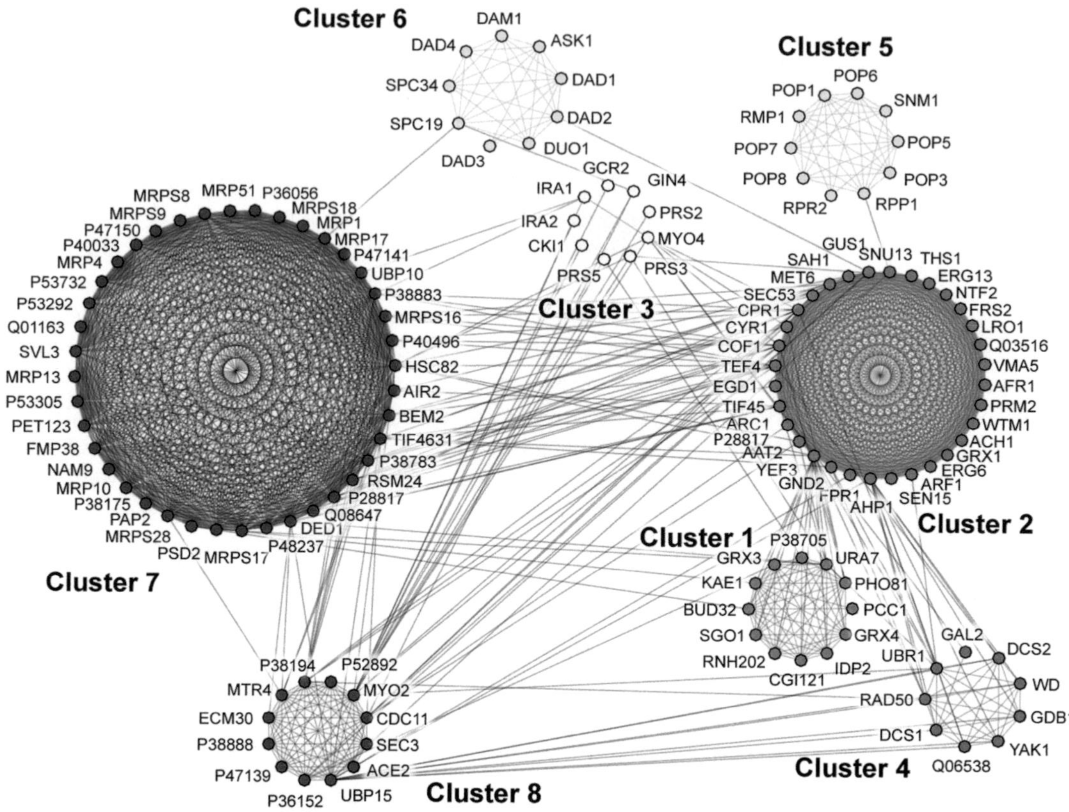

Fig. 1. Eight dense putative complexes identified by RNSC in the yeast protein interaction data. RNSC was run on the I2D yeast interaction database ver. 1.71 (http://ophid.utoronto.ca/i2d), a network of 6,027 proteins and 120,974 interactions (15, 16). Of the 1,276 multi-protein complexes output by RNSC, 702 had density at least 0.8, 93 had size at least 9, and 36 satisfied both criteria. This figure shows the first 8 of these 36 predicted complexes and the interactions between all member vertices. The largest consists of 40 proteins and has density 0.84. For simplicity, individual complexes and interactions within them are highlighted with different colour for nodes and edges. The network was visualized using NAViGaTOR 2.1.14 (http://ophid.utoronto.ca/navigator) (17). A NAViGaTOR XML file and a colour version of the figure are available at http://www.cs.utoronto.ca/~juris/rnsc.

Cluster #4 with 51 nodes, density 0.520784; and Cluster #5 with 15 nodes, density 0.523810. Two pairs of the putative complexes are further interconnected (shown with dashed edges): Clusters #2 and #4 are linked by 41 edges between 27 proteins, and clusters #2 and #5 are linked by 2 edges between 3 proteins.

2. Materials

2.1. Software

RNSC and its associated support programs run on a Unix platform. A basic knowledge of the Unix shell is required for its use. The main software necessary to apply our approach is the RNSC implementation available at http://www.cs.utoronto.ca/~juris/data/rnsc/. This .zip/.tar.gz archive also contains a program for parsing and analyzing output data. Installation instructions are

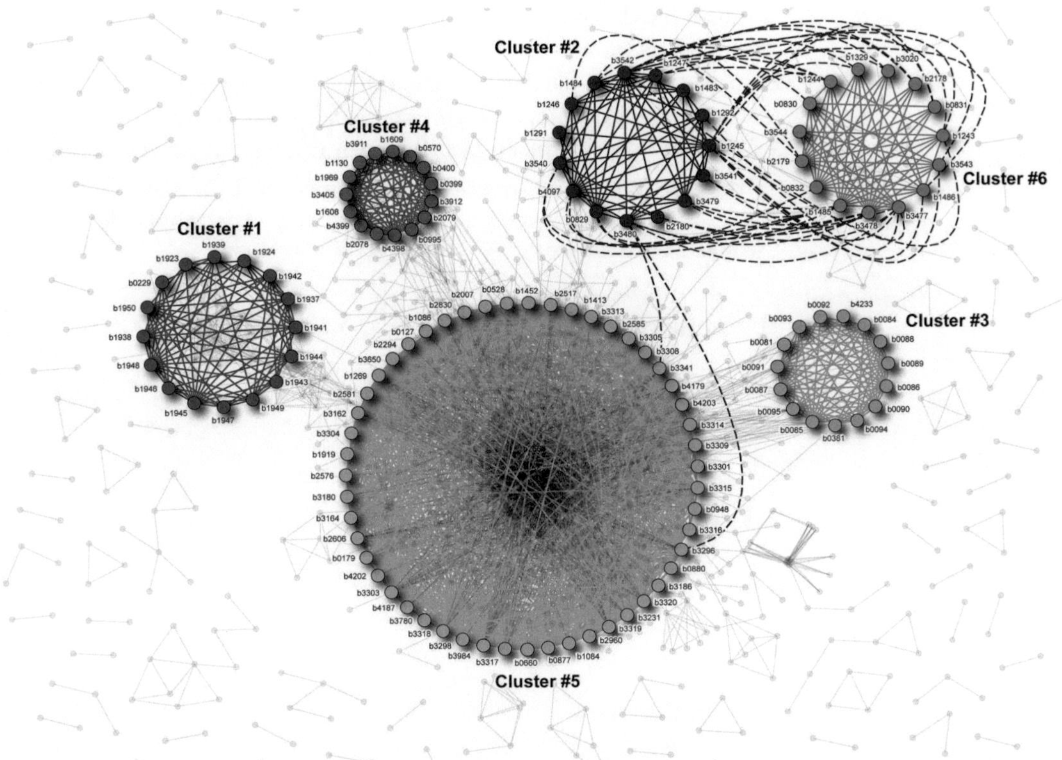

Fig. 2. Five dense putative complexes identified by RNSC in the *Escherichia coli* protein interaction data from ref. 19. RNSC was run on a network with 2,283 proteins and 7,613 interactions. Five large and dense putative complexes have been identified: *Cluster #0* with 16 nodes, density 0.675000; *Cluster #1* with 15 nodes, density 0.590476; *Cluster #2* with 15 nodes, density 0.780952; *Cluster #3* with 15 nodes, density 0.733333; *Cluster #4* with 51 nodes, density 0.520784; and *Cluster #5* with 15 nodes, density 0.523810. Two of the putative complexes are further interconnected (shown with *dashed edges*). *Clusters #2* and *#4* are linked by 41 edges between 27 proteins, and *Clusters #2* and *#5* are linked by 2 edges between 3 proteins. For simplicity, individual putative complexes and interactions within them are *highlighted* with *different shades* for nodes and edges, and other nodes and edges are kept partially transparent. The network was visualized using NAViGaTOR 2.1.14 (http://ophid.utoronto.ca/navigator) (17). A NAViGaTOR XML file and a colour version of the figure are available at http://www.cs.utoronto.ca/~juris/rnsc.

given in the next section; they require compilation with gcc version 3.0.0 or later. The other necessary software is an included program called *rnscfilter*, which filters the RNSC output clusters by size, density, and functional homogeneity, and then outputs the predicted complexes in a convenient human-readable format.

2.2. File Formats

The current RNSC implementation follows a strict but simple input/output file format protocol. There are two types of files: graph files and clustering files. In these files, each of the n nodes in the graph is represented by an integer between 0 and $n - 1$. The program *rnscfilter* uses an additional file type in which the names and functional groups of the proteins are stored. The *rnsc.zip* archive includes a program that can convert a verbal list of interactions and protein functions to an RNSC graph file and a protein name list.

2.2.1. Graph File (Input)

A graph file describes the graph to be partitioned using RNSC. It is described as an adjacency list, which lists each edge once. For each node in turn, the file lists: (1) the node label, followed by (2) each of the node's neighbours *with a greater label*, followed by (3) the number −1, used to signify the end of the given node's *greater neighbour list*.

Example: *The graph* C_5 consists of five nodes 0, 1, 2, 3, 4, with edges (0,1), (1,2), (2,3), (3,4), (4,0), respectively. The graph file for C_5 would be as follows:

```
<beginning of file>

0 1 4 -1

1 2 -1

2 3 -1

3 4 -1

4 -1

<end of file>
```

Notice that the edge (0,4) must appear in the list for 0, not the list for 4, because 4 is greater than 0.

2.2.2. Clustering File (Input and Output)

Clustering files represent a fixed clustering of a corresponding graph. This is the format of RNSC's output. A clustering file can also be loaded into RNSC as an initial clustering, which RNSC then works to improve. A clustering file describes a partitioning of the nodes of the graph into *k* clusters by listing, for each cluster in turn, the nodes of the cluster (in any order) followed by the terminating token −1. Thus, for the graph C_5, supposing that a clustering has 0, 1, 3 in one cluster and 2, 4 in another cluster, the clustering file might be:

```
<beginning of file>

0 3 1 -1

4 2 -1

<end of file>
```

2.2.3. Name File (Input)

RNSC inputs and outputs a network on n nodes using only numerical labels from 0 to $n-1$. When filtering and reporting predicted complexes with *rnscfilter*, the user may optionally include a *name* file. The name file consists of $2 \times n$ lines. For each node in turn, there is a line giving the name of the protein corresponding to the node on one line, and an uppercase letter corresponding to the protein's functional group on the next line (e.g. T for transport). Functional groups are used for determining functional homogeneity *P*-values of predicted complexes, as described in ref. 2, so if no functions are known or desired, arbitrary dummy values can be used without trouble. Each protein name must be alphanumeric, unique, and at most 15 characters in length. Note: If no name file is specified, or if a protein's functional group is not an uppercase letter, then the protein's functional group will be specified as U for unknown.

Example: If the nodes 0, 1, 2, 3, and 4 of the network C_5 correspond to proteins named ABC, ABC2, PROTEIN.34, 2647, and PROTEIN_X, with functional assignments A, U, T, T, and O, respectively, the name file would be as follows:

```
<beginning of file>
ABC
A
ABC2
U
PROTEIN.34
T
2647
T
PROTEIN_X
O
<end of file>
```

2.2.4. Verbal Interaction List

One common and easily manageable file format for PPI networks is a list of interactions, one on each line, in which an interaction is expressed as the name of the first protein, then a tab or space, and the name of the second protein. Given a two-column spreadsheet of this type, such a file can easily be generated by exporting the spreadsheet as a tab-delimited text file.

The program *rnscconvert*, included in *rnsc.zip*, can convert such a file into an RNSC graph file and a name file. As input, *rnscconvert* takes a file in which every line has the format:

Protein1 Function1 Protein2 Function2

The protein names are strings of alphanumeric characters and punctuation with no spaces. Protein names and functions must be separated by spaces, tabs, or line breaks. The function names, which are optional, are single letters denoting the protein's function. Protein names cannot be a single letter, although single digits are allowed. They are case sensitive. If the same protein is given two different functions, *rnscconvert* chooses the latest function. If a protein receives no function, *rnscconvert* gives it function U for unknown. Usage of *rnscconvert* is covered in the next section.

Subsequent versions of the program support a broader variety of functional labels, i.e. multi-character and case-sensitive labels.

Example: The network C_5, described in the previous examples, could be expressed as a verbal interaction list as follows:

```
<beginning of file>

ABC          A    ABC2

PROTEIN.34 T      ABC2

PROTEIN.34 T      2647         T

2647              PROTEIN_X O

ABC               PROTEIN_X O

<end of file>
```

3. Methods

3.1. Installation of RNSC, rnscfilter, and rnscconvert on a Unix Machine

We first describe the installation method for RNSC, *rnscfilter*, and *rnscconvert* on a Unix machine using the command line.

1. Make a directory called *PPI/*. Download the .zip archive found at http://www.cs.utoronto.ca/~juris/data/rnsc/rnsc.zip and put it in the *PPI/* directory. In the terminal command window, in the *PPI/* directory, run the command *unzip rnsc*.

zip. This should create directories *PPI/data/* and *PPI/source/* as well as files *README.TXT* and *makefile*. If you have trouble with the zip archive, you may prefer the .tar.gz archive http://www.cs.utoronto.ca/~juris/data/rnsc/rnsc.tar.gz, which is extracted with the command *tar xvzf rnsc.tar.gz*.

2. Compile the three programs by running the command *make* in the *PPI/* directory. This should compile all three programs, creating the executables *rnsc*, *rnscconvert*, and *rnscfilter* in the *PPI/* directory.[1] Compilation requires gcc version 3.0.0 or later, which can be installed by standard methods, for example the *apt-get* tool under Ubuntu or *yum* under Fedora.

3.2. Preparing the Data

As discussed in the previous section, RNSC graph and clustering files must follow a particular protocol. Graph and clustering files can be located anywhere relative to the *rnsc* file, but for the sake of simplicity let us assume that all input graphs are in a directory called *PPI/data/*, in the proper format. Example data sets are provided in the .zip archive and should appear in the *PPI/data/* directory upon extraction.

3.3. Converting a Verbal Interaction List

If the PPI network is not in RNSC format, you must first convert it to RNSC format from a verbal interaction list. The interaction list must be of the format specified in the previous section. For a list with filename *test.txt* in the *PPI/data/* directory, conversion would be done by running the following command from the *PPI/* directory:

```
./rnscconvert -i ./data/test.txt -g ./data/test.graph -n ./data/test.names
```

This runs *rnscconvert* on the input file *PPI/data/test.txt* (chosen with the *-i* parameter). There are two output files. The *-g* parameter specifies the filename to which the RNSC graph file is written, and the *-n* parameter specifies the filename to which the names file is written. In this case, the graph output file and the names output file are *PPI/data/test.graph* and *PPI/data/test.names*, respectively.

3.4. Running the RNSC Algorithm

We run RNSC from the command line in the *PPI/* directory as in the following example:

```
./rnsc -g ./data/test.graph -e50 -o ./data/test.clustering -n5 -N100 -D40

-d10 -t2
```

In this example, RNSC is run on the graph *PPI/data/test.graph*. Fifty experiments are run, each of which has naïve stopping tolerance 5, scaled stopping tolerance 100, diversification period

[1] Users sufficiently familiar with Unix systems can copy the binary executables to a directory in their path.

40, diversification length 10, and tabu length 2. The output, i.e. the best clustering from the 50 experiments, is written to the file *PPI/data/test.clustering*. (see Note 2 for more information on parameters for both RNSC and *rnscfilter*.)

3.5. Filtering the Output with rnscfilter

Once the output of RNSC is obtained, we must filter the predicted complexes. This involves discarding clusters that are too small, too sparse, or with insufficient functional homogeneity. This can be done using the *rnscfilter* program.

The *rnscfilter* program has two mandatory input files: a graph file (i.e. the RNSC input) and a corresponding clustering file (i.e. the RNSC output). In addition, it can input a protein name file, and cut-off values for minimum size, minimum density, and maximum functional homogeneity. Continuing the example, we run *rnscfilter* using the terminal command line from the directory *PPI/* as follows:

```
./rnscfilter -g ./data/test.graph -c ./data/test.clustering -n
./data/test.name -s3 -d 0.5 -o ./data/test.complexes
```

In this example, *rnscfilter* is run on the graph *./data/test.graph* and the RNSC output clustering *./data/test.clustering* with protein name file *./data/test.name*. Only clusters of size at least 2 with density at least 0.5 are output. No restriction is put on the functional homogeneity *P*-value. Predicted complexes are written to the output file *./data/test.complexes.*

3.6. Application of the Method to Study Case 1

We now run through an application of our method to the Υ_{11k} network. After saving the *rnsc.zip* archive in the directory *PPI/*, we first compile *rnsc* and *rnscfilter* with the following terminal command sequence. Note that the $ and preceding characters are not part of the command; they simply mark the current working directory:

```
~/PPI$ unzip rnsc.zip
~/PPI$ make
```

We have now compiled the necessary programs. We then place the input graph file *yeast11.graph* and its name file *yeast11.name* in the directory *PPI/data/* (these files are already included in this directory as examples). The next step is to cluster the network using RNSC:

```
~/PPI$ ./rnsc -g ./data/yeast11.graph -o ./data/yeast11.clustering -t10
-e3 -d10 -D40
```

This takes just a second or two, and writes the clustering file to *PPI/data/yeast11.clustering*. The next step is to filter the candidate complexes to generate our predicted complexes.

```
~/PPI$ ./rnscfilter -g ./data/yeast11.graph -c ./data/yeast11.clustering
-n ./data/yeast11.name -s5 -d 0.7 -p1e-3 -o ./data/yeast11.complexes
```

The resulting 48 predicted complexes were written to the file *PPI/data/yeast11.complexes.* Note that in ref. 2, only 45 complexes were predicted using the same method and cut-offs. This is explained by the fact that RNSC is a stochastic algorithm and does not always give the same output, so the number of predicted complexes is not necessarily equal between two runs of RNSC.

3.7. Application of the Method to Study Case 2

We now walk through the application of our method to the worm network, step by step. First, we extract and compile RNSC and *rnscfilter* as in the previous case. Note: This only needs to be done once, before the first application of RNSC. We run RNSC on the network file, stored at *PPI/data/worm.graph.*

```
~/PPI$ ./rnsc -g ./data/worm.graph -o ./data/worm.clustering -t10 -e3
-D40 -d10
```

This only takes about a second. Next, we run *rnscfilter* on the RNSC clustering. We did not have any functional data for the *worm* network, so in *worm.name*, every protein is assigned functional symbol "U". This means that we do not want to filter by functional homogeneity. We use size and density cut-offs 4 and 0.67, respectively:

```
~/PPI$ ./rnscfilter -g ./data/worm.graph -c ./data/worm.clustering -n
./data/worm.name -s4 -d 0.67 -o ./data/worm.complexes
```

This results in 32 predicted complexes being written to the file *PPI/data/worm.complexes.*

4. Notes

1. *Scalability and resource requirements.* RNSC can handle relatively large PPI networks with ease. Memory requirements are on the order of the number of interactions in the network, which is typically small in a PPI network. For example,

running RNSC on a network of 4,687 proteins and 45,000 interactions requires 17 MB, and on a network of 5,321 proteins and 78,390 interactions requires 26 MB of memory. Thus, the memory requirements for a PPI network can be estimated as about 1 MB per 3,000 interactions. Time requirements show slightly poorer scaling. Per experiment, on a computer with 2 GB of memory and a 2.26 GHz processor, PPI networks on 2,455, 11,000, 45,000, and 78,390 interactions required about 0.03, 0.35, 6.63, and 18.79 s, respectively. The scalability of RNSC is demonstrated by analyzing an extremely large network (15,447 nodes and 1,722,708 edges; "MouseNet" from http://www.functionalnet.org/mousenet/). RNSC ran the naive scheme in 6 min using 560 MB of RAM. Then, it ran the scaled scheme in 30 s using 280 MB of RAM. The drop in memory requirements comes from the fact that one key data structure (the list of moves being considered) has size proportional to the number of edges with endpoints in different clusters. For this large network, most of the edges get put in clusters straight away.

There is a clique of 1,448 nodes, which itself contains more than half of the edges in the graph. The experiment resulted in many very dense and quite large clusters. Of the 265 of size 10 or greater, 201 were cliques.

These tests were performed with the command:

```
~/PPI$ ./rnsc -g ./data/worm.graph -o ./data/worm.clustering -t10 -e3
-D40 -d10
```

Lower values of the scaled stopping tolerance N would result in significantly faster trials (see ref. 10 for empirical data on the relationship between parameter values and performance).

Future development of RNSC will likely include improved integration with standard file types and tools, support for edge-weighted graphs, and user-defined *sparsity bias*. In a PPI network, confidence values for the interactions, modelled as edge weights, reduce the impact of false positives in high-throughput datasets (4, 19). Sparsity bias affects the relative values of high cluster density versus low cluster intra-connectedness. This would correct RNSC's tendency to favour very small, dense clusters in a sparse network (13) and allow it to identify larger clusters with less emphasis on density. At present, this issue can be addressed by using the "*-c*" flag to specify a maximum number of clusters.

2. *Appropriate parameter choices.* RNSC is run with several parameters that affect its performance. These affect the mechanics of the search algorithm, generally trading a deeper and more

diverse search for speed. One parameter in particular, the *-c* option, allows the user to specify a maximum number of clusters (normally, there is no maximum). This can be used to force larger clusters in very sparse networks. Future versions of RNSC will implement more sophisticated bias parameters that handle this situation more effectively.

A list of parameters appears when RNSC is run with the *-h* flag or with no options; see the README.TXT file included with the source for a brief discussion of appropriate values. The *rnscfilter* program also requires several parameter choices, which are discussed in the same file.

When running *rnsc*, *rnscconvert*, and *rnscfilter*, make sure to choose output filenames that do not match important files. All three programs overwrite the output files, erasing the files' previous contents.

Acknowledgments

AK is supported by an NSERC Postdoctoral Fellowship. NP is supported by the NSF CAREER IIS-0644424, the NSF Cyber-Enabled Discovery and Innovation OIA-1028394, and the Serbian Ministry of Education and Science Project III44006 grants. IJ is supported in part by the Natural Sciences Research Council (NSERC #104105), Genome Canada via the Ontario Genomics Institute, Canada Foundation for Innovation (CFI #12301 and CFI #203383), Canada Research Chair Program, CIHR Catalyst grant (BIO-99745), and IBM. This research was funded in part by the Ontario Ministry of Health and Long Term Care. The views expressed do not necessarily reflect those of the OMOHLTC. We thank Tijana Milenković for insightful comments.

References

1. Sharan R, Ulitsky I, Shamir R. (2007) Network-based prediction of protein function. Mol Syst Biol, 3:88.
2. King AD, Przulj N, Jurisica I. (2004) Protein complex prediction via cost-based clustering. Bioinformatics 20:3013–3020.
3. von Mering C, Krause R, Snel B, Cornell M, Oliver SG, Fields S, Bork P. (2002) Comparative assessment of large-scale data sets of protein–protein interactions. Nature, 417:399–403.
4. Yu H, Braun P, Yildirim MA, Lemmens I, Venkatesan K, Sahalie J, Hirozane-Kishikawa T, Gebreab F, Li N, Simonis N, Hao T, Rual JF, Dricot A, Vazquez A, Murray RR, Simon C, Tardivo L, Tam S, Svrzikapa N, Fan C, de Smet AS, Motyl A, Hudson ME, Park J, Xin X, Cusick ME, Moore T, Boone C, Snyder M, Roth FP, Barabasi AL, Tavernier J, Hill DE, Vidal M. (2008) High-quality binary protein interaction map of the yeast interactome network. Science, 322:104–110.
5. Li S, Armstrong CM, Bertin N, Ge H, Milstein S, Boxem M, Vidalain PO, Han JD, Chesneau A, Hao T, Goldberg DS, Li N, Martinez M, Rual JF, Lamesch P, Xu L, Tewari M, Wong SL, Zhang LV, Berriz GF, Jacotot L, Vaglio P, Reboul J, Hirozane-Kishikawa T,

Li Q, Gabel HW, Elewa A, Baumgartner B, Rose DJ, Yu H, Bosak S, Sequerra R, Fraser A, Mango SE, Saxton WM, Strome S, Van Den Heuvel S, Piano F, Vandenhaute J, Sardet C, Gerstein M, Doucette-Stamm L, Gunsalus KC, Harper JW, Cusick ME, Roth FP, Hill DE, Vidal M. (2004) A map of the interactome network of the metazoan *C. elegans*. Science, 303:540–543.

6. Sharan R, Ideker T. (2006) Modeling cellular machinery through biological network comparison. Nat Biotechnol, 24:427–433.
7. de Lichtenberg U, Jensen LJ, Brunak S, Bork P. (2005) Dynamic complex formation during the yeast cell cycle. Science, 307:724–727.
8. Schaeffer SE. (2007) Graph clustering. Comput Sci Rev, 1:27–64.
9. King AD (2004) *Graph clustering with restricted neighbourhood search*. Department of Computer Science, University of Toronto, Toronto.
10. King AD. (2005) *An efficient cost-based graph clustering algorithm*. McGill University, Montreal.
11. Van Dongen S. (2000) A Cluster Algorithm for Graphs. In *Information Systems*. National Research Institute for Mathematics and Computer Science in the Netherlands, Amsterdam, 1–40.
12. Brohee S, van Helden J. (2006) Evaluation of clustering algorithms for protein–protein interaction networks. BMC Bioinformatics, 7:488.
13. Moschopoulos CN, Pavlopoulos GA, Schneider R, Likothanassis SD, Kossida S. (2009) GIBA: a clustering tool for detecting protein complexes. BMC Bioinformatics, 10(Suppl 6):S11.
14. Milenkovic T, Przulj N. (2008) Uncovering biological network function via graphlet degree signatures. Cancer Informatics, 6:257–273.
15. Brown KR, Jurisica I. (2005) Online predicted human interaction database. Bioinformatics, 21:2076–2082.
16. Brown KR, Jurisica I. (2007) Unequal evolutionary conservation of human protein interactions in interologous networks. Genome Biol, 8:R95.
17. Brown KR, Otasek D, Ali M, McGuffin MJ, Xie W, Devani B, Toch IL, Jurisica I. (2009) NAViGaTOR: network analysis, visualization and graphing Toronto. Bioinformatics, 25:3327–3329.
18. Peregrin-Alvarez JM, Xiong X, Su C, Parkinson J. (2009) The modular organization of protein interactions in *Escherichia coli*. PLoS Comput Biol, 5:e1000523.
19. Bader GD, Hogue CW. (2003) An automated method for finding molecular complexes in large protein interaction networks. BMC Bioinformatics, 4:2.
20. Giot L, Bader JS, Brouwer C, Chaudhuri A, Kuang B, Li Y, Hao YL, Ooi CE, Godwin B, Vitols E, Vijayadamodar G, Pochart P, Machineni H, Welsh M, Kong Y, Zerhusen B, Malcolm R, Varrone Z, Collis A, Minto M, Burgess S, McDaniel L, Stimpson E, Spriggs F, Williams J, Neurath K, Ioime N, Agee M, Voss E, Furtak K, Renzulli R, Aanensen N, Carrolla S, Bickelhaupt E, Lazovatsky Y, DaSilva A, Zhong J, Stanyon CA, Finley RL Jr, White KP, Braverman M, Jarvie T, Gold S, Leach M, Knight J, Shimkets RA, McKenna MP, Chant J, Rothberg JM. (2003) A protein interaction map of *Drosophila melanogaster*. Science 302:1727–1736.

Chapter 17

Network Analysis and Protein Function Prediction with the PRODISTIN Web Site

Anaïs Baudot, Ouissem Souiai, and Christine Brun

Abstract

Interactions between macromolecules are deciphered to gain information about biological processes and protein function. This information is hidden in large interaction networks, yet very complicated to dissect. In this context, the PRODISTIN Web site is dedicated to the clustering of network proteins according to the identity of their interaction partners, and to the subsequent functional annotation of these clusters. It allows analysing functionally networks and eventually leads to the prediction of function for uncharacterized protein based on their belonging to protein clusters. PRODISTIN analyses also provide an overview of the different biological processes existing in a given interactome. Here, we present a step-by-step procedure to analyse interaction networks using the PRODISTIN Web site. The protocol is illustrated by an application to the *Campylobacter jejuni* interactome.

Key words: Interaction networks, Maps, Graph, Partitioning, Functional classification, Function prediction

1. Introduction

Protein function can be viewed at different levels, from molecules to organisms, depending on the organisational level under consideration. However, in practice, protein function(s) are mainly defined at the molecular level (i.e. a protein is a kinase or a transcription factor). Nevertheless, proteins do not function isolated but rather in combination in order to achieve their goals at the cellular level. These roles may correspond to molecular machines, signalling pathways or biological processes in which proteins are involved. Hence, to fully understand the function(s) of one or several proteins, as well as to have a system-level overview of the cellular mechanisms, the ensemble of the molecular interactions, among which protein–protein interactions are of paramount importance, have to be deciphered.

Jacques van Helden et al. (eds.), *Bacterial Molecular Networks: Methods and Protocols*, Methods in Molecular Biology, vol. 804, DOI 10.1007/978-1-61779-361-5_17,

1.1. Protein Interaction Maps

The complete set of protein–protein interactions of an organism – also called its *interactome* (1) – is often represented as a graph, or a network, in which the nodes correspond to the proteins and the edges the interactions between the proteins. Thanks to the development of high-throughput screenings, these maps are nowadays available for many organisms but still incomplete. Indeed, different techniques have been applied to organism complete genomes to search for protein complexes and binary protein–protein interactions (see Chapter 2 by Bouveret and Brun). In bacteria, large-scale affinity purification screenings have identified complexes in *Escherichia coli* (2–4) and *Mycoplasma pneumoniae* (5), and 2-hybrid screenings have been successfully applied to *Helicobacter pylori* (6), *Synechocystis* (7), *Mesorhizobium meloti* (8), *Campylobacter jejuni* (9), and *Treponema pallidum* (10) (for an overview of the protein interaction data available in bacteria, see Chapter 2 by Bouveret and Brun). These experiments have generated a large number of protein–protein interaction data (for instance, 3,174 protein–protein interactions of *C. jejuni* are represented as a network in Fig. 1). However, despite the importance and abundance of the information contained in these large interaction networks, their complexity hinders any intuitive understanding.

1.2. Network Partitioning Methods for Interactome Analyses

Bioinformatics methods have been proposed to extract information from interaction networks, and to predict protein functions. In particular, some of them focused on the partitioning of the interaction network into functional modules (see Chapter 15 by van Dongen and Abreu-Goodger and Chapter 16 by King et al.). These modules may correspond to groups of proteins involved in the same pathway, the same protein complex or the same cellular process. Based on the belonging of a protein of unknown function to some of these modules, a putative function can then be inferred for the protein (for a review on function prediction, (11)).

A number of these network analysis methods have been grounded on principles that derive from graph analyses. Noticeably, a functional module or a group of proteins that are functionally related can be deduced from: (1) a search for graph regions particularly densely populated by interactions (12, 13); (2) the similarity between the shortest paths in the graph (14); (3) the progressive pruning of the graph until disconnection according to the "betweenness" (15, 16); (4) the sharing of interactors (17, 18) or a combination thereof (19) (for a review on methods, (20); for a comparison of the performances of such methods, (21)). Overall, these methods use network topological criteria to identify modules.

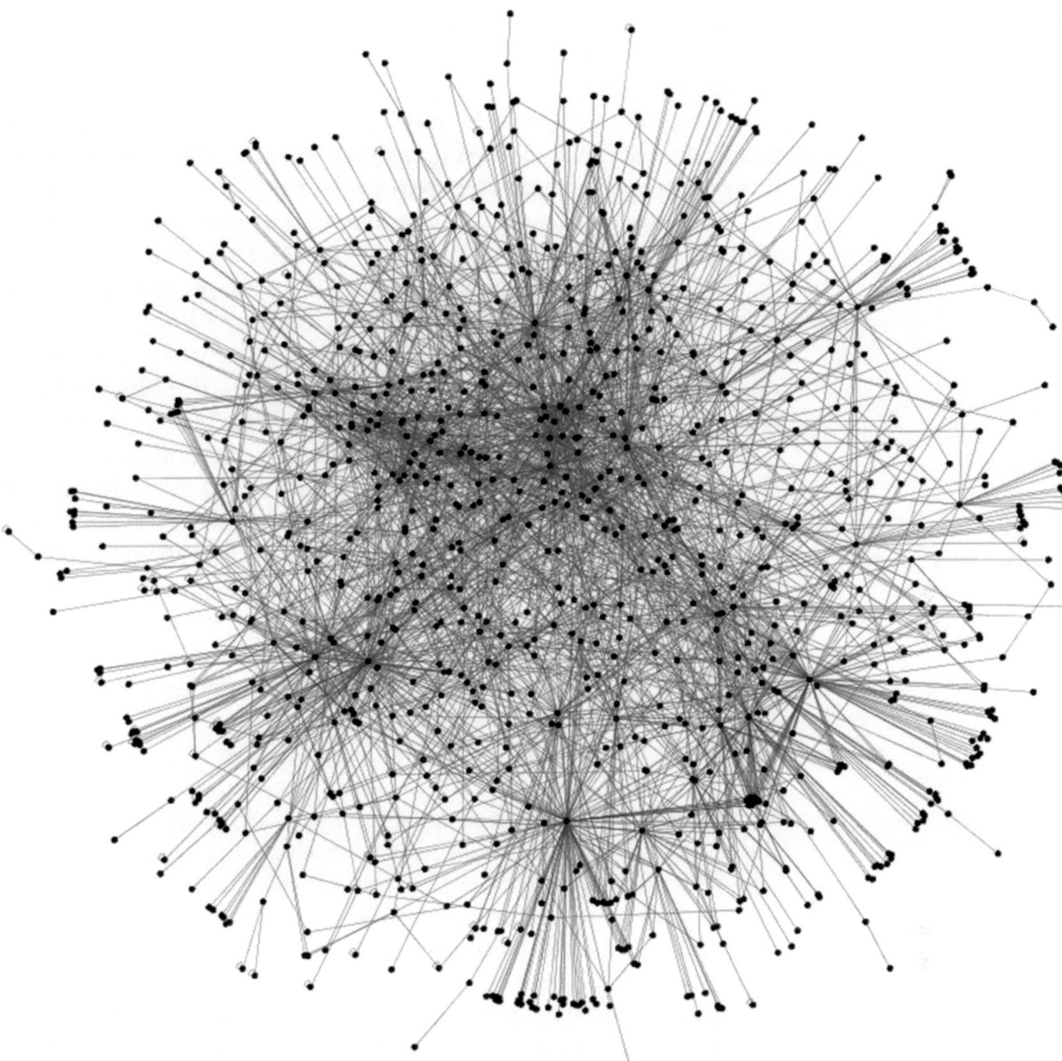

Fig. 1. A high quality protein–protein interaction map of the Campylobacter jejuni, after ref. 8.

1.3. The PRODISTIN Method

It was soon noticed that proteins of similar cellular functions tend to lie within a short distance in the interaction network (22, 23). This assertion underlined the development of the PRODISTIN method, which is based on the biological hypothesis that the more two proteins share interactors, the more likely they should be functionally related. Following this principle, the method computes a functional dissimilarity index between all protein pairs in the interaction network with regard to the number of interactors they share. This index corresponds to the Czekanowski-Dice formula (24):

$$D(i,j)=\frac{|Int(i)\Delta Int(j)|}{|Int(i)\cup Int(j)|+|Int(i)\cap Int(j)|}$$

in which i and j denote two proteins, Int(i) and Int(j) are the lists of their interactors plus themselves and Δ is the symmetrical difference between the sets.

A classification tree is then built from the computed values using the BIONJ clustering algorithm (25). This tree allows representing the relative relatedness of the proteins according to the similarity of their list of interactors. The proteins classified in the tree will be further referred as "classified proteins".

In the final step, the method identifies classes of functionally related proteins by using their Gene Ontology biological process (GOBP) annotations (26) as well as the topology of the classification tree. Throughout, we refer to the output of the PRODISTIN method as "functional or PRODISTIN classes".

It has been shown that the PRODISTIN method is able to (1) cluster proteins participating in the same cellular processes in the same functional classes, (2) predict function for unknown proteins, and (3) is statistically valid (17). In addition, the method has been successfully used to investigate the evolutionary fates of the yeast duplicated genes (27), the modularity of the drosophila signalling network (28), and the genetic network of *Caenorhabditis elegans* (29). PRODISTIN was implemented as a Web server and is available to the community (30).

1.4. Study Case : The Interactome of C. jejuni

C. jejuni is a pathogenic bacterium causing gastroenteritis in human. Despite its importance as a pathogen, little is known about its biology, neither about the mechanisms causing diseases. The first complete genome of this species, from strain NCTC 11168, has been released in 2000 (31), and a second strain, RM1221, in 2005. Nevertheless, similarly to other newly sequenced genomes, the prediction of function solely based on sequence similarities with known proteins in other organisms is only successful for a subset of the proteins (see Chapter 2 by Bouveret and Brun). Indeed, although 77.8% of the NCTC11168 strain proteins have recognisable orthologs in other species (31), *C. jejuni* proteins lack functional annotations, in particular at the biological process level. This is inherently due to the inference of function(s) based on sequence similarities which usually only give information about the molecular functions of the proteins. In this context, the protein interaction map released in 2007 (Fig. 1), that implicates ~80% of *C. jejuni* proteins of the NCTC11168 strain (9), could provide new insights into protein cellular functioning. For this, we have applied the PRODISTIN method to the *C. jejuni* protein–protein interaction network.

Noticeably, in order to circumvent the lack of annotations of the NCTC11168 strain, which will have impaired the use of PRODISTIN on this interactome, we further used the Gene Ontology annotations available for the RM1221 strain. For this, we have transformed the protein names into their orthologs in the strain RM1221.

2. Material

2.1. Software

The PRODISTIN Web site is available at http://crfb.univ-mrs.fr/webdistin/. The Web site application is mainly written in PHP with interfaces to Perl scripts and C programs.

2.2. File Formats

2.2.1. Input File

The protein–protein interaction file is a text-formatted file respecting the following instructions:

1. The first line must contain the number of binary interactions described in the file.
2. Each line contains a couple of interacting proteins separated by spaces or tabs. Because the network interactions are considered as non-oriented, the order of the protein names in each row does not matter.
3. For protein name requirements, see the dedicated "Gene/Protein names" paragraph below.

   Example of input protein interaction file:

```
12012
CJE_1626    CJE_1174
CJE_1626    CJE_0043
..........
CJE_0043    CJE_0661
```

2.2.2. Output Files (Fig. 2)

Several files are built during the different steps of the PRODISTIN Web site processing. The program successively generates the files described below, which are downloadable for further analyses (see Note 1).

1. The "Classified gene list" file (.txt) is the list of proteins from the original interaction dataset that are recovered in the classification tree (see Note 3).
2. The newick file (.nw) contains the description of the classification tree in a standard tree format.
3. The "Biological Process annotations for classified proteins" file (.tlf2) contains for each tree protein, the GOBP terms and their ancestor terms.

4. The "PRODISTIN classes" file (.txt) describes each PRODISTIN class. (CA), for Class Annotations, precedes the list of GOBP terms annotating the class, (P#) denotes the number of proteins in the class and (PN) the name of proteins composing the class.
5. The "Biological Process annotations for class" file (.tlf) contains, for each protein, the GOBP annotation of the class it belongs to.
6. The "Annotated tree for TreeDyn" file (.tgf) can be directly loaded into the TreeDyn tree editor tool ((32), http://www.treedyn.org/).
7. Image files of the coloured classification tree, final results of the PRODISTIN Web site are available in gif and PDF formats.

2.2.3. Gene/Protein Names

The classification tree can be computed for any protein interaction network file. However, in order to recover the GOBP annotations of each protein, the appropriate gene/protein names, matching the ones used in the Gene Ontology annotation files have to be used (http://www.geneontology.org/GO.current.annotations.shtml). For bacterial gene names, they correspond to the locus name used by the TIGR Comprehensive Microbial Resources (http://cmr.jcvi.org/tigr-scripts/CMR/CmrHomePage.cgi).
Currently, 25 species are available in the PRODISTIN Web site, among which nine are bacterium: *Silicibacter pomeroyi dss-3*; *Shewanella oneidensis mr-1*; *Pseudomonas syringae dc3000*; *Methylococccus capsulatis bath*; *Listeria monocytogenes 4b*; *Geobacter sulfurreducens pca*; *Dehalococcoides ethenogenes 195*; *Campylobacter jejuni rm1221*; *Coxiella burnett rsa493* (see Note 2).

2.3. Sample Files

The list of 3,174 high quality protein–protein interactions detected in *C. jejuni* by Parrish et al. (9) is available as Supplementary Material (http://tagc.univ-mrs.fr/SupplementaryFile_Chapter17_MiMB.htm).

3. Methods

3.1. General Indications

The method proceeds in two successive steps. First, the similarity between protein pairs is computed from the protein–protein interaction list. Then, the obtained values are clusterized in a classification tree. Secondly, the GO annotations of each classified proteins are retrieved from a local database and used to identify the subtrees corresponding to functional classes.

3.2. Step-by-Step Description of the Method

Step 1: From a binary interaction list to a classification tree (Fig. 2a)

1. Open a connection to the Protistin Web site (http://crfb.univ-mrs.fr/webdistin/).
2. On the top of the screen, click on the "Start" link.
3. Choose the model organism from the "species" selection list (see Note 2).
4. Upload your interaction file using the "list of interactions" file input field proposed by the PRODISTIN Web site. The uploaded file must comply with the format described in the previous section.
5. Set the minimal "gene/protein connectivity" to be classified. This corresponds to the minimum number of interactions a protein should have in the network to appear in the final tree. This cut-off is set by default to 3, but can range from 1 to 12 interactors (see Note 3).
6. Click on the "Compute" button to build the classification tree. After this first step, the first two files described in the previous "output files" paragraph can be downloaded by clicking on the links in the box "Output files".

Step 2: Functional class identification and visualisation (Fig. 2b–d)

7. On the bottom of the first step result page, click on the button "*Proceed with step 2*", and in the next page, click "*Proceed with Class identification*". This starts the gene ontology annotation retrieval process: the GOBP annotations of the classified proteins are recovered from a local database in order to build the PRODISTIN classes (see Note 4).
8. A summary of the number of annotated and non-annotated proteins is given (Fig. 2b). At this step, the third output file containing protein annotations is generated (see Note 5).
9. Then, the identification of functional classes is ready to proceed. The user has to choose two parameters that will be used to define PRODISTIN classes. The variable X sets the minimum number of proteins sharing the same GOBP annotation per PRODISTIN class (by default, a PRODISTIN class has to contain a minimum of three proteins sharing the same annotation) (see Note 6). The variable Y sets the percentage of class members which should be annotated to the same GO term. Finally, it is possible to set a GOBP term as a root; only child annotation terms in the GO hierarchy are then used to identify the PRODISTIN classes (see Note 7).
10. Click on the "*Draw*" button to proceed with the "*Functional class identification*" with your customised parameters (see Note 8).

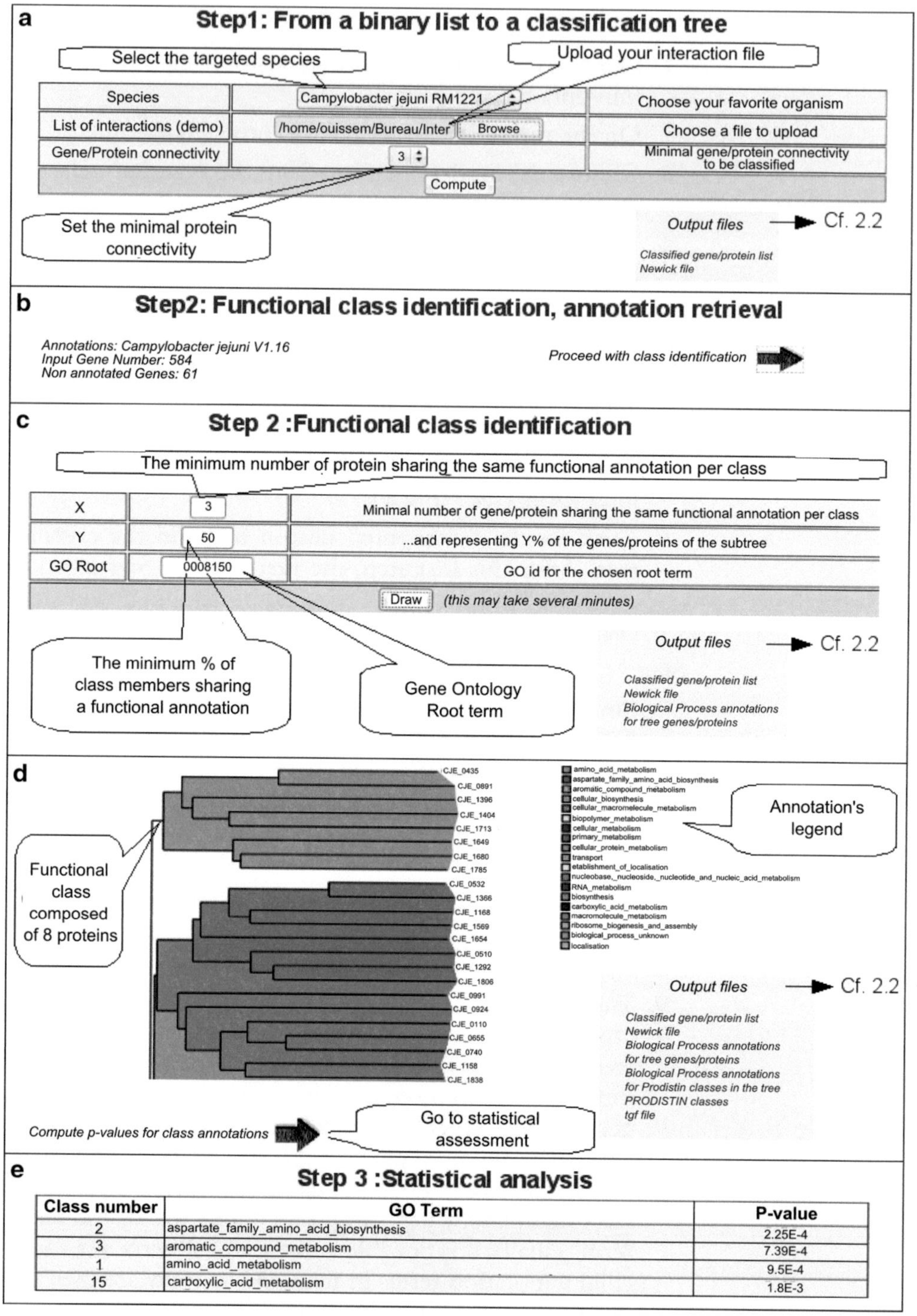

Class number	GO Term	P-value
2	aspartate_family_amino_acid_biosynthesis	2.25E-4
3	aromatic_compound_metabolism	7.39E-4
1	amino_acid_metabolism	9.5E-4
15	carboxylic_acid_metabolism	1.8E-3

Fig. 2. The different steps of the PRODISTIN Web site methods. (**a**) Starting with the computation of a classification tree, (**b**) the method retrieves the GOBP annotations of the proteins. (**c**) The PRODISTIN classes are defined and identified and (**d**) a coloured classification tree is obtained. Statistical significance of the functional classes can be computed (**e**).

11. Once identified, the functional classes are reported onto the classification tree which is coloured according to the annotations of classes. A specific colour is dedicated to each class annotation (*i.e.*, 1 colour per term of the GOBP ontology). Classes with multiple annotations appear with the colour of the last annotation term in the alphabetical ordered-list of the class annotation terms. An example of the resulting classification tree is showed in Fig. 2d.
12. Statistical assessment (Fig. 2e).

The PRODISTIN Web site permits to statistically assess the quality of the class annotations. It provides for each PRODISTIN class a *p*-value corresponding to the over-representation of the GOBP term(s) annotating the class as compared to the tree (see Note 9).

3.3. Application to the C. jejuni Interactome

Parrish et al. (9) recently identified 12,012 direct interactions in a high-throughput yeast two-hybrid screen of the *C. jejuni* strain NCTC11168. Among all these interactions, a subset of 3,174 interactions has a high confidence score (Fig. 1). We have created a protein–protein interaction file (available as Supplementary Material) containing all these interactions, after having transformed the protein names into their orthologs in the strain RM1221. This transformation was made necessary by the fact that the gene ontology annotations, used in the second part of the PRODISTIN method, are available only for this second strain.

The dataset of 3,174 interactions was loaded in the PRODISTIN Web site, and submitted with the default parameters, after having selected the *C. jejuni* strain RM1221 in the species box. The first step of the method permitted to obtain a classification tree of 584 proteins having at least three interactors. After the second step, which retrieves the GO annotations for the proteins, 46% of the proteins do not have any gene ontology annotation, and are considered as “biological process unknown”. The functional classes were identified according to the tree topology and the proteins being annotated in GOBP, with the default parameters (Fig. 3).

The PRODISTIN method permitted to obtain 34 annotated functional classes containing 244 proteins. Among all the proteins contained in the classes, 65 do not have any annotation in the Gene Ontology database, and we can propose a possible involvement in some biological process for them. *C. jejuni* proteins are poorly annotated, hence we obtain a small number of functional classes, and many of them have a broad annotation, such as “localization” or “primary metabolism”. However, the PRODISTIN Web site

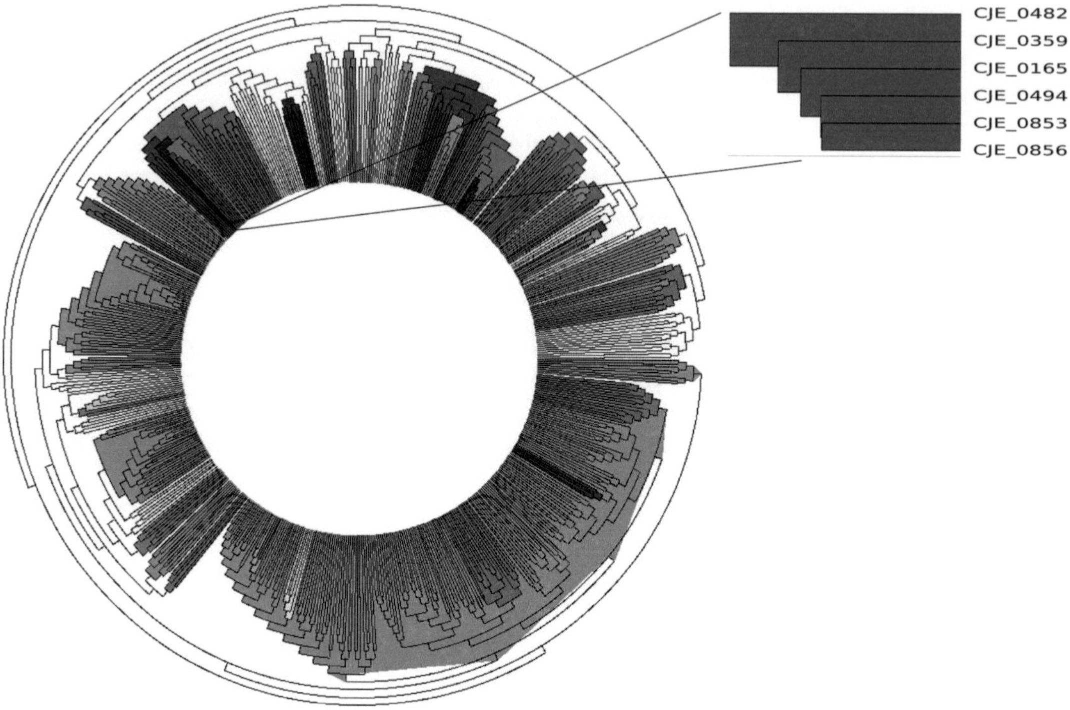

Fig. 3. PRODISTIN Web site output for the *Campylobacter jejuni* high quality interactome. The coloured classification tree is represented as a *circle*, and a functional class of interest annotated "aspartate family amino acid biosynthetic process", and containing six proteins is shown.

also permits to compute a *p*-value reflecting the significance of a given class annotation, in comparison with the rest of the classification tree. In the *C. jejuni* classification tree, seven classes have a *p*-value lower than 0.01. The more significant class is annotated "aspartate family amino acid biosynthetic process" (Fig. 3, subtree), the pathway of formation of aspartate amino acids. This class contains six proteins among which three are known to be involved in aspartate biosynthesis. Among the other proteins, one is involved in the binding of amino acids to their corresponding tRNA, and the two others do not have any precisely defined role. For these three proteins, we can therefore propose their participation to the aspartate biosynthesis pathway, based on their PRODISTIN classification.

4. Notes

1. The (.nw), (.tlf), (.tlf2), and (.tgf) files can be used by advanced users in tree editor Treedyn (32) to visualise and manipulate the resulting trees. Newick files can be used with

other tree editors accepting this format (NJPlot (33) http://pbil.univ-lyon1.fr/software/njplot.html, TreeGraph http://treegraph.bioinfweb.info/, ...).

2. If the species with which you work is not proposed, the classification tree can still be computed and be informative. However, the second step of the method (annotation retrieval and functional class identification) will not be achieved. An uncoloured classification tree will be obtained, and the corresponding Newick file will be available for download and further use.
3. The existence of false positive and false negative interactions in dataset weights more in the analysis of poorly connected genes/proteins than of others. Then, the system offers the opportunity to set a minimal connectivity threshold for genes/proteins to be classified. Only genes/proteins displaying a connectivity equal to or above the set threshold are classified, whereas the less connected genes/proteins are not classified but taken into account for the computation of the Czekanowski-Dice dissimilarity.
4. Gene Ontology has a hierarchical organisation, from broad to specific annotations. Hence, for each annotated protein, its annotations, as well as all the ancestor terms, are recovered. Note that only annotations contained within the Biological Process ontology are retrieved because the PRODISTIN method is usually used to cluster protein–protein interaction networks, therefore leading to groups of proteins involved in the same biological process.
5. If the number of "Not annotated genes/proteins" is very large though you are working with an annotated model organism, check that the gene/protein names fit with Standard Gene Names used by Gene Ontology. By default, a protein without any annotation found will be considered to be annotated "biological process unknown".
6. A PRODISTIN functional class is the largest possible subtree composed of at least X proteins sharing the same functional annotation and representing at least Y % of the individual class members for which a functional annotation is available. X and Y are set to 3 and 50% by default but can be user-defined.
7. A GO term ID can be selected by the user as a root term in order to filter out a part of the GOBP annotations. The default root value is "0008150", the identification for the ontology root term "Biological Process". Hence, the complete GOBP hierarchy is used to identify functional classes by default. But one may want to interpret a classification tree only with the annotations relative to "development", for instance. In this case, the user has to enter the GOBP identification number for "development", 0007275, as root term. Consequently, the

functional classes will be built by only using the descendant terms of the chosen root. Available GO tools permit to visualise GO hierarchy (e.g., AmiGO!, http://amigo.geneontology.org/cgi-bin/amigo/go.cgi) and choose your root GO identification number.

8. As we mentioned in Note 4, gene ontology has a hierarchical organisation, from broad to specific annotations. Hence, to construct PRODISTIN classes, the system takes into account for each protein, it GOBP terms plus the complete hierarchy of ancestor terms. Among all these terms, terms shared by at least X proteins in a same class are selected, and the more precise ones (= the deepest in the ontology) are kept to annotate the class. Note that a same class may be annotated with several terms. For the sake of clarity, very large classes annotated with broad annotation terms, such as "Biological Process" or "Cellular Process" are withdrawn from the classification.

9. The p-value is computed using the hypergeometric distribution:

$$\mathrm{Prob}\{X = k\} = \frac{C_m^k C_{N-m}^{m-k}}{C_N^n}$$

where k is the number of annotated genes for a given GO term among a dataset of size n, knowing that the reference dataset (all annotated tree proteins) contains m such annotated genes out of N genes.

Acknowledgments

This work was supported by the CNRS and the Inserm. A.B. is supported by a Juan de la Cierva Spanish fellowship and O.S. by a fellowship from the French Institute of Cooperation (IFC).

References

1. Sanchez C, Lachaize C, Janody F, Bellon B, Roder L, Euzenat J, Rechenmann F, Jacq B. (1999) Grasping at molecular interactions and genetic networks in *Drosophila melanogaster* using FlyNets, an Internet database. Nucleic Acids Res, 27:89–94.
2. Arifuzzaman M, Maeda M, Itoh A, Nishikata K, Takita C, Saito R, Ara T, Nakahigashi K, Huang HC, Hirai A, Tsuzuki K, Nakamura S, Altaf-Ul-Amin M, Oshima T, Baba T, Yamamoto N, Kawamura T, Ioka-Nakamichi T, Kitagawa M, Tomita M, Kanaya S, Wada C, Mori H. (2006) Large-scale identification of protein–protein interaction of *Escherichia coli* K-12. Genome Res, 16:686–691.
3. Butland G, Peregrin-Alvarez JM, Li J, Yang W, Yang X, Canadien V, Starostine A, Richards D, Beattie B, Krogan N, Davey M, Parkinson J, Greenblatt J, Emili A. (2005) Interaction network containing conserved and essential protein complexes in *Escherichia coli*. Nature, 433:531–537.

4. Hu P, Janga SC, Babu M, Diaz-Mejia JJ, Butland G, Yang W, Pogoutse O, Guo X, Phanse S, Wong P, Chandran S, Christopoulos C, Nazarians-Armavil A, Nasseri NK, Musso G, Ali M, Nazemof N, Eroukova V, Golshani A, Paccanaro A, Greenblatt JF, Moreno-Hagelsieb G, Emili A. (2009) Global functional atlas of *Escherichia coli* encompassing previously uncharacterized proteins. PLoS Biol, 7:e96.
5. Kuhner S, van Noort V, Betts MJ, Leo-Macias A, Batisse C, Rode M, Yamada T, Maier T, Bader S, Beltran-Alvarez P, Castano-Diez D, Chen WH, Devos D, Guell M, Norambuena T, Racke I, Rybin V, Schmidt A, Yus E, Aebersold R, Herrmann R, Bottcher B, Frangakis AS, Russell RB, Serrano L, Bork P, Gavin AC. (2009) Proteome organization in a genome-reduced bacterium. Science, 326:1235–1240.
6. Rain JC, Selig L, De Reuse H, Battaglia V, Reverdy C, Simon S, Lenzen G, Petel F, Wojcik J, Schachter V, Chemama Y, Labigne A, Legrain P. (2001) The protein–protein interaction map of *Helicobacter pylori*. Nature, 409:211–215.
7. Sato S, Shimoda Y, Muraki A, Kohara M, Nakamura Y, Tabata S. (2007) A large-scale protein–protein interaction analysis in *Synechocystis* sp. PCC6803. DNA Res, 14:207–216.
8. Shimoda Y, Shinpo S, Kohara M, Nakamura Y, Tabata S, Sato S. (2008) A large scale analysis of protein–protein interactions in the nitrogen-fixing bacterium *Mesorhizobium loti*. DNA Res, 15:13–23.
9. Parrish JR, Yu J, Liu G, Hines JA, Chan JE, Mangiola BA, Zhang H, Pacifio S, Fotouhi F, DiRita VJ, Ideker T, Andrews P, Finley RL Jr. (2007) A proteome-wide protein interaction map for *Campylobacter jejuni*. Genome Biol, 8:R130.
10. Titz B, Rajagopala SV, Goll J, Hauser R, McKevitt MT, Palzkill T, Uetz P. (2008) The binary protein interactome of *Treponema pallidum* – the syphilis spirochete. PLoS ONE, 3:e2292.
11. Sharan R, Ulitsky I, Shamir R. (2007) Network-based prediction of protein function. Mol Syst Biol, 3:88.
12. Spirin V, Mirny LA. (2003) Protein complexes and functional modules in molecular networks. Proc Natl Acad Sci U S A, 100:12123–12128.
13. Bader GD, Hogue CW. (2003) An automated method for finding molecular complexes in large protein interaction networks. BMC Bioinformatics, 4:2.
14. Rives AW, Galitski T. (2003) Modular organization of cellular networks. Proc Natl Acad Sci U S A, 100:1128–1133.
15. Girvan M, Newman ME. (2002) Community structure in social and biological networks. Proc Natl Acad Sci U S A, 99:7821–7826.
16. Dunn R, Dudbridge F, Sanderson CM. (2005) The use of edge-betweenness clustering to investigate biological function in protein interaction networks. BMC Bioinformatics, 6:39.
17. Brun C, Chevenet F, Martin D, Wojcik J, Guénoche A, Jacq B. (2003) Functional classification of proteins for the prediction of cellular function from a protein–protein interaction network. Genome Biol, 5:R6.
18. Samanta MP, Liang S. (2003) Predicting protein functions from redundancies in large-scale protein interaction networks. Proc Natl Acad Sci U S A, 100:12579–12583.
19. Brun C, Herrmann C, Guénoche A. (2004) Clustering proteins from interaction networks for the prediction of cellular functions. BMC Bioinformatics, 5:95.
20. Aittokallio T, Schwikowski B. (2006) Graph-based methods for analysing networks in cell biology. Brief Bioinform, 7:243–255.
21. Brohee S, van Helden J. (2006) Evaluation of clustering algorithms for protein–protein interaction networks. BMC Bioinformatics, 7:488.
22. Schwikowski B, Uetz P, Fields S. (2000) A network of protein–protein interactions in yeast. Nat Biotechnol, 18:1257–1261.
23. Jacq B. (2001) Protein function from the perspective of molecular interactions and genetic networks. Brief Bioinform, 2:38–50.
24. Dice L. (1945) Measures of the amount of ecologic association between species. Ecology, 26:297–302.
25. Gascuel O. (1997) BIONJ: an improved version of the NJ algorithm based on a simple model of sequence data. Mol Biol Evol, 14:685–695.
26. Ashburner M, Ball CA, Blake JA, Botstein D, Butler H, Cherry JM, Davis AP, Dolinski K, Dwight SS, Eppig JT, Harris MA, Hill DP, Issel-Tarver L, Kasarskis A, Lewis S, Matese JC, Richardson JE, Ringwald M, Rubin GM, Sherlock G. (2000) Gene ontology: tool for the unification of biology. The Gene Ontology Consortium. Nat Genet, 25:25–29.
27. Baudot A, Jacq B, Brun C. (2004) A scale of functional divergence for yeast duplicated genes revealed from analysis of the protein–protein interaction network. Genome Biol, 5:R76.

28. Baudot A, Angelelli JB, Guénoche A, Jacq B, Brun C. (2008) Defining a modular signalling network from the fly interactome. BMC Syst Biol, 2:45.

29. Zhong W, Sternberg PW. (2006) Genome-wide prediction of *C. elegans* genetic interactions. Science, 311:1481–1484.

30. Baudot A, Martin D, Mouren P, Chevenet F, Guenoche A, Jacq B, Brun C. (2006) PRODISTIN Web Site: a tool for the functional classification of proteins from interaction networks. Bioinformatics, 22:248–250.

31. Parkhill J, Wren BW, Mungall K, Ketley JM, Churcher C, Basham D, Chillingworth T, Davies RM, Feltwell T, Holroyd S, Jagels K, Karlyshev AV, Moule S, Pallen MJ, Penn CW, Quail MA, Rajandream MA, Rutherford KM, van Vliet AH, Whitehead S, Barrell BG. (2000) The genome sequence of the food-borne pathogen *Campylobacter jejuni* reveals hypervariable sequences. Nature, 403:665–668.

32. Chevenet F, Brun C, Banuls AL, Jacq B, Christen R. (2006) TreeDyn: towards dynamic graphics and annotations for analyses of trees. BMC Bioinformatics, 7:439.

33. Perriere G, Gouy M. (1996) WWW-query: an on-line retrieval system for biological sequence banks. Biochimie, 78:364–369.

Chapter 18

Using the NeAT Toolbox to Compare Networks to Networks, Clusters to Clusters, and Network to Clusters

Sylvain Brohée

Abstract

In this chapter, we present and interpret some operations on biological networks that can easily performed with NeAT, a set of Web tools aimed at studying biological networks (or graphs) and classifications.

These approaches are of particular interest for biologists and scientists who need to assess the reliability of new datasets (either experimental or predicted) by comparing them to established references.

Firstly, we describe the steps that will allow a nonspecialist user to compare two networks to compute their union and the statistical significance of their intersection. Next, we show how to map functional classes (e.g., GO categories, sets of regulons or complexes) onto a biological network. A third protocol explains how to compare two sets of functional classes, e.g., to assess statistically the biological relevance of some computationally returned groups of genes (clustering).

The metrics as well as the results obtained by following the different protocols are extensively described and explained.

NeAT is available at the following URL: `http://rsat.bigre.ulb.ac.be/rsat/index_neat.html`.

Key words: Network biology, Network comparison, Clustering assessment, Network visualization, Graph clustering, Neat

1. Introduction

Throughout this book, we had the occasion to study a large variety of networks implying bacterial proteins, genes, or processes that illustrate the large variety of existing types of networks. This chapter describes methods to compare networks to other networks and to classifications.

In the following sections, the term network (or graph) is considered in its simpler meaning, i.e., a set of *nodes* (elements), which are connected by *edges*. Graphs can be directed, in which case edges are named *arcs* (an arc going from node *a* to node *b* is not equivalent to an arc from node *b* to node *a*). Accordingly, the two nodes characterizing an arc are denoted *source* and *target*

Jacques van Helden et al. (eds.), *Bacterial Molecular Networks: Methods and Protocols*, Methods in Molecular Biology, vol. 804, DOI 10.1007/978-1-61779-361-5_18,

nodes, respectively. The terms *incoming* and *outgoing* are used to denote the direction of arcs respective to a node. Note that in a nondirected graph, the source and target node denomination are often used even if they may be swapped.

A classification consists in a set of classes (or clusters) containing elements. The classes may overlap, i.e., some elements may belong to more than one class or not.

The methods described in the following sections are applied to three different bacterial study cases. However, these methods are applicable to any kind of network and of classification (biological or not).

1.1. Study Case 1: Network–Network Comparison: Comparing Two Inferred Regulatory Networks

Network comparison presents many application domains. For example, the statistical merits of predictive methods can be calculated by comparing an inferred network to a well-established dataset. Another example is the computation of the intersection of two networks (i.e., the common edges). By doing this, we can increase the reliability of the interactions discovered several times by differing means. Note that if this manipulation increases the specificity of the network, it lowers its sensitivity.

In this first study case, we mainly focus on two networks obtained from the STRING database. This database contains many types of weighted links between genes. These interactions can be functional (e.g., cotranscription, phylogenetic cooccurrence) as well as physical (experimentally described protein–protein interaction). A score is attributed to each interaction, reflecting its reliability (1).

In this first study case, we compare two inferred regulatory networks in *Escherichia coli* and visualize their union using the popular *Cytoscape* network tool. The first graph is a coexpression network built using the experimental data of a large number of microarray experiments. The second one is a network inferred from neighborhood data (sets of genes that are found repeatedly in close neighborhood within different genomes, and that may belong to the same operon). The network comparison allows us to consider the level of overlap between these two regulatory network inference approaches as well as its statistical significance. In a second step, we randomize one of the two networks and apply the same graph comparison procedure to a randomized network to calculate the random expectation.

1.2. Study Case 2: Network–Clusters Comparison: Mapping and Extracting Clusters from a Coexpression Network

Many classifications or predefined clusters are available to biologists: functional classes (Gene Ontology (2)), list of complexes or of regulons (sets of genes regulated by the same transcription factor). Moreover, various computational methods produce clusters (microarray or network clustering, supervised classification methods, etc). It is of interest to visualize whether groups of nodes belonging to the same class are localized together in the network.

To this end, in the second study case, we extract and map groups of annotated coregulated genes (regulons) obtained from the RegulonDB database (3) onto the microarray cotranscription network described in the previous section. We are then able to visualize the annotated regulons together with the predicted cotranscription edges. Indeed, for each edge of the coexpression network, we determine whether it connects two nodes belonging to the same class (regulated by the same factor) or not. We distinguish the terms *mapping* (annotation of each edge as being intra- or an extracluster) and *extraction* (removing all intracluster edges).

1.3. Study Case 3: Clusters–Clusters Comparison: Comparing Annotated to Predicted Regulons

Besides the various annotated classifications, there exist many algorithms that produce clusters or classes according to varying criteria. A classical example are the groups of coexpressed genes computed by comparing expression profiles. Another common example consists in constituting clusters from groups of highly interconnected subparts of the graph. Indeed, for example, in a protein–protein interaction network, these highly connected parts of the network might correspond to protein complexes (e.g., see refs. 4–7 and the chapters of this book dedicated to network clustering (Chapters 15 and 16)).

Having two classifications, each consisting in a set of classes, we compare each class of the first classification to each class of the second one and compute the statistical significance of their intersection size as the probability of reaching an intersection of at least that size to be obtained by chance only. This approach is widespread in bioinformatics (e.g., see refs. 8–10). Indeed, in a large collection of clusters, it allows for the easy retrieval of the clusters constituted of a significant number of elements sharing at least one characteristic. The elements of these clusters, unannotated for these characteristics, can then be investigation candidates. For example, comparing a set of highly connected proteins of a protein interaction network to a set of annotated complexes enable the identification of biologically relevant clusters (those having a large intersection with the annotated complexes) and detecting new members of well established complexes.

In the following sections, we compare an annotated set of regulons (obtained from the regulonDB database) to clusters constituted of the highly interconnected subparts of the STRING coexpression network (computed with the MCL network clustering algorithm (see Chapter 15; (5)). It is interesting to see whether the clusters returned by MCL do actually correspond to existing regulons.

2. Material

2.1. Software

1. Neat
   The NeAT suite (Network Analysis Tools) is a set of tools allowing users to perform statistical operations on networks (network comparison, node statistics computation, network randomization, etc.) and clusters (mapping of clusters onto a network, classification comparison, etc.). In the present chapter, we only describe the Web server (`http://rsat.bigre.ulb.ac.be/rsat/index_neat.html`) but the tools are also accessible using SOAP Web services, as well as simple command line tools (11). Note that NeAT also presents path finding applications and strategies, but these are discussed in another chapter of this volume (Chapter 7).

   The NeAT home page displays two frames. The left frame presents a menu of the available tools, the right frame displays the forms and result pages. The tools are modular, and the output of one tool can easily be used as the input of another tool.

   To analyze the study cases described above, we mainly use the following programs:

   (a) *Compare-graphs*, which performs comparison between two networks, by computing the edges that are common to both or found in only one of the two networks.
   (b) *Graph-get-clusters*, which performs a mapping of a classification on a network by computing the intracluster and extracluster edges.
   (c) *Random-graph*, which randomizes a network and will be used to compute random controls.
   (d) *Compare-classes*, which takes as input two classifications and computes the statistical significance of the intersection of each pair of classes.

2. Cytoscape
   NeAT is not focused on the problem of network visualization. This is why we use *Cytoscape* for a proper visualization of the network.

   Cytoscape is the most known platform dedicated to the study of networks in bioinformatics. In its basic version, it is mainly devoted to visualization and includes a set of network layout algorithms. Among those, the spring embedding layout is commonly used for network visualization. Its principle is quite intuitive: linked nodes are attracted (edges are seen as springs) while two nonconnected nodes repell each other (like similarly charged particles). Starting from a random solution, at each iteration, the algorithm calculates the forces acting on each node and moves it according to them. After many iterations, an equilibrium state is reached where the force applied on each node is close to zero.

Additional plugins can be added to bring new functionalities (clustering, topological analysis, . . .). This program must be installed and runs on all major platforms. It is downloadable from http://www.cytoscape.org (12).

2.2. Intersection Statistics

In several sections of this protocol, we try to detect significant intersections between two classifications or between two graphs. In both cases, we can consider that we have a finite set of N elements. For *compare-classes*, N is the total number of elements that can be member of any reference or query class (e.g., all the genes of an organism). For *compare-graphs*, the N elements are all the possible edges between any pair of nodes (of both graphs).

Let us then define N, the total number of elements, R the reference set containing N_r elements, Q a query set containing N_q elements, C the intersection between a query and the reference set, and N_c the number of elements in this intersection.

1. Calculating the maximum number of edges in a graph
Knowing if it is directed and may contain self-loops (edges where the source node is also the target node), we can easily calculate the maximum number of edges in a network of n nodes with no duplicated edges.

Directed	Self-loops	(N) Max. number of edges
Yes	Yes	n^2
Yes	No	$(n^2 - n) = n(n-1)$
No	Yes	$(n + n^2)/2 = n(n+1)/2$
No	No	$(n^2 - n)/2 = n(n-1)/2$

2. Jaccard coefficient
The Jaccard coefficient (index) simply consists in the quotient of the size of the intersection between the classes by the size of the union. Although this index seems very intuitive, its main disadvantage is that is does not reflect the sizes of the compared sets. Indeed, an intersection of 1 between two sets of 2 and 1 element can have the same Jaccard index as an intersection of 100 between two sets of 200 and 100 elements, even if the latter is intuitively more significant. For this reason, we prefer the hypergeometric significance described below.

The Jaccard index J is calculated as follows:

$$J = \frac{R \cap Q}{R \cup Q}$$

3. Hypergeometric significance
   The hypergeometric distribution is generally used to estimate the statistical significance of the intersection between two random selections in a set. Its classical example of application is the random selection without replacement in an urn containing a set of white and black balls. The reference set (classes or graph) corresponds to the black balls of the urn example. The query set corresponds to the selection without replacement (indeed, a member cannot appear several times in the same class, and an edge cannot appear several times in the same graph). The hypergeometric P-value (p-val) indicates the probability to observe by chance at least x elements at the intersection between the query set and the reference set.

$$p\text{ - val} = P(X \geq x) = \sum_{i=N_c}^{N_q} \frac{C_{N_r}^{i} C_{N-N_r}^{N_q-i}}{C_N^{N_q}}$$

   where C_n^k corresponds to the number of ways that k things can be "chosen" from a set of n things and is computed as follows:

$$C_n^k = \frac{n!}{k!(n-k)!}$$

   When the number of comparisons is large, a correction for multitesting must be introduced. Indeed, when comparing many classes together (like in the third study case) the number of comparisons is the number of reference classes multiplied by the number of query classes, a low P-value is expected to be found by chance. A classical way to apply this correction is to compute the E-value (e-val) as:

$$e\text{ - val} = P\text{ - val} \cdot T$$

   The E-value corresponds to the expected number of false positives at a given P-value for a battery of T tests. The hypergeometric significance is then commonly calculated as:

$$sig_H = -log_{10}(e\text{ - val})$$

   The significance score gives an intuitive perception of the exceptionality of the intersection between sets: a negative significance indicates that an intersection of at least that size is likely to occur by chance while a positive value means that it is significant.

2.3. File Formats

1. Networks can be saved in a variety of formats. The NeAT program *convert-graph* facilitates the handling of these

formats, by supporting interconversions between some input and output formats.

The *tab-delimited* format is a convenient and intuitive way to encode a graph, where each row represents an edge and each column contains its attributes. Generally, the two first columns contain the nodes linked by an edge. Logically, in undirected graphs, the columns containing the source and the target nodes may be swapped. Orphan nodes (nodes without neighbor) can be included by specifying a source node without target. Some additional edge attributes (weight, label, color) can be placed in additional columns.

The *GML* format (`http://www.infosun.fim.uni-passau.de/Graphlet/GML/`) allows one to specify layout characteristics of the network (node size, color, label). GML is made up of nested key-value pairs. The most popular graph editors (such as Cytoscape and yEd) support GML as input format.

2. Classifications used by NeAT are files composed of two columns delimited by tabulation. The first column indicates the name of an element while the second column specify the cluster this element belongs to.

2.4. Sample Files

The files that are used for our study cases can be downloaded from the sample file repository on `http://rsat.bigre.ulb.ac.be/rsat/data/published_data/MiMB_BMN_2011/Brohee_supp_material/`.

`string_ecoli_coexpression.tab` and `string_ecoli_neighborhood.tab` contain the STRING network inferred from microarray data and from gene neighborhood, respectively. These networks are encoded in the tab delimited format. The two first columns contain the node names and the third column the weight of the edge linking both nodes.

`regulonDB.tab` contains a list of regulons extracted from the RegulonDB database. This is a NeAT classification file as the first column contains the name of the regulated gene and the second column the transcription factor that regulates them.

`string_ecoli_coex_mcl.tab` is the tab-delimited file obtained when applying the MCL graph clustering algorithm on the `string_ecoli_coexpression.tab` described above. The first column contains the node names and the second column the clusters they belong to

2.5. Supplementary Material

Supplementary figures and data are stored on the following Web page `http://rsat.bigre.ulb.ac.be/rsat/data/published_data/MiMB_BMN_2011/Brohee_supp_material/neat_supmater.html`

3. Methods

3.1. Network Comparisons and Visualization

In this subsection, we show how to calculate the union of two networks. This procedure takes as input two networks (called the query and the reference networks), and returns a larger colored network corresponding to their union. Each edge is colored according to the fact it belongs only to the query graph (red), to the reference graph (violet) or to both (green).

1. Download the sample networks from the data repository described in Subheading 2.4. To this end, with your Web browser, go to the repository Web site (`http://rsat.bigre.ulb.ac.be/rsat/data/published_data/MiMB_BMN_2011/Brohee_supp_material/`) and right click on the file `string_ecoli_coexpression.tab`, choose *Save link as* and save it on your computer under the same name. Do the same for the classification file `string_ecoli_neighborhood.tab`
2. Open a connection to the NeAT Web server: `http://rsat.ulb.ac.be/neat/`.
3. Click on "Network comparison" in the left frame.
4. Select the reference and the query networks that you want to compare by clicking on the *Browse* button. Select the previously downloaded file `string_ecoli_neighborhood.tab` as query and `string_ecoli_coexpression.tab` as reference.
5. The source and target node columns are correctly attributed by default (columns 1 and 2). However, as these graphs are weighted, specify for both networks column containing the weight is the third one.
6. Select *GML* as output format (so that we can open the resulting file with Cytoscape).
7. Check that union is selected as *Output type*. In this menu, you could have chosen *Intersection*, *Difference* (which would remove all the common edges from the query graph), *Query graph only* or *Reference graph only*.
8. There are different types of weights for the resulting union graph. By default, on common edges, the weight of both edges is returned. You can also select the sum, the mean, the minimal and maximal values of both edges, or only the weight of the query or of the reference.
9. Finally, you can indicate whether or not the graph may admit self-loops (edges having the same node as source and target), or whether it is directed. In our study case, the graph is

undirected and has no self-loop. Networks that may contain loops orwhich are directed contain a larger number of edges. The maximum number of edges is used for the computation of the intersection size significance.

10. Click on the *GO* button. The operation might take up to 1 min depending on how many people are connected and working on the NeAT server.
11. The result page displays some comparison statistics (that is discussed in the Results section) as well as a link pointing to the resulting network (GML file). Below the link to the file, there are some buttons displaying the names of some other graph tools, which allow us to use the union graph that we just computed as input for these tools.
12. To visualize the network with Cytoscape, right-click on the link to the union graph and store it on your computer.
13. Open Cytoscape.
14. In the File menu, select Import and Network (Multiple file types) and select the GML file you just copied on your computer. The union graph appears on the screen, the nodes being placed on a oblique line.
15. In the *Layout* menu, select *yFiles* > *organic* for a more intuitive visualization of the network. The network is now displayed in the classical spring embedding layout.
16. Back to the NeAt Web page, to build the negative control, in the NeAT left menu, click on the ***Randomize network*** link. The form of the ***random-graph*** tool appears.
17. Upload one of the two graphs used in the graph comparison via the *Browse* button (e.g., `string_ecoli_neighborhood.tab`)
18. Choose the type of randomization. We recommend you to use the node degree conservation that consists in shuffling the edges between the nodes, each node keeping the same number of neighbors as in the original graph. This procedure is specially designed to avoid duplicating edges, unless you check the option "Allow duplicated edges" (which should not be done here).
19. Click on the *GO* button. After some seconds of computation, a link to the results is displayed as well as buttons proposing to use the result file as input to other operations.
20. Click on the button *Compare this graph to another one*, as we want to compare the randomized neighborhood network to the nonrandomized coexpression network. The *compare-graphs* form described above is displayed with a notable

difference. The query graph part of the form is prefilled with the randomized network. To run the comparison, follow again steps 4–11 of this section.

3.2. Network–Clusters Comparison and Visualization

In this section, we describe how to compare a classification (set of classes) to a network.

1. Download the sample network and classification files from the data repository described in Subheading 2.4. To this end, with your Web browser, go to the repository Web site (`http://rsat.bigre.ulb.ac.be/rsat/data/published_data/MiMB_BMN_2011/Brohee_supp_material/`) and right click on the file `string_ecoli_coexpression.tab`, choose *Save link as* and save it on your computer under the same name. Do the same for file `regulonDB.tab`
2. Click on the *Map clusters onto network* link in the NeAT left menu *Map clusters onto network*. The form of *graph-get-clusters* is displayed on the screen
3. Select the coexpression graph that you want to compare by clicking on the *Browse* button next to the *Upload graph from file* field.
4. The source and target node columns are correctly attributed by default (1 and 2). However, as these graphs are weighted, specify that the third column contains the weight.
5. Select the clusters file, by clicking on the *Browse* buttons next to the *Upload clusters from file* field.
6. The main choice for the tool *graph-get-clusters* is the output type. The program supports different types of operations between the network and the clusters. We test the two major ones. (1) The option *annotated graph* labels each edge according to its intra- or intercluster nature, while (2) the *intra-cluster edges* option selects a subnetwork restricted to the intracluster edges (intercluster edges are deleted from the network). Let us start by selecting the latter option "intracluster edges."
7. Select the GML output format to visualize the output network with Cytoscape.
8. Click on the *GO* button. The operation might take up to 1 min depending on how many people are connected and working on the NeAT server.
9. As usual, the result page contains a blue hypertext link to the result network, as well as some buttons proposing to use the result network as input for other operations.
10. To visualize the network, perform operations 12–15 of Subheading 3.1.

11. To map the clusters, perform again steps 2–10 of this section. However, at step 6, select the option *annotated graph (all edges)*.

3.3. Clusters–Clusters Comparison

Here, we show how to compare two classifications and find clusters presenting a statistically larger intersection.

1. Download the classification files from the data repository described in Subheading 2.4. To this, with your Web browser go to repository Web site (`http://rsat.bigre.ulb.ac.be/rsat/data/published_data/MiMB_BMN_2011/Brohee_supp_material/`) and right click on the file `regulonDB.tab`, then choose *Save link as* and save it on your computer under the same name. Do the same for file `string_ecoli_coexpression.tab`
2. Click on the *Compare classes/clusters* link in the NeAT left menu. The input form of the *compare-classes* tool appears.
3. Upload the classifications files by clicking on the *Browse* buttons. Query classification corresponds to clusters returned by the MCL algorithm (`string_ecoli_coexpression.tab`) while the reference file is the file containing the regulons annotated in the RegulonDB database (`regulonDB.tab`).
4. As there is no score for these classification files, the *score column* field may remain empty.
5. The option *Thresholds on return* fields allows one to combine various constraints to select the most significant intersections between query and reference clusters/classes. The default threshold on significance ($sig\_H > 0$) usually gives satisfactory results. For our study case, the other thresholds do not need to be changed.
6. Click on the *GO* button. The operation might take up to 1 min depending on how many people are connected and working on the NeAT server.
7. Here, the result page contains two hypertext links to the results. The first link will redirect you to a tab-delimited text file and the second one to an HTML conversion of the first file. These files consist mainly of a table where each line reports the result of one comparison. The HTML format is useful for inspecting and handling the result on the Web browser. The table can be sorted according to the data of any column simply by clicking on the header of these columns. The first lines of the result file indicate the parameters used for the analysis, and document the content of the columns. This is followed by a table, where each line reports the comparison between one MCL cluster and one complex (Fig. 3).

3.4. Study Case 1: Network–Network Comparison: Comparing Two Inferred Regulatory Networks

The first study case consisted of computing the union of two inferred regulatory networks downloaded from the STRING database. The results of the procedure can be seen on Fig. 1 and Supplementary Figure S4.

Figure 1 shows the result page of the NeAT *compare-graphs* tool. On this page, some network characteristics are summarized (number of edges, nodes in the query or reference network, etc.). The coexpression network used as reference contains 1,609 genes and 16,267 edges, while the query graph (neighborhood) contained 3,699 nodes and 34,138 edges. The intersection between these graphs is apparently weak: 1,344 edges only. The Jaccard coefficient indicates that this intersection represents no more than 2.7% of the union. However, the number of edges expected by chance at the intersection is even smaller: $E(Q \cap R) = 68$. The hypergeometric *P*-value indicates the probability to observe at least 1,344 edges at the intersection when 68 are expected by chance. In this study case, we observe that even with no more than 0.68% of edges at the intersection, the *P*-value is extremely low ($<10^{-300}$). In other terms, the number of edges at the intersection is much too high to be explained by chance, and is more likely to result from the biological relevance of the datasets. By contrast, the number of edges in common between the randomized query and the reference only delivers about

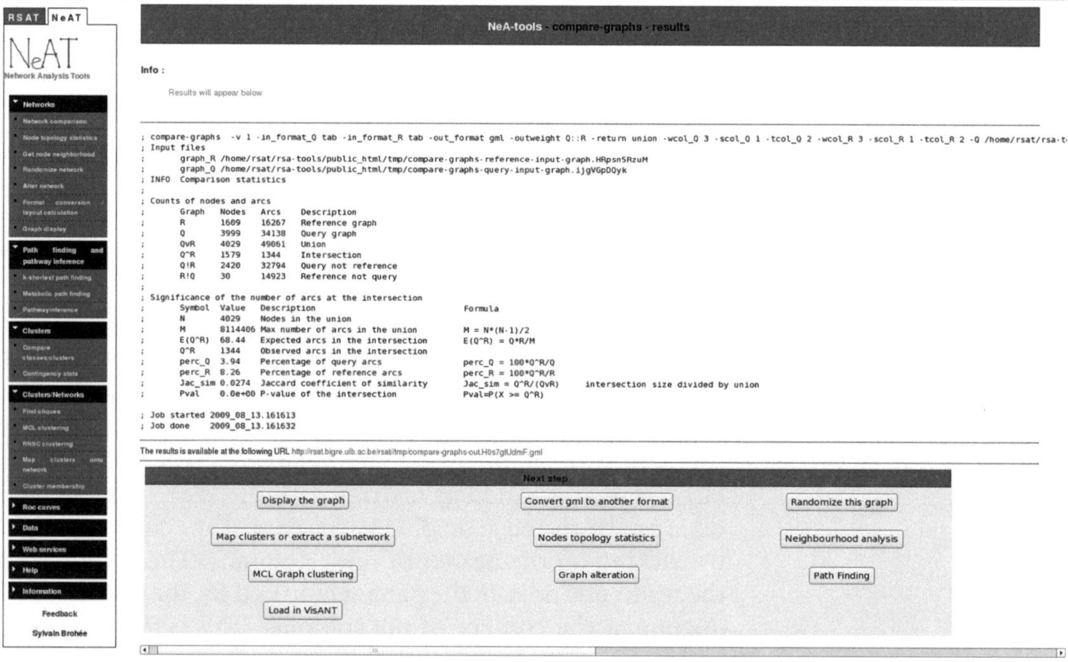

Fig. 1. Study case 1. Comparison between two inferred regulatory networks downloaded from the STRING database (query: neighborhood and reference: microarray data). Result page of the NeAT tool *compare-graphs* detailing the network characteristics and the intersection statistics.

100 edges which means a Jaccard index of ~0.2% and a much higher *P*-value (2.3×10^{-5}, i.e., more than 10^{295} times lower). The results obtained with the random control may vary, this is why generally, random controls are computed multiple times to obtain an estimate of their score distribution.

Supplementary Figure S4 contains the union of both reference and query graph plotted in Cytoscape with the *Organic* layout. This figure and the network itself can be downloaded from the supplementary material Web page.

3.5. Study Case 2: Network–Clusters Comparison: Mapping and Extracting Clusters from a Coexpression Network

The network resulting from the cluster extraction methodology described in Subheading 3.2 consists in the STRING coexpression network, where each intercluster edge, i.e., the edges connecting a source and a target node not belonging to the same annotated regulon has been removed (Supplementary Figure S5). It is thus only composed of intracluster edges, the edges being colored according to the cluster they belong to. Some edges may belong to more than one cluster. In this case, the program duplicates the edges so that the color of each cluster is represented (Fig. 2).

Figure 2(a) shows an enlarged part of the intracluster edges network consisting of the genes implicated in sulfur metabolism which are all regulated by the cysB factor. This representation also shows that genes cysH, cysI, and cysJ are also annotated as regulated by another factor (the global repressor FIH) as these three genes are linked by an edge presenting a different color (here, dashed and wider for visualization purposes).

Figure 2(b) shows a part of the network resulting from the cluster mapping where all the intercluster edges have been colored in gray and intracluster edges have been assigned a different color. For visualization purposes, the intracluster edges are wider and

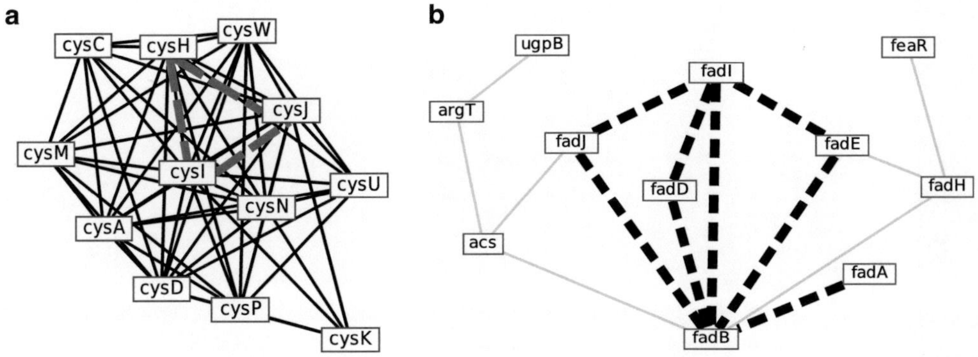

Fig. 2. Graphical representation of the cluster extraction (**a**) and mapping (**b**) of coregulated genes clusters (regulons) on a regulatory network. Each intercluster edge is removed (in case of extraction) or colored in grey (mapping) while intracluster edges are colored differently according to the cluster they belong to. (**a**) Enlargement of the CysB regulon (sulfur metabolism) showing that some edges may belong to more than one cluster. (**b**) Enlargement of the representation of the cluster mapping consisting in the fadR regulon (fatty acid metabolism). It shows that some genes are not annotated as regulated by fadR (light grey edges) but are found nevertheless to be coexpressed with fadR genes.

dashed. With this representation, the user can compare the annotated data (here, the different regulons from regulonDB) to the inferred network and directly correct them. Indeed, if a set of genes connected by edges of the same colors (belonging to the same regulon) are also linked to one or more genes of similar function (predicted as coexpressed), this might mean that the annotations are not complete. This is what we observe, indeed, the gene fadH is predicted as coexpressed with the genes of the fadR regulon (implied in fatty acid metabolism) but is not annotated as such in regulonDB.

3.6. Study Case 3: Clusters–Clusters Comparison: Comparing Annotated to Predicted Regulons

Figure 3 shows the comparison results between the dense clusters (obtained by applying the MCL clustering algorithm on a coexpression network) and the regulons of regulonDB, ranked by the hypergeometric significance. The header gives a short explanation for the content of each column of the results table. Each row describes the comparison between one MCL cluster and one regulon. For example, the first row compares the MCL cluster "cl_12," which contains 17 genes, with a set of 24 genes regulated by the CysB factor (implied in cysteine metabolism regulation). The intersection contains 12 genes, which represents a high fraction of both the MCL cluster, and the annotated complex. The probability to observe such an intersection by chance is 2.4×10^{-22}. The *E*-value, obtained with the correction for multitesting, indicates that the number of

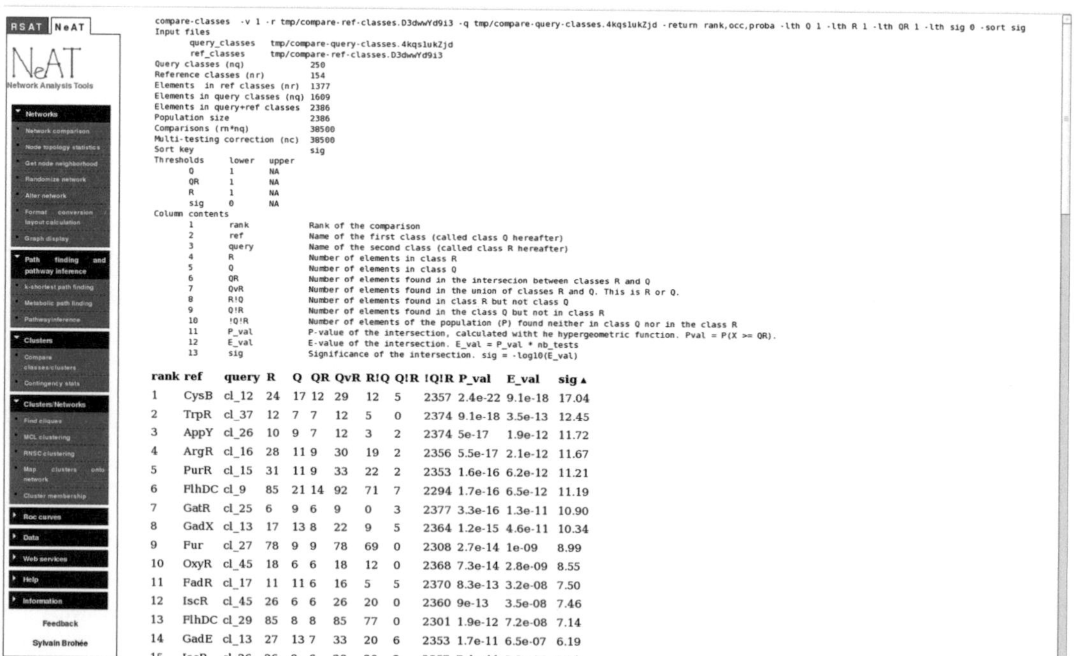

rank	ref	query	R	Q	QR	QvR	R!Q	Q!R	!Q!R	P_val	E_val	sig
1	CysB	cl_12	24	17	12	29	12	5	2357	2.4e-22	9.1e-18	17.04
2	TrpR	cl_37	12	7	7	12	5	0	2374	9.1e-18	3.5e-13	12.45
3	AppY	cl_26	10	9	7	12	3	2	2374	5e-17	1.9e-12	11.72
4	ArgR	cl_16	28	11	9	30	19	2	2356	5.5e-17	2.1e-12	11.67
5	PurR	cl_15	31	11	9	33	22	2	2353	1.6e-16	6.2e-12	11.21
6	FlhDC	cl_9	85	21	14	92	71	7	2294	1.7e-16	6.5e-12	11.19
7	GatR	cl_25	6	9	6	9	0	3	2377	3.3e-16	1.3e-11	10.90
8	GadX	cl_13	17	13	8	22	9	5	2364	1.2e-15	4.6e-11	10.34
9	Fur	cl_27	78	9	9	78	69	0	2308	2.7e-14	1e-09	8.99
10	OxyR	cl_45	18	6	6	18	12	0	2368	7.3e-14	2.8e-09	8.55
11	FadR	cl_17	11	11	6	16	5	5	2370	8.3e-13	3.2e-08	7.50
12	IscR	cl_45	26	6	6	26	20	0	2360	9e-13	3.5e-08	7.46
13	FlhDC	cl_29	85	8	8	85	77	0	2301	1.9e-12	7.2e-08	7.14
14	GadE	cl_13	27	13	7	33	20	6	2353	1.7e-11	6.5e-07	6.19
15	IscR	cl_26	26	9	6	29	20	3	2357	7.4e-11	2.9e-06	5.54

Fig. 3. Study case 3. Most significant associations between MCL clusters versus regulonDB regulons. The figure only shows the top of the table returned by the program *compare-classes*. Each row represents the comparison between one regulon (reference) and one MCL cluster (query).

false positives expected with such a p-value would be 9.1×10^{-18}. In other terms, the correspondence between this MCL cluster and the cysteine metabolism gene is too high to be explained by chance.

Acknowledgments

KULeuven SCD-SISTA lab is funded by the Research Council KUL (GOA MaNet, GOA AMBioRICS, CoE EF/05/007 SymBioSys, PROMETA, START 1, several PhD/postdoc and fellow grants, GOA 2006/12), FWO [PhD/postdoc grants, projects G.0241.04 (Functional Genomics), G.0499.04 (Statistics), G.0232.05 (Cardiovascular), G.0318.05 (subfunctionalization), G.0553.06 (VitamineD), G.0302.07 (SVM/Kernel), research communities (ICCoS, ANMMM, MLDM)], G.0733.09 3UTR; G. 082409 (EGFR), G.0254.05 (Genetics of human heart develoment), IWT (PhD Grants, GBOU-McKnow-E (Knowledge management algorithms), GBOU-ANA (biosensors), TAD-BioScope-IT, Silicos; SBO-BioFrame, SBO-MoKa, TBM Endometriosis), the Belgian Federal Science Policy Office [IUAP P6/25 (BioMaGNet, Bioinformatics and Modeling: from Genomes to Networks, 2007–2011), IUAP P5/25 (Molecular Pathology of Genetic Diseases) and the EU-RTD (ERNSI: European Research Network on System Identification; FP6-NoE Biopattern; FP6-IP e-Tumours, FP6-MC-EST Bioptrain, FP6-STREP Strokemap). Sylvain Brohée is Chargé de Recherches at the Fonds National de la Recherche Scientifique (FNRS) de la Communauté Française de Belgique and was supported by a post-doc grant of the CheartED project (in the SISTA lab).

References

1. Jensen LJ, Kuhn M, Stark M, Chaffron S, Creevey C, Muller J, Doerks , Julien P, Roth A, Simonovic M, Bork P, von Mering C. (2009) STRING 8—a global view on proteins and their functional interactions in 630 organisms. Nucleic Acids Res, 37(Database issue): D412–D416. [http://www.dx.doi.org/10.1093/nar/gkn760].
2. Ashburner M, Ball CA, Blake JA, Botstein D, Butler H, Cherry JM, Davis AP, Dolinski K, Dwight SS, Eppig JT, Harris MA, Hill DP, Issel-Tarver , Kasarskis A, Lewis S, Matese JC, Richardson JE, Ringwald M, Rubin GM, Sherlock G. (2000) Gene ontology: tool for the unification of biology. The Gene Ontology Consortium. Nat Genet, 25:25–29. [http://www.dx.doi.org/10.1038/75556].
3. Gama-Castro S, Jiménez-Jacinto V, Peralta-Gil M, Santos-Zavaleta A, Peñaloza-Spinola MI, Contreras-Moreira B, Segura-Salazar J, Muñniz- Rascado L, Martínez-Flores I, Salgado H, Bonavides-Martínez C, Abreu-Goodger C, Rodríguez-Penagos C, Miranda-Ríos J, Morett E, Merino E, Huerta AM, Treviño-Quintanilla L, Collado-Vides J. (2008) RegulonDB (ver- sion 6.0): gene regulation model of Escherichia coli K-12 be- yond transcription, active (experimental) annotated promoters and Textpresso navigation. Nucleic Acids Res, 36(Database issue): D120–D124. [http://www.dx.doi.org/10.1093/nar/gkm994]. 26.
4. King AD, Przulj N, Jurisica I. (2004) Protein complex prediction via cost- based

clustering. Bioinformatics, 20(17):3013–3020. [http://www.dx.doi.org/10.1093/bioinformatics/bth351].

5. Enright AJ, Dongen SV, Ouzounis CA. (2002) An efficient algorithm for large-scale detection of protein families. Nucleic Acids Res, 30(7):1575–1584.
6. Blatt M, Wiseman S, Domany E. (1996) Superparamagnetic clustering of data. Phys Rev Lett, 76(18):3251–3254.
7. Bader GD, Hogue CWV. (2003): An automated method for finding molecular complexes in large protein interaction networks. BMC Bioinformatics, 4:2.
8. Naamane N, van Helden J, Eizirik DL. (2007) In silico identification of NF-kappaB-regulated genes in pancreatic beta-cells. BMC Bioinformatics, 8:55. [http://www.dx.doi.org/10.1186/1471-2105-8-55].
9. Dessailly BH, Lensink MF, Wodak SJ. (2007) Relating destabilizing regions to known functional sites in proteins. BMC Bioinformatics, 8:141. [http://www.dx.doi.org/10.1186/1471-2105-8-141].
10. Barriot R, Sherman DJ, Dutour I. (2007) How to decide which are the most pertinent overly-represented features during gene set enrichment analysis. BMC Bioinformatics, 8:332. [http://www.dx.doi. rg/10.1186/1471-2105-8-332]. 27.
11. Brohee S, Faust K, Lima-Mendez G, Sand O, Janky R, Vanderstocken G,Deville Y, van Helden J. (2008) NeAT: a toolbox for the analysis of bio- logical networks, clusters, classes and pathways. Nucleic Acids Res, 36(Web Server issue):W444–W451. [http://www.dx.doi.org/10.1093/ nar/gkn336].
12. Shannon P, Markiel A, Ozier O, Baliga NS, Wang JT, Ramage D, Amin N, Schwikowski B, Ideker T. (2003) Cytoscape: a software environment for integrated models of biomolecular interaction networks. Genome Res, 13(11):2498–2504. [http://www.dx.doi.org/10.1101/gr.1239303].

Chapter 19

Analyzing Biological Data Using R: Methods for Graphs and Networks

Nolwenn Le Meur and Robert Gentleman

Abstract

R is a powerful language and widely used software tool for the analysis and visualization of data. Its core capabilities can be extended through many different add-on packages. Among the many packages are some which offer a broad range of facilities for analyzing statistical properties of graphs. This chapter provides a practical tutorial covering the use of R methods for graphs and networks to examine biological data and analyze their topological and statistical properties.

Key words: Graph, Random graph, Network, Statistic, Systems biology

1. Introduction

1.1. Context

Recent advances in molecular biology and associated technologies have expanded our opportunities to explore and understand the complexity of biological systems at different levels: from DNA to proteins to signaling networks and beyond. To integrate these data, to build and explore biological networks, we now need computational tools, efficient flexible data structures, and statistical methods designed with the particular challenges of systems biology and the analysis of high-throughput data in mind.

For the past 10 years, the Bioconductor community has worked on developing such infrastructure. The tools come in the form of different packages for R (1), a free software environment for statistical computing and graphics. Each package provides software, or data, to carry out a specific set of tasks. Developing workflows that rely on multiple packages is an effective strategy for the rapid development of tools for analyzing biological data. Packages are being developed to address the specific needs of systems biology approaches, such as network modeling, simulation, formal verification, and graph visualization (2). For instance,

Jacques van Helden et al. (eds.), *Bacterial Molecular Networks: Methods and Protocols*, Methods in Molecular Biology, vol. 804, DOI 10.1007/978-1-61779-361-5_19, © Springer Science+Business Media, LLC 2012

Castelo and Roverato developed an R package providing reverse engineering procedures for molecular network discovery from microarray expression data (3). Similarly, using graph theoretical concepts, Le Meur and Gentleman (4) identified multiprotein complexes and pairs of multiprotein complexes that share an unusually high number of synthetic genetic interactions. Network data computed and analyzed in R can directly be visualized within R using for instance the ***graph*** and ***Rgraphviz*** Bioconductor libraries that are discussed in this chapter, or ***igraph***, an interesting and valuable R library (5). Additionally, the network data generated by R can easily be exported to other visualization tools. For example, Cytoscape (6) can be seen as an analyst-friendly and interactive alternative. Among many features, Cytoscape notably allows zooming in and out or using *bird's eye view* for browsing large graphs. It also offers classical network analysis such as selecting subnetworks based on graph properties (e.g., node or edge annotation).

In this chapter, we review methods available in R for working with graphs. We discuss how to retrieve and integrate biological data to build graphs of gene networks. We also consider some of the techniques available to explore and analyze the topological and statistical properties of these graphs. This chapter is written using the Sweave system (7), and all the codes used to produce the outputs, graphics, etc. can be retrieved from our online supplements, located at `http://www.bioconductor.org/pub/MiMB`.

1.2. Case Studies

1.2.1. Case Study 1: Bacillus subtilis Protein–Protein Interaction Data

Bacillus subtilis, known as the hay bacillus or grass bacillus, is a Gram-positive bacterium commonly found in soil. It may contaminate food but rarely causes food poisoning. It is not considered a human pathogen. *B. subtilis* and the enzymes it produces have many applications such as immunostimulatory agents in alternative medicine or as additives in laundry detergents. In this chapter, we explore some of the protein–protein interaction (PPI) data available for *B. subtilis.*

1.2.2. Case Study 2: Response of Escherichia coli During an Oxygen Shift

Escherichia coli is a Gram-negative bacterium that is commonly found in the lower intestine of warm-blooded organisms. Most *E. coli* strains are harmless, but some, such as serotype O157:H7, can cause serious food poisoning in humans. *E. coli* can grow in aerobic as well as in anaerobic conditions (e.g., the intestine). In this case study, we explore gene expression results of various knockout (KO) mutants under aerobic conditions to characterize elements of the transcriptional network.

2. Material

We assume that the reader is acquainted with R and only give a quick overview on the installation procedures and help functionalities. For more details about R and its programming language, we refer the reader to the many manuals and courses freely available on the R and Bioconductor Web sites (`http://www.r-project.org`; `http://bioconductor.org/`). In addition, there are many good books on R and Bioconductor such as (8–11).

The Graphviz package (`http://www.graphviz.org/`) and the Boost graph library (`http://www.boost.org/`) are additional tools required to run graph examples of this chapter. Their installation might not be trivial; it depends on the operating system. One should be especially careful to the location of those libraries and the definition of their path in the environmental variables. For details, we refer the reader to the project Web sites or the Bioconductor packages that depend on them, namely ***Rgraphviz*** and ***RBGL***.

2.1. R and Bioconductor

R is freely available and runs on a wide variety of commonly used operating systems. To install R, download the most recent version of R from `http://www.r-project.org/`. Detailed instructions of installation are available in the R FAQ and the *R Installation and Administration Manual* on the R project Web site. The base installation of R consists of the programming language itself together with a number of packages that are mainly oriented to statistical analyses. R can be extended and its capabilities expanded through the use of add-on packages that are notably available through the R and Bioconductor Web sites. Bioconductor is comprised of several hundred R packages that provide solutions to a large number of bioinformatic and computational biology problems. The scope of Bioconductor packages is broad covering many aspects of the analysis of microarrays but also including tools for genomic annotation, high-throughput genotyping, cell-based assays (e.g., RNAi, cytometry), handling graphs, and more recently high-throughput sequencing.

2.2. Using R and Bioconductor Packages

Starting R on Windows or OS X is usually accomplished by double-clicking on the R application icon. On UNIX-like systems users typically type `R` at a shell prompt. Because there are many complex dependencies between packages and the version of R being used the recommended way to install Bioconductor packages is to use the `biocLite.R` installation script in a R session:

```
> source("http://bioconductor.org/biocLite.R")
> biocLite()
```

To find, download, and install add-on packages, we recommend that you use built-in capabilities, either in the GUI or through the command line. On Windows and OS X packages can be installed *via* a menu. You can also use the R function `install.packages` or the `biocLite` script to download a specified list of packages and their dependencies. There is a difference between installing and using a package. Typically, you install a package once, using the tools described above, and use it multiple times. For example, we could install the ***graph*** and ***Rgraphviz*** packages using the code below.

```
> biocLite(c("graph", "Rgraphviz"))
```

To use those packages, you must load them using the `library` function:

```
> library("graph")
> library("Rgraphviz")
```

2.3. R Documentation

To get started, it is important to know where you can find documentation such as help for functions, classes, and concepts. The `?` operator followed by the name of a function will display the manual page of this function. `class?foo` will show the manual page for the `foo` class and most often the available methods for this class. The function `apropos` can be used to find objects in the search path with names that partially match the character string given as an argument to `apropos`. The function `help.search` will search the help system for documentation matching a given character string, and again partial matching will be used. You should compare the output for the three commands below on your system.

```
> class?graphNEL
> apropos("graphNEL")
> help.search("graphNEL")
```

In addition, each Bioconductor package contains at least one *vignette*, which is a document that provides a task-oriented description of the functionalities provided by the package. Vignettes should contain executable examples and are intended to be used interactively.

2.4. R Data Structures

The basic R data structure is the atomic vector. It contains an indexed set of values that are all of the same type: logical, numeric, complex or character. The numeric type can be further broken down into integer, single or double types. In the code below we create a vector named x, and while it may seem that we have mixed integer and real types, the integers have been coerced to their real representation, which can be seen in the output of the call to `typeof`.

```
> x <- c(10, 5, 6, 1.5, 9)
> typeof(x)
```

```
[1] "double"
```

Additional data structures are lists, arrays and data frames. An `array` is a vector with a dimension attribute and matrices are a special variant where there are exactly two dimensions. Arrays are like atomic vectors in that all elements must be of the same basic type. A `list` is an object consisting of an ordered collection of objects. It is different from atomic vectors because its elements can be of any type, and for example a list could contain a numeric vector, a logical value, and a matrix. A `data.frame` is much like a matrix in that it is two dimensional. Data frames are widely used for holding sample level data, since they are typically rectangular (some variables for each sample or individual) but the data types (columns) can be of mixed types (e.g., age, gender, and so on).

In the code below, we create a matrix, `y`, of integers with two rows and five columns. We also create a `data.frame` with two columns, one contains numeric values and the other contains character values.

```
> y <-c(1:10)
> dim(y) <- c(2, 5)
> age <- c(33, 44, 55)
> gender <- c("M", "F", "F")
> myDF <- data.frame(age, gender)
```

While creating the `data.frame` with the default parameters, we allow the conversion of the character vector to factors. Factors are used to encode vectors as a category (or enumerated type). Levels are then optional vectors of the values that in our case might be taken by `gender`. The default is the unique set of values taken by `as.character`(gender), sorted into increasing order of `gender`.

```
> is.factor(myDFgender)

[1] TRUE

> levels(myDFgender)

[1] "F" "M"
```

Subsets of the elements of a vector may be selected using the square bracket operator. If the object being subscripted has multiple dimensions, then each dimension can be subscripted independently with arguments separated by commas. If a subscript is not supplied (see the examples for y and myDF in the code below) then it is presumed that all elements are selected. In the code below we select the first two elements of the vector x, then the first two columns of the matrix `y`, and finally the first row of the data.frame myDF.

```
> x[1:2]
> y[,1:2]
> myDF[1,]
```

R also supports a number of paradigms for object oriented programming. Object-oriented programming provides a coherent way to describe and manipulate rich data structures. For instance in the ***Biobase*** package, the `ExpressionSet` class, designed to standardize data structures to represent DNA microarray data, has become the basic data structure for most gene expression data analysis packages. Similarly, recent developments propose classes and basic methods to support many other types of high-throughput data such as cell-based assays (12, 13), genotyping, or DNA sequencing (14).

More generally, we distinguish packages that are mainly infrastructure (sets of tools that can be used to create other pieces of software) and packages that are designed to provide an end-user application. For instance, the ***graph*** and ***Rgraphviz*** packages are infrastructure, while the ***GOstats*** and ***apComplex*** are specific applications built on that infrastructure. In addition there is a large set of packages that provide annotation for whole organisms and a number of real data sets are also available as self-contained packages suitable for method development and testing.

In the following sections, we make use of the packages listed in Table 1. We focus on data structures suited to graphs and demonstrate their usefulness to manipulate and represent biological networks.

Table 1
Bioconductor's graph-related packages. Packages used in this chapter

Package	Description	Type
graph	Basic class definitions and graph handling capabilities	Infrastructure
Rgraphviz	Interfaces R with the AT and T Graphviz library for plotting R graph objects from the graph package	Infrastructure
RBGL	Interface to the graph algorithms contained in the *BOOST* library	Infrastructure
RpsiXML	R interface to PSI-MI 2.5 files	Infrastructure
ppiStats	Tools for the analysis of protein interaction data	Application
limma	Data analysis, linear models and differential expression for microarray data	Application
biomaRt	Interface to the BioMart software suite	Annotation
GOstats	Basic manipulation tools for graphs, hypothesis testing and other simple calculations on Gene Ontology	Application
Category	Statistical methods for performing category analysis	Application
org.EcK12.eg.db	Genome wide annotation for *E. coli* strain K12, primarily based on mapping using ENTREZ GENE identifiers from NCBI	Annotation

3. Methods

3.1. Working with Graphs in R

3.1.1. Graph Types

There are many different types of graphs and their definitions primarily depend on their edge and node properties (see Notes 1 and 2). An edge is used to represent a particular type of relationship (interaction) between two nodes. R offers to build, compute, and represent many types of graphs from classical interaction graphs to specially developed graphs for answering specific biological questions (see Note 3).

3.1.2. Building and Exploring Graphs

There are many ways to computationally represent graphs. Some R data structure are designed for efficiency of different types of access or manipulation while others are efficient for handling very large, but sparse graphs (see Note 4).

The ***graph*** package offers three basic strategies for building random graphs. The `randomEGraph` function generates a random edge graph. In this graph, edges are either randomly generated according to a specified probability, or the user can specify a number of edges that are then randomly assigned. `randomNodeGraph` builds a random graph with a pre-specified node degree distribution. And for the `randomGraph` function, the user provides both the node labels and a factor with some fixed number of levels. Each node is randomly assigned levels of the factor and edges are created between nodes that share the same levels of the factor.

1. As an example, we create a random ***graph*** `g1` using the `randomEGraph` function from the ***graph*** package (Fig. 1). `g1` is an undirected graph composed of 15 nodes (labeled by the first 15 letters of the alphabet) and 20 edges (see Note 1).

```
> set.seed(123)
> g1 <- randomEGraph(LETTERS[1:15], edges= 20)
> class(g1)

[1] "graphNEL"
attr(,"package")
[1] "graph"
> g1
A graphNEL graph with undirected edges
Number of Nodes = 15
Number of Edges = 20
```

The ***graph*** package also offers a collection of functions and methods to convert various forms of matrices into graph objects (e.g., `ftM2graphNEL`, `aM2bpG`).

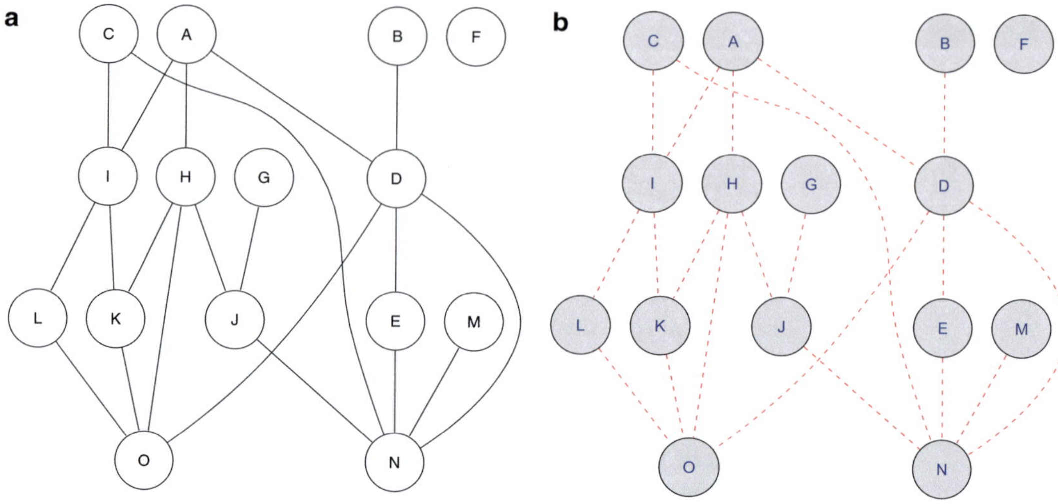

Fig. 1. Flexible graph rendering with R. The graph was generated by the graph package. In panel (**a**), the default parameters of the `layoutGraph` and `renderGraph` functions were used. In panel (**b**), the node and edge parameters were modified using the `graph.par` function. This affected the node color, line color and type, etc.

2. Next, we can determine graph properties of our random graph `g1` such as the connected components using the `connectedComp` function (see Note 2).
3. Once we know which nodes are connected we can ask about paths in the graph, from one node to another.

```
> library("RBGL")
> s1 <- connectedComp(g1)
> s1

$ '1'
 [1] "A" "B" "C" "D" "E" "G" "H" "I" "J" "K" "L"
[12] "M" "N" "O"

$ '2'
[1] "F"
> sp.between(g1, "A", "O")

$ 'A:O'
$ 'A:O'$length
[1] 2

$`A:O`$path_detail
[1] "A" "D" "O"
$`A: O`$length_detail
$`A: O`$length_detail[[1]]
A--D D--O
   1    1
```

4. Finally, we can extract the largest connected component using the `subGraph` function and thus perform other computations using just that subgraph. For instance, we can look for a solution to the minimum cut problem (see Note 2 and the `minCut` main page for more details).

```
> gsub <- subGraph(s1[[1]], g1)
> minCut(gsub)

Mincut
[1] 1

$S [1] "B"

'V-S'
[1] "A" "C" "D" "E" "G" "H" "I" "J" "K" "L" "M"
[12] "N" "O"
```

3.2. Graph Layout

The main graph layout algorithms are provided by the ***Rgraphviz*** package (see Note 5), which is based on the *Graphviz project* (15).

1. A typical workflow of a graph layout is to pass a graph object to the layout function (`layoutGraph`), which returns another graph object containing all the necessary information for subsequent rendering.

```
> g1 <- layoutGraph(g1)
> renderGraph(g1)
```

Figure 1a shows the result of the rendering function (with default parameters) used below to represent our random graph `g1`.

There is a hierarchy to set rendering parameters for a graph. The levels of this hierarchy are the session, the rendering operations, and the individual nodes and edges. At the session level, defaults parameters are used if not set somewhere further down the hierarchy. For single rendering operation, i.e., a single call to `renderGraph`, defaults can be set using its `graph.par` argument.

2. All parameters set using renderGraph's `graph.pars` argument will be transient.

```
> renderGraph(g1,
+              graph.pars=list(edges= list(col= "lightblue",
+                                          lty= 1)))
```

3. Setting session-wide parameters, using the function `graph.par` will affect all subsequent rendering operations. Figure 1b is the result of modifying the `graph.par` as follow:

```
> graph.par(list(nodes = list(fill= "lightgray",
+                     textCol= "blue", cex= 1),
+               edges=list(col= "red", lty= 2)))
> renderGraph(g1)
```

4. Parameters for individual nodes or edges in a graph object can also be set using the `nodeRenderInfo` and `edgeRenderInfo` functions. These changes will be retained in all subsequent layout or rendering operations of that particular graph.

```
>nodeRenderInfo(g1)<-list(fill=c(A= "lightyellow",
+                               B= "lightblue"))
> renderGraph(g1)
```

There are many options for layout type, node color, node shape, edge representation, etc., and we refer the reader to the vignette of the **Rgraphviz** package for further details.

4. Case Studies

4.1. Case Study 1: B. subtilis Protein–Protein Interaction Data

In this case study, we are interested in *B. subtilis* PPI data. We focus on physical binary interactions between protein pairs obtained via a Yeast-2-Hybrid (Y2H) assay (16) and complex comembership between sets of proteins detected via Affinity Purification–Mass Spectrometry (AP–MS) (17) (see Note 6). Curated, PSI-MI 2.5 XML, *B. subtilis* PPI data are available from the IntAct database (see Note 7). In Bioconductor, the ***RpsiXML*** package serves as a programmatic interface between the R statistical environment and the XML files. It implements the PSI-MI 2.5 standard and currently supports eight databases (unfortunately each database has its own implementation of the PSI-MI 2.5 standard).

1. To download and parse the *B. subtilis* PPI data, we use the RpsiXML. First, the URL is specified, then that is used to open the connection and download the text and finally that text is parsed to extract the different components.

```
>library ("RpsiXML")
>site <- ftp://ftp.ebi.ac.uk/pub/databases/intact/current/psi25"
> burl <- paste(site, "species/bacsu_small.xml" , sep="/")
>file <- paste(readLines(url(burl)), collapse= "")
>bacsu <- parsePsimi25Interaction(bfile,
+                                  INTACT.PSIMI25,
+                                  verbose= FALSE)
```

The R object `bacsu` is of class `psimi25Interaction Entry`. You can access its attributes using the associated methods like interactions, that lists each interaction, or organism `Name`, that returns the name of the organism (see man page for the complete list of methods).

2. To explore the data, we might first want to understand what types of interactions are present and in what quantities, as well as how many different experiments are represented (see Note 7).

```
> ii = interactions(bacsu)
> table(sapply(ii, function(x) x@interactionType))
                 2 hybrid  adenylate cyclase
                       41  1
           affinity chrom| anti bait coip
                        7  1
             anti tag coip          confocal microscopy
                        3                             1
                Crosslink          electron microscopy
                        1                             4
                     emsa          fluorescence imaging
                        4                             3
                     fret              light scattering
                        1                             2
     protein kinase assay                     pull down
                        5                            12
                       tp           transcription compl
                        4                             3
         x-ray diffraction
                        4
> table(sapply(ii, function(x) x@expPubMed))
 10997907    11722727    12065423    12100548    12359875
        1           4           5           3           1
 12644235    12738765    14711369    14993308    15045029
        1           1           3           9           4
 15104138    15362849    15491161    16796675    17662947
       13           1           4          19           7
 17803906    17994626    18854156    19450516    19800110
        8           3           4           1           5
```

We can see that 41 of the interactions are from Y2H screens and 12 are from AP–MS experiments. From the second call to table we see that 18 different PubMed citations are involved and that the two largest sources of data are 13 interactions from PMID:15104138 and 19 interactions from PMID:16796675.

3. We next can verify the quality of the downloaded interactions. For instance, we can create a function that test whether Y2H dataset have all their bait and prey information and apply this function to the entire list of interactions.

```
> filterfun <- function(x){
+  x@interactionType=="2hybrid"  &&  is.na(x@bait)&&
is.na(x@prey)
+  }
> table(sapply(ii, filterfun))

FALSE  TRUE
   91     6
```

The result indicates that there may be some quality issues: 6 out of 41 have missing information. More generally, high-throughput experiments are prone to biases and special care must be taken to evaluate and take into account experimental errors in data analysis (see Note 8). In the following paragraph, we provide an overview of the ***ppiStats*** package that provides quality assessment (QA) tools for *PPI* experiments.

4. Upon reading the PPI data into R, one intuitive exploratory approach is to represent it as a graph. To extract a graph from XML data, **RpsiXML** provides two methods: `psimi25XML2-Graph` that generates one large graph composed of all the bait-prey data across all experiments while `separateXMLDataByExpt` yields a list of graphs indexed by the PubMed ID. Here we choose the latter as the data originate from many various experimental techniques (mostly low throughput) and it might be misleading to mix them together. For similar reasons, the separate `XMLData-ByExpt` function does not mix interaction data based on direct, e.g. binary physical interactions, with that obtained from indirect methods, such as AP–MS. Here we ask for `indirect`, e.g. AP–MS, interactions.

```
>graphs<-separateXMLDataByExpt(bfile, INTACT.PSIMI25,
+                               type="indirect",
+                               directed= TRUE,
+                               abstract= TRUE,
+                               Verbose= FALSE)
> names(graphs)
```

5. This subset of the PPI data originated from six different works. One way of getting a descriptive summary of these experiments is to use the `createSummaryTables` function of the ***ppiStats*** package. This function takes a list of directed graph objects and creates a summary table based on the directed connectivity, as shown in Table 2.

Table 2
An overview of summary statistics for all AP–MS data sets

	VB	VP	VBP	VBP/VB	VP/VB	TI	TI/VB
12100548	1	1	0	0.00	1.00	1	1.00
15045029	4	22	3	0.75	5.50	57	14.25
15491161	3	3	2	0.67	1.00	3	1.00
16796675	6	13	4	0.67	2.17	32	5.33
17994626	2	2	1	0.50	1.00	2	1.00
19450516	1	1	0	0.00	1.00	1	1.00

Each row displays the result reported in each paper (PMID). *VB* viable baits, *VP* viable prey, *VBP* viable bait/prey, *TI* the total number of directed interactions. The statistics within this table can differentiate between experiment type. For example, the last column details the number of interactions per VB; the experiment reported in paper 15045029 and 16796675 have remarkably high numbers of average interactions per VB, and so would produce markedly distinct features (i.e., protein complexes) from the others

```
> library("ppiStats")
> directedGraph <- sapply(graphs, isDirected)
>aSummary<-
createSummaryTables(graphs[directedGraph])
```

6. We next consider the paper 16796675 (18) in more detail. First, we can read the abstract information.

```
> abstract(graphs[["16796675"]])

An object of class 'pubMedAbst':
Title: A new FtsZ-interacting protein, YlmF,
     complements the activity of FtsA during
     progression of cell division in
     Bacillus subtilis.
PMID: 16796675
Authors: S Ishikawa, Y Kawai, K Hiramatsu, M
     Kuwano, N Ogasawara
Journal: Mol Microbiol
Date: Jun 2006
```

7. Then, we verify the number and type of interactions reported.

```
> Ishikawa <- sapply(ii, function(x) x@expPubMed ==
16796675)
> sum(Ishikawa)
[1] 19
> IshikawaInts <- ii[Ishikawa]
>table(sapply(IshikawaInts,function(x)
x@interactionType))
2 hybrid pull down
     8    11
```

They report 19 interactions, 11 detected by pull down (most likely AP–MS technology) and 8 detected via Y2H.

8. We next extract the graph for their data from the `graphs` object. Recall that in creating this we only accessed indirect interactions, so only the AP–MS-like interactions are represented.

```
> IshikawaGraph <- graphs$ '16796675'
> IshikawaGraph

A graphNEL graph with directed edges
Number of Nodes = 15
Number of Edges = 32
[1] "psimi25Graph"
attr(,"package")
[1] "RpsiXML"
```

The resulting graph, `IshikawaGraph`, is a directed graph composed of 15 nodes (representing proteins) and 32 edges (4representing observed indirect interactions). The IntAct data reported 11 interactions, but recall that each such interaction is between one bait and one or more prey. The nodes in the graph correspond to the unique set of proteins detected as either bait or prey, while the edges indicate which baits detected which prey. Thus, there were 32 interactions detected among 15 different proteins.

9. To assess the quality of the reported information we can compute the degree of the nodes in the graph (see Note 2). Since the graph is directed the in-degree and out-degree are reported separately.

```
> degree(IshikawaGraph)

$inDegree
 031728   034894   P17865   P28264   P37469   P94542   P54420
      4        4        4        4        0        0        2
 P33166   P45694   P80868   P39751   P06574   P24327   P21465
      2        2        2        1        2        2        2
 P06567
      1

$outDegree
 031728   034894   P17865   P28264   P37469   P94542   P54420
      3        3        3       11        8        4        0
 P33166   P45694   P80868   P39751   P06574   P24327   P21465
      0        0        0        0        0        0        0
 P06567
      0
```

Two proteins, P28264 and P37469 have large out-degrees, 11 and 8 respectively. However these relationships are not reciprocated (the in-degrees are 4 and 0, respectively). In fact, in PPI experiments a major issue is that not all proteins are employed as both bait and prey and we cannot observe interactions that were not tested (see Note 8 and ref. 19 for terminology).

10. To count the proteins that are observed as both bait and prey (VBP) in an experiment, we use the `idViableProteins` function.

```
> idViableProteins(IshikawaGraph) $VBP

[1] "O31728" "O34894" "P17865" "P28264"
```

The `assessSymmetry` function from the ***ppiStats*** package can be used to test the symmetry of your data by calculating the reciprocated degree (nr), the nonreciprocating out (no) and in (ni) degrees. Using this information and a Binomial error model you can assess the *p*-value for the in and out degree of each protein (20).

11. In protein interaction graphs, another question of special interest is the identification of functionally related modules (e.g., protein complexes). In graph analysis, this might be achieved by searching for cliques (see Note 2). Here, we search for the maximum clique, i.e., the largest subgraph(s), using the `maxClique` function implemented in *RBGL*. Though this algorithm needs to be applied on an undirected graph. We thus extract the underlying structure of `IshikawaGraph` using the `ugraph` function from graph.

```
> library("RBGL")
> xu <- ugraph(IshikawaGraph)
> cls <- maxClique(xu)maxCliques
> cs <- sapply(cls, length)
> maxC <- cls[cs== max(cs)]
> maxC

[[1]]
[1] "P28264" "O34894" "P17865" "O31728" "P94542"

> selectedProt<- unlist(maxC[1])
```

We identify a subgraph that contains the four VBPs plus one additional protein. Figure 2 illustrates the result of the maximum clique algorithm. Panel (a) displays all the interactions observed by (18) using the AP–MS technology. Panel (b) renders only the maximum clique. Note, that P9542 only has out edges indicating it was a bait.

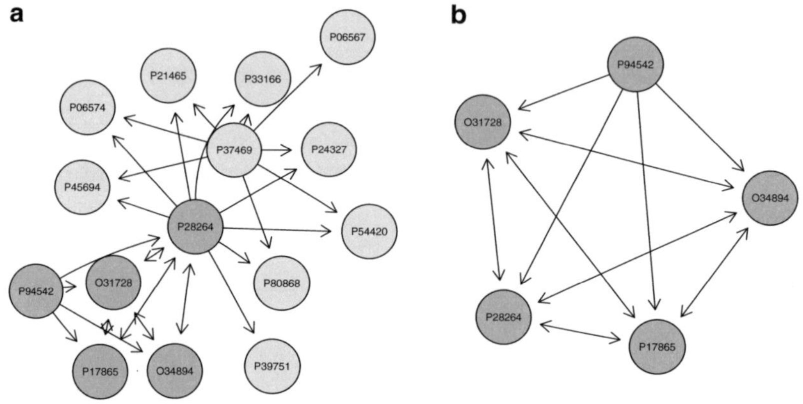

Fig. 2. Graphs of interacting proteins in *B. subtilis*, during cell division, as reported by Ishikawa et al. (18). Only those interactions detected by AP–MS are rendered here. Panel (**a**) shows all interactions and their directions (the *arrows* go from a bait protein to a prey protein). In *orange* is the maximum clique found by the `maxClique` function of the *RBGL* package. In panel (**b**) only the maximal clique is rendered. The proteins 031728, P17865, and P28264 are part of the divisome complex; P94542 is a divisome activator and 034894 is a divisome inhibitor.

12. The next step in an analysis could be to identify and understand the roles and function of the proteins in the clique. The protein interaction data are recorded using UNIPROT IDs but most papers will use common names, or symbols, so we need some way to navigate between these (see Note 9). To this aim we use the ***biomaRt*** package. Every analysis with ***biomaRt*** starts with selecting a BioMart database and one of the associated datasets (see the ***biomaRt*** vignette for details). Here, we query the *B. subtilis* dataset of the "*bacterial_mart_8*" database with the five selected proteins. We define the list of filters and attributes we are interested in and use the `getBm` to retrieve annotation information.

```
> library("biomaRt")
> ensembl <- useMart("bacterial_mart_8" )
> bsubtilis <- useDataset("bac_6_gene ,mart=ensembl)
> filters <- listFilters(bsubtilis)
> attributes <- listAttributes(bsubtilis)
> getBM(attributes = c(attributes[c(4,12), 1]),
+     filters = "uniprot_swissprot_accession",
+     values = selectedProt,
+     mart = bsubtilis)
```

The name and comments columns of the results show us that the selected proteins are indeed tightly related physically or functionally. The proteins O31728 (*sepF*), P17865 (*ftsZ*) and P28264 (*ftsA*) are part of the divisome complex (or *FtsZ complex*), while P94542 (*zapA*) is a divisome activator and O34894 (*ezrA*) is a divisome inhibitor (Fig. 2b). We also remark that the protein *sepF* corresponds to the gene name *YlmF* used in ref. 18.

4.2. Case Study 2: Response of E. coli *During an Oxygen Shift*

In this study case, the data consist of 43 microarrays for six mutants in two conditions aerobic and anaerobic. The mutants are knockout strains (KO) of five key transcriptional regulators of the oxygen response network: Δ*arcA*, Δ*appY*, Δ*fnr*, Δ*oxyR*, Δ*soxS*, and the double knockout Δ*arcA*Δ*fnr*. These data are part a genome-scale analysis (21). We restrict our attention to the effect of the single transcription factor knockout data in aerobic conditions. We realize that a comparison of aerobic to anaerobic conditions would be of some importance, but such an analysis would be well beyond the scope of this chapter, where our primary goal is to inform the reader of the available tools for graph analysis and to demonstrate their use in a setting that is not especially complex. Even with that simplification, our analysis is incomplete and quite cursory. However, the reader could easily extend our example in many interesting directions.

Here, we consider a model that will help to identify the set of genes that are regulated by our transcription factors. The nodes in our network will be the genes in *E. coli* that were assayed. We will define an edge to exist between one of the five transcription factors listed above and any other gene if the gene expression between the wild type condition and the condition with the transcription factor knocked out is significantly different. Expression of the second gene can either increase (in which case the TF was likely a repressor) or decrease (in which case the TF was likely an enhancer). We note that the biological effect need not to be direct; that is, there is no reason to assume that the TF which acts in *cis*, could not affect an intermediate regulator of a second gene. One could determine which TFs are acting in *cis* by using either known TF binding sites or by performing a ChIP-seq experiment. We use the genes identified by the knockout experiments to identify cellular modules regulated by those transcription factors. We use the biological process component of GO (22) and KEGG pathways (http://www.genome.jp/kegg/) to identify the candidate pathways and sets of genes for this part of the analysis.

1. To get started, we create a vector with the names of the KO genes.

```
> KOgene <- c("arcA", "appY", "oxyR", "soxS", "fnr")
```

2. Next we load the expression data that we formatted from the online supplementary information of *Covert et al.* (21) (see Note 10).

```
> load("data/bES.rda")
> bES

ExpressionSet (storageMode: lockedEnvironment)
assayData: 4345 features, 37 samples
    element names: exprs
protocolData: none
phenoData
     sampleNames: aerobic_wild_rep1
        aerobic_wild_rep2 ...
        anaerobic_soxS_rep3 (37 total)
     varLabels: strain growth genotype name
     varMetadata: labelDescription
featureData
   featureNames: thrl_b0001_at thrA_b0002_at
     ...las T_b4403_at (4345 total)
   fvarLabels: genename
   fvarMetadata: labelDescription
experimentData: use 'experimentData(object)'
Annotation:

> pData(bES)[1:5,1:3]

                      strain growth genotype
aerobic_wild_rep1 wild type aerobic     wt
aerobic_wild_rep2 wild type aerobic     wt
aerobic_wild_rep3 wild type aerobic     wt
aerobic_appY_rep1   mutant aerobic    appY
aerobic_appY_rep2   mutant aerobic    appY
```

The object bES is an object of class ExpressionSet, a container that holds both gene expression data and the corresponding, per sample, metadata.

3. Using the bES data set, we subset the data so that only the arrays corresponding to the aerobic condition are retained (in code not shown we have also removed the double knockout arrays).

```
> library("limma")
> aerobia <- bES[, bESgrowth=="aerobic"]
```

4. Then we evaluate contrasts between experimental conditions using statistical testing. Since there are relatively few arrays we use the ***limma*** package to test for differentially expressed genes between KO and wild-type strains. To this aim, we first define a design matrix to describe which samples have been applied to each array (see Note 11).

```
> aGeno <- as.factor(pData(aerobia)genotype)
> aDesign <- model.matrix(~0 + aGeno)
> colnames(aDesign) <- levels(aGeno)
```

5. In the following code we fit the model, define the contrast to test, and apply the moderated empirical Bayes test with Benjamini Hocheberg *p*-value adjustment. This last step is to adjust for multiple testing concerns.

```
> afit <- lmFit(aerobia, aDesign)
> a.cont.matrix <- makeContrasts ("appY"=appY-wt,
+                                  "arcA"=arcA-wt,
+                                  "fnr"=fnr-wt,
+                                  "oxyR"=oxyR-wt,
+                                  "soxS"=soxS-wt,
+                                  levels=aDesign)
> afit2 <- contrasts.fit(afit, a.cont.matrix)
> afit2 <- eBayes(afit2)
> aRes <- c()
> for(i in 1:ncol(a.cont.matrix)){
+ aDE <- topTable(afit2, coef= i, n=500, adjust="BH",
+                 p.value=0.001)
+ intRes <- cbind(colnames(a.cont.matrix)[i],
+                 aDEgenename , round(aDElogFC, 2))
+ aRes <- rbind(aRes, intRes)
+  }
> colnames(aRes) <- c("KO","targets","logFC")
```

6. We define as significantly differentially expressed the genes (probes) with *p*-value ≤ 0.001. The number of significantly differentially expressed genes for each TF KO condition is:

```
> table(aRes[,1])

appY arcA fnr oxyR soxS
  33    1  65  147    1
```

We note that three of the TFs have a large number of targets identified, while two (*arcA* and *soxS*) have only one target each. This may be due to the rather stringent *p*-value threshold we used or the biological functions of those genes that depend on the oxygen condition. Another interesting feature of the data is that the transcription factor KOs Δ*appY*, Δ*fnr*, Δ*oxyR* seem to have similar effects on fairly large sets of their targets suggesting that perhaps they affect similar sets of intermediate (Fig. 3).

7. We can display the graph of these interactions using the ***graph*** and ***Rgraphviz*** packages. The type of graph that is most helpful is one that is essentially a bipartite graph (see Note 3). In practical terms, we construct this graph using the information from ***limma*** which tells us both the TFs and their targets. This is what can be referred to as a *from-to* representation of a graph and we use the function `ftM2graphNEL` to construct an instance of the ***graphNEL*** class. In the call we specify that the log fold change should be included in the graph as edge weights.

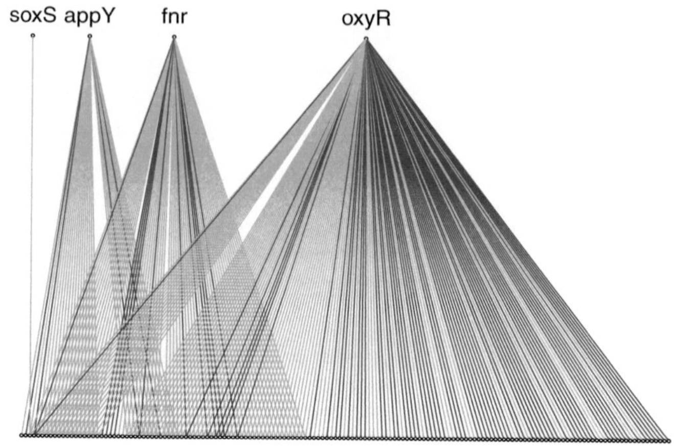

Fig. 3. Unbalanced effect of the deletion of *Escherichia coli* key transcriptional regulators of the oxygen response, in aerobic condition. In this figure, the TF knockouts (*top*) are linked to their targets (*bottom*). Self-loops have been removed (autoregulation of the TFs) as well as the node for the TF *arcA*, which only regulates itself and affects no other gene in the aerobic conditions. The edges are colored according to whether there is higher expression in WT than in KO (*blue*) or higher expression in KO than WT (*orange*). There are a surprising number of orange edges. We can also see the high level of overlap in targets between TFs.

```
>graphAERO<-ftM2graphNEL(aRes[, c("KO","targets")],
+                         W=as.numeric(aRes[,"logFC"]),
+                           edgemode="directed")
```

8. Our graph might not be a bipartite graph per se as it may contain self-loops (TF can autoregulate themselves). We thus remove self-loops. Only *arcA* regulates itself (and has no other target) which brings us to remove the *arcA* node. The fact that *arcA* only regulates itself is intriguing. However *arcA* is known to act primarily as a negative transcriptional regulator under anaerobic conditions thus might not be expressed (or at very low level) in aerobic condition, explaining its auto-regulation and the mild effect of its deletion (23).

```
> aeroG <- removeSelfLoops(graphAERO)
> aeroG <- removeNode("arcA", aeroG)
```

9. We use the fold change (FC) between KO and wild type to color the edges (orange means higher expression in KO and blue means higher expression in wild type). First, the graph is laid out, then we determine which of the edges correspond to positive log FC and which to negative. We then construct a vector named `edgeColor` that contains the value `"orange"` for those edges that are higher in KO, and `"blue"` otherwise. We apply that information to the graph using the `edgeRenderInfo`

function. Finally, for readability we remove the labels from the TF targets using the `nodeRenderInfo` function and we render the `graph`.

```
> aeroG <- layoutGraph(aeroG, layoutType="dot")
> edgeColor <- rep("orange", numEdges(aeroG))
> names(edgeColor) <- edgeNames(aeroG)
> low <- unlist(lapply(edgeWeights(aeroG), function(x) x < 0))
> edgeColor[low] <- "blue"
> edgeRenderInfo(aeroG) <- list(col=edgeColor)
> KOlabel    <-    nodes(aeroG)>    nodeRenderInfo(aeroG)$labelY[!KOlabel] <- NA
> renderGraph(aeroG)
```

There is a preponderance of orange edges and indeed 158 genes have higher expression in the KO strains while only 85 have lower expression levels. Suggesting that these TFs were largely repressing gene expression, or perhaps that their deletion caused other systems to upregulate. Had these TFs been largely activating transcription then we would have seen blue edges, showing that their deletion corresponded to a decrease in expression.

To better understand this, we turned to the literature and found evidence that these TFs have dual roles. *fnr* is the primary transcriptional regulator mediating the transition from aerobic to anaerobic growth through the regulation of hundreds of genes. Generally speaking, this protein activates genes involved in anaerobic metabolism and represses genes involved in aerobic metabolism, which may partly explain the upregulation observed in aerobic conditions (24). In *E. coli*, *oxyR* was until now mostly described as an activator (except for the *flu* gene (25)), but it has recently been characterized as a repressor in several other bacteria (such as *Neisseria gonorrhoeae* and *Deinococcus radiodurans* (26)). *appY* induces the expression of energy metabolism genes under anaerobiosis, stationary phase, and phosphate starvation (27). Finally, *soxS* is a transcriptional activator of oxidative stress genes, which might explain that in physiological aerobic condition its depletion has a minor effect. Of course, we cannot rule out the possibility that there was somewhat more variation in some of the experiments than in others. This would make it harder to detect differential expression under some conditions and hence could reduce the target set. A thorough analysis would engage this idea and ensure that there were no obvious differences between the conditions.

We can also use KEGG and GO categories to try to better understand whether there are particular pathways or processes that are affected. This analysis consists of testing for overrepresentation of categories among genes in a selected gene set based upon the Hypergeometric distribution (28) (see Note 12). To this aim, we make use of the methods in the ***GOstats*** and ***Category*** packages.

10. First, we select the *E. coli* gene name from the `bES ExpressionSet` and retrieve the unique Entrez Gene Identifiers from the **org.EcK12.eg.db** annotation package (see Note 9).

```
> library("GOstats")
> library("Category")
> library("org.EcK12.eg.db")
> universe <- unique(fData(bES)$genename)
> univEG <- unlist(mget(universe,
+                org.EcK12.egSYMBOL2EG, ifnotfound=NA))
> univEG <- univEG[!is.na(univEG)]
```

11. We next apply the same procedure to annotate the genes associated with each KO and simultaneously (in a *for* loop) perform the GO and KEGG `hyperGTest` function to search for overrepresentation with a *p*-value cut-off of 0.001.

```
> pCutoff <- 0.001
> GO <- vector("list", length=length(KOgene))
> KEGG <- vector("list", length=length(KOgene))
> names(GO) <- names(KEGG) <- KOgene
> for(i in 1:length(KOgene))
+ adjnode <- unlist(adj(graphAERO, KOgene[i]))
+ genelist <- c(KOgene[i], adjnode)
+ genelist <- unique(genelist)
+       genelist        <-        unlist(mget(genelist,
org.EcK12.egSYMBOL2EG,
+                     ifnotfound=NA))
+ genelist <- genelist[!is.na(genelist)]
+   a.GOparams    <-   new("GOHyperGParams",    geneIds    =
genelist,
+                     universeGeneIds = univEG,
+                     annotation = "org.EcK12.eg.db",
+                     ontology = "BP",
+                     pvalueCutoff = pCutoff,
+                     conditional=TRUE,
+                     testDirection = "over")
+ GO[[i]] <- hyperGTest(a.GOparams)
+
+   a.Kparams    <-   new("KEGGHyperGParams",    geneIds    =
genelist,
+                     universeGeneIds = univEG,
+                     annotation = "org.EcK12.eg.db",
+                     pvalueCutoff = pCutoff,
+                     testDirection = "over")
+ KEGG[[i]] <- hyperGTest(a.Kparams)
+ }
```

We first consider the results for the GO category analysis. We restricted our attention to the biological process (BP) ontology.

```
> sigGO <- lapply(GO, sigCategories)
> sapply(sigGO, length)

arcA appY oxyR soxS fnr
   0   10   12    0   9
> lapply(GO["oxyR"], function(x) summary(x)[,c(2,7)])
$oxyR
           Pvalue
1     2.316280e-17
2     2.316280e-17
3     3.010293e-13
4     6.792395e-11
5     4.578291e-09
6     4.578291e-09
7     4.578291e-09
8     3.475130e-07
9     1.510304e-04
10    3.125890e-04
11    3.125890e-04
12    7.966690e-04

                                                   Term
1                                   iron ion transport
2                                  metal ion transport
3                                        ion transport
4                         establishment of localization
5                     enterobactin biosynthetic process
6                             catechol metabolic process
7                     siderophore biosynthetic process
8                         peptide biosynthetic process
9                                   aerobic respiration
10                         thiamin biosynthetic process
11             thiamin and derivative metabolic process
12    cellular aromatic compound metabolic process
```

Very similar categories are identified for *fnr*, *oxyR*, and *appY*. The set of GO categories for *fnr* was a subset of those identified for *appY*, with `GO:0006066` being associated with *appY*, but not *fnr*. Similarly, the set of GO categories for *appY* was essentially a subset of those for *oxyR*, with `GO:0006066` and `GO:0006725` being associated with *appY* but not *oxyR*. These GO categories are labeled *alcohol metabolic process* and *cellular aromatic compound metabolic process*, respectively. We list the complete set of GO categories for *oxyR* using the code below.

12. Next, we consider the results for the KEGG pathways.

```
> sigKEGG <-lapply(KEGG, sigCategories)
> sapply(sigKEGG, length)

arcA appY oxyR soxS fnr
   0    1    3    0   2
```

We see that there is four significant KEGG pathways for *oxyR* and one for each of *appY* and *fnr*. All 3 TFs share the one significant pathway, *Biosynthesis of siderophore group nonribosomal peptides*, which is in accordance with the GO categories identified.

13. We can then return to the original data to get a sense of which of the TFs had genes involved in this pathway.

```
>gnodes <- nodes(graphAERO)
> geneInKEGG <- geneIdsByCategory(KEGG[[2]])$'01053'
> symbol <- unlist(mget(geneInKEGG,
+                      org.EcK12.egSYMBOL, ifnotfound=NA))
> siderophore <- gnodes[gnodes%in%symbol]
> as.data.frame(aRes[aRes[,2]%in%siderophore,])
```

```
      KO targets logFC
1    appY   entC  2.16
2    appY   entE  2.06
3    appY   entB  2.09
4    appY   entF  2.36
5     fnr   entC  2.72
6     fnr   entE   2.6
7     fnr   entB  2.56
8     fnr   entF  2.93
9    oxyR   entC   3.1
10   oxyR   entE   2.9
11   oxyR   entB  2.91
12   oxyR   entF  3.43
```

We note that several genes involved in iron metabolism are up regulated between WT and KO. And indeed it seems that KOs of *appY*, *fnr*, and *oxyR* seem to have similar effects on a largely overlapping set of genes. Microorganisms growing under aerobic conditions need iron for a variety of functions including reduction of oxygen for synthesis of ATP, reduction of ribotide precursors of DNA, for formation of heme, and for other essential purposes. Accumulation of the intracellular siderophore has also been shown to be upregulated by oxidative stress in *Aspergillus nidulans* (29), which underscores the intertwining of iron metabolism and oxidative stress.

5. Discussion

We have provided a fairly brief and concise overview of how some of the many Bioconductor packages can be used to help analyze network data. The tools are quite varied, and certainly much remains to be done, but most of the building blocks needed for a comprehensive tool suite now exist and can fairly easily be used to accomplish complex analyses.

We believe that the current capabilities are extensive, but that perhaps the most compelling aspect of these tools is the ease with which they can be extended and adapted to different analysis problems. Interested readers can obtain our data and scripts from the accompanying Web site. They can verify our results, perhaps point out errors or bugs in our code, and then go on to carry out much more detailed analyses of these. We encourage them to do so.

6. Notes

1. *Graph orientation* A graph can be undirected or directed. For an undirected graph the relationship determined by the edges is symmetric, and hence does not have direction assigned to its edges like `g1` in Fig. 1. For a directed graph (or digraph) the relationship is asymmetric, and node A can have an edge to node B, but B need not have an edge to A. These relationships are thought of as being directed and are often represented by edges with arrowheads indicating the direction, see panel (a) of Fig. 2 for an example. A directed graph with no directed cycles is called a *directed acyclic graph*, or DAG.
2. *Graph properties.*
   (a) *Node degree* The node degree of a node is the number of edges that are incident on that node. If the graph is directed then we can consider in-degree and out-degree separately. In a biological network it is often useful to know the number of interactors associated with a node. This information can be used for a variety of purposes, such as monitoring flux between components of the network, characterizing protein complexes, or in assessing the quality of proteomics bait-prey technologies by modeling experimental errors on node degree (20).
   (b) *Complete graph* A complete graph is a graph in which every pair of distinct vertices is connected by a unique edge.
   (c) *Cut problem* In graph theory, a cut is a partition of the vertices of a graph into two non-empty but disjoint subsets. The cut-set of the cut is the set of edges whose end points are in different subsets of the partition. Edges are said to be crossing the cut if they are in its cut-set. A cut is minimum if the disjoints subets are build by cutting the minimum number of edges. Inversely a cut is maximum if the disjoints subets are build by cutting the maximum number of edges.

(d) *Clique* A clique is a complete subgraph, i.e., there is an edge between every pair of vertexes. For instance, in PPI network a clique might represent a multiprotein complex. Although, we note that a clique is a very extreme notion of a cohesive subgroup. There are many algorithms implemented in RBGL that might be used and compute different type of cliques. We mentioned the maximum clique, i.e., the largest subgraph(s) (`maxClique function`), but RBGL also provides implementations of *k-cliques*, *k-cores* and many other concepts described in the social network literature (30). These latest might be particularly useful for high-throughput PPI data with high error rates (see Note 8). Also, (31) consider this problem and provide both an algorithm and an implementation via the **apComplex** package.

(e) *Shortest-path* As an example, if a graph represents a metabolic network, we might be interested in computing the shortest-path to generate a product downstream of a cascade of interlacing reactions.

3. *Graph types.*

(a) *Directed graphs* are useful for representing directional interactions, or experiments, that are asymmetric such as protein interaction experiments where one protein is the *bait* and other proteins are *prey.* In many cases the proteins need to be modified to perform the assay and it is not uncommon to observe that protein A interacts with B, when A is the bait, but B fails to interact with A when B is the bait.

(b) *Bipartite graphs* are graphs where nodes can be separated into two sets and all edges go from a member of one set to a member of the other set. For instance, bipartite graphs may be helpful in modeling the relationship between transcription factors (TFs) and their targets. The mappings between GO terms and genes may form a bipartite graph, as do the mappings between KEGG pathways and genes.

(c) *Directed acyclic graphs* or DAGs are used by the Gene Ontology consortium (22) to store and explore gene annotation data.

(d) *Cluster graph* represents the output of a clustering algorithm. They have the property that nodes are separated into groups or clusters where within a cluster all nodes are connected (a complete graph) but between clusters there are no edges.

(e) *Network* is a type of graph that can be used to represent flow. In particular a network is a directed graph with a vertex, called the source, whose in-degree is 0, and a vertex, called the sink, whose out-degree is 0, and there can be a numerical capacity associated with each edge in a network.

4. *Building graph in R*. The **graph** package provides different representations such as matrix and list. A number of methods are available to coerce graphs between the different representations. The hypergraph handle more specialized operations and types of graphs. For instances, it offers specialized tools for bipartite graphs analysis and representation. Other graph-related packages provide methods for computing graph properties. Some packages are aimed at specific areas of application such as the **ppiStats** package which is aimed at PPI data and the **rsbml** package which provides support for SBML modeling (Systems Biology Markup Language http://www.sbml.org). The **RBGL** package is an interface from R to the Boost Graph Library (32). It is the primary source of algorithms to explore and analyze graphs. For instance, it contains methods to search for shortest paths, connected components, and cliques (see Note 2).
5. *Graph layout with R*. In Bioconductor, we have separated the process of drawing graphs into two distinct phases. The first is *layout* which involves determining the locations of the nodes and edges, typically in a two dimensional space. This can then be followed by the process of rendering the graph, which is the actual plotting of the graph. There are two reasons for separating these tasks. First, we would like to use R's graphics capabilities for rendering, so that graphs can easily be intermingled with other plots and so that the full range of R's advanced graphics can be used. Second, we often want to plot graphs based on the same set of nodes (and sometimes edges) multiple times. Being able to control the location of the nodes and edges helps us achieve that goal. The main graph layout algorithms are provided by the **Rgraphviz** package but some can also be found in the **RBGL** package.
6. *Protein–protein interaction: Y2H and AP–MS experiments.* Experimental data issued from PPI experiments are most often incomplete. For yeast 2-hybrid assay (Y2H) both bait and prey proteins need to be modified. One of the two is modified to contain a DNA binding domain, while the other is modified to contain an activation domain. A typical experiment involves a generally smaller set of bait proteins crossed with a generally larger set of prey. The resulting data are usually processed by the experimenter to be binary and usually only positive results are reported. Thus the data generally

consist of a set of bait-prey pairs. For these experiments both the set of baits that were used and the set of prey they were targeted to are known, but are seldom reported if not interacting. For AP–MS like experiments only the bait is modified and hence the potential set of prey is very large. Each *pull-down* provides data on the set of interactors with the bait protein (the protein that was pulled-down). For these experiments the set of baits is known, but the set of prey is typically the set of expressed proteins in the tissue being assayed.

7. *IntAct database.* IntAct is an open-source, open data molecular interaction database (http://www.ebi.ac.uk/intact/). Interaction data are abstracted from the literature or from direct data depositions by expert curators. The interaction types refer to the methods used to detect the interactions as stored in the IntAct database like "Y2H", representing binary physical interactions, or "pull-down", representing complex comembership (see the IntAct annotation rules for more details at http://www.ebi.ac.uk/~intact/site/doc/IntActAnnotationRules.pdf). Data are stored and downloadable as curated PSI-MI 2.5 XML format. The PSI-MI 2.5 XML schema is a recent initiative by the *Protein Standardization Initiative* to formally standardize the way biologists should report molecular interaction data.
8. *Quality assessment of PPI high-throughput experiments.* More generally quality assessment (QA) is an crucial processing step in data analysis and exploration. Experimental data (especially high-throughput) are often biased due to systematic or stochastic errors (33, 34). QA allows identifying biases (and potentially their causes) and minimizing the effects by correcting or removing erroneous data. Several Bioconductor packages offers QA methods specific to a variety technology. For example, the ppiStats provides tools for the analysis of Y2H and AP–MS data. The package is specifically designed to summarize graph data, identify systematically biased baits, and estimate stochastic error probabilities (19, 35). In particular, its authors (19) introduced the concepts of viable baits (VBs) and viable prey (VP) to help take into account the problem of untested interactions versus tested but not observed interactions. Experimental results are typically only reported for proteins that are observed in at least one interaction, leaving uncertainty as to which proteins were actually tested. VB is thus the set of bait proteins that successfully detect at least one interaction partner, VP is the set of all proteins detected by at least one bait, and VBP is the set of proteins that are both viable baits and viable preys.
9. *Metadata and annotation.* For many important organisms the Bioconductor project offers database-oriented packages that

provide maps between the different identifiers used for that organism. For example, the **org.EcK12.eg.db** annotation package allows mapping between gene symbols and Entrez Gene identifiers along with various annotations such as GO categories, REFSEQ or PMID. In addition one can rely on external resources such as BioMart (36) which can be accessed from within R via the **biomaRt** package.

10. *E. coli dataset.* We downloaded at http://www.nature.com/nature/journal/v429/n6987/suppinfo/nature02456.html. *Covert et al.* provided expression estimates for each probe based on dChip (37). This is slightly problematic for our purposes since dChip can give negative expression values, so we renormalized the data using the VSN algorithm (38) from the **vsn** package. The normalized data are provided on the Web page that accompanies this article (http://www.bioconductor.org/pub/MiMB). The data were saved in a *.rda file (specific binary format for external representation of R objects). Once downloaded from the Web site, it can be loaded in R using the function `load` and the location of the file (e.g., `load(path/to/bES.rda)`).
11. *Limma analysis.* **limma** is a R package for the analysis of gene expression microarray data, especially the use of linear models for analyzing designed experiments and the assessment of differential expression. Every ***limma*** analysis starts with the definition of a design matrix. A design matrix is a matrix of explanatory variables, which here allows you to formalize your experimental design (observed data) and biological question for statistical analysis. Complex experimental design implies complex definition of design matrix, requiring some statistical knowledge (for more details we refer the reader to the well-documented vignette of the ***limma*** package).
12. *Gene Set Enrichment Analysis.* GSEA is a computational method that determines whether an a priori defined set of genes (e.g., based on biological function, process, pathways...) shows statistically significant, concordant differences between two biological states (e.g., phenotypes). In expression data, we do note that this approach, while widely used, is slightly problematic. Among the many issues that arise is the fact that we are combining genes that are both upregulated and downregulated.

Acknowledgments

Funding for this research was provided by grants from La Ligue Contre Le Cancer (RAB08008NSA; project R08010NS). We also gratefully acknowledge the Instituts national de la santé et de la recherche médicale (INSERM) for supporting this project. Funding for RG was provided by NIH grant P41 HG004059.

The authors would like to thank Wolfgang Huber, Tony Chiang, Shailesh Date and Denise Scholtens for helpful comments and for providing a stimulating research environment that has led to the many tools used here and to the ideas that underlie this research.

References

1. R Development Core Team. (2009) *R: A Language and Environment for Statistical Computing*. R Foundation for Statistical Computing, Vienna, Austria, [http://www.R-project.org]. [ISBN 3-900051-07-0].
2. Huber W, Carey VJ, Long L, Falcon S, Gentleman R. (2007) Graphs in molecular biology. BMC Bioinformatics, 8(6):S8.
3. Castelo R, Roverato A. (2009) Reverse engineering molecular regulatory networks from microarray data with qp-graphs. J Comput Biol, 16(2):213–227.
4. Le Meur N, Gentleman R. (2008) Modeling synthetic lethality. Genome Biol, 9(9):R135.
5. Csardi G, Nepusz T. (2006) The igraph software package for complex network research. InterJournal, Complex Systems:1695.
6. Shannon P, Markiel A, Ozier O, Baliga N, Wang J, Ramage D, Amin N, Schwikowski B, Ideker T. (2003) Cytoscape: a software environment for integrated models of biomolecular interaction networks. Genome Res, 13(11):2498.
7. Leisch F. (2002) Sweave: Dynamic Generation of Statistical Reports Using Literate Data Analysis. In *Compstat 2002 Proceedings in Computational Statistics*. Edited by Härdle W, Rönz B, Physika Verlag, Heidelberg, Germany, 575–580, [http://www.ci.tuwien.ac.at/~leisch/Sweave]. [ISBN 3-7908-1517-9].
8. Venables WN, Ripley BD. (2002) *Modern Applied Statistics with S (4e)*. Springer, New York.
9. Gentleman R. (2008) *R Programming for Bioinformatics*. CRC Press, Boca Raton.
10. Chambers JM. (2008) *Software for Data Analysis: Programming with R*. Springer, New York.
11. Hahne F, Huber W, Gentleman R, Falcon S. (2008) *Bioconductor Case Studies*. Springer, New York.
12. Boutros M, Bras L, Huber W. (2006) Analysis of cell-based RNAi screens. Genome Biol, 7(7):R66.
13. Hahne F, Le Meur N, Brinkman R, Ellis B, Haaland P, Sarkar D, Spidlen J, Strain E, Gentleman R. (2009) FlowCore: a bioconductor package for high throughput flow cytometry. BMC Bioinformatics, 10:106.
14. Morgan M, Anders S, Lawrence M, Aboyoun P, Pagès H, Gentleman R. (2009) ShortRead: a bioconductor package for input, quality assessment and exploration of high-throughput sequence data. Bioinformatics, 25(19):2607–2608.
15. Ellson J, Gansner E, Koutsofios E, North S, Woodhull G. (2004) Graphviz and Dynagraph – Static and Dynamic Graph Drawing Tools. In *Graph Drawing Software*. Edited by Junger M, Mutzel P, Springer, Berlin/Heidelberg, 127–148.
16. Brückner A, Polge C, Lentze N, Auerbach D, Schlattner U. (2009) Yeast two-hybrid, a powerful tool for systems biology. Int J Mol Sci, 10(6):2763–2788.
17. Wingren C, James P, Borrebaeck C. (2009) Strategy for surveying the proteome using affinity proteomics and mass spectrometry. Proteomics, 9(6):1511–1517.
18. Ishikawa S, Kawai Y, Hiramatsu K, Kuwano M, Ogasawara N. (2006) A new FtsZ-interacting protein, YlmF, complements the activity of FtsA during progression of cell division in *Bacillus subtilis*. Mol Microbiol, 60(6):1364–1380.

19. Chiang T, Scholtens D, Sarkar D, Gentleman R, Huber W. (2007) Coverage and error models of protein-protein interaction data by directed graph analysis. Genome Biol, 8(9):R186.
20. Scholtens D, Chiang T, Huber W, Gentleman R. (2008) Estimating node degree in bait-prey graphs. Bioinformatics, 24(2):218–224.
21. Covert M, Knight E, Reed J, Herrgard M, Palsson B. (2004) Integrating high-throughput and computational data elucidates bacterial networks. Nature, 429(6987):92–96.
22. The Gene Ontology Consortium. (2000) Gene ontology: tool for the unification of biology. Nat Genet, 25:25–29.
23. Iuchi S, Lin E. (1988) arcA(dye), a global regulatory gene in *Escherichia coli* mediating repression of enzymes in aerobic pathways. Proc Natl Acad Sci U S A, 85(6):1888–1892.
24. Salmon K, Hung S, Mekjian K, Baldi P, Hatfield G, Gunsalus R. (2003) Global gene expression profiling in *Escherichia coli* K12. The effects of oxygen availability and FNR. J Biol Chem, 278(32):29837–29855.
25. Correnti J, Munster V, Chan T, Woude M. (2002) Dam-dependent phase variation of Ag 43 in *Escherichia coli* is altered in a seqA mutant. Mol Microbiol, 44(2):521–532.
26. Chen H, Xu G, Zhao Y, Tian B, Lu H, Yu X, Xu Z, Ying N, Hu S, Hua Y. (2008) A novel OxyR sensor and regulator of hydrogen peroxide stress with one cysteine residue in *Deinococcus radiodurans*. PLoS ONE, 3(2):e1602.
27. Brondsted L, Atlung T. (1996) Effect of growth conditions on expression of the acid phosphatase (cyx-appA) operon and the appY gene, which encodes a transcriptional activator of *Escherichia coli*. J Bacteriol, 178(6):1556.
28. Falcon S, Gentleman R. (2007) Using GOstats to test gene lists for GO term association. Bioinformatics, 23(2):257–258.
29. Eisendle M, Schrettl M, Kragl C, Muller D, Illmer P, Haas H. (2006) The intracellular siderophore ferricrocin is involved in iron storage, oxidative-stress resistance, germination, and sexual development in *Aspergillus nidulans*. Eukaryot Cell, 5(10):1596.
30. Wasserman S, Faust K. (1994) *Social Network Analysis, Methods and Applications*. Cambridge University Press, Cambridge.
31. Scholtens D, Vidal M, Gentleman R. (2005) Local dynamic modeling of global interactome networks. Bioinformatics, 21:3548–3557.
32. Siek JG, Lee LQ, Lumsdaine A. (2002) *The Boost Graph Library*. Addison Wesley, Boston.
33. Schuchhardt J, Beule D, Malik A, Wolski E, Eickhoff H, Lehrach H, Herzel H. (2000) Normalization strategies for cDNA microarrays. Nucleic Acids Res, 28(10):E47.
34. Gentleman R, Huber W. (2007) Making the most of high-throughput protein-interaction data. Genome Biol, 8(10):112.
35. Chiang T, Scholtens D. (2009) A general pipeline for quality and statistical assessment of protein interaction data using R and Bioconductor. Nat Protoc, 4(4):535–546.
36. Haider S, Ballester B, Smedley D, Zhang J, Rice P, Kasprzyk A. (2009) BioMart Central Portal–unified access to biological data. Nucleic Acids Res, 37:W23–W27.
37. Li C, Wong WH. (2001) Model-based analysis of oligonucleotide arrays: expression index computation and outlier detection. Proc Natl Acad Sci U S A, 98:31–36.
38. Huber W, von Heydebreck A, Sueltmann H, Poustka A, Vingron M. (2002) Variance stabilization applied to microarray data calibration and to the quantification of differential expression. Bioinformatics, 18 (Suppl. 1):S96–S104.

Part III

Dynamical Modelling

Chapter 20

Detecting Structural Invariants in Biological Reaction Networks

Jörn Behre, Luís Filipe de Figueiredo, Stefan Schuster, and Christoph Kaleta

Abstract

The detection and analysis of structural invariants in cellular reaction networks is of central importance to achieve a more comprehensive understanding of metabolism. In this work, we review different kinds of structural invariants in reaction networks and their Petri net-based representation. In particular, we discuss invariants that can be obtained from the left and right null spaces of the stoichiometric matrix which correspond to conserved moieties (P-invariants) and elementary flux modes (EFMs, minimal T-invariants). While conserved moieties can be used to detect stoichiometric inconsistencies in reaction networks, EFMs correspond to a mathematically rigorous definition of the concept of a biochemical pathway. As outlined here, EFMs allow to devise strategies for strain improvement, to assess the robustness of metabolic networks subject to perturbations, and to analyze the information flow in regulatory and signaling networks. Another important aspect addressed by this review is the limitation of metabolic pathway analysis using EFMs to small or medium-scale reaction networks. We discuss two recently introduced approaches to circumvent these limitations. The first is an algorithm to enumerate a subset of EFMs in genome-scale metabolic networks starting from the EFM with the least number of reactions. The second approach, elementary flux pattern analysis, allows to analyze pathways through specific subsystems of genome-scale metabolic networks. In contrast to EFMs, elementary flux patterns much more accurately reflect the metabolic capabilities of a subsystem of metabolism as well as its integration into the entire system.

Key words: Elementary flux modes, Elementary flux patterns, *K*-Shortest elementary flux modes, Petri nets, Signal transduction, Stoichiometric network analysis, Structural robustness

1. Introduction

In systems biology, the analysis of reaction networks and their integration into other cellular processes is of central importance (1, 2). In this field, structural analysis methods have seen particular success (3, 4). These methods abstract from the precise structure of metabolites as well as kinetic laws of reactions and focus on the stoichiometric structure of the network (Fig. 1a).

Jacques van Helden et al. (eds.), *Bacterial Molecular Networks: Methods and Protocols*, Methods in Molecular Biology, vol. 804, DOI 10.1007/978-1-61779-361-5_20,

a

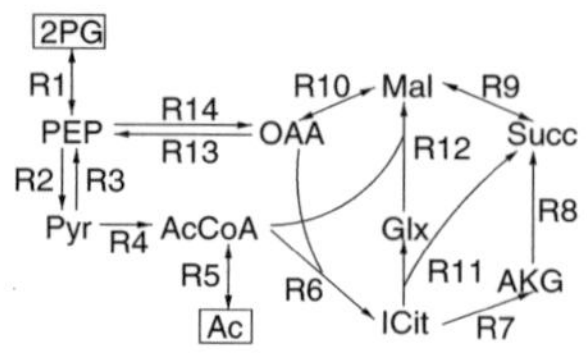

b

	R1	R2	R3	R4	R5	R 6	R7	R8	R9	R10	R11	R12	R13	R14
AcCoA	0	0	0	1	-1	-1	0	0	0	0	0	-1	0	0
AKG	0	0	0	0	0	0	1	-1	0	0	0	0	0	0
Glx	0	0	0	0	0	0	0	0	0	0	1	-1	0	0
ICit	0	0	0	0	0	1	-1	0	0	0	-1	0	0	0
Mal	0	0	0	0	0	0	0	0	1	-1	0	1	0	0
OAA	0	0	0	0	0	-1	0	0	0	1	0	0	-1	1
PEP	1	-1	1	0	0	0	0	0	0	0	0	0	1	-1
Pyr	0	1	-1	-1	0	0	0	0	0	0	0	0	0	0
Succ	0	0	0	0	0	0	0	1	-1	0	1	0	0	0

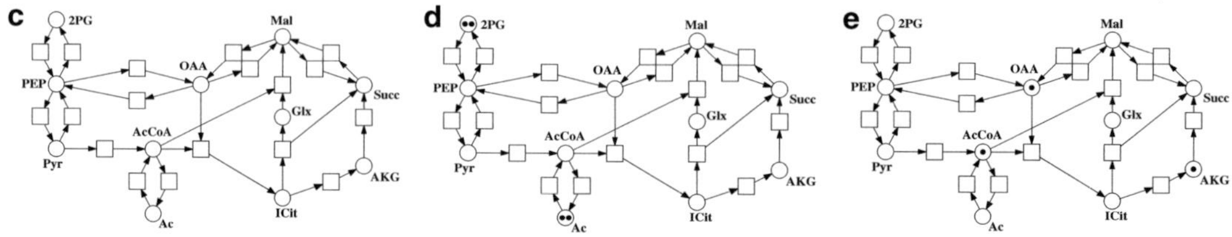

Fig. 1. Different representations of a reaction network. (**a**) Simple model of the tricarboxylic acid cycle (TCA cycle) and some adjacent reactions in *Escherichia coli*. External metabolites are *boxed*. A list of abbreviations can be found in the supplemental material of (67). (**b**) Stoichiometric matrix of the system. (**c**) Representation as a Petri net. (**d**) Dynamic representation of the Petri net with four tokens: two 2PG and two Ac. (**e**) Instance of the dynamic of the Petri net after several transitions have fired. Currently, the transitions connecting OAA, AcCoA to ICit as well as the transition connecting AKG to Succ are able to fire.

The stoichiometric structure of a reaction network is commonly summarized by its stoichiometric matrix that indicates the consumption and production of metabolites in each reaction. In this matrix, each row corresponds to a metabolite and each column to a reaction. An entry $N_{i,j}$ corresponds to the coefficient with which metabolite i is consumed or produced in reaction j: it is negative if i is an educt of reaction j and positive otherwise (Fig. 1b). The advantage of structural analysis methods lies in that most of the stoichiometric structure can be identified from the sequenced genome of an organism (5, 6), whereas the precise kinetics of reactions are often partially or totally unknown.

One important theoretical framework for the analysis of reaction networks is represented by Petri nets ((7), cf. (8, 9)). Petri nets are bipartite graphs in which the two types of nodes, called places and transitions, correspond to metabolites and reactions, respectively (Fig. 1c–e). Since transitions in Petri net theory are unidirectional, reversible reactions are decomposed into forward and reverse reaction steps. A second special feature of Petri nets is the concept of token which correspond to a discrete representation of a molecule. A token is an entity that can flow along transitions from one place to another. This could, for example, correspond to a molecule in a reaction network. The distribution of tokens in a given net at a given time point is called a marking.

Each arc (directed edge) connecting a place and a transition is weighted by a coefficient that is equivalent to the stoichiometric coefficient in the stoichiometric matrix. Whenever the magnitude of that coefficient is greater than one, it is indicated by a number next to the arc in the graphical representation. The matrix comprising all these coefficients is called incidence matrix, **N**, and is equivalent to the stoichiometric matrix of a reaction network. For the analysis of Petri nets a well-elaborate apparatus of theoretical and software tools is available (Subheading 3). Moreover, they have been extended to continuous versions within the so-called Hybrid Functional Petri Nets (10, 11). For a more comprehensive description of Petri net analysis in the context of biological networks, see (12, 13).

The central focus of this review is structural invariants of reaction networks that can be derived from their stoichiometric matrix. These structural invariants, elementary flux modes (EFMs) and moiety conservation, called minimal T-invariants and P-invariants in Petri net modeling, respectively, correspond to fundamental properties of reaction networks. Thus, in the first part, we outline important properties of these invariants, the kind of insights into reaction networks they provide and which software frameworks can be used to analyze them. In the second part, we focus on EFM analysis that has recently been used to devise strategies for strain improvement, to analyze the robustness of metabolic networks subject to perturbations and to characterize signaling networks. In the third part, we discuss challenges arising in the analysis of pathways in form of EFMs in genome-scale metabolic networks. In particular, we discuss two recently introduced approaches that allow to cope with such large networks. First is an algorithm to compute the K-shortest EFMs, that is, those with the least number of reactions. Second is the concept of elementary flux patterns that allows for the analysis of pathways through subsystems of genome-scale metabolic networks.

2. Structural Invariants in Reaction Networks

Many fundamental properties of a reaction network can be derived from its stoichiometric matrix **N**. In particular the right and the left null spaces of **N** are of interest.

2.1. The Right Null Space of N

Given the stoichiometric matrix **N**, the right null-space (Kernel) **V** of **N** is defined by the equation

$$\mathbf{NV} = 0. \tag{1}$$

This condition is often referred to as the steady-state condition. It implies that the consumption and production of metabolites must be balanced and hence the metabolite concentrations should not change. The columns of **V** define a sub-space of the n-dimensional flux space $\mathbb{R}^n$ containing all fluxes $\mathbf{v} \in \mathbb{R}^n$ satisfying the steady-state condition. That is, every flux vector

v satisfying the steady-state condition can be written as linear combination of columns of **V**. The steady-state condition is usually relaxed by differentiating between internal and external metabolites to take into account the consumption of nutrients and the production of biomass (i.e. growth) and waste metabolites. While internal metabolites need to be balanced in consumption and production, external metabolites need not. In consequence, external metabolites define the system's boundary. Metabolites often considered as external are metabolites of the growth media, for instance, the carbon or nitrogen source. Additionally, "energy currency" metabolites like ATP or NADH are often considered as external.

Since all flux distributions satisfying Eq. 1 can be written as linear combinations of columns of **V** it allows to derive important conclusion about fluxes at steady state. For instance, sets of reactions that are always operating together in fixed proportions, so-called reaction subsets (enzyme subsets), can be identified from rows in **V** that are multiples of each other (14). However, the analysis of the flux vectors contained in **V** is limited in two important points.

First, **V** is not unique. That is, there are an infinite number of matrices that represent a right null-space of **N**. Second, analyzing **V** implicitly underlies the assumption that all reactions are reversible. However, due to thermodynamic constraints some reactions can be practically irreversible at physiological conditions.

Thus, for every reaction in the set of irreversible reactions, irr, any flux **v** has to fulfill the additional condition

$$\mathbf{v}_{\mathrm{irr}} \geq 0 \tag{2}$$

to be thermodynamically feasible. Since **V** usually does not fulfill this condition, elaborate methods that allow one to analyze and characterize the solution space of Eqs. 1 and 2 have been developed. The most important methods are represented by elementary flux modes (15) in biochemical networks and minimal T-invariants (7) in Petri nets. One aim of this section is to outline both concepts and the relationship between them.

2.2. Elementary Flux Modes

A central method in the analysis of reaction networks is represented by the concept of elementary flux modes (EFMs, Fig. 2, (15)). EFMs correspond to minimal feasible steady-state fluxes through a reaction network (15). They are minimal in the sense that removing a reaction from an EFM prohibits any steady-state flux through the remaining reactions. The feasibility of EFMs refers to that only the thermodynamically feasible direction of irreversible reactions can be used. The EFMs of the example network (Fig. 1a) are displayed in Fig. 2. Consider for instance the EFM displayed in Fig. 2d. This EFM converts 2PG (2-phosphoglycerate) into Ac

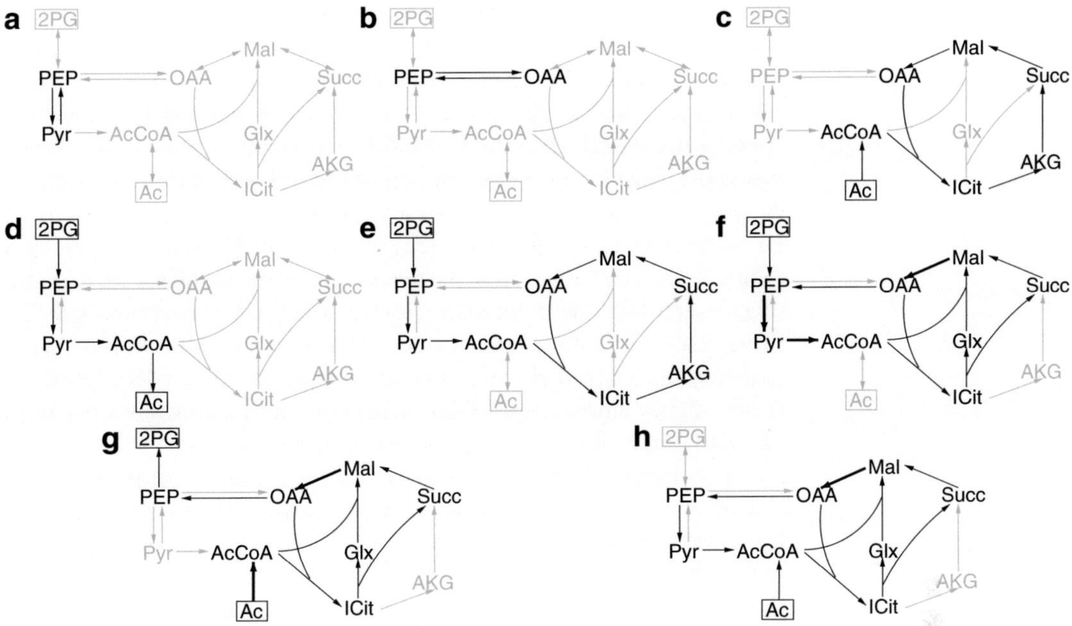

Fig. 2. EFMs of the example network. Reactions of the EFMs are drawn in *black*. *Arrow thickness* corresponds to relative fluxes. External metabolites are *boxed*. (**a**, **b**) Futile cycles. (**c**) Acetate (Ac) oxidized in the TCA cycle. (**d**) Conversion of 2-phosphoglycerate (2PG) into acetate used during fast growth on glucose (overflow metabolism). (**e**) Phosphoglycerate in the TCA cycle. (**f**) Phosphoglycerate in the PEP-glyoxylate cycle used during growth on low glucose concentrations (15, 33). (**g**) Lower part of gluconeogenesis starting from acetate. (**h**) Acetate oxidized in the PEP-glyoxylate cycle. A list of abbreviations can be found in the supplemental material of (67).

(acetate). In the sense of the steady-state condition (Eq. 1), 2PG and Ac are external metabolites and hence need not be balanced. For the other metabolites of the EFM, by contrast, the steady-state condition requires that they are balanced. Thus, the rate of production of PEP (phosphoenolpyruvate), Pyr (pyruvate) and AcCoA (acetyl-coenzyme A) needs to equal the rate of consumption of these metabolites. Furthermore, as can be easily seen, this EFM is minimal since there is no steady-state flux through the remaining reactions if removing any of those used by the EFM. A concept similar to EFMs is represented by extreme pathways (16) and minimal metabolic behaviors (17). Similarities and differences between extreme pathway analysis and EFM analysis have been discussed (18).

EFM analysis has been successful in various biotechnological applications. For example, Friedrich Srienc's group at the University of Minnesota used it for predicting optimal genetic manipulations in yeast and *Escherichia coli* (for a review, also covering the work by other groups, see (19)). First, a metabolic model of a genetically engineered strain able to produce poly-β-hydroxybutyrate (PHB), a bacterial storage compound that can be used as a biodegradable thermoplastic, had been studied (20). Adding four reactions required for PHB synthesis and considering only acetate, ethanol, glycerol, and succinate as possible substrates and products, 20 EFMs, seven of which produce PHB, were obtained. Adding the natively absent ATP citrate-lyase to the system, 496 EFMs were identified. The maximum theoretical PHB-to-carbon yield was thereby increased from 0.67 to 0.83. Inverse metabolic engineering based on EFM analysis was applied to maximize the biomass yield of *E. coli* MG1655 (21). After an analysis of the wild-type system an EFM with optimal biomass yield was identified. Knocking out six genes such that only this pathway remained resulted in a strain whose growth characteristics closely matched theoretical predictions. Here, it is important to note that it can be shown that every pathway with optimal yield corresponds to an EFM (15). Furthermore, if there are several EFMs with optimal yield, such a pathway can also correspond to a combination of EFMs (15).

The biosynthesis of the essential amino acid lysine has been the subject of a number of *in silico* pathway analyses. As an animal food supplement, lysine is of great economic importance. The main objective of these studies has been to compute the maximal theoretical lysine-over-glucose yield. Considering all cofactors including ATP as external, Schuster (22) computed a yield of 1:1 for *E. coli*. This result may have been guessed because both lysine and glucose involve six carbon atoms. However, the lysine synthesis pathway includes both carboxylating and decarboxylating reactions. When ATP is considered internal, part of the glucose is needed to produce the ATP that is necessary for lysine synthesis and only the carbons of the remainder can be used to form the skeleton of lysine. For *Corynebacterium glutamicum*, a yield of 3:4 has been computed, both by linear algebra (23) and EFM analysis (24, 25). Using a slightly different network (in which malate dehydrogenase is reversible), a set of EFMs involving one with a yield of 9:11 has been computed (26). Similar studies have been performed for other amino acids, such as tryptophan (15) and methionine (27).

Besides biotechnological applications, EFM analysis is of interest for medical studies. For example, the clinical effect of enzyme deficiencies can be assessed. When there are redundancies in the network due to bypasses, enzyme deficiencies can be tolerated in metabolism. There are some pathway analysis studies on red blood cells in this direction, for example, for energy metabolism (28) and nucleotide metabolism (29). Moreover, the abovementioned tools have been used in functional genomics (30, 31). Finally, EFM analysis detected a pathway alternative to the TCA cycle (15, 32) that was later found in experiment (33).

2.3. The Left Null Space of N

The left null space $\mathbf{K}$ of $\mathbf{N}$ is defined by the equation

$$\mathbf{KN} = 0. \tag{3}$$

This relationship describes the fundamental principle of the conservation of mass that requires that the atomic composition on both sides of a reaction must be identical. The rows of **K** correspond to linear combinations of metabolite concentrations that always stay constant, referred to as conserved moieties (34). In Petri nets, conserved moieties correspond to P-invariants (7, 35). If all entries for a specific metabolite are zero (i.e., a row of **K** only containing zeros) this indicates that the network contains reactions that do not fulfill the conservation of mass. That is, there are reactions that are consuming or producing mass. For this type of analysis it is important to note that external metabolites need to be included in the stoichiometric matrix. Otherwise, reactions that contain external metabolites would not fulfill the law of conservation of mass.

Considering, for instance, the set of reactions

$$\mathrm{ATP} \rightarrow \mathrm{ADP} + Pi$$
$$\mathrm{ADP} + Pi \rightarrow \mathrm{ATP}$$

the stoichiometric matrix reads

$$N = \begin{pmatrix} -1 & 1 \\ 1 & -1 \\ 1 & -1 \end{pmatrix}$$

The left null space of is

$$K = \begin{pmatrix} 1 & 1 & 0 \\ 1 & 0 & 1 \end{pmatrix}$$

which translates into the conserved moieties

$$\mathrm{ATP} + \mathrm{ADP} = \mathrm{const.}$$
$$\mathrm{ATP} + Pi = \mathrm{const.}$$

These relationships correspond to the conservation of the ADP backbone and the phosphate in the above set of reactions.

Furthermore, since **K** can also contain negative entries, which is difficult to interpret in the sense of atomic compositions, the concept of minimal semipositive conserved moieties has been introduced (34). These correspond to minimal sets of conserved moieties that only contain positive entries.

2.3.1. Detecting Stoichiometric Inconsistencies in Reaction Networks

As described above, a network might contain metabolites without a nonzero entry in the matrix of conserved moieties **K**. These metabolites point to stoichiometric inconsistencies in some reactions of a reaction network. One example is the metabolic network depicted in Fig. 1. In this network, **K** is of dimension zero, indicating that there is no single conserved moiety. In order to detect reactions that do not fulfill the conservation of mass, elementary leakage modes can be computed (36). Such a mode indicates a set

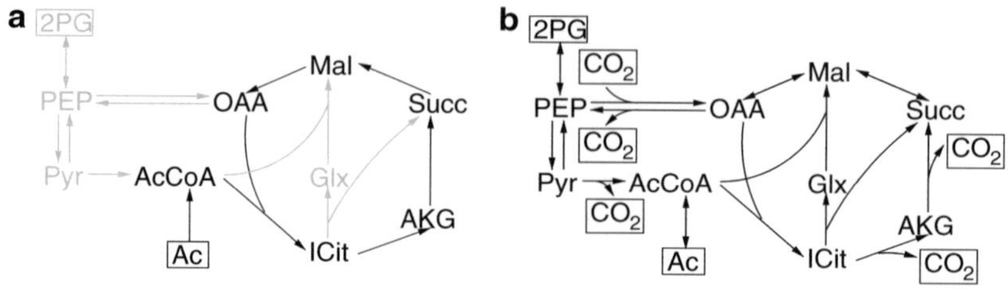

Fig. 3. Correction of stoichiometric inconsistencies in the example network (Fig. 1). (**a**) Leakage mode consuming the atoms in acetate. (**b**) Corrected network after adding carbon dioxide, which is produced and consumed in several reactions.

of reactions by which mass can be produced or is consumed. They can be obtained by computing the EFMs of a modified version of the stoichiometric matrix **N** that additionally includes the rows corresponding to the external metabolites. In large-scale networks this might not be feasible due to the combinatorial explosion in the number of EFMs in such networks (see Subheading 5). However, in this case an Integer Linear Programming formulation similar to the formulation used to compute the *K*-shortest EFMs (Subheading 5.1) can be used (36).

An example for an elementary leakage mode is depicted in Fig. 3a. Using this mode, the atoms of an acetate molecule are consumed. By inspection of the reactions used in this leakage mode it is easy to deduce that two carbon dioxide molecules that are released in the TCA cycle are missing. After modifying the corresponding reactions and taking into account CO_2 consumption and production in three other reactions, we find no further stoichiometric inconsistency (Fig. 3b). However, this example shows that detected stoichiometric inconsistencies not automatically indicate errors in a model, but can also be due to simplifications and abstractions often used in modeling. Thus, the carbon dioxide was left out of the initial model, since it was not considered as an important component of the model.

Solving Eq. 3 for the stoichiometric matrix of the corrected network, we now find a single conserved moiety

$$6\,\mathrm{ICit} + 4\mathrm{Mal} + 2\mathrm{Glx} + 4\ \mathrm{OAA} + 2\ \mathrm{Ac} + 1\ \mathrm{CO_2} + 3\ \mathrm{Pyr} \\ + 3\ \mathrm{PG} + 5\ \mathrm{AKG} + 4\ \mathrm{Succ} + 2\ \mathrm{AcCoA} + 3\ \mathrm{PEP} = \mathrm{const.}$$

The coefficients exactly correspond to the number of carbon atoms in each metabolite. The only exception is AcCoA. Here, the carbon atoms in the coenzyme A (CoA) moiety are not directly taken into account, since neither its binding nor release is modeled.

2.4. Structural Invariants in Petri Nets

In Petri net theory, the interest in structural properties arose independently of the structural analysis of reaction networks and started relatively early. In particular, T-invariants and P-invariants have been defined (7, cf. (9)]. A vector **v** is called a T-invariant if it fulfills Eq. 1. Intuitively speaking, T-invariants are distributions of token flows that allow the Petri net to regain the initial marking. True T-invariants **v** have the property that all components in **v** are nonnegative and at least one is positive. For the following definition of minimal T-invariants the mathematical concept of support is used. The support of **v** is the set of positions where **v** is not zero. A true T-invariant **v** is called a minimal T-invariant if there is no other true T-invariant, $\mathbf{v}^*$, the support of which would be a proper subset of the support of **v**:

$$\mathrm{supp}(\mathbf{v}) = \{i : v_i \neq 0\}$$
$$\nexists \mathbf{v}^* : \mathrm{supp}(\mathbf{v}^*) \subset \mathrm{supp}(\mathbf{v})$$

From the definition of EFMs given in Subheading 2.2, it can be seen that in the case where all reactions are irreversible, EFMs and minimal T-invariants are equivalent concepts. In the frequent case where several reactions are reversible, the two concepts differ to some extent. In EFM analysis, these reactions are left as they are. While this makes the algorithm for computing EFMs somewhat more complex, a decomposition or reversible reactions as usual in Petri net theory has the drawback that spurious minimal T-invariants consisting of forward and reverse transitions resulting from the same reaction occur, which need to be discarded after being computed.

Similar to conserved moieties in biochemical networks, P-invariants (7) have been defined for Petri nets. They describe relations of marking conservation, that is, a linear combination of token numbers remains constant irrespective of any firing of transitions. The counterpart in biochemical networks are conserved moieties. Analogously as for T-invariants, minimal P-invariants were defined.

It is worth noting that the concepts of EFMs (37) and extreme currents (38) were established independently of the concept of minimal T-invariants, which dates back to 1973 (7). Note that Petri nets were used originally for describing production, transportation and similar processes rather than (bio)chemical reaction systems. Hofestädt (39) and Reddy *et al.* (40) were the first to use them in that context. The similarity between EFMs and minimal T-invariants as well as between minimal semipositive moiety conservations and minimal P-invariants was first discussed by Schuster *et al.* (35) and Voss *et al.* (41). Previously, the development of algorithms ran in parallel. For example, Colom and Silva (42) presented an algorithm for computing minimal P-invariants (without dealing with biological networks). It shares certain features with the algorithm for computing EFMs proposed by Schuster *et al.* (43) in that it involves row operations on the incidence matrix augmented with an identity matrix. By transposition of the incidence matrix, the Colom-Silva algorithm can be used also for calculating minimal T-invariants.

Sackmann *et al.* (44) introduced the concept of maximal common transition sets (MCT sets). An MCT set is a maximal set of transitions that operate together in each steady state. Hence, they participate in exactly the same minimal T-invariants. Each Petri net can be decomposed into MCTs, that is, disjoint subpathways, which can be interpreted as building blocks of the net. Earlier, the concept of reaction subsets discussed above had been introduced (14) which is stricter than MCT sets because two reactions belong to the same reaction subset if they always function together and do so in fixed flux proportions. By contrast, a constant ratio of fluxes is not required for MCT sets. Therefore, all reaction subsets correspond to MCT sets, while the opposite does not hold. In consequence, MCT sets are identical to the concept of partially coupled reaction sets (45) that represent a generalization of the concept of reaction subsets.

An increasing number of applications of Petri nets to modeling metabolic and signaling networks can be found in the literature (for a review see (12, 13)). Sackmann *et al.* (44) studied the mating pheromone response pathway in yeast. In this publication, they demonstrated the systematical building and validation process of discrete Petri-net models, which reflect provably the qualitative biological behavior of a biological system without any knowledge of its kinetic parameters. Thereby, the mating pheromone response pathway in *Saccharomyces cerevisiae* served as case study. Furthermore, they introduced two new concepts: the feasible minimal T-invariants and the above mentioned MCT sets. Sackmann *et al.* (46) analyzed the homeostasis of iron in humans. Many components of this complex process are already described in the literature. Sackmann *et al.* compiled and validated a Petri net model of this crucial process and conducted a comprehensive structural analysis. The 85 feasible minimal T-invariants that were obtained from the model were grouped into ten clusters. Sackmann *et al.* showed that all clusters of feasible minimal T-invariants correspond to known biological processes in the system. Kielbassa *et al.* (47) analyzed the assembly of one subunit of the spliceosome relevant in alternative mRNA splicing. The simulation of the Petri net provided insights into the importance of the stability of complexes during the formation of the spliceosome.

3. Software for Analyzing Structural Invariants in Reaction Networks

There exist a large number of software packages for the analysis of reaction networks. Here, we give just a short and incomplete list of the most prominent tools.

3.1. Software with a Focus on Reaction Networks

- *CellNetAnalyzer* (48) (http://www.mpi-magdeburg.mpg.de/projects/cna/cna.html) offers a graphical user interface for EFM analysis from within *Matlab*. For the computation of EFMs, *efmtool* or Metatool serve as back-ends. Besides a graphical display of EFMs within a pre-drawn representation of the reaction network, many additional tools including selection of specific EFMs, computation of minimal cut sets (49, 50) and analysis of regulatory networks are included.
- *efmtool* (51) (http://www.csb.ethz.ch/tools/efmtool/) implements the currently fastest algorithm for the computation of EFMs (51). Reaction networks can be imported in the SBML (Systems Biology Markup Language, (52)) format. A *Matlab* integrations is available.
- *Metatool* (53) (http://pinguin.biologie.uni-jena.de/bioinformatik/networks/index.html) is available in two versions: as stand-alone executable and as *Matlab* library. The stand-alone version implements an improved version of the canonical base approach and can be used for networks represented in the so-called *Metatool* format. EFMs in terms of used enzymes and net conversions as well as further information concerning the network structure are returned as a text-file. This version of *Metatool* is also available from the Systems Biology Workbench (54). As *Matlab* library, *Metatool* can be called with a network in the SBML-format if the SBML Toolbox is installed (55).
- *ScrumPy* (56) (http://mudshark.brookes.ac.uk/ScrumPy) is a software package for the reconstruction and analysis of reaction networks. It is written in Python and implemented as a user interface. Network models can be stored as SBML files or in a proprietary format. The program includes methods for both kinetic and structural modeling, but with an emphasis on the latter. One important feature is the computation of elementary leakage modes to detect stoichiometric inconsistencies in reaction networks (36).
- *YANAsquare* (57) (http://yana.bioapps.biozentrum.uni-wuerzburg.de/) offers a Java-based graphical user interface that integrates the process of network reconstruction facilitated by an import of reactions and metabolites from KEGG (58), automated network visualization and EFM analysis. Reaction networks can be imported and saved in *Metatool* or SBML format.

3.2. Software with a Focus on Petri Nets

- *Charlie* (http://www-dssz.informatik.tu-cottbus.de/index.html?/software/charlie.html) is a software package to analyze Petri nets. It is implemented in Java and offers a graphical user interface, a textual user interface and more features, such as a marking editor. *Charlie* is able to import all Petri-net classes that are supported by *Snoopy*. The tool can be used to analyze and validate technical systems, as well as biological networks. Among its features are functions for the analysis of structural properties, such as deadlocks/traps or P- and T-invariants, for determining shortest paths, and for the visualization of reachability/coverability graphs.

- *Integrated Net Analyzer (INA)* (http://www2.informatik.hu-berlin.de/lehrstuehle/automaten/ina/) is a package for the analysis of Petri Nets and Colored Petri nets. *INA* comprises a textual editor to compile and to edit nets, a simulation function, and an analysis function to derive structural information such a P- and T-invariants as well as reachability/coverability-graphs. The analysis can be carried out under different transition rules, with or without priorities or time restrictions, and under firing of single transitions or maximal sets of concurrently enabled transitions. Furthermore, minimal pathways can be computed, and the (non)reachability of a certain marking can be determined. Some external graphical editors and tools can export nets to *INA*, such as *Snoopy.*
- *Snoopy* (59, 60) (http://www-dssz.informatik.tu-cottbus.de/index.html?/software/snoopy.html) is a software package to construct and animate hierarchical graphs with a focus on Petri nets. *Snoopy* can be used to validate technical systems, as well as biochemical networks, such as metabolic, signal-transduction, or gene-regulatory networks. Snoopy offers an intuitive graphical user interface to design Petri nets and to simulate token flows. The support for hierarchical graphs (hierarchical nodes, logical nodes) is particularly useful for larger models, or models with a higher connectivity degree. *Snoopy* provides import from the several other Petri net tools and file formats (e.g. SBML (52)). For a more detailed description of Snoopy, see (Marwan *et al.* Chapter 21).

For a comprehensive list of available Petri-net tools see http://www.informatik.uni-hamburg.de/TGI/PetriNets/tools/db.html.

4. Recent Applications of EFM Analysis

4.1. Robustness

Robustness is a general feature of living cells being exposed to varying conditions because biological systems must be robust against environmental and genetic perturbations to be evolvable. Kitano (61) describes robustness to be "fundamental feature of evolvable complex systems". He distinguishes different mechanisms to ensure the robustness of a system: system control (covering positive and negative feedback) to attain a robust dynamic response in a wide range of environmental conditions, alternative mechanisms (redundancy) to compensate, for instance, knockouts and modularity plus decoupling to serve for a higher and faster adaptability of organisms to evolutionary changing environmental conditions.

To analyze these different aspects of robustness several modeling approaches can be used. Most of them are dynamic approaches, focused on the analysis of parametric sensitivities (see (62)). With the help of structural approaches just topological aspects of robustness such as redundant pathways or bottlenecks can be analyzed. However, there is also a study by Jacobsen and Cedersund (63) where a combination of structural and dynamic approaches is used. They stated that perturbations such as knockouts cannot be represented by changes in the model parameters, because they are affecting the model structure itself and consequently, they studied the change in the dynamic behavior of a system after removing entire interactions and the corresponding equations.

Stelling *et al.* (64) studied the structural robustness of central metabolism of *E. coli* using EFM analysis. They compared the wild type with different knockout mutants. If, after deletion of an enzyme, at least one EFM with a positive biomass yield remained, the mutant was considered to be viable. Thereby, they could show that the biomass yield remains quite constant even if the number of EFMs is reduced significantly. This result demonstrates quite well that redundancy is an important factor for the structural robustness of biological systems. But it is also apparent that the number of EFMs of a system is not a sufficient measure for its structural robustness.

By considering only the number of EFMs, it remains unclear what happens to the topology of the network if an enzyme is knocked out. For instance, Fig. 4 shows two small networks having the same number of EFMs but different robustnesses. Thus, a more sophisticated approach to calculate the structural robustness of reaction networks on the basis of EFMs was introduced (65, 66). To handle also the structural robustnesses of subnetworks, different variants of robustness measures (R_1, R_2, and R_3) that were distinguished by

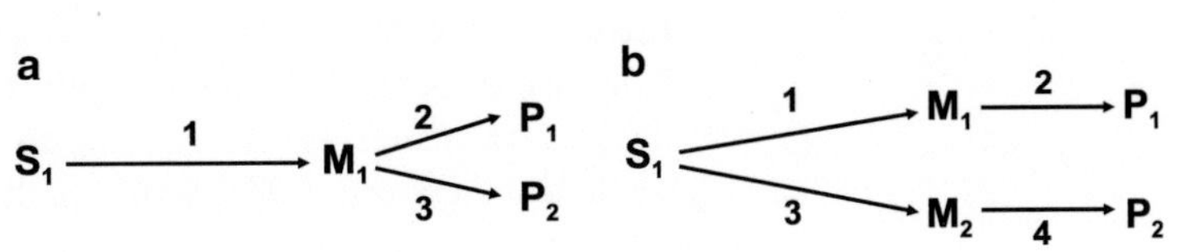

Fig. 4. Example networks with the same number of EFMs but different robustnesses. Network A is less robust than B, since a single knockout can cut off any EFM in A, while at least two knockouts are needed in B to remove all EFMs. However, to be more robust, an additional reaction is required indicating the cost of robustness. According to the robustness measures introduced in (65, 66), network A has a structural robustness of **1/3**, whereas network B has a value of **1/2**. M_i are internal metabolites, S_k substrates and P_m products.

their indices were introduced by Wilhelm *et al.* (65). Here, we only focus on the global measure R_1, but we keep its index for the sake of consistency.

To characterize the structural robustness against knockout of enzyme E_i, the ratio $z^{(i)}/z$, between the number of EFMs remaining after knockout, $z^{(i)}$, and the number, z, in the unperturbed network is used. To quantify the global robustness of the entire network, the arithmetic mean of all these ratios over all possible knockouts, is determined. The result is a normalized quantity between 0 and 1.

$$R_1(d) = \frac{\sum_{i=1}^{c(d)} z^{(i)}}{c(d) \cdot z} \qquad \text{with} \qquad c(d) = \binom{r}{d} \tag{4}$$

Here, $c(d)$ denotes the total number of knockout combinations with cardinality d, out of r reactions that can be knocked out (no spontaneous reactions). The number of EFMs remaining after knockout i is denoted by $z^{(i)}$, and z is the total number of EFMs. $R_1(d)$ is the robustness with respect to knockouts of exactly d enzymes.

In (66) this framework was applied to the amino acid anabolism of *E. coli* and human hepatocytes, and to the central metabolism in human erythrocytes, showing that *E. coli* has the most robust metabolism of those cell types, followed by the human hepatocyte and the human erythrocyte. From an evolutionary viewpoint, these results are intuitively understandable because *E. coli*, hepatocytes, and erythrocytes live, in that order, in increasingly homeostatic environments requiring less and less metabolic robustness.

As a further example to demonstrate this concept of structural robustness, we compiled an additional medium-scale metabolic network of *E. coli* comprising glycolysis, TCA cycle, pentose phosphate pathway, and arginine synthesis. Altogether, the network consists of 62 reactions, 52 internal, and 9 external metabolites, and gives rise to 5,736 EFMs. Table 1 shows the structural robustnesses of our arginine synthesis network depending on the combinatorial depth *d* of knockouts.

From the values in Table 1, it is apparent that the robustness score decreases as the knockout depth increases. This is an inherent property of the concept, which was generally proven by Behre *et al.* (66). For the different amino acid synthesis networks of *E. coli* that were under study in (66), we obtained robustness scores for $R_1(1)$ of in the range of 0.776 to 0.623, whereas the corresponding value of the present network is of medium size (0.419). This is due to a principle phenomenon in structural analysis that always emerges when cut-outs of larger networks are considered. In the present model, we only consider arginine synthesis, whereas the former models cover the synthesis of all amino acids. This leads to a pruned and thus less interconnected network which consequently shows a

Table 1
Robustness values $R_1(d)$ for knockouts up to a combinatorial depth (cardinality) of $d = 5$

Cardinality d	Robustness $R_1(d)$
1	0.419
2	0.186
3	0.087
4	0.042
5	0.021

lower structural robustness. On the contrary, the maximum arginine yield is at the level of 0.93 molecules of arginine per molecule of glucose and thus significantly increased compared to a genome-scale model of *E. coli* (67), discussed below, that shows an optimal yield of 0.88. The reason for this increased yield is the absence of uptake reactions that are usually energy-consuming. In the genome-scale model, this issue is considered, whereas in the medium-scale model the necessary input metabolites are just set as external. This is also the reason for the reduced length of the EFMs in the present model compared with the genome-scale model. Hence, in principle only the structural robustnesses of genome-scale models should be discussed. This is unfortunately still not feasible due to the exponentially growing number of EFMs in large networks. But keeping this shortcoming in mind, our concept of structural robustness is a useful tool for a comparison of different network topologies.

4.2. Modeling of Signal Transduction

Signal transduction is fundamentally important for the regulation of biological functions in living cells. The modeling of signaling networks has recently gained appreciable interest (68–70). While in metabolic networks mass flow is the important aspect, in signal transduction networks the flow of information matters. Furthermore, metabolic networks usually subsist in steady states, whereas in signaling systems, time-dependent signals such as short pulses occur.

Several approaches to model structural properties of signal transduction networks have been already presented, for instance, by Papin and Palsson (68) using extreme pathways. A further method that is applicable for modeling signaling systems is the concept of Petri nets, which has already been discussed above. A more abstract approach for modeling signal transduction networks are logical interaction hypergraphs (LIHs), which were introduced by Klamt *et al.* (70). This approach reduces the processes on the molecular level to a Boolean perspective and offers a useful framework to model interactions in biological networks. However, it is not useful for all applications to reduce the level of systems states to *on* and *off*.

Difficulties in modeling signal transduction networks generally arise because the direction of information flow is not necessarily the same as the direction of the mass flow that is required to transport the information. Whereas, for instance, in the phosphotransferase system (PTS) the phosphate moiety is transferred along the reaction chain, in kinase cascades the mass flow consists of cycles of enzymes being phosphorylated and dephosphorylated, while the information is transduced along the cascade. In the case when the flow of information corresponds to the mass flow, methods from metabolic modeling can be used in a straightforward manner as shown by Papin and Palsson (68). Although in most signaling systems mass flow does not correspond to information flow, several authors have applied methods from metabolic modeling formally to such systems (e.g., (44, 46, 71)). This approach has led to meaningful results, although its theoretical foundation is still not completely clear.

In a recent publication (72), we show that it is possible to apply the concept of EFMs to signal transduction networks composed of enzyme cascades. We propose a theoretical justification for applying methods of structural analysis of metabolic systems also to enzyme cascades based on the following reasoning. First, such networks usually allow signal amplification but no signal diminution. Second, after the signal has been transmitted, the system must return to its original state so that a pseudo-steady state can be assumed in the sense of an average over the time span that includes the signal flow and the recovery phase. When discussing enzyme cascades we take phosphorylation with *NTP* and *NDP* as unspecified placeholders for any covalent modification that can switch *on* and *off* the activity of enzymes.

Owing to the lack of kinetic data no conclusion about the signal amplification along the cascade is possible. But based on the irrelevance of signal diminution a minimal stoichiometry can be derived requiring that at least one kinase of the next cascade level must be activated before the kinase of the former level is inactivated again. Thus, in the case of a single-phosphorylation cascade for all levels of the cascade, reaction equations of the following form can be written.

$$E_iP + E_{i+1} + \mathrm{NTP} = E_i + P + E_{i+1}P + \mathrm{NDP}$$

with E being a kinase of the considered cascade level. With that approach, kinase cascades are modeled as chains of reactions with each reaction containing the phosphorylation of kinase E_{i+1} together with the dephosphorylation of kinase E_i. The resulting EFMs go through the cascades and have thus the same direction as the flows of information. The overall reactions seem to be repeated handovers of phosphate groups but as the consumption of *NTPs* indicates, the phosphate groups are renewed at each level of the cascades. So, the systems are modeled as a kind of relay race with changing batons.

This approach holds for single-phosphorylation and double-phosphorylation cascades including convergent and divergent branching. Divergent cascades fork to two or more branches at their branching point, whereas in convergent cascades several branches are merged. In Fig. 5, a diverging double-phosphorylation cascade is shown as an example. At the branching point, the cascades may have an OR-connection or an AND-connection. For an OR-connection, the branching point is modeled as a set of two reactions, whereas in the case of an AND-connection these two reaction equations are merged (for details see (72)). Both, divergent and convergent OR-connected cascades give rise to two EFMs each, as expected, whereas the AND-connected variants result in one EFM each.

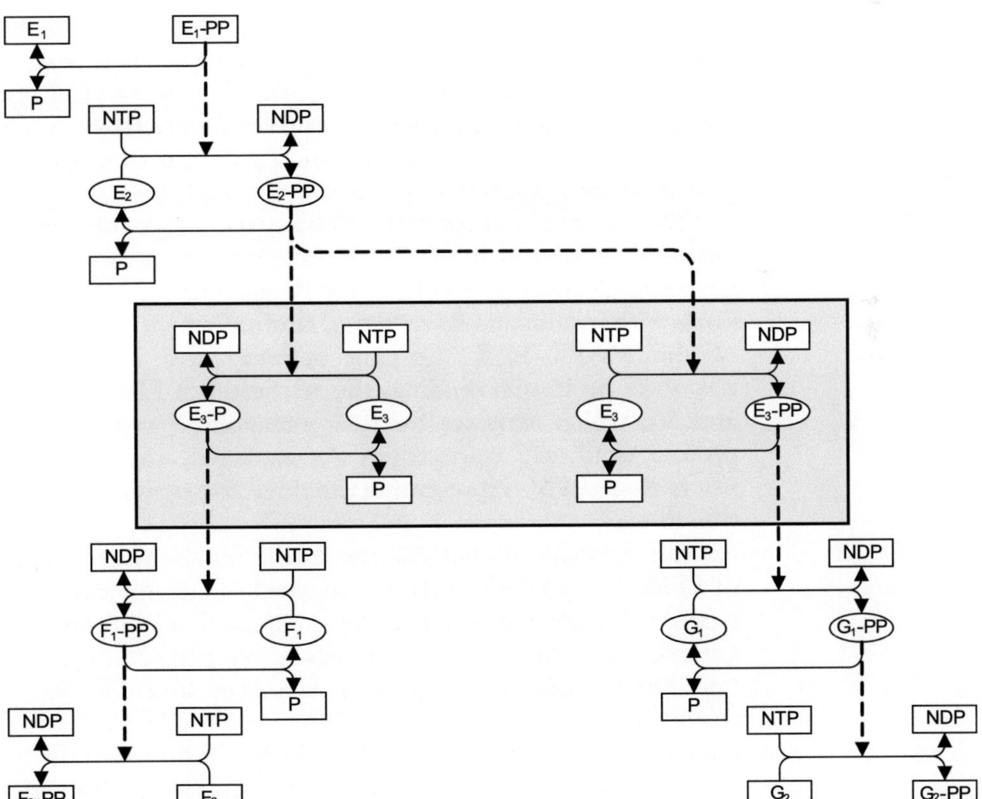

Fig. 5. Simplified diverging double-phosphorylation cascade, comprising the kinases E_1–E_3, F_1–F_2 and G_1–G_2. All double-phosphorylation steps within the three branches (kinases E_1–E_2, F_1–F_2, and G_1–G_2) are merged, whereas at the branching point (*gray box*), the phosphorylations of E_3 to E_3P and E_3PP, respectively, are separate reactions. External and internal species are represented by rectangles and ellipses, respectively.

A further motif in signal transduction networks are double-phosphorylation cascades. In such cascades an enzyme E_{i+1} needs to be phosphorylated twice to be activated. These two phosphorylation steps and also the subsequent dephosphorylation steps are often catalyzed by the same preceding kinase and subsequent phosphatase, respectively. Without kinetic data, the exact chronology of phosphorylation and dephosphorylation and thus the amplification of the signal along the cascade cannot be identified. But again, a minimal stoichiometry can be deduced, leading to reaction equations for each level of double-phosphorylation cascades, where both phosphorylation steps are merged. The remaining modeling process is analogous to that of single-phosphorylation cascades, but with one exception. In the case that a single-phosphorylated kinase $E_{i+1}P$ has a different substrate specificity than its double-phosphorylated variant $E_{i+1}PP$, $E_{i+1}P$ may catalyze, for instance, the phosphorylation of a kinase F_1, while $E_{i+1}PP$ phosphorylates G_1. This results in a diverging double-phosphorylation cascade. Here, obviously the two phosphorylation steps at the branching point may not be merged. From the diverging double-phosphorylation cascade, shown in Fig. 5, it can be seen that all double-phosphorylation steps within the three branches, represented by the kinases E_1–E_2, F_1–F_2, and G_1–G_2, respectively, are merged, whereas at the branching point (emphasized by a gray box), the phosphorylations of E_3 to E_3P and E_3PP, respectively, are modeled as separate reactions.

We evaluated our approach with a real example for a diverging cascade including complex formations, that is part of the insulin signaling pathway in humans (for details see (72)). The model starts with insulin and its receptor, continuing with the main part of the MAPK/ERK signaling pathway and leads to three responses on insulin binding, the activation of PFKFB-2, CREB and S6, which activates fructose metabolism, cell growth and protein synthesis, respectively. As expected, the model gives rise to three EFMs representing the three corresponding information flows.

Our approach is a suitable framework for applying the concept of EFMs to signaling systems composed of enzyme cascades and complex formations. It combines the detection of nondecomposable routes in signaling systems with a modeling concept close to "molecular reality". Nevertheless, in analyzing inhibitions, a difference to Boolean approaches shows up. Modeling inhibitions close to the molecular level leads to EFMs ending with the inhibited component as dead-end metabolite. This makes it impossible to sum up two consecutive inhibitions to an activation as it is feasible with Boolean concepts.

Another concept suggested by Gianchandani *et al.* (73) offers a quasi-stoichiometric approach to handle regulatory systems. This so-called "Matrix Formalism" defines interactions between components of regulatory systems as pseudo-reactions with a minimal stoichiometry necessary for information transport. Participating enzymes are considered to be "consumed" and "produced". Moreover, Gianchandani *et al.* introduce "absent compounds". Thus, they model, for instance, an inhibition of enzyme D by inhibitor C as $C_{\text{Presence}} = D_{\text{Absence}}$. Altogether, the concept of Gianchandani *et al.* is a Boolean-like approach with the same level of abstraction and a similar set of possibilities to analyze signaling and regulatory systems as the LIH approach by Klamt *et al.* (70).

5. Going Genome Scale: Pathway Analysis in Large-Scale Reaction Networks

In recent years an increasing number of models covering the metabolism of entire organisms have been reconstructed (4). These models differ from classical models usually covering a single or only few pathways in that they try to give a holistic picture of the metabolic capabilities of an organism. While constraint-based methods, building on linear programing, like flux balance analysis (74), can be readily applied to this kind of networks, the enumeration of all EFMs is not possible (75).

There are two ways to circumvent this problem. On the one hand, the network under consideration can be reduced to a small subsystem of a size at which EFMs can still be computed by abstracting parts of the system that are of no interest (Fig. 6). On the other hand, algorithms that allow to enumerate a subset of EFMs can be used. While the first approach is commonly used, only recently a new algorithm that allows an iterative enumeration of EFMs in large-scale networks has been developed (76).

There are three important problems associated to the analysis of only specific subsystems of metabolism. These problems arise due to the need to define metabolites as external and to introduce

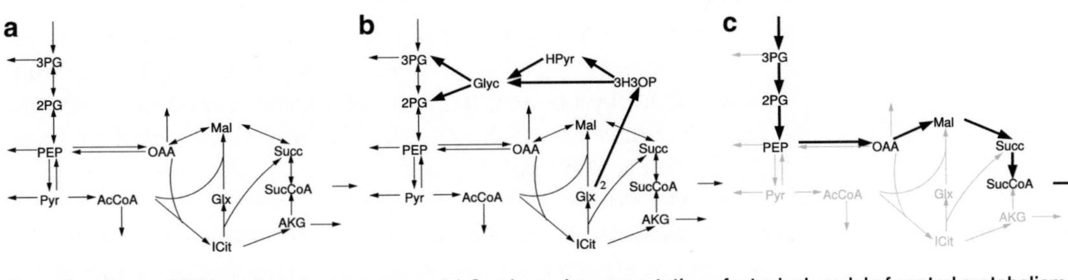

Fig. 6. Problems of EFM analysis in subsystems. (**a**) Condensed representation of a typical model of central metabolism in *E. coli*. (**b**) Glycolate pathway (*bold arrows*) that is often not considered in the analysis of central metabolism. (**c**) *Bold reactions* denote an EFM that fulfills the steady-state condition on the local scale but is not part of any steady-state flux on the global scale, since every reaction consuming succinyl-CoA in the remaining system requires further metabolites that can only be produced by the subsystem (79). A list of abbreviations can be found in the supplemental material of (67).

exchange reactions that model production and consumption of metabolites of a subsystem by the underlying large-scale system.

First, even though systematic approaches for the definition of metabolites as external have been put forward (77, 78), there is no unique rule for the definition of a metabolite as internal or external. However, this definition heavily impacts the EFMs that are found (79).

Second, small reconstructed networks often only represent the modeler's knowledge of the network in contrast to genome-scale networks which often contain many more pathways. For instance, the glycolate pathway which is active during growth on acetate is often not considered in analyses of central metabolism (80) (Fig. 6b). In consequence, the results are biased toward the parts of the metabolic network known by the modeler.

Third, the definition of metabolites as external and the introduction of exchange reactions often do not take into account dependencies between the production and consumption of metabolites by the remaining system. This leads to EFMs that satisfy the steady-state condition on a local, but not on a global scale (79) (Fig. 6c).

The aim of this section is to illustrate two new approaches that allow to circumvent these problems. First, an algorithm allowing to iteratively enumerate EFMs in genome-scale metabolic networks is discussed. Second, the concept of elementary flux patterns which allows a more realistic assessment of the metabolic capabilities of a subsystem within a genome-scale system is presented.

5.1. K-Shortest Elementary Flux Modes

As mentioned above, the enumeration of the full set of EFMs in genome-scale networks is impossible due to computational constraints. Nevertheless, this issue does not preclude the enumeration of a subset of EFMs of interest, like the *K*-shortest EFM (76). These correspond to the *K* EFMs with the least number of reactions.

The motivation for the "*K-shortest*" concept stems from graph-theory and has been used to detect the *K*-shortest paths in metabolic networks (81, 82). However, as mentioned in (Arita, Chapter 6), not all connected reactions form a metabolic pathway. Thus, the inclusion of stoichiometric information brings more meaning to the metabolic routes (83).

Experimentally, the "*K-shortest*" concept is also of interest. First, it is laborious to over-express a large number of enzymes. Additionally, shorter pathways are better suited as a target for genetic manipulation, since the highest increase in pathway flux is achieved if all enzymes (84) or (at least) a considerable number of enzymes in a pathway (85) are overexpressed. Moreover, it has been theoretically shown that a structure characterized by the least number of reaction steps emerges when flux maximization through a pathway is considered (86).

The enumeration of a subset of EFMs starts with the computation of a shortest EFM in the input network. This problem can be formulated as an integer linear program (ILP). The inclusion of a binary variable which has the value of 1 if a particular reaction j is active and, 0 otherwise, allows to use the minimization of the number of reactions as objective function and consequently, to compute the shortest EFM. Another important aspect of the mathematical model is the inclusion of a constraint that removes the previous solution from the solution space before each new iteration step. This way, it is possible to enumerate EFMs in increasing order of length starting from the shortest, the second shortest and so on. Please note, however, that there might be several EFMs with the least number of reactions. In this case, the order in which they are found depends on the implementation. For a more detailed outline on the mathematical formulation of the K-shortest EFMs, see (76).

This approach is very efficient for large-scale metabolic networks because classical methods have to compute all EFMs to obtain a single EFM, whereas the K-shortest EFM method can compute a single EFM from the solution space. Moreover, the flexibility of this method allows to direct the EFM computation to a solution space of interest by adding more information, that is, more constraints, to the optimization problem. For example, one can compute EFMs that produce or consume a given metabolite through the inclusion of constraints forcing the overall flux associated to this metabolite to be bigger or lower than zero, respectively. This approach was used in the study of lysine synthesis in two genome-scale networks of *E. coli* and *C. glutamicum* (76).

5.1.1. Application of K-Shortest Elementary Flux Modes to Arginine Synthesis

To exemplify the application of K-shortest EFMs, we will compute the 10-shortest EFMs producing arginine in the genome-scale metabolic network of *E. coli* (67). The exchange reactions in the original metabolic network were removed as well as the biomass equation. The stoichiometric coefficients from the respiratory chain reactions were scaled to become integers. The *in silico* growth medium, that is, the set of metabolites that can be consumed from the extra cellular space ([e]), was defined as: potassium (k[e]), sulfate (so4[e]), sodium (na1[e]), iron (fe2[e]), ammonium (nh4[e]), inorganic phosphate (pi[e]), water (h2o [e]), oxygen (o2[e]), and glucose (glc-D[e]). The target metabolite was arginine (arg[e]). The ILP was solved using ILOG CPLEX® 11.0.

From Fig. 7 it becomes evident that most variability of the 10-shortest EFMs lies in glucose catabolism which supplies glutamate, a precursor of arginine, and cofactors. Some variability can also be found in the regeneration of the oxidized ubiquinol (q8).

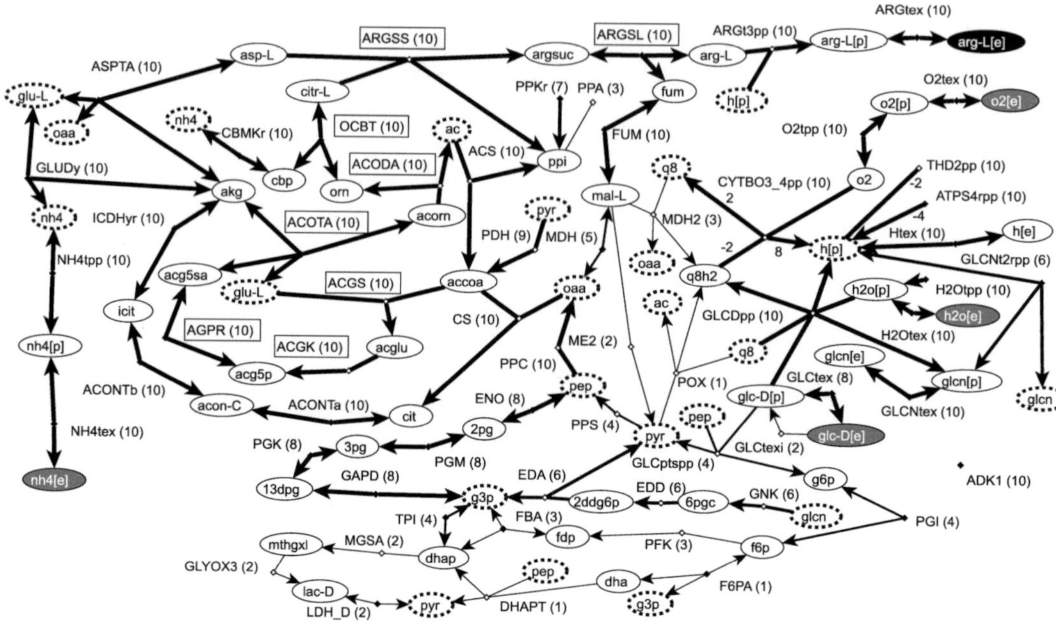

Fig. 7. 10-shortest EFMs producing arginine merged into a subnetwork. The number in parenthesis after the reaction label corresponds to the number of EFMs in which the reaction is present and the *arrow thickness* is proportional to this value. Arginine, *node in black*, is synthesized from glutamate using a pathway described in the literature (*boxed reactions labels*). The metabolites taken up from the medium are represented by *gray nodes* and copied metabolite nodes have a *dashed border*. Some metabolite nodes concerning cofactors, such as ATP and NADH, were removed from the visualization for clarity. A list of abbreviations can be found in the supplemental material of (67).

This compound is reduced in the oxidation of glucose (glc-D) to gluconate (glcn). The oxidation of glc-D to glcn is not very efficient in generating a proton gradient and consequently has to carry high fluxes. This is the shortest route for generating NADPH, which is required in the anabolic part of the EFMs.

Other alternative routes for $NADP^+$ reduction are the pentose phosphate pathway alone and the tricarboxylic acid cycle coupled with oxidative phosphorylation and the NAD^+ transhydrogenase (THD2pp). These pathways have a higher efficiency in NADPH generation. However, their inclusion into a pathway synthesizing arginine would require high costs in enzyme synthesis. Nevertheless, these pathways are tightly integrated in *E. coli* metabolism and consequently act in other parts of metabolism, thus reducing the enzyme costs.

From Table 2, it can be seen that these EFMs do not have high arginine over glucose yield. Indeed, the maximum yield is 0.18 and corresponds to EFM number 5 in which glycolysis is the glucose catabolic pathway. The maximal theoretical yield (discarding biomass production) for this network is 0.88 and can be computed using linear programming (74). Thus, it is evident that other EFMs with higher yield exist as is also indicated by

Table 2
Order (*K*), length (*L*), and overall reaction of the 10-shortest EFMs producing arginine in *E. coli*

K	*L*	Overall reaction
1	45	26 glc-D[e] + 16 nh4[e] + 13 o2[e] → 4 arg-L[e] + 22 glcn[e] + 20 h2o[e] + 34 h[e]
2	45	26 glc-D[e] + 16 nh4[e] + 13 o2[e] → 4 arg-L[e] + 22 glcn[e] + 20 h2o[e] + 34 h[e]
3	45	34 glc-D[e] + 16 nh4[e] + 17 o2[e] → 4 arg-L[e] + 30 glcn[e] + 20 h2o[e] + 42 h[e]
4	45	34 glc-D[e] + 16 nh4[e] + 17 o2[e] → 4 arg-L[e] + 30 glcn[e] + 20 h2o[e] + 42 h[e]
5	46	22 glc-D[e] + 16 nh4[e] + 11 o2[e] → 4 arg-L[e] + 18 glcn[e] + 20 h2o[e] + 30 h[e]
6	46	8 glc-D[e] + 4 nh4[e] + 4 o2[e] → 1 arg-L[e] + 7 glcn[e] + 5 h2o[e] + 10 h[e]
7	46	26 glc-D[e] + 16 nh4[e] + 13 o2[e] → 4 arg-L[e] + 22 glcn[e] + 20 h2o[e] + 34 h[e]
8	46	46 glc-D[e] + 16 nh4[e] + 23 o2[e] → 4 arg-L[e] + 42 glcn[e] + 20 h2o[e] + 54 h[e]
9	46	50 glc-D[e] + 16 nh4[e] + 25 o2[e] → 4 arg-L[e] + 46 glcn[e] + 20 h2o[e] + 58 h[e]
10	46	10 glc-D[e] + 4 nh4[e] + 5 o2[e] → 1 arg-L[e] + 9 glcn[e] + 5 h2o[e] + 12 h[e]

the analysis of a medium-scale system of arginine synthesis in the previous section.

5.2. Elementary Flux Patterns

In order to overcome the limitations of the application of EFMs to subsystems of large-scale reaction networks as discussed in the beginning of this section, the concept of elementary flux patterns (EFPs) has been introduced (79). They correspond to basic routes that all feasible pathways within a large reaction network can take through a specific subsystem (that is, a specific set of reactions within the large reaction network). Thus, if a steady-state flux of the large-scale system uses reactions of the subsystem, they correspond to combinations of elementary flux patterns. An example for elementary flux patterns is given in Fig. 8.

A formal definition of elementary flux patterns will be given next. Given the stoichiometric matrix **N**, we assume for simplicity that the first *k* reactions make up the subsystem of interest. A flux pattern *s* is defined as a set of reactions within the subsystem that can be associated to at least one steady-state flux v of the entire system. A set of reaction indices *s* is called a flux pattern if it fulfills the following conditions:

$$\mathrm{v} \geq 0 \tag{5}$$

$$\mathrm{N} \cdot \mathrm{v} = 0 \tag{6}$$

$$\forall i \in s : v_i > 0 \tag{7}$$

$$\forall j \in \{1..k\} \setminus s : v_j = 0. \tag{8}$$

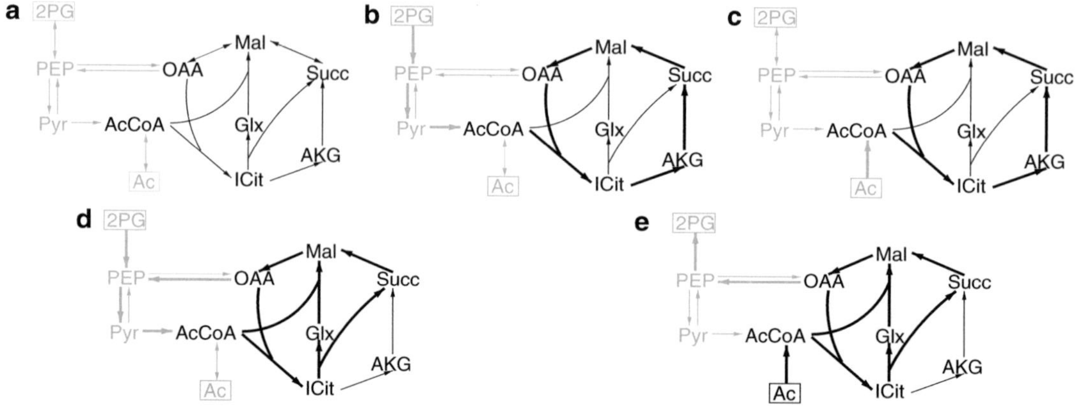

Fig. 8. Example for EFPs in a subsystem of the example reaction network in Fig. 1. The subsystem is defined as the reactions of the TCA cycle and the glyoxylate shunt (**a**). The reactions of the subsystem are displayed in *black*. Fluxes are indicated by *thick arrows*: fluxes within the subsystem are drawn in *black* and the global pathway of which each EFP forms part is indicated by *gray arrows*. (**b**, **c**) EFPs corresponding to the TCA cycle. In (**b**) 2PG (2-phosphoglycerate), an intermediate of glycolysis, is oxidized, in (**c**), Ac (acetate) is oxidized. (**d**, **e**) EFPs corresponding to the glyoxylate cycle. In (**d**), the phosphoenolpyruvate-glyoxylate cycle (15, 33) is used to oxidize 2PG, in (**e**) Ac is converted into 2PG that can be used for gluconeogenesis.

Similarly to EFMs we can define the basic routes of each steady-state flux through the subsystem by constraining the analysis to elementary flux patterns. Hence, a flux pattern is called elementary if it cannot be derived as combination, that is, as set union, of other flux patterns.

By definition, each elementary flux pattern is associated to at least one EFM through the entire system. In consequence, when analyzing the fluxes within a subsystem, the fluxes through the entire system are explicitly taken into account. Another advantage of elementary flux patterns lies in that many approaches building on EFMs can also be applied using elementary flux patterns (79).

5.2.1. Elementary Flux Patterns of Arginine Metabolism

As an example we will analyze the elementary flux patterns within a subsystem of arginine and proline metabolism in the same genome-scale system that has been used for the computation of the *K*-shortest EFMs. EFM analysis of the subsystem would require the definition of the set of reactions necessary for arginine and proline synthesis as well as the definition of exchange reactions with the remaining system. For elementary flux pattern analysis only a set of reactions is required which can be obtained from the subsystem annotation of the model. Additionally, to determine whether arginine can be produced by a particular elementary flux pattern, an outflow reaction of arginine present in the entire system has been added to the subsystem. In consequence, the subsystem contains 45 reactions. Computations have been performed with *Elementary Flux Pattern Analyzer* available from http://users.minet.uni-jena.de/~m3kach/EFPA/.

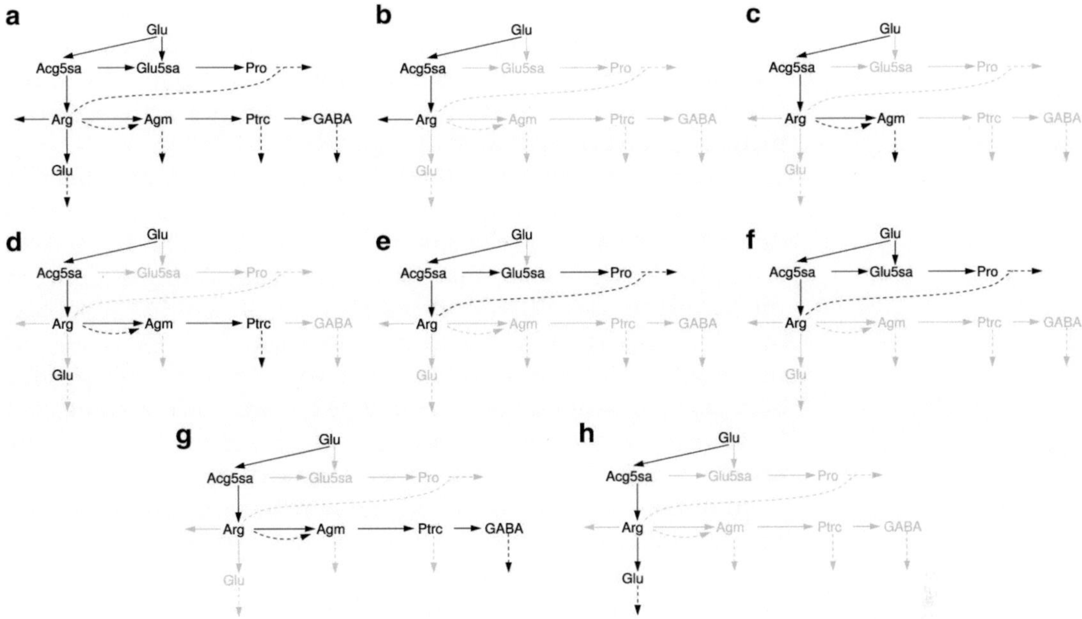

Fig. 9. Pathways in arginine metabolism. (**a**) Condensed system of arginine metabolism. Reactions of the remaining system are drawn with *dashed lines*. The reaction consuming arginine and proline corresponds to the biomass reaction. Elementary flux patterns (reactions drawn in *black*): (**b**) EFP 18; (**c**) combined EFPs 19 and 20, decarboxylation of arginine into agmatine can proceed either in the cytosol (*bold line*) or in the periplasm (*dashed line*); (**d**) combined EFPs 21 and 22; (**e**) EFP 23; (**f**) EFP 26; (**g**) combined EFPs 22 and 24; (**h**) EFP 27. Abbreviations: *Acg5sa N*-acetyl-L-glutamate 5-semialdehyde, *Agm* agmatine, *Arg* arginine, *Glu* glutamate, *Glu5sa* glutamate 5-semialdehyde, *Pro* proline, *Ptrc* putrescine.

This system gives rise to 27 elementary flux patterns, five of which correspond to futile cycles of forward and backward directions of split reversible reactions. Here, we will focus on the ten elementary flux patterns producing arginine. Apart from variability in co-factor recycling reactions, there exists only a single arginine biosynthetic pathway in *E. coli*. However, since the subsystem also includes reactions for arginine degradation, most variability is found in these pathways. Accordingly, the elementary flux patterns can be classified into six different ways by which arginine can be disposed (Fig. 9). The indices of the corresponding elementary flux patterns are indicated in brackets.

The first and simplest way is the disposal by excreting arginine (EFP 18). Second, arginine can be incorporated into proteins which are modeled as a sink in form of a biomass reaction in the model. However, another product of the subsystem, proline, is also required for protein synthesis. In consequence, elementary flux patterns disposing arginine through this mechanism additionally need to produce proline. Indeed, there are two such elementary flux patterns, one for each of the two proline producing pathways (EFPs 23 and 26). Third, arginine can be degraded into agmatine which can be excreted. The decarboxylation of arginine can either be performed by the protein SpeA in the cytoplasm (EFP 18) or by Adi in the periplasm (EFP 20). While the former protein is used during growth on arginine (87), the latter is active during acid stress to increase the pH of the environment by decarboxylation of arginine (88).

Here, we also find an interesting case showing that defining a reaction as not belonging to the subsystem does not represent a drawback when using elementary flux patterns. The EFM associated to the elementary flux pattern decarboxylating arginine in the periplasm uses a transport reaction not present in the subsystem to transport arginine into the periplasm. In an EFM analysis of the subsystem one would just have found that the periplasmic transport reaction cannot carry any flux, since it is not connected to the subsystem. Fourth, agmatine can be converted into putrescine which can also be either excreted or converted into 4-aminobutyric acid (GABA) in the remaining system (EFPs 21 and 22). Fifth, putrescine can be converted into GABA which can be further degraded to succinate in the subsystem yielding two further elementary flux patterns (EFPs 24 and 25). Finally, a pathway converting arginine into glutamate can be used (EFP 27).

6. Conclusion

The analysis of structural invariants that can be derived from the stoichiometric matrix of a reaction network has proved fruitful in many studies. They span from theoretical analyses to biotechnological applications. Conserved moieties (P-invariants) derived from the left null-space of the stoichiometric matrix can be used to detect stoichiometric inconsistencies in reaction networks. Elementary flux modes (EFMs, minimal T-invariants), by contrast, are a most useful tool in pathway analysis. Thus, EFM analysis has shown its value in guiding experimental work aimed at strain improvement in production processes, to analyze the robustness of metabolic networks against perturbations and to characterize signaling networks.

Since Petri nets have their origin in engineering and computer sciences and were originally devised for the description of production and transportation processes, the framework of Petri nets comprises (besides the classical place/transition nets) many different variants. There are, for instance, stochastic Petri nets, continuous Petri nets, or hybrid functional Petri nets (10, 11). Thus, concepts developed for the analysis of Petri nets have a broader range of applications than concepts such as EFM analysis that is mainly used in the context of metabolic networks. Nevertheless, concerning biochemical topics, EFM analysis is the more prominent tool.

But due to the close relationship between EFMs and T-invariants, conserved moieties and P-invariants, and reaction subsets and maximal common transition sets (MCT sets) (44), results, determined with one of the two concepts can be directly compared with those, identified with the other one. And also extensions, established for one of the two concepts (e.g., the measures for structural robustness, that were introduced on the basis of EFMs), can be (at least partially) adapted to the other concept.

However, EFM analysis (and therefore also Petri nets) meets its limitations in the size of the network to which it can be applied, the effort that has to be put into model reconstruction prior to analysis, and problems involved in the analysis of over-simplified subsystems of reaction networks. Thus, the extension of concepts from EFM analysis to genome-scale metabolic networks, which are now available for many organisms, has received particular attention. These approaches try to enumerate either a subset of EFMs that are of interest (e.g., the *K*-shortest) or focus on minimal routes that steady-state fluxes of a large-scale network can take through a particular subnetwork (elementary flux patterns).

An important challenge that lies ahead is the analysis of the relationship between large-scale experimental data sets, such as transcriptomic, proteomic, metabolomic, and fluxomic data, and the structural invariants discussed in this work. Some first studies into this direction show promising results (6, 78, 89, 90). However, a holistic integration of such data sets into structural metabolic networks has yet to be achieved to pass from "genome-scale" to truly "organism-scale" models.

Acknowledgments

We thank Ines Heiland for the reconstruction of the medium-scale model of arginine metabolism in *E. coli* and for valuable discussions. Financial support from the German Ministry for Research and Education (BMBF) to C.K. and J.B. within the frameworks of the Forsys Partner initiative and the HepatoSys project, respectively, and from the Fundação Calouste Gulbenkian, Fundação para a Ciência e a Tecnologia (FCT) and Siemens SA Portugal (PhD grant SFRH/BD/32961/2006) to L.F.F. is gratefully acknowledged.

References

1. Schuster S, Fell DA, Dandekar T, (2000) A general definition of metabolic pathways useful for systematic organization and analysis of complex metabolic networks. Nat Biotechnol, 18(3):326–332.
2. Palsson BØ. (2006) *Systems Biology – Properties of Reconstructed Networks.* Cambridge University Press, New York.
3. Price ND, Reed JL, Palsson BØ. (2004) Genome-scale models of microbial cells: evaluating the consequences of constraints. Nat Rev Microbiol, 2(11):886–897.
4. Feist AM, Palsson BØ. (2008) The growing scope of applications of genomescale metabolic reconstructions using *Escherichia coli.* Nat Biotechnol, 26(6):659–667.
5. Paley SM, Karp PD. (2002) Evaluation of computational metabolic-pathway predictions for *Helicobacter pylori.* Bioinformatics, 18 (5):715–724.
6. Notebaart RA, Teusink B, Siezen RJ, Papp B. (2008) Co-regulation of metabolic genes is better explained by flux coupling than by network distance. PLoS Comput Biol, 4(1):e26.
7. Lautenbach K. (1973) Exakte Bedingungen der Lebendigkeit für eine Klasse von Petri-Netzen (in German). GMD Report, 82.
8. Murata T. (1989) Petri nets: properties, analysis and applications. Proc IEEE, 77 (4):541–580.
9. Starke PH. (1990) *Analyse von Petri-Netz-Modellen.* Teubner Verlag, Leipzig.
10. Matsuno H, Doi A, Nagasaki M, Miyano S. (2000) Hybrid Petri net representation of gene regulatory network. Pac Symp Biocomput, 5:341–352.
11. Wu J, Voit E. (2009) Hybrid modeling in biochemical systems theory by means of functional Petri nets. J Bioinform Comput Biol, 7(1):107–134.
12. Hardy S, Robillard PN. (2004) Modeling and simulation of molecular biology systems using Petri nets: modeling goals of various approaches. J Bioinform Comput Biol, 2 (4):595–613.
13. Chaouiya C. (2007) Petri net modelling of biological networks. Brief Bioinform, 8 (4):210–219.
14. Pfeiffer T, Sánchez-Valdenebro I, Nuño JC, Montero F, Schuster S. (1999) METATOOL: for studying metabolic networks. Bioinformatics, 15(3):251–257.
15. Schuster S, Dandekar T, Fell DA. (1999) Detection of elementary flux modes in biochemical networks: a promising tool for pathway analysis and metabolic engineering. Trends Biotechnol, 17(2):53–60.
16. Schilling CH, Letscher D, Palsson BØ. (2000) Theory for the systemic definition of metabolic pathways and their use in interpreting metabolic function from a pathway-oriented perspective. J Theor Biol, 203(3):229–248.
17. Larhlimi A, Bockmayr A. (2009) A new constraint-based description of the steady-state flux cone of metabolic networks. Discrete Appl Math, 157(10):2257–2266.
18. Papin JA, Stelling J, Price ND, Klamt S, Schuster S, Palsson BØ. (2004) Comparison of network-based pathway analysis methods. Trends Biotechnol, 22(8):400–405.
19. Trinh CT, Wlaschin A, Srienc F. (2009) Elementary mode analysis: a useful metabolic pathway analysis tool for characterizing cellular metabolism. Appl Microbiol Biotechnol, 81(5):813–826.
20. Carlson R, Fell D, Srienc F. (2002) Metabolic pathway analysis of a recombinant yeast for rational strain development. Biotechnol Bioeng, 79(2):121–134.
21. Trinh CT, Carlson R, Wlaschin A, Srienc F. (2006) Design, construction and performance of the most efficient biomass producing *E. coli* bacterium. Metab Eng, 8(6):628–638.
22. Schuster S. (2004) Metabolic Pathway Analysis in Biotechnology. In *Metabolic Engineering in the Post Genomic Era.* Edited by Kholodenko BN, Westerhoff HV, Horizon Scientific, Wymondham, 181–208.
23. de Graaf AA. (2000) Metabolic Flux Analysis of *Corynebacterium glutamicum.* In *Bioreaction Engineering, Modelling and Control.* Edited by Schügerl K, Springer, New York, 506–555.
24. Gayen K, Venkatesh KV. (2006) Analysis of optimal phenotypic space using elementary modes as applied to *Corynebacterium glutamicum.* BMC Bioinformatics, 7:445.
25. Schuster S, von Kamp A, Pachkov M. (2007) Understanding the roadmap of metabolism by pathway analysis. Methods Mol Biol, 358:199–226.
26. Wittmann C, Weber J, Betiku E, Krömer J, Böhm D, Rinas U. (2007) Response of fluxome and metabolome to temperature-induced recombinantprotein synthesis in *Escherichia coli.* J Biotechnol, 132(4):375–384.
27. Krömer JO, Wittmann C, Schröder H, Heinzle E. (2006) Metabolic pathway analysis

for rational design of L-methionine production by *Escherichia coli* and *Corynebacterium glutamicum*. Metab Eng, 8(4):353–369.

28. Çakir T, Tacer CS, Ülgen KO. (2004) Metabolic pathway analysis of enzyme-deficient human red blood cells. BioSystems, 78 (1–3):49–67.
29. Schuster S, Kenanov D. (2005) Adenine and adenosine salvage pathways in erythrocytes and the role of S-adenosylhomocysteine hydrolase. A theoretical study using elementary flux modes. FEBS J, 272(20):5278–5290.
30. Förster J, Gombert AK, Nielsen J. (2002) A functional genomics approach using metabolomics and *in silico* pathway analysis. Biotechnol Bioeng, 79(7):703–712.
31. Pachkov M, Dandekar T, Korbel J, Bork P, Schuster S. (2007) Use of pathway analysis and genome context methods for functional genomics of *Mycoplasma pneumoniae* nucleotide metabolism. Gene, 396(2):215–225.
32. Liao JC, Hou SY, Chao YP. (1996) Pathway analysis, engineering, and physiological considerations for redirecting central metabolism. Biotechnol Bioeng, 52(1):129–140.
33. Fischer E, Sauer U. (2003) A novel metabolic cycle catalyzes glucose oxidation and anaplerosis in hungry *Escherichia coli*. J Biol Chem, 278 (47):46446–46451.
34. Schuster S, Höfer T. (1991) Determining all extreme semi-positive conservation relations in chemical reaction systems: a test criterion for conservativity. J Chem Soc Faraday Trans, 87(16):2561–2566.
35. Schuster S, Pfeiffer T, Moldenhauer F, Koch I, Dandekar T. (2000) Structural analysis of metabolic networks: elementary flux modes, analogy to Petri nets, and application to In Bornberg-Bauer E, Rost U, Stoye J (Eds.): *Proceedings of the German Conference on Bioinformatics (GCB 2000)*, Logos Verlag 115–120.
36. Gevorgyan A, Poolman MG, Fell DA. (2008) Detection of stoichiometric inconsistencies in biomolecular models. Bioinformatics, 24 (19):2245–2251.
37. Schuster S, Hilgetag C,Woods JH, Fell DA. (2002) Reaction routes in biochemical reaction systems: algebraic properties, validated calculation procedure and example from nucleotide metabolism. J Math Biol, 45 (2):153–181.
38. Clarke BL. (1981) Complete set of steady states for the general stoichiometric dynamical system. J Chem Phys, 75(10):4970–4979.
39. Hofestädt R. (1996) A Petri net application to model metabolic processes. Syst Anal Model Sim, 16(2):122.
40. Reddy VN, Liebman MN, Mavrovouniotis ML. (1996) Qualitative analysis of biochemical reaction systems. Comput Biol Med, 26(1):9–24.
41. Voss K, Heiner M, Koch I. (2003) Steady-state analysis of metabolic pathways using Petri nets. In Silico Biol, 3(3):367–387.
42. Colom JM, Silva M. (1991) Convex Geometry and Semiflows in P/T Nets: A Comparative Study of Algorithms for Computation of Minimal P-semiflows. In *APN 90: Proceedings on Advances in Petri nets 1990*. Springer-Verlag, New York, 79–112.
43. Schuster S, Pfeiffer T, Moldenhauer F, Koch I, Dandekar T. (2002) Exploring the pathway structure of metabolism: decomposition into subnetworks and application to *Mycoplasma pneumoniae*. Bioinformatics, 18(2):351–361.
44. Sackmann A, Heiner M, Koch I. (2006) Application of Petri net based analysis techniques to signal transduction pathways. BMC Bioinformatics, 7:482.
45. Burgard AP, Nikolaev EV, Schilling CH, Maranas CD. (2004) Flux coupling analysis of genome-scale metabolic network reconstructions. Genome Res, 14(2):301–312.
46. Sackmann A, Formanowicz D, Formanowicz P, Koch I, Blazewicz J. (2007) An analysis of the Petri net based model of the human body iron homeostasis process. Comput Biol Chem, 31(1):1–10.
47. Kielbassa J, Bortfeldt R, Schuster S, Koch I. (2009) Modeling of the U1 snRNP assembly pathway in alternative splicing in human cells using Petri nets. Comput Biol Chem, 33(1):46–61.
48. Klamt S, Saez-Rodriguez J, Gilles ED. (2007) Structural and functional analysis of cellular networks with CellNetAnalyzer. BMC Syst Biol, 1:2.
49. Klamt S, Gilles ED. (2004) Minimal cut sets in biochemical reaction networks. Bioinformatics, 20(2):226–234.
50. Klamt S. (2006) Generalized concept of minimal cut sets in biochemical networks. BioSystems, 83(2–3):233–247.
51. Terzer M, Stelling J. (2008) Large-scale computation of elementary flux modes with bit pattern trees. Bioinformatics, 24 (19):2229–2235.
52. Hucka M, Finney A, Sauro HM, Bolouri H, Doyle JC, Kitano H, Arkin AP, Bornstein BJ, Bray D, Cornish-Bowden A, Cuellar AA, Dronov S, Gilles ED, Ginkel M, Gor V, Goryanin II, Hedley WJ, Hodgman TC, Hofmeyr JH, Hunter PJ, Juty NS, Kasberger JL, Kremling A, Kummer U, Le Novère N, Loew LM,

Lucio D, Mendes P, Minch E, Mjolsness ED, Nakayama Y, Nelson MR, Nielsen PF, Sakurada T, Schaff JC, Shapiro BE, Shimizu TS, Spence HD, Stelling J, Takahashi K, Tomita M, Wagner J, Wang J. (2003) The Systems Biology Markup Language (SBML): A medium for representation and exchange of biochemical network models. Bioinformatics, 19(4):524–531.

53. von Kamp A, Schuster S. (2006) Metatool 5.0: fast and exible elementary modes analysis. Bioinformatics, 22(15):1930–1931.

54. Sauro HM, Hucka M, Finney A, Wellock C, Bolouri H, Doyle J, Kitano H. (2003) Next generation simulation tools: the Systems Biology Workbench and BioSPICE integration. OMICS, 7(4):355–372.

55. Keating SM, Bornstein BJ, Finney A, Hucka M. (2006) SBMLToolbox: an SBML toolbox for MATLAB users. Bioinformatics, 22 (10):1275–1277.

56. Poolman MG. (2006): ScrumPy: metabolic modelling with Python. IEE Proc Syst Biol, 153(5):375–378.

57. Schwarz R, Liang C, Kaleta C, Kühnel M, Hoffmann E, Kuznetsov S, Hecker M, Griffiths G, Schuster S, Dandekar T. (2007) Integrated network reconstruction, visualization and analysis using YANAsquare. BMC Bioinformatics, 8:313.

58. Kanehisa M, Araki M, Goto S, Hattori M, Hirakawa M, Itoh M, Katayama T, Kawashima S, Okuda S, Tokimatsu T, Yamanishi Y. (2008) KEGG for linking genomes to life and the environment. Nucleic Acids Res, 36(Database issue):D480–D484.

59. Heiner M, Richter R, Schwarick M, Rohr C. (2008) Snoopy – a tool to design and execute graph-based formalisms. Petri Net Newsletter, 74:8–22.

60. Rohr C, Marwan W, Heiner M. (2010) Snoopy – a unifying Petri net framework to investigate biomolecular networks. Bioinformatics, 26(7):974–975.

61. Kitano H. (2004) Biological robustness. Nat Rev Genet, 5(11):826–837.

62. Wolf J, Becker-Weimann S, Heinrich R. (2005) Analysing the robustness of cellular rhythms. IEE Proc Syst Biol, 2:35–41.

63. Jacobsen EW, Cedersund G. (2008) Structural robustness of biochemical network models–with application to the oscillatory metabolism of activated neutrophils. IET Syst Biol, 2(1):39–47.

64. Stelling J, Klamt S, Bettenbrock K, Schuster S, Gilles ED. (2002) Metabolic network structure determines key aspects of functionality and regulation. Nature, 420(6912):190–193.

65. Wilhelm T, Behre J, Schuster S. (2004) Analysis of structural robustness of metabolic networks. IEE Proc Syst Biol, 1(1):114–120.

66. Behre J, Wilhelm T, von Kamp A, Ruppin E, Schuster S. (2008) Structural robustness of metabolic networks with respect to multiple knockouts. J Theor Biol, 252(3):433–441.

67. Feist AM, Henry CS, Reed JL, Krummenacker M, Joyce AR, Karp PD, Broadbelt LJ, Hatzimanikatis V, Palsson BØ. (2007) A genome-scale metabolic reconstruction for *Escherichia coli* K-12 MG1655 that accounts for 1260 ORFs and thermodynamic information. Mol Syst Biol, 3:121.

68. Papin JA, Palsson BØ. (2004) Topological analysis of mass-balanced signaling networks: a framework to obtain network properties including crosstalk. J Theor Biol, 227 (2):283–297.

69. Zevedei-Oancea I, Schuster S. (2005) A theoretical framework for detecting signal transfer routes in signalling networks. Comput Chem Eng, 29(3):597–617.

70. Klamt S, Saez-Rodriguez J, Lindquist JA, Simeoni L, Gilles ED. (2006) A methodology for the structural and functional analysis of signaling and regulatory networks. BMC Bioinformatics, 7:56.

71. Xiong M, Zhao J, Xiong H. (2004) Network-based regulatory pathways analysis. Bioinformatics, 20(13):2056–2066.

72. Behre J, Schuster S. (2009) Modeling signal transduction in enzyme cascades with the concept of elementary flux modes. J Comput Biol, 16(6):829–844.

73. Gianchandani EP, Papin JA, Price ND, Joyce AR, Palsson BØ. (2006) Matrix formalism to describe functional states of transcriptional regulatory systems. PLoS Comput Biol, 2(8 e101):0902–0917.

74. Varma A, Palsson BØ. (1994) Metabolic flux balancing: basic concepts, scientific and practical use. Bio/Technology, 12(10):994–998.

75. Acuña V, Marchetti-Spaccamela A, Sagot MF, Stougie L. (2010) A note on the complexity of finding and enumerating elementary modes. BioSystems, 99(3):210–214.

76. de Figueiredo LF, Podhorski A, Rubio A, Kaleta C, Beasley JE, Schuster S, Planes FJ. (2009) Computing the shortest elementary flux modes in genome-scale metabolic networks. Bioinformatics, 25(23):3158–3165.

77. Dandekar T, Moldenhauer F, Bulik S, Bertram H, Schuster S. (2003) A method for classifying metabolites in topological pathway analyses based on minimization of pathway number. BioSystems, 70(3):255–270.

78. Schwartz JM, Gaugain C, Nacher J, de Daruvar A, Kanehisa M. (2007) Observing metabolic functions at the genome scale. Genome Biol, 8(6):R123.
79. Kaleta C, de Figueiredo LF, Schuster S. (2009) Can the whole be less than the sum of its parts? Pathway analysis in genome-scale metabolic networks using elementary flux patterns. Genome Res, 19(10):1872–1883.
80. Oh MK, Rohlin L, Kao KC, Liao JC. (2002) Global expression profiling of acetate-grown *Escherichia coli.* J Biol Chem, 277 (15):13175–13183.
81. Croes D, Couche F, Wodak SJ, van Helden J. (2006) Inferring meaningful pathways in weighted metabolic networks. J Mol Biol, 356(1):222–236.
82. Planes FJ, Beasley JE. (2009) Path finding approaches and metabolic pathways. Discrete Appl Math, 157(10):2244–2256.
83. de Figueiredo LF, Schuster S, Kaleta C, Fell DA. (2009) Can sugars be produced from fatty acids? A test case for pathway analysis tools. Bioinformatics, 25:152–158.
84. Kacser H, Acerenza L. (1993) A universal method for achieving increases in metabolite production. Eur J Biochem, 216(2):361–367.
85. Niederberger P, Prasad R, Miozzari G, Kacser H. (1992) A strategy for increasing an *in vivo* flux by genetic manipulations. The tryptophan system of yeast. Biochem J, 287(Pt 2):473–479.
86. Meléndez-Hevia E, Waddell TG, Montero F. (1994) Optimization of metabolism: the evolution of metabolic pathways toward simplicity through the game of the pentose phosphate cycle. J Theor Biol, 166(2):201–220.
87. Wu WH, Morris DR. (1973) Biosynthetic arginine decarboxylase from *Escherichia coli.* Purification and properties. J Biol Chem, 248 (5):1687–1695.
88. Blethen SL, Boeker EA, Snell EE. (1968) Argenine decarboxylase from *Escherichia coli.* I. Purification and specificity for substrates and coenzyme. J Biol Chem, 243(8):1671–1677.
89. Schuster S, Klamt S, Weckwerth W, Pfeiffer T. (2002) Use of network analysis of metabolic systems in bioengineering. Bioproc Biosyst Eng, 24(6):363–372.
90. Reed JL, Palsson BØ. (2004) Genome-scale *in silico* models of *E. coli* have multiple equivalent phenotypic states: assessment of correlated reaction subsets that comprise network states. Genome Res, 14(9):1797–1805.

Chapter 21

Petri Nets in Snoopy: A Unifying Framework for the Graphical Display, Computational Modelling, and Simulation of Bacterial Regulatory Networks

Wolfgang Marwan, Christian Rohr, and Monika Heiner

Abstract

Using the example of phosphate regulation in enteric bacteria, we demonstrate the particular suitability of stochastic Petri nets to model biochemical phenomena and their simulative exploration by various features of the software tool Snoopy.

Key words: Petri nets, Stochastic Petri nets, Non-Markovian Petri nets, Deterministic delay, Animation, Gillespie simulation, Stochastic simulation analysis, Continuous time Markov chain

1. Introduction

1.1. Executable Petri Nets as a Unifying Framework for Systems Biology

Systems biology is based on a transdisciplinary joint effort to understand the complex mechanisms of life at the molecular systems level. For good reasons, experimentalists and theoreticians traditionally use different languages specific to their respective disciplines, thinking and expressing themselves in different ways. Experimentalists think in molecules and molecular mechanisms, which they illustrate with pictures, qualitative schemes, and biochemical reaction pathways. Theoreticians communicate by using equations and mathematical symbols, which are difficult to read for the majority of experimentalists, who frequently do not have any significant mathematical background. A true understanding of each other would be greatly facilitated by establishing a communication platform (and language) which is equally easy to use for both experimentalists and theoreticians. Such common language should be of unequivocal expressiveness and directly refer to the traditional languages of both.

Jacques van Helden et al. (eds.), *Bacterial Molecular Networks: Methods and Protocols*, Methods in Molecular Biology, vol. 804, DOI 10.1007/978-1-61779-361-5_21, © Springer Science+Business Media, LLC 2012

With Snoopy, we provide a tool that supports the use of Petri nets as

- a common communication platform (modelling language) for experimentalists and theoreticians, together with
- a unifying framework for the graphical display, computational modelling, simulation, and bioinformatic annotation of biochemical networks, such as bacterial regulatory networks.

Petri nets as executable models. A Petri net is a mathematical graph with a strictly defined syntax. A Petri net is directly executable, if it is represented with the help of an appropriate tool like Snoopy. Snoopy translates automatically and thus reproducibly the graphical scheme into a set of equations used by the program to run simulations. In other words, a graphical representation of a Petri net drawn in Snoopy can be executed, i.e. simulations can be run with a mouse click; no special additional encoding is required.

Petri nets represent molecular and other mechanisms with a strictly defined syntax. The syntax of the Petri net language is simple and therefore very easy to learn (see below). Because the syntax is strictly defined, there is no ambiguity in what a graphical representation of the Petri net model means in terms of reaction mechanism. The Petri net formalism is ideal to naturally represent chemical reactions and their mechanisms, and any type of biochemical interactions. As defined by the user, Petri net components may represent molecules and reactions, or even more complex entities and processes. This allows to describe and represent biological processes at arbitrary level of abstraction (i.e. with arbitrary resolution in detail), but always mechanistically unambiguous, and to join those processes into a coherent, executable model. This option is extremely useful for signal transduction networks or gene regulatory networks as usually not all processes involved in a phenomenon of interest are and will be known with the same resolution in detail.

Petri nets provide a unifying framework for modelling and simulation in systems biology. Petri nets can be used to perform all major modelling and simulation approaches central to systems biology. Briefly, there are several reasons suggesting the deployment of Petri nets to investigate biochemical networks.

1. Petri nets enjoy an intuitive bipartite graphical representation, which makes them easily comprehensible and facilitates the communication between wet-lab experimentalists and computational theoreticians (1).
2. Petri nets allow the unambiguous representation of various types of biological processes at different levels of abstraction (with different resolution of detail) in one and the very same model – ranging from the conformational change of a single molecule to the macroscopic response of a cell (2), the development of a tissue, or even the behaviour of a whole

organism; for a recent survey of case studies employing Petri nets see ref. 3.

3. Petri nets come along with an explicit notion of concurrency. This allows to distinguish unmistakably between alternative and concurrent behaviour and means an adequate representation of inherently concurrent behaviour, such as the concurrent dephosphorylation in the individual stages of a signalling cascade (4) (see also Note 2).
4. Petri nets are directly executable by an appropriate tool like Snoopy (5). This execution allows a time-free animation in the case of qualitative Petri nets (these are the standard Petri nets), and time-dependent animation and simulation in the case of quantitative Petri nets (in our case, stochastic Petri nets (SPN)).
5. Petri nets can be explored by a substantial body of mathematically founded analysis techniques, covering structural and behavioural properties as well as their relations (4, 6); see also Notes 3–5.
6. Petri nets may serve as a kind of umbrella formalism integrating qualitative and quantitative (i.e. stochastic, continuous, or hybrid) modelling and analysis techniques. A related unifying framework is demonstrated in refs. 4, 6.
7. Petri nets in all their flavours and related analysis techniques are supported by several reliable tools, developed by an international community of computer scientists (7–10).

As we will show below, Snoopy provides several features that allow to comfortably handle large, complex models.

2. Material

2.1. Software

In this chapter we use Snoopy (5), a software tool to design and animate or simulate hierarchical graphs, among them the Petri net classes: standard Petri nets (PN), extended Petri nets (xPN), and extended stochastic Petri nets (xSPN). The Snoopy software has three distinguished features.

1. It is extensible; its generic design facilitates the implementation of new graph types.
2. It is adaptive by supporting the simultaneous use of several graph types, while the GUI adapts dynamically to the graph type in the active window.
3. It is platform-independent. The tool runs on Mac OS X, Windows, and Linux.

The software is freely available for non-profit academic research purposes at "http://www-dssz.informatik.tu-cottbus.de/software/snoopy.html."

2.2. File Formats

Petri nets which have been designed with Snoopy can be saved and reloaded in a Snoopy-specific file format applying XML technology. The file formats vary slightly for the individual Petri net classes and are typically indicated by specific file extensions, such as spped (PN), spept (xPN), and spstochpn (xSPN). Snoopy offers build-in animation and stochastic simulation, which are used in this chapter. Supplementary, Snoopy provides continuous simulation of continuous Petri nets and export to various analysis tools as well as import and export of the standard exchange format SBML, Level 2, Version 3 (http://sbml.org). The Petri nets and simulation plots can be saved in Encapsulated Postscript (eps) format as well. For more information see Snoopy's website, and Notes 3 and 4.

2.3. Sample Files

All Petri nets used in this chapter are available at "http://www-dssz.informatik.tu-cottbus.de/examples/mimb/." To repeat the computational experiments discussed in this chapter, you just need to download Snoopy, to open these example files, and to follow the steps described in Subheading 3.

3. Methods

In this section, we describe briefly three net classes and the main features of their animation and simulation, which will then be used to perform the case study, the phosphate network of enteric bacteria. See also Note 1.

3.1. Petri Nets

The standard notion of Petri nets (PN) does not involve any timing aspects. Those nets are purely qualitative, i.e. time-free models. Technically speaking, Petri nets are weighted, directed, bipartite graphs which consist of the following basic ingredients (compare Figs. 1 and 2).

1. There are two types of nodes, which are called *places*, graphically represented by circles, and *transitions*, graphically represented by rectangles. Places usually model passive system components like species (substrates) playing the role of precursors and products of chemical reactions, while transitions stand for active system components like chemical reactions, transforming precursors into products or simply transporting species from a site A to a site B. Reversible chemical reactions are modelled by two opposite transitions (see Fig. 2b). Due-to their strong correspondence, we often use the two terms "transition" and "reaction" interchangeably.
2. The directed *arcs* (edges) connect always nodes of different type. They go from precursors to reactions, and from reactions to products. In other words, the pre-places of a

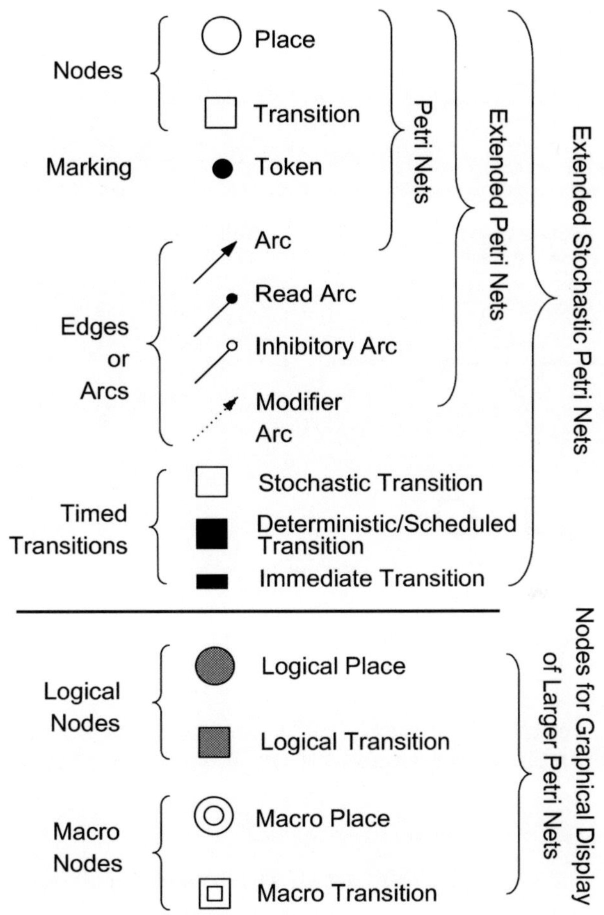

Fig. 1. Petri net elements and their graphical representation. Petri nets are weighted, directed, bipartite graphs consisting of nodes and arcs. The nodes of a Petri net, the places and transitions, are interconnected by arcs. An arc always connects a place with a transition or vice versa, but never two places or two transitions with each other. Places may contain (be marked with) tokens, while transitions cannot contain any tokens. Standard Petri nets are composed of places, transitions, and (standard) arcs. The other Petri net components are introduced in the text.

transition correspond to the reaction's precursors, and its post-places to the reaction's products.

3. Arcs are weighted by natural numbers. The arc weight may be read as the *multiplicity* of the arc, reflecting known stoichiometries or other semi-quantitative assumptions. The arc weight 1 is the default value and is usually not given explicitly.
4. A place carries an arbitrary amount of *tokens*, represented as black dots or a natural number. The number zero is the default value and usually not given explicitly. Tokens can be interpreted as the available amount of a given species in number of molecules or any abstract, i.e. discrete concentration

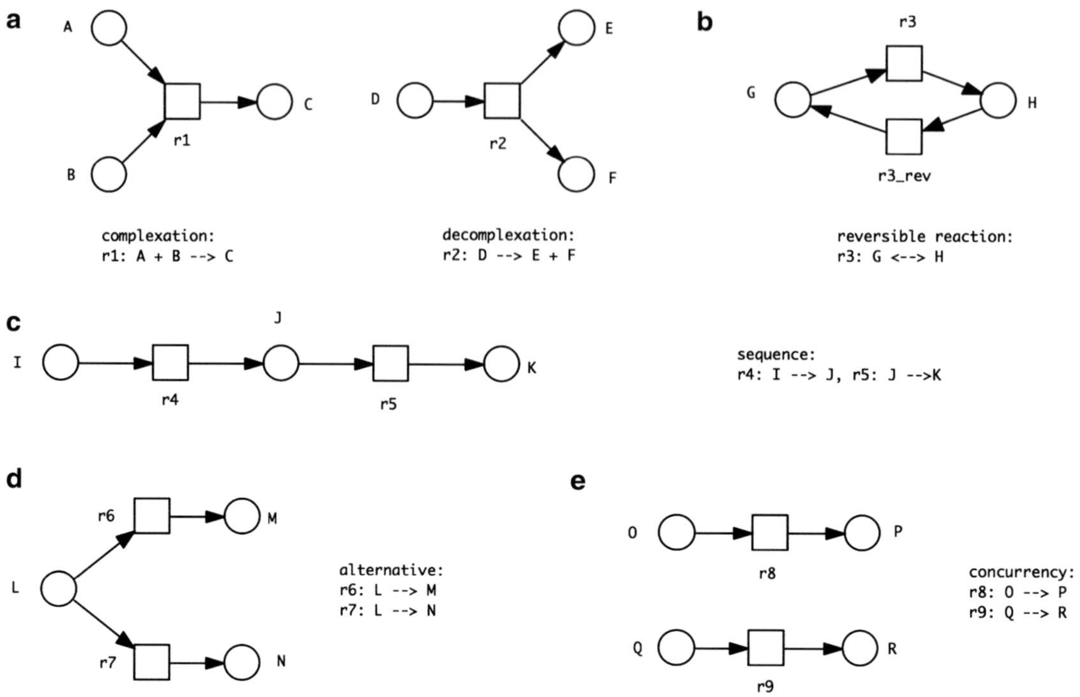

Fig. 2. Some typical basic structures of biochemical networks. (**a**) formation and decay of a macro-molecular complex; (**b**) reversible reactions; (**c**) sequential reactions; (**d**) alternative reactions; (**e**) concurrent, i.e. independent reactions.

level. The tokens on all places establish together the marking of the net, which represents the current state of the system.

How to draw a Petri net with Snoopy. To draw your first Petri net corresponding to Fig. 3, repeat the following steps.

1. Create a new net and chose the appropriate net class. You get a new drawing window with the name "unnamed".
2. Select the graph element *Place* in the menu panel on the left-hand side, and make a left mouse click on the drawing window at those positions where you want to get a place. Each mouse click creates a new place.
3. Similarly, select the graph element *Transition* in the menu panel, and create transitions.
4. To connect two nodes, select *Edge* in the menu panel, click on the source node, and move the mouse pointer, while keeping the mouse button pressed, to the target node, where the mouse button is released.
5. A double click on a graph element (place, transition, edge) opens an attribute window, which allows to edit the element-specific properties.

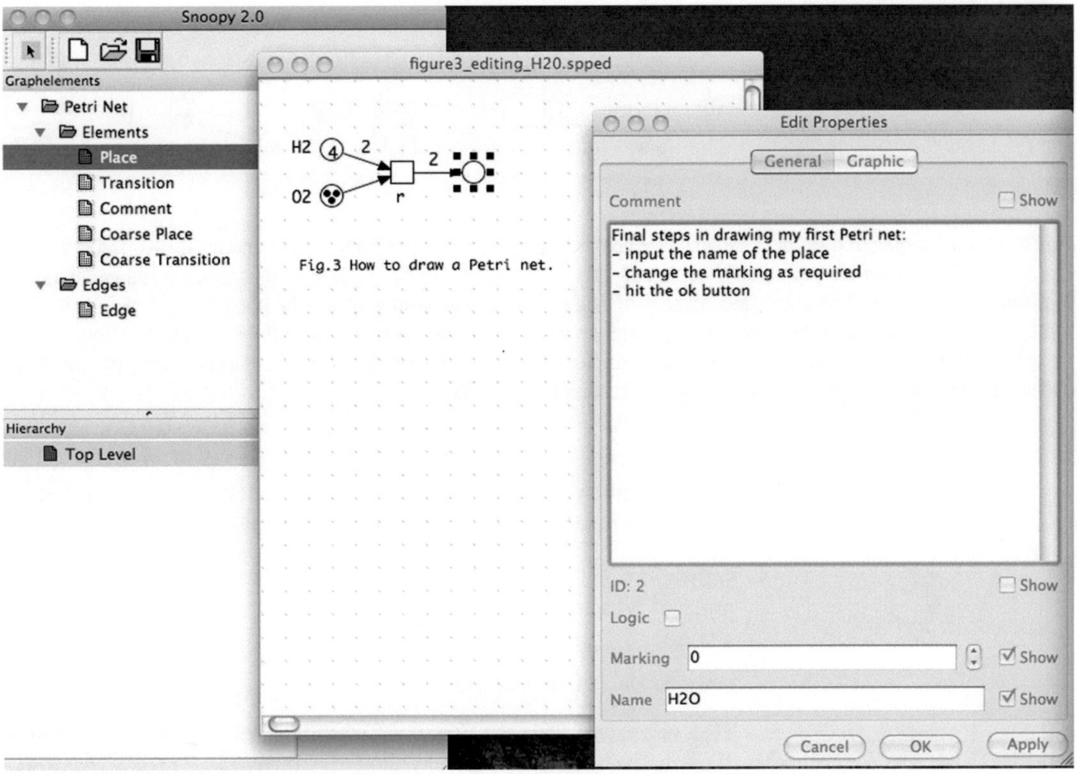

Fig. 3. How to draw a Petri net with Snoopy. The menu panel on the *left* allows to select the type of graph elements to be created, the hierarchy panel to open macro-nodes to show their sub-nets (not used here). The drawing window in the *middle* shows the Petri net under development. The window on the *right* belongs to the place that is selected in the drawing window and allows to edit the properties of this place.

6. Finally, do not forget to save your work under a new name (File → Save as).

The tokens may move through the net driven by the firing of transitions (see Fig. 4). The rules of the token game are defined by the firing rule. It consists of two parts: the precondition and the firing itself.

1. A transition is *enabled*, if all its pre-places carry at least as many tokens as required by the weights of the corresponding ingoing arcs.
2. An enabled transition *may fire* (may occur), i.e. an enabled transition is never forced to fire. Upon firing a transition removes from all its pre-places as many tokens as specified by the weights of the ingoing arcs, and adds to all its post-places as many tokens as specified by the weights of the outgoing arcs. The firing happens atomically (i.e. there are no states in between), and firing does not consume any time.

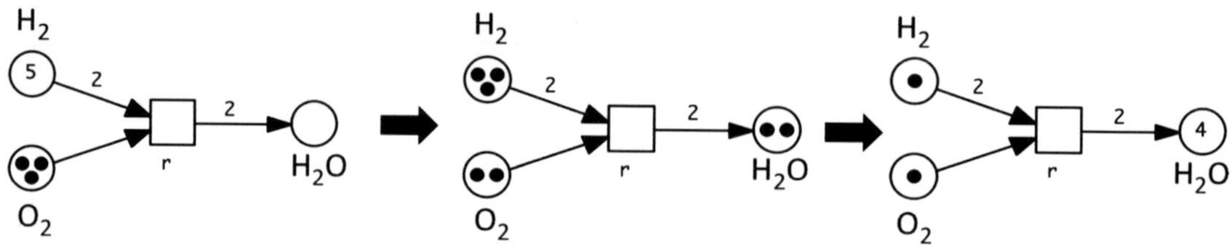

Fig. 4. Petri net model of a simple chemical reaction: the formation of water from oxygen and hydrogen. The reaction's reactants and products are represented as places, the number of available molecules of each chemical compound is represented by tokens that mark the respective place. The stoichiometry of the reaction 2 $H_2 + O_2 \rightarrow 2H_2O$ is represented by the arc weights, indicated as numbers proximal to the respective arc. If no number is given, the arc weight is always 1. Transition *r* can fire two times to produce four molecules of water. Subsequently, the transition cannot fire anymore as the stoichiometry of the reaction is not fulfilled.

How to animate a Petri net with Snoopy. Snoopy visualises the token flow, allowing for an easy comprehension. To animate your first Petri net, repeat the following steps.

1. Open the net, and open the animate window (View → Animation Mode).
2. A left mouse click on a transition triggers the firing of this transition, if it is enabled.

The repeated atomic firing of enabled transitions establishes the discrete behaviour of the qualitative Petri net. This behaviour may include (compare Fig. 2c–e):

1. *Sequential* reactions reflecting causality (e.g., reaction r5 cannot happen before r4 happened).
2. *Alternative* reactions competing for tokens on shared pre-places and therefore branching into alternative behaviour; in Petri net terminology, the transitions are said to be in conflict (e.g. reactions r6 and r7 share the pre-place L; a token on L can either be consumed by r6 or by r7).
3. *Concurrent* reactions, which are neither causal nor conflict relation; thus, they are independent and can fire in any order or even concurrently (e.g. reactions r8 and r9).

In addition to a basic toolkit for drawing Petri nets, Snoopy supports two distinguished features for the design and systematic construction of larger Petri nets (see Figs. 1 and 5).

1. Places and transitions can be specified as *logical nodes* (also called fusion nodes). They are automatically coloured in grey. Logical nodes with the same name are identical, i.e. graphical copies of a single node (place or transition). They are often used for compounds involved in several reactions or reactions involving distributed compounds.
2. There are two types of *macro-nodes*. Macro-transitions (drawn as two centric squares) help to hide transition-bordered

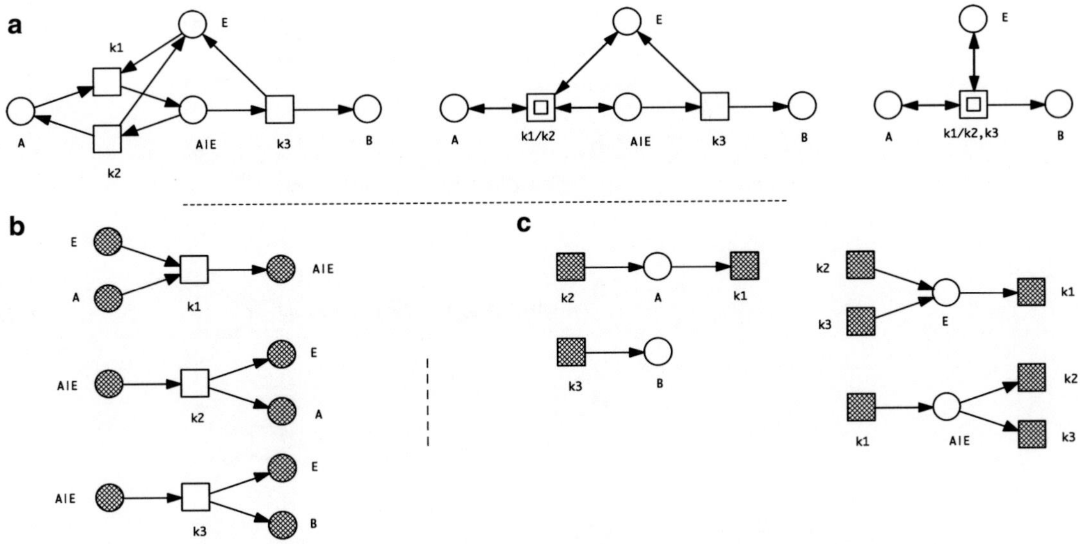

Fig. 5. The use of macro-transitions and logical nodes. Different graphical representations of one and the same model representing the enzymatic reaction A + E ↔ AIE → B + E, where AIE is the enzyme–substrate complex. (**a**) Macro-transitions yield hierarchically structured models; (**b**) reaction-centric representation by the use of logical places; (**c**) species-centric representation by the use of logical transitions.

subnets (i.e. subnets having only transitions as interface to the super-net). Likewise, macro-places (drawn as two centric circles) can be used to hide place-bounded subnets (i.e. subnets having only places as interface to the super-net). Both types of macro-nodes allow a hierarchical structuring of Petri nets.

Logical nodes and macro-nodes help to deal with larger networks. They may contribute to a net's readability and, therefore, are crucial for non-trivial net examples. However, they do not extend the expressiveness of the modelling language. Contrary, the features introduced in the next section do extend the expressiveness.

How to draw an hierarchical Petri net with Snoopy. To hierarchically structure a Petri net, repeat the following steps.

1. Draw your flat net (or portion of it) as you wish to have it.
2. Select the sub-graph which you want to be abstracted by a macro-node, and select Hierarchy → Coarse. Chose the appropriate coarse element and hit the OK button.
3. A double click on the macro-node opens its attribute window and allows to assign a suitable name, a click on the entry with this name in the hierarchy panel on the left-hand side opens the sub-graph in a separate window. The blue net parts have been automatically generated and represent the connection of the sub-graph (macro-node) with the neighbouring nodes on the next higher hierarchy level.

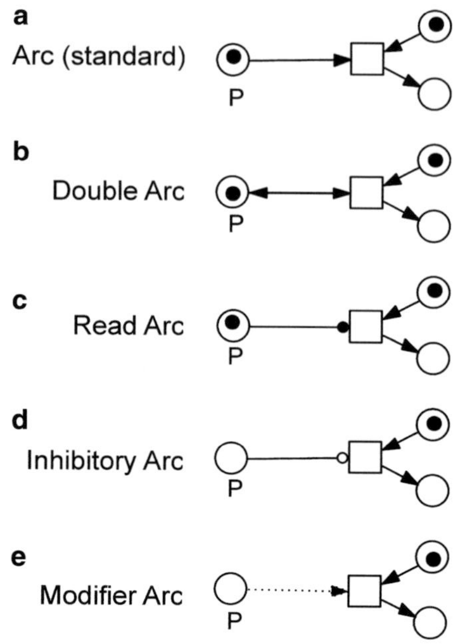

Fig. 6. Different types of arcs and their influence on the firing of a transition. (**a**) Standard arc; the transition fires whereby the token in P is consumed. (**b**) Double arc; the transition fires if there is at least one token in P; upon firing, the token is transiently consumed and put back into P. (**c**) Read arc; the transition fires only if there is at least one token in P; the token is not consumed upon firing of the transition. (**d**) Inhibitory arc; the transition fires only if there is no token in place P. The different types of arcs in (**a–d**) may also carry weights. (**e**) Modifier arc; used with stochastic transitions only and always without weights; the probability per unit of time for the transition to fire depends on the number of tokens in place P; the transition can even fire if there is no token in P. These different types of arcs can be used to model phenomena that typically occur in biochemical networks: (**b**) Double arcs can be used to model a catalytic reaction by neglecting formation and decay of the enzyme–substrate complex. In dynamic models, the effect of substrate saturation will not occur if this representation is chosen. (**c**) Read arcs can be used, e.g. to model the influence of the conformation of a protein molecule on the occurrence or rate of a biochemical reaction. (**d**) Inhibitory arcs can be used, e.g. to model the inhibitory influence of the conformation of a protein molecule on the occurrence or rate of a biochemical reaction, e.g. upon binding of an inhibitory subunit to an enzyme. (**e**) Modifier arcs can be used, e.g. to model the influence of a covalent modification on the activity of a protein. The covalent modification may be represented by a token in P. If there is no token in P, the protein displays its basal activity.

4. Finally, do not forget to save your work under a new name (File → Save as).

3.2. Extended Petri Nets

xPN build on Petri nets and additionally provide special arc types, which go always from a place to a transition, see Figs. 1 and 6. They can also carry multiplicities. The two most popular special arcs are:

1. *Read arcs* (also called test arcs), graphically represented by a black dot as arc head, query for tokens in a place representing positive side-conditions, e.g. the conformation of a protein complex that may determine whether a reaction can occur, or a specific physiological state of a cell that may determine whether a cell is responsive to a certain stimulus. The tested place needs at least as many tokens as given by the read arc's multiplicity to enable a transition. The firing of the transition does not change the number of tokens on the tested place.

   A read arc may be simulated by two opposite arcs. Although a read arc and two opposite arcs have the same total effect, there is a subtle semantic difference, which is sometimes useful to precisely represent molecular mechanisms within a biochemical network. Assume that an enzyme E catalyses the reaction of substrate X to product Y. Then E is transiently consumed by forming the enzyme–substrate complex and reformed as the enzyme–product complex decays to form Y. Such an enzymatic reaction is represented in a mechanism-oriented style by two opposite arcs if formation and decay of the enzyme–substrate complex are not explicitly modelled. On the other hand, using the read arc is appropriate if, for example an unphosphorylated receptor (Y) autophosphorylates (Y-P) as the consequence of a conformational change to R (the token in R represents the active conformation of the receptor). Upon autophosphorylation, the active conformation of the receptor is not transiently consumed and accordingly, the token stays in R (compare the two panels of Fig. 7; see also Note 2).

2. *Inhibitory arcs*, graphically represented by an hollow dot as arc head, indicate negative side-conditions in an abstract way, e.g. if the presence of a given protein (inhibitory subunit of a

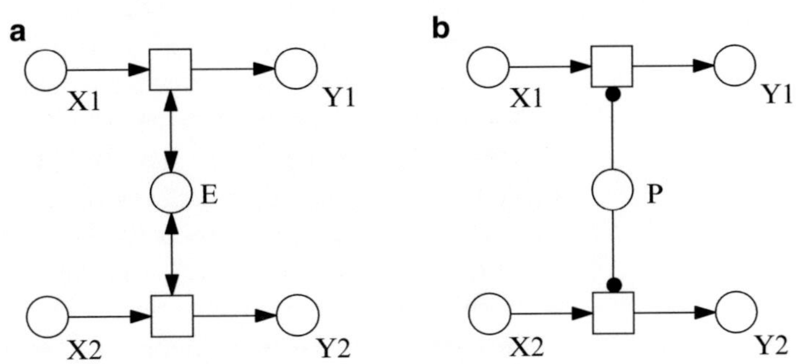

Fig. 7. Semantic differences between two opposite arcs and a read arc. (**a**) E catalyses the reactions r1 and r2. Thus, r1 and r2 compete for the tokens on E and can only use them alternatively. (**b**) P stands for an active confirmation, allowing r3 and r4 to happen concurrently, thus sharing the tokens on P. The two arc types are not distinguishable in interleaving semantics, but in partial order semantics (see also Note 2).

protein complex) or condition inhibits a specific reaction. The inhibiting place must have less tokens than given by the inhibitory arc's multiplicity to enable a transition. The firing of the transition does not change the number of tokens on the inhibiting place. Inhibitory arcs can only be simulated by standard Petri nets, if the inhibiting place is bounded (the number of tokens never exceeds a given finite number). So they strictly increase the expressiveness.

There are two other special arc types, not used in this chapter: equal arcs and reset arcs (for more details, see ref. 11).

3.3. Animation of Qualitative Petri Nets

Having obtained a new Petri net, the next step often aims at a better understanding of the net behaviour. As we have already seen, we can execute the Petri net by playing the token game to experience the net behaviour. Each execution exemplifies some possible net behaviour. The animation can be triggered manually by clicking on an enabled transition. It can also be done in automatic mode by clicking on the control panel. In each step of the automatic mode the set of all enabled transitions is determined. There are three strategies to choose the transition(s) to be fired in the next execution step among all enabled transitions.

1. *Single step* – one single transition is randomly chosen.
2. *Intermediate step* – an arbitrary subset of concurrent transitions is randomly chosen.
3. *Maximal step* – a maximal set of concurrent transitions is randomly chosen.

Steps just done can be played backwards up to a depth of 10; this default value can be changed in the Global Preferences dialogue.

By playing the token game, we can produce and observe any reachable state. All states, which can be reached from a given state by any firing sequence of arbitrary length, constitute the set of reachable states. Each execution run of an xPN corresponds to a walk through the state space defined by this xPN. The set of states reachable from the initial state is said to be the state space of a given system. Often, this qualitative state space is infinite, caused by considering all possible behaviour of a structurally unbounded Petri net under any timing constraints. We call a Petri net bounded, if the number of tokens on all places is bounded by a constant independently of what happens. Then we get a finite state space, which – however – can still be too huge to be explored exhaustively.

Executing a Petri net generally needs to make decisions between alternative behaviour. Encountered alternatives (conflicts) are taken non-deterministically (automatic mode) or user-guided (manual mode). To get an exhaustive picture, we have to consider all possible execution sequences. Obviously, that is not possible for systems with infinite behaviour, which can be caused by cyclic behaviour and/or infinite state space. Thus, we have to confine ourselves to a subset of execution sequences, which are considered to be representative.

In the next section, we introduce time and we see how specific kinetic assumptions typically restrict the qualitatively infinite state space to a quantitatively finite state space, if we are only interested in states with a probability above a certain threshold.

How to change the net class in Snoopy. A (qualitative) Petri net can be converted into a stochastic Petri net, which allows to re-use the structure. Only the additional attributes have to be set.

1. Open the qualitative Petri net, and go to the export Window (File → Export), chose the appropriate target(s) and hit the OK button.
2. Open the stochastic Petri net, just generated. It looks exactly the same as the qualitative net. Now you are ready to specify the attributes specific for SPN, which are discussed in the next section.

3.4. Stochastic Petri Nets

SPN build on standard Petri nets. As in the qualitative case, a stochastic Petri net maintains a discrete number of tokens on its places. But contrary to the time-free case, a stochastic firing rate is associated with each transition, determining a stochastic waiting time before an enabled transition actually fires, provided it did not loose its license to fire in between. The waiting times are random variables following an exponential probability distribution (with the parameter lambda). Therefore, all transition firing sequences of the qualitative Petri net can theoretically still occur, but the probability depends on the individual firing rates of the transitions involved (see Fig. 8). Snoopy provides the following features to specify state-dependent firing rates (i.e. of the parameter lambda of the exponential probability distribution).

1. *Rate functions* are arbitrary mathematical functions, which are stored in lookup tables, if necessary. However, to keep a close relation to the net structure, only the transition pre-places are allowed as variables in the transition rate function. Popular kinetics (mass-action semantics, level semantics, see ref. 4) are supported by pre-defined function patterns. Of course, each transition gets its own rate function, making up together a list of rate functions. Moreover, several of such rate functions lists can be maintained, allowing for quite flexible models.
2. *Parameters*, in figures represented by ellipses, are real-valued constants, which are used for the rate functions. Parameters can be collected in macro-parameter nodes, in figures

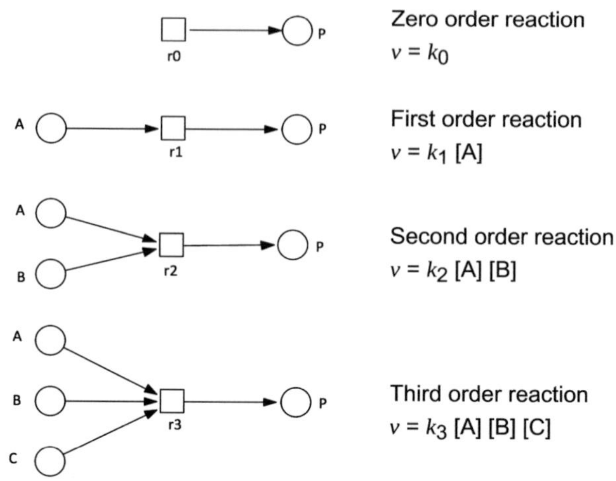

Fig. 8. Automatic generation of reaction rate equations with stochastic transitions in extended stochastic Petri nets. Using the "MassAction" function, Snoopy automatically generates the correct reaction rate equation depending on whether a reaction is zero, first, second, or third order. Here, [x] stands for the concentration of the place x in the case of continuous Petri nets, and for the number of tokens on x in the case of stochastic Petri nets.

represented as two centric ellipses. Again, several sets of parameters can be maintained.

3. *Modifiers* are particular arcs, in figures represented by dashed lines, which always go from a place to a transition. Pre-places connected with a transition by a modifier arc may modify the transition's firing rate, but do not have an influence on the transition's enabledness (contrary to the special arcs).

The firing itself of a stochastic transition does not consume time and follows the standard firing rule of qualitative Petri nets. SPN with exponentially distributed firing delays for all transitions fulfil the Markov property; thus their semantics is described by a continuous time Markov chain (CTMC); see (4) for more details. However, in this chapter we confine ourselves to simulative approaches. Each simulation run of an SPN corresponds to a walk through the CTMC defined by this SPN, see Subheading 6.

3.5. Extended Stochastic Petri Nets

xSPN combine SPN with the special arc types of xPN. Additionally, there are deterministically timed transitions, which come in three flavours.

1. *Deterministic transitions* have – contrary to stochastic transitions – a deterministic firing delay which is specified by an integer constant. The delay is always relative to the time point where the transition gets enabled. The transition may lose its enabledness while waiting for the delay to expire. Then, the transition will not fire. This behaviour is also known as the

so-called pre-emptive firing rule. As for stochastic transitions, the firing itself of a deterministic transition does not consume time and follows the standard firing rule of qualitative Petri nets. Deterministic transitions may be useful to reduce networks, and thus speed-up simulations, e.g. by replacing a linear sequence of stochastic transitions by one deterministic transition with the delay set to the expectation value of the sum of the sequence.

2. *Immediate transitions* are a popular special case of deterministic transitions. They have a zero delay and always highest priority. The latter creates a subtle difference between an immediate transition and a deterministic transition with zero firing delay: if there is a conflict between the two, i.e. a situation, where two transitions compete for tokens, the immediate transition gets priority. Immediate transitions may help to avoid stiff systems by using them for transitions with extremely high rates (non-significant delay), compared to the other transitions in the system.
3. *Scheduled transitions* are another special case of deterministic transitions. The deterministic firing occurs according to a schedule specifying absolute points of the simulation time. A schedule can specify just a single time point, or equidistant time points within a given interval, triggering the firing once or periodically. However, transitions only fire at their scheduled time points if they are enabled at this time. Scheduled transitions can dramatically restrict the (qualitative) net behaviour. The crucial point is that they allow to disturb the core model at well-defined time points as it is done experimentally with the actual biological system under investigation in the wet-lab.

An unrestricted use of deterministic transitions destroys the Markov property, which precludes an analytical evaluation by constructing and analysing the CTMC. If we consider SPN without deterministic transitions, the probability of two transitions to fire at the same time is practically zero. Contrary, in SPN with deterministic transitions, it is possible that two transitions want to fire simultaneously. To analyse such a system, all possible choices have to be considered, while in the simulation a random choice takes place, see next section (for more details and related examples see ref. 12).

In most practical cases, xSPN are the obvious choice. Therefore, Snoopy does not distinguish between SPN and xSPN.

3.6. Simulation of Quantitative Petri Nets

The simulation of SPN can be read as a timed version of the qualitative token game, taking into account the stochastic and deterministic delays of enabled transitions. Snoopy builds upon the Gillespie algorithm (13). This exact method does a step-by-step simulation of possible states of the stochastic Petri net. Consequently, the result represents always a valid state of the underlying stochastic process at any time point of the simulation. The standard stochastic simulation algorithm works in the following way:

1. Initialise the SPN network with the chosen initial marking.
2. Calculate the firing rates of all enabled transitions using their rate functions.
3. Calculate the combined rate by summing up all transition rates.
4. Determine the time interval until the next state change takes place. This is done by computing an exponentially distributed random number depending on the current combined transition rate of the net.
5. Increase the simulation time by this time interval.
6. Determine the next system state change. For this purpose, a weighted random selection of the transition is made which gets the license to fire.
7. Let the selected transition fire and update the marking of the SPN.
8. Go back to step 2, if the simulation time has not yet reached its end point.

The simulation of xSPN requires some straightforward modifications of the standard simulation algorithm. Deterministic transitions require a higher priority than stochastic transitions to ensure their correct handling. Immediate transitions need to have the highest priority (over deterministic and scheduled transitions) because they have to fire instantaneously when they get enabled. Consequently, the simulation algorithm has to check for enabled immediate transitions after any change of the marking. In the case of more than one enabled immediate transitions, the selection is done randomly, but uniformly distributed. The system may run into a time deadlock, if always an immediate transition is enabled. Then, time will not progress anymore. Time deadlocks are an indication of inconsistent time assignments. The user has to take care of avoiding them.

Deterministic and scheduled transitions are treated differently. If a deterministic transition gets enabled, its timer starts running until it expires. If the deterministic transition is still enabled, it fires. The timer will be simply switched-off, if the transition should have lost in between the enabledness. The firing of a transition in conflict does not change the timer value. The scheduled transitions fire at the specified absolute time points, if they are enabled at that time. Otherwise the intention to fire is not used. All conflicts between deterministic and/or scheduled transitions are resolved randomly, by uniformly distributed selection.

Stochastic animation. The progress of a stochastic simulation run can be observed in the automatic animation mode, which shows the corresponding token flow.

Repeated stochastic simulation. Each simulation run is one possible trace of state changes over time. Therefore, when doing stochastic simulation, a significant number of simulation runs is usually required to get meaningful results. For this purpose, the traces of the simulation runs are averaged.

The simulation can be parametrised in four ways.

1. Different sets of markings, rate functions and parameters can be chosen.
2. The simulation time of interest can be specified, specifically the interval start and interval end, e.g. from 0 to 100 or from 250 to 300.
3. The number of output step points in the specified simulation interval can be declared. They define together with the interval start and end values the grid of the recorded simulation data.
4. The number of simulation runs can be specified.

Having started the simulation, the progress bar indicates how far the simulation has been gone until now, and the time consumed by the simulation is displayed too.

The user can select one out of three export functions:

1. *Direct export*: the averaged result of the selected places or transitions is exported.
2. *Single trace export*: the result of every simulation run is exported separately.
3. *Exact trace export*: every change of the marking of the selected places or the transition rates at any time point is exported.

The simulation results can be shown as

1. *Table*, where each column represents a selected place (transition) and each row shows the averaged marking (firing times) at one output step point.
2. *Plot*, where each curve stands for one selected place (transition); the x-axis holds the time and the y-axis holds the number of tokens (firing times).

Different tables and plots can be created to switch conveniently between different views on the simulation results. Each table is characterised by a set of selected places (transitions). If a node was coloured, its corresponding curve gets the same colour in thc plot. Simulation plots can also be saved in eps format.

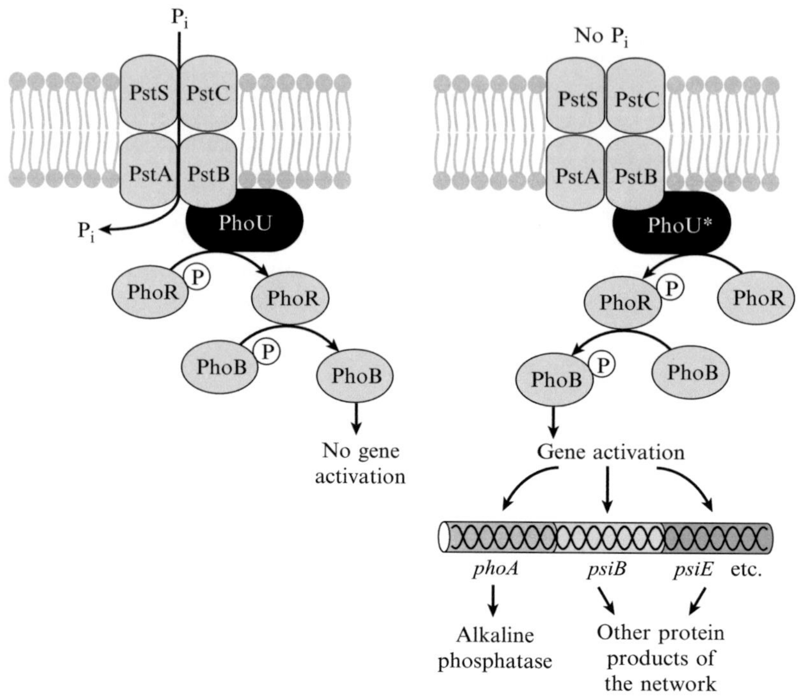

Fig. 9. Biochemical model of the phosphate regulatory network in enteric bacteria. The scheme is adapted from Neidhardt et al. (14).

3.7. Case Study

3.7.1. A Case Study: The Phosphate Regulation Network in Enteric Bacteria

Let us now consider the phosphate utilisation gene regulatory network in *Escherichia coli* (see Fig. 9 discussed in ref. 14). Note that this case study is neither meant as a scientific contribution to the understanding of the gene regulatory network of phosphate utilisation nor does it provide any comprehensive representation of the knowledge on this subject; for a recent survey see ref. 15. Instead, the case study is intended to show how a typical biochemical model, a gene regulatory network, may be formally translated into a Petri net. As we will show, the Petri net can serve as a qualitative scheme representing the molecular reaction mechanisms, but it can also be used for dynamic simulations.

For different reasons, inorganic phosphate may become a growth-limiting factor for a bacterial cell population. Under these conditions cells synthesise alkaline phosphatase (PhoA), an enzyme which is secreted into the periplasm (i.e. into the space between cytoplasmic membrane and outer membrane of a Gram-negative bacterium), where it degrades organic phosphate esters into inorganic phosphate to be taken up and recycled by the cell. Transport of inorganic phosphate is mediated by an uptake system composed of four proteins, PstS, PstC, PstA, and PstB that form a transmembrane protein complex. Experimental evidence suggests that phosphate transport is sensed by the PhoU protein. If the phosphate transport system is active, the PhoU protein resides in

Table 1
Molecular components involved in phosphate regulation

Abbreviation	Molecular component
PhoA	Alkaline phosphatase enzyme degrading organic phosphate compounds to inorganic phosphate
Pi	Inorganic phosphate
Po	Organic phosphate
PstSCAB	Transmembrane protein complex, transporter of inorganic phosphate
PhoU	Signal transducer relaying the PstSCAB complex activity
PhoR	Phosphorylatable regulatory protein
PhoB	Phosphorylatable regulatory protein

its inactive state. Phosphate limitation renders the transport system inactive, which activates PhoU and causes phosphorylation of the PhoR protein which in turn phosphorylates PhoB. Phosphorylated PhoB (PhoB-P) is a positive regulator which binds to the promotor region of certain operons. Upon PhoB-P binding, among others, the *phoA* gene is expressed and alkaline phosphatase (PhoA protein) is synthesised and exported into the periplasm. While active PhoU causes the phosphorylation of PhoR, inactive PhoU is thought to act as a PhoB-P phosphatase which switches off the DNA binding activity of the PhoB protein. A summary of the molecular components involved in the regulation of the phosphate utilisation network is shown in Table 1.

3.7.2. Application of xSPN to Study Case: The Phosphate Network in Enteric Bacteria

Qualitative modelling (xPN). The regulatory mechanism illustrated in Fig. 9 is implemented in the Petri net shown in Fig. 10. The molecular components in their different states (e.g. active, inactive, phosphorylated, unphosphorylated) are represented as places (Table 2) and their biochemical reactions or functional interactions are represented as stochastic transitions (Table 3). Additionally, there are deterministically timed transitions (immediate and scheduled transitions) to model the experimental addition of inorganic/organic phosphate. There are different arc types. Standard arcs represent the mass flow. Read arcs and inhibitory arcs are used to model regulatory interactions between proteins through physical protein–protein interaction. Double arcs (a short-hand notation for two opposite arcs) represent catalytic reactions which may be complex.

The graphical representation of the cytoplasmic membrane and of some proteins of interest is put underneath the Petri net in order to explain the modular structure of the net and to highlight which

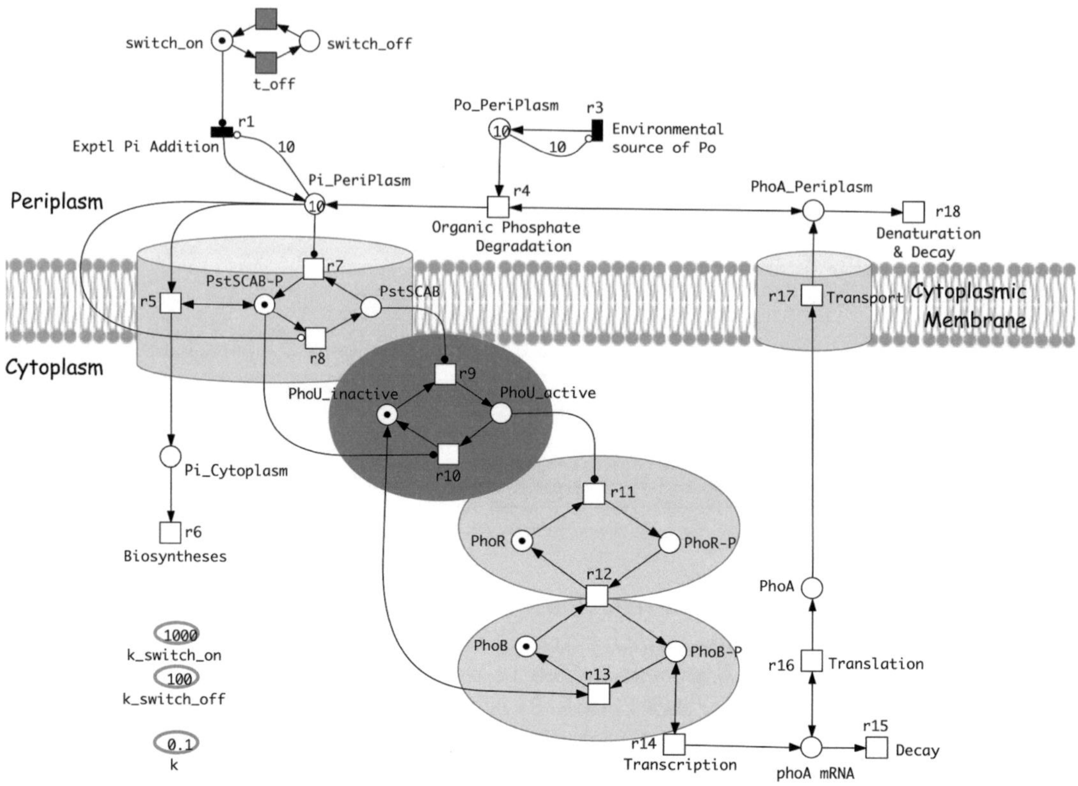

Fig. 10. Petri net model of the phosphate regulatory network, along with a schematic representation of the membrane and of some relevant proteins. Biomolecular components and their functional states are represented as places, (bio-) chemical reactions are represented as transitions. Stoichiometric reactions are represented as standard arcs, allosteric interactions of proteins by read arcs and inhibitory arcs. Organic and inorganic phosphate is supplied by the environment of the cell at constant concentration. In the simulations, the constant concentration of these phosphate compounds is obtained by using transitions with immediate firing behaviour (r1, r3) delivering tokens to the corresponding places (Pi_PeriPlasm; Po_PeriPlasm). The firing of r1 and r3 is controlled by inhibitory arcs. These inhibitory arcs shut the transitions off if the pre-defined number of tokens is in the pre-place of the respective transition. Experimental addition and removal of inorganic phosphate to the cell is modelled with the help of the switch_on place which is connected by a read arc to r1. Therefore, r1 delivers inorganic phosphate as long as (1) there is a token in the switch_on place and (2) the number of tokens in the Pi_PeriPlasm place is less than ten. The supply with inorganic phosphate may be switched off and on at defined time points by switching the token in the switch_on place with the deterministically timed (scheduled) transitions t_on and t_off. In the simulations, organic phosphate in the periplasm remains constant at all times. The Petri net represents the processes involved in phosphate regulation with different resolution of details. Only part of the net represents molecular processes in terms of interactions of individual molecules. Complex reaction mechanisms, like biosyntheses using inorganic phosphate (r6), transcription of the *phoA* gene (r14), or translation of the *phoA* mRNA (r16) are represented as individual transitions that condense multiple individual steps into one single reaction, assuming first order rate constants in the example provided. Note, however, that the quantitative behaviour of a transition in the xSPN version of Snoopy can be programmed individually so that even complex and non-linear kinetic mechanisms can be modelled.

reactions occur inside or outside the cell, respectively. Note that these graphics are not a functional element of the Petri net as generated in Snoopy. The PstSCAB transmembrane protein complex has a dual function as transporter and as a sensor of inorganic phosphate. When inorganic phosphate is present in the periplasm,

Table 2
Petri net places represent molecular components in their different states

Place	Molecular component
Pi_PeriPlasm	Inorganic phosphate in the periplasm
Pi_Cytoplasm	Inorganic phosphate in the cytoplasm
Po_PeriPlasm	Organic phosphate in the periplasm
PhoA_Periplasm	PhoA protein in the periplasm
PhoA	PhoA protein in the cytoplasm
PhoA mRNA	PhoA gene mRNA
PstSCAB	Transporter of inorganic phosphate, inactive form
PstSCAB-P	Transporter of inorganic phosphate, phosphorylated, active form
PhoU_inactive	Signal transducer, inactive form
PhoU_active	Signal transducer, active form
PhoR	PhoR protein, dephosphorylated form
PhoR-P	PhoR protein, phosphorylated form
PhoB	PhoB protein, dephosphorylated form
PhoB-P	PhoB protein, phosphorylated form
switch_on, switch_off	Auxiliary places to model the experiments

the PstSCAB complex is phosphorylated through reaction r7. The phosphorylated form, PstSCAB-P is active in transporting inorganic phosphate through the cytoplasmic membrane into the cytoplasm (r5) where it is used for biosynthetic reactions (r6). The PstSCAB protein allosterically controls the activity of the PhoU protein, modelled by read arcs that control the transitions r9 and r10 representing the activation (r9) and the deactivation (r10) of the PhoU protein, respectively. As long as inorganic phosphate is in the periplasm, the PstSCAB complex is kept in its phosphorylated, transport-active state, modelled by the inhibitory arc that prevents dephosphorylation via transition r8. In its active form, the PhoU protein causes the phosphorylation of PhoR (r11) which in turn phosphorylates PhoB. This direct transfer reaction of the phosphate group between the two protein molecules is considered as a second order reaction and represented by transition r12. PhoB-P then binds to DNA to promote the transcription of several genes including the *phoA* gene. The complex process of regulator binding and gene transcription is represented by transition r14. Translation of the *phoA* mRNA into the PhoA protein occurs through transition r16 which also summarises a complex set of biochemical reactions

Table 3
Petri net transitions represent biochemical reactions or functional interactions

Transition	Reaction or process
r1	Experimental addition of inorganic phosphate (immediate transition)
r2a, r2b	Switch on/off the experimental addition by r1 (scheduled transitions)
r3	Environmental source of organic phosphate (immediate transition)
r4	Degradation of organic phosphate
r5	Transport of inorganic phosphate into the cytoplasm
r6	Consumption of inorganic phosphate by biosyntheses
r7	Phosphorylation of PstSCAB complex
r8	Dephosphorylation of PstSCAB complex
r9	Activation of PhoU by PstSCAB complex
r10	Deactivation of PhoU by PstSCAB-P complex
r11	Phosphorylation of PhoR by active PhoU
r12	Phosphorylation of PhoB by PhoR-P, involving dephosphorylation of PhoR-P
r13	Dephosphorylation of PhoB-P by inactive PhoU
r14	Transcription of the *phoA* gene upon binding of PhoB-P to its promoter
r15	Decay of the *phoA* mRNA
r16	Translation of the *phoA* mRNA into PhoA protein
r17	Transport of the PhoA protein into the cytoplasm
r18	Denaturation and decay of the periplasmic PhoA protein

The transitions are stochastic if not specified otherwise.

involving many additional components. The double arrow linking r16 and the *phoA* mRNA place indicates that from each mRNA molecule many copies of the PhoA protein can be synthesised. The PhoA protein is finally transported into the periplasm (r17) where it degrades organic phosphate to supply the cell with inorganic phosphate which in turn switches off the signalling cascade and subsequently the biosynthesis of the PhoA protein, as long as sufficient inorganic phosphate enters the cell.

We have already seen how to draw a simple Petri net, so you will not have any trouble in drawing the Petri net for your first case study. Coarse transitions and coarse places may be used to structure large networks into modules in order to improve the readability of the net. Since the Petri net presented for the case study is rather small, we did not use coarse nodes here.

Animation of the xPN. To explore the net behaviour, do the following steps:

1. Load the net (File → Open).
2. Start the animation (View → Start Animation Mode).
3. Run the net in manual or automatic mode.

Specifically you should check the following scenarios:

1. Which reactions are necessary in which (partial) order to fire the transitions Biosynthesis (r6), Decay (r15), or Denaturation&Decay (r18)?
2. Which net behaviour is possible if the inorganic supply is switched on (place switch_on is marked), which net behaviour is triggered if the inorganic supply is switched off?
3. Try to figure out the maximal token numbers you can get on each place!
4. Is it possible to reach a marking where none of the transitions is enabled?
5. Having played with the net for a while, is it always possible to come back to the given initial marking?

Quantitative modelling (xSPN). All stochastic reactions follow mass-action kinetics with all parameters initially set to 0.1. Afterwards, the parameters were adjusted so that the steady-state concentration of inorganic phosphate in the periplasm is approximately the same no matter whether the external source is inorganic or organic phosphate, respectively: the parameter of r4 is set to 0.2, and the parameter of r15 to 0.075.

The transition "t_off" is scheduled to fire at time point 100, and the transition "t_on" is scheduled to fire at time point 1,000.

Stochastic animation of the xSPN. Follow the stochastic token flow in the automatic animation mode. Which reaction sequences do you observe while the inorganic supply is switched on? What is the difference to the reaction and state sequences, which you have observed in the animation of the corresponding xPN? The answer lies in the immediate transitions (here r1), which have now always highest priority.

Stochastic simulation of the xSPN. In the stochastic simulation, the reaction of each individual molecule is considered in the form of individual tokens that move through the net. When the number of molecules of each biochemical component per cell is known, one can obtain realistic traces of how the number of molecules develops over time. The result may be a time-dependent change in the concentration of the compound if the number of molecules per cell is sufficiently high or the number of molecules may be subjected to stochastic fluctuations over time if it is low. When a stochastic simulation is run many times and the simulation results are averaged, one can approach the result of a deterministic simulation as obtained, for example, by solving a set of ordinary differential equations (ODEs). Each way of simulation, stochastic as performed here, or deterministic, e.g. by solving ODEs, may be of particular advantage dependent on the question under consideration. For studying and simulating the behaviour of individual cells, stochastic simulations may be essential.

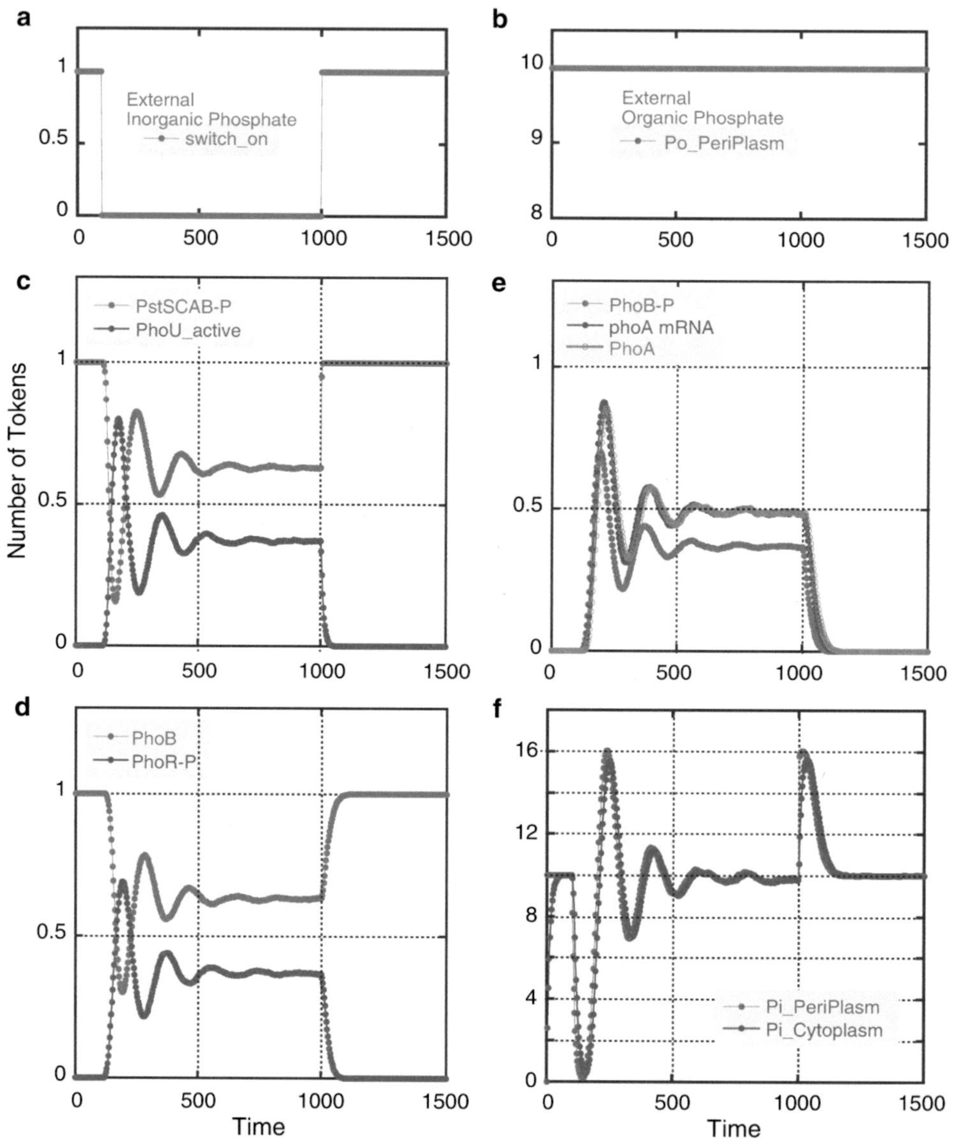

Fig. 11. Response of the phosphate regulatory network to step-wise changes in the supply of external inorganic phosphate. Traces were obtained by averaging 50,000 simulation runs. The rate constants of the stochastic transitions are r4 – 0.2, r15 – 0.075, else 0.1. Initial marking as given in Fig. 9. Simulation results were exported from Snoopy as csv file and imported into *Kaleidagraph*, *see* "http://www.kaleidagraph.com/," for display.

In order to provide an example of a simulation, we have run the model 50,000 times to show how the phosphate regulation network may behave dynamically. The concentration of the sources, inorganic and organic, was assumed to be constant over time. The system was first allowed to equilibrate in the presence of constant external inorganic phosphate (Fig. 11). After the steady state was reached, external inorganic phosphate was removed (switched off) by a step-down to zero concentration and the system was allowed to approach the new steady state. The oscillations in the concentration of the components obtained in the simulation are due to the feed-back loops in the system. Note that the precise dynamic behaviour of the system depends on the ratio of the rate constants which we do not know. Finally, the external source of inorganic phosphate is switched on again and the system equilibrates into its pre-stimulus state. The results of the stochastic simulation are shown as a panel obtained by importing the simulation results into the plot program *Kaleidagraph* (Fig. 11) and as screen shots of the simulation results window of Snoopy (Figs. 12 and 13).

You can continue by trying the following simulation scenarios:

1. Which rates influence amplitudes, frequencies, and damping of the oscillations caused by feedback regulation?
2. Check how the dynamic system behaviour is influenced by the decay rates of mRNA and PhoA protein.

4. Notes

1. In this chapter, we deliberately omitted formal definitions. The interested reader finds them together with accompanying examples of the basic notions of biochemically interpreted Petri nets in ref. 4; for xPN see specifically (11), for xSPN (12), and for general textbooks introducing into Petri nets and SPN the references therein.
2. Petri nets represent an inherently concurrent modelling paradigm. They allow to distinguish precisely between alternative and concurrent (i.e. independent) behaviour. However for analysis purposes, the partial order (true concurrency) semantics is often reduced to the interleaving semantics, where concurrency of reactions is described by all interleaving sequences of these reactions. We get a state transition system (known in the Petri net community as reachability graph or marking graph), which is basically a finite automaton, i.e. a sequential model, and as such does not support the differentiation between alternative and concurrent behaviour. This disadvantage is (partially) compensated by the numerous

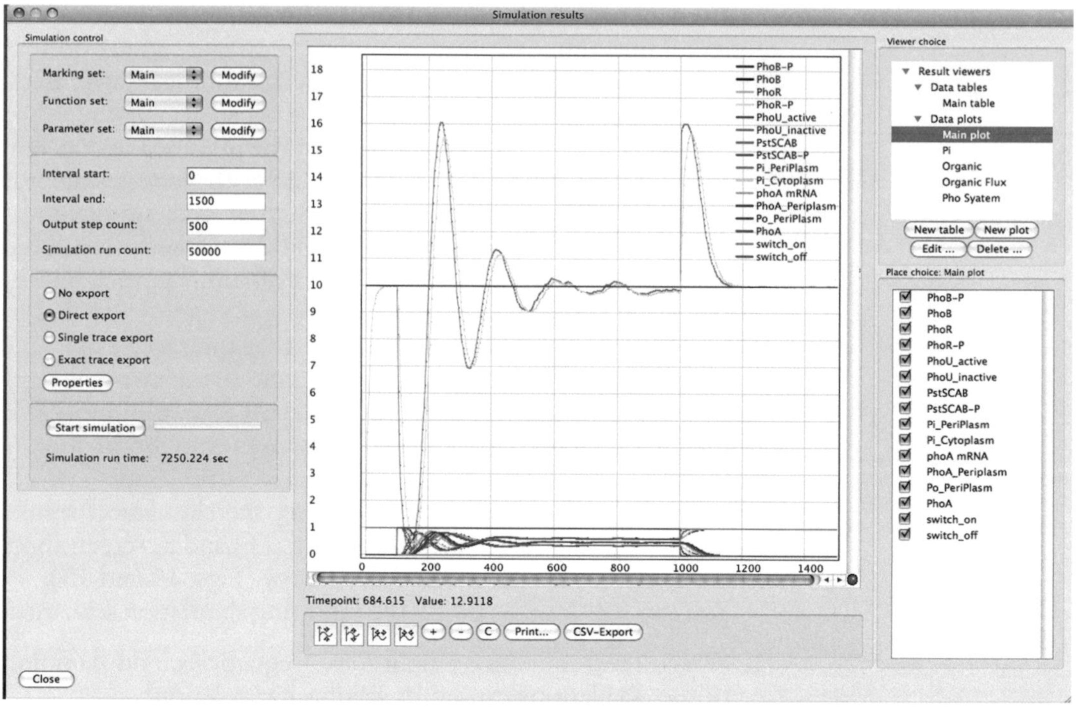

Fig. 12. Screen shot of the simulation dialogue window of Snoopy displaying simulation results of the Petri net shown in Fig. 10. Simulation run parameters: Interval start: 0; Interval end: 1,500; Output step count: 500; Simulation run count: 50,000. Snoopy is a modelling and simulation tool. Petri nets can be graphically edited, parameter and function lists can be edited and simulation runs and results can be graphically displayed without leaving the program.

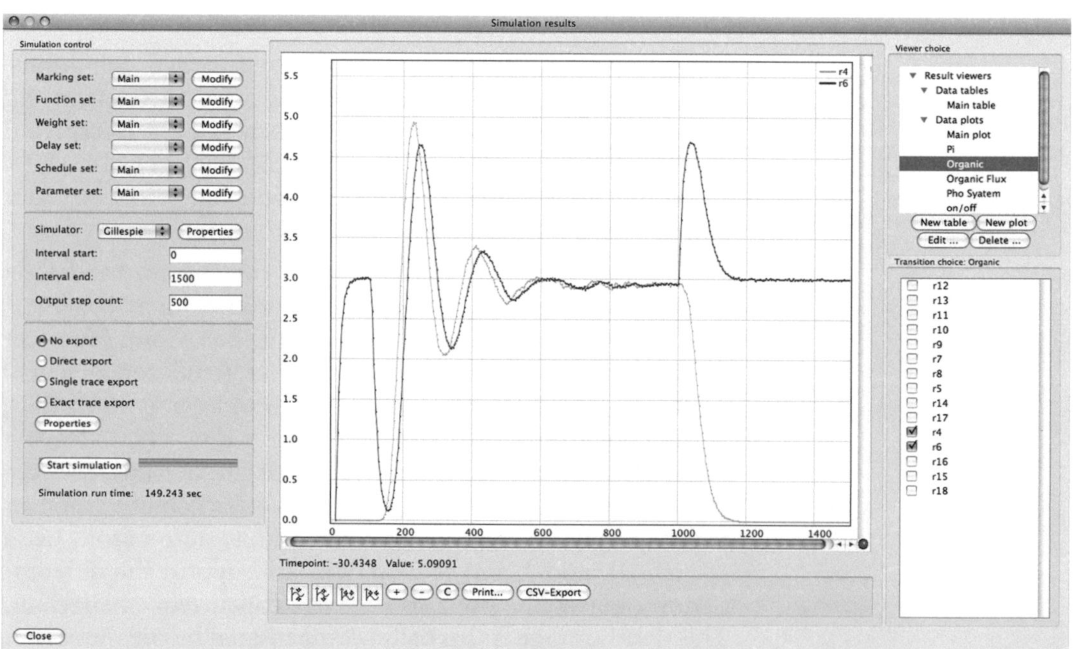

Fig. 13. Simulation dialogue showing transition firing times (amount of firing in the last grid interval).

sophisticated algorithms and advanced data structures which have been developed for their efficient analysis.

Nevertheless, often it is advisable to preserve all the concurrency information, which a biochemical network model enjoys, e.g. to distinguish the subtle difference between read arcs and two opposite arcs (Fig. 7). The partial order semantics yields a behaviour description comprising all (infinite) partial order runs. It is often given as labelled condition/event net. Here, transitions represent events, labelled by the name of the reaction taking place, while places stand for binary conditions, labelled by the name of the species, set or reset by the event, respectively. Partial order runs, which can be read as the unfolding of a net structure, give further insight into the dynamic behaviour of a network, which may not be apparent from the standard net representation. For examples see refs. 4, 6, where the behaviour of subnets induced by T-invariants is given as a partial order run.

3. Snoopy supports standard analysis techniques of Petri net theory, such as static and dynamic analyses, by export to various analysis tools, see (5). Additionally, Snoopy's Petri nets are directly read by the Petri net analyser Charlie (16). One of the standard static analysis techniques comprises the computation of place and transition invariants, which are often helpful for network validation, see e.g. (4, 6, 17–19), and (Behre et al. Chapter 20). Sets or multi-sets of nodes, such as place and transition invariants, Parikh vectors, structural deadlocks, or traps (see ref. 4 for explanations of these terms), which have been produced with Charlie, are read by Snoopy to visualise the sub-networks induced by them.

   The Petri net of our case study is covered by transition invariants. There are five place invariants, and such conservation laws, which are obvious in the given case: (switch_on, switch_off), (PstSCAB, PstSCAB-P), (PhoU_inactive, PhoU_active), (PhoR, PhoR-P), and (PhoB, PhoB-P). The following five places are not covered by place invariants: Pi_PeriPlasm, Po_PeriPlasm, Pi_CytoPlasm, PhoA_Periplasm, PhoA, phoA mRNA; their maximal token numbers are determined by the timing constraints.

4. An established analysis technique for special behavioural properties is model checking. It checks the Petri net behaviour against properties formally specified in temporal logics, see (Batt et al. Chapter 22). Charlie provides some standard explicit model checkers, basically for teaching purposes. Snoopy allows also to export the designed Petri nets to quite a number of model checkers, among them the symbolic model checker MARCIE (20, 21). Simulation traces generated by Snoopy's (stochastic or continuous) simulation engines can be checked against PLTLc properties with the

Monte Carlo Model Checker MC2 (22). *See* refs. 4, 6 for case studies demonstrating a systematic and seamless model checking approach in the qualitative, stochastic, and continuous modelling paradigms.

5. A given Petri net may also be read as a continuous Petri net, if it is amenable to continuization and the population semantics allows to consider just the averaged case. Continuous Petri nets define uniquely a system of ODEs (4), but not vice versa. The paper (23) demonstrates the structured design of ODEs by the step-wise composition of hierarchically structured continuous Petri nets. A family of related Petri net models builds the core of an integrative approach comprising qualitative, stochastic and continuous Petri nets, which is demonstrated by a running example each in ref. 4, 6. It also provides an adequate framework to reason about the behavioural relation of models, sharing structure, but having different kinetics; see, e.g., (24). For a general outline of BioModel Engineering see ref. 25.

Acknowledgement

Christian Rohr is funded by the International Max Planck Research School for Analysis, Design, and Optimisation in Chemical and Biochemical Process Engineering, Magdeburg.

References

1. Petri CA, Reisig W. (2008) Petri net. Scholarpedia, 3(4):6477 [http://www.scholarpedia.org/article/Petri_net].
2. Marwan W, Wagler A, Weismantel R. (2011) Petri nets as a framework for the reconstruction and analysis of signal transduction pathways and regulatory networks. J Nat Comput, 10(2):639–654.
3. Baldan P, Cocco N, Marin A, Simeoni M. (2010) Petri nets for modelling metabolic pathways: a survey. J Nat Comput, 955–989.
4. Heiner M, Gilbert D, Donaldson R. (2008) Petri nets in systems and synthetic biology. Lect Notes Comput Sci, 5016:215–264.
5. Rohr C, Marwan W, Heiner M. (2010) Snoopy – a unifying Petri net framework to investigate biomolecular networks. Bioinformatics, 26(7):974–975.
6. Heiner M, Donaldson R, Gilbert D. (2010) Petri Nets for Systems Biology. In *Symbolic Systems Biology: Theory and Methods, Chapter 3*. Edited by Iyengar MS, Jones & Bartlett Publishers, LLC.
7. Petri Nets World: Online Services for the International Petri Nets Community, [http://www.informatik.uni-hamburg.de/TGI/PetriNets/].
8. SBML Software Summary, [http://sbml.org/SBML_Software_Guide/SBML_Software_Summary].
9. Bitesize Bio – Free Online Bioinformatics tools, [http://bitesizebio.com/2009/01/06/free-online-bioinformatics-tools/].
10. Systems Biology – Software, [http://systems-biology.org/software/].
11. Tovchigrechko A. (2008) Efficient symbolic analysis of bounded Petri nets using Interval

Decision Diagrams. PhD thesis, BTU Cottbus, Germany.

12. Heiner M, Lehrack S, Gilbert D, Marwan W. (2009) Extended stochastic petri nets for model-based design of wet-lab experiments. Lect Notes Bioinformatics, 5750:138–163.
13. Gillespie DT. (1977) Exact stochastic simulation of coupled chemical reactions. J Phys Chem, 81(25):2340–2361.
14. Neidhardt FC, Ingraham JL et al. (1990) Physiology of the bacterial cell – a molecular approach. Sinauer Associates, Sunderland, 370.
15. Yi-Ju Hsieh Y-J, Wanner BL (2010) Global regulation by the seven-component Pi signaling system. Curr Opin Microbiol, 13:198–203.
16. Franzke A. (2009) Charlie 2.0 – a multithreaded Petri net analyser. Diploma thesis, BTU Cottbus, Germany.
17. Heiner M, Koch I. (2004) Petri net based system validation in systems biology. Lect Notes Comput Sci, 3099:216–237.
18. Palsson BO (2006) *Systems Biology: Properties of Reconstructed Networks*. Cambridge University Press, Cambridge.
19. Heiner M. (2009) Understanding Network Behaviour by Structured Representations of Transition Invariants – A Petri Net Perspective on Systems and Synthetic Biology. In *Algorithmic Bioprocesses*. Edited by Condon A, Harel D, Kok JN, Salomaa A, Winfree E. Springer, Natural Computing Series, 367–389.
20. Heiner M, Schwarick M, Tovchigrechko A. (2009) DSSZ-MC – a tool for symbolic analysis of extended petri nets. Lect Notes Comput Sci, 5606:323–332.
21. Schwarick M, Heiner M. (2009) CSL model checking of biochemical networks with Interval Decision Diagrams. Lect Notes Bioinformatics, 5688:296–312.
22. MC2(PLTLc) – Monte Carlo Model Checker for PLTLc properties, University of Glasgow, [http://www.brc.dcs.gla.ac.uk/software/mc2/].
23. Breitling R, Gilbert D, Heiner M, Orton R. (2008) A structured approach for the engineering of biochemical network models, illustrated for signalling pathways. Brief Bioinformatics, 9:404–421.
24. Heiner M, Sriram K. (2010) Structural analysis to determine the core of hypoxia response network. PLoS ONE, 5(1):e8600.
25. Breitling R, Donaldson RA, Gilbert D, Heiner M. (2010) Biomodel engineering – from structure to behavior. Lect Notes Bioinformatics, 5945:1–12.

Chapter 22

Genetic Network Analyzer: A Tool for the Qualitative Modeling and Simulation of Bacterial Regulatory Networks

Grégory Batt, Bruno Besson, Pierre-Emmanuel Ciron, Hidde de Jong, Estelle Dumas, Johannes Geiselmann, Regis Monte, Pedro T. Monteiro, Michel Page, François Rechenmann, and Delphine Ropers

Abstract

Genetic Network Analyzer (GNA) is a tool for the qualitative modeling and simulation of gene regulatory networks, based on so-called piecewise-linear differential equation models. We describe the use of this tool in the context of the modeling of bacterial regulatory networks, notably the network of global regulators controlling the adaptation of *Escherichia coli* to carbon starvation conditions. We show how the modeler, by means of GNA, can define a regulatory network, build a model of the network, determine the steady states of the system, perform a qualitative simulation of the network dynamics, and analyze the simulation results using model-checking tools. The example illustrates the interest of qualitative approaches for the analysis of the dynamics of bacterial regulatory networks.

Key words: Genetic Network Analyzer, Qualitative modeling and simulation, Gene regulatory networks, Piecewise-linear differential equations, Model checking, Carbon starvation, *Escherichia coli*

1. Introduction

1.1. Bacterial Regulatory Networks

In order to survive, in the short term as well as on an evolutionary time scale, bacteria constantly have to adapt to changing environmental conditions. For instance, under conditions of nutrient deprivation, the Gram positive soil bacterium *Bacillus subtilis* initiates a range of responses to deal with an increasingly hostile environment, such as the production of antibiotics, the development of motility and competence, and as a final resort, the formation of dormant, environmentally resistant spores. Even bacteria such as *Escherichia coli*, which do not differentiate into multiple cell types, dramatically modify their properties (membrane permeability, cell surface, metabolic flux distribution, etc.) in response to environmental stresses.

Jacques van Helden et al. (eds.), *Bacterial Molecular Networks: Methods and Protocols*, Methods in Molecular Biology, vol. 804, DOI 10.1007/978-1-61779-361-5_22, © Springer Science+Business Media, LLC 2012

Bacteria are able to monitor and respond to changes in their environment by adjusting the synthesis and degradation of regulatory proteins controlling growth and differentiation. These processes involve large and complex networks of interactions encompassing genes, mRNA, proteins, signaling molecules, and metabolites. For many model bacteria, such as the examples mentioned in the previous paragraph, decades of work in genetics, biochemistry, and molecular biology have led to detailed maps of the underlying regulatory networks. However, we still know very little about how the response of a bacterium to a particular environment emerges from the interactions between the molecular components of the network.

The study of bacterial regulatory networks obviously requires experimental tools for characterizing the individual interactions and measuring the dynamics of intracellular processes. However, experimental tools alone are not enough when dealing with such large and complex systems. When trying to predict the effect of the downregulation of a gene on the overall response of the system, intuitive reasoning alone may quickly lead one into error. We need mathematical modeling and computer simulation to integrate available biological data, and understand and predict the dynamics of the system under various environmental and physiological conditions.

1.2. Quantitative Versus Qualitative Models

In theory, it is possible to write down kinetic models of biochemical networks, and study these by means of classical analysis and simulation tools for ordinary differential equations (ODEs) (1, 2). In practice, this is not easy to achieve though, as kinetic parameters are often only constrained to within a range spanning several orders of magnitude for most systems of biological interest. Moreover, the models consist of a large number of variables, are strongly nonlinear, and include different time scales, which make them difficult to handle both mathematically and computationally. This has motivated the interest in *qualitative models* which, from incomplete knowledge of the system, are able to provide a coarse-grained picture of its dynamics (see also the chapters by Marwan et al., Chapter 21 and by Chaouiya et al., Chapter 23).

Qualitative models are based, explicitly or implicitly, on approximations of the rate functions found in kinetic models, which allow the latter to be replaced by easier-to-analyze, often discrete models. The approximations are justified by the assumption that important dynamical properties of living systems are not expected to depend on precise numerical values or a specific reaction mechanism. In many situations such qualitative dynamical properties, invariant for a range of reaction mechanisms and values of kinetic constants, are more interesting than quantitative dynamical properties. The qualitative properties express the intimate connection between the behavior of the system and the structure of the regulatory network, independently of the quantitative details of the latter. The modeler must assure that the approximations are applicable in the situation at hand, however, to assure that the predictions from the qualitative model are valid.

A variety of qualitative models for regulatory networks have been proposed in the literature (3–5), such as logical models (6), Petri nets (7), constraint-based models (8), and process algebras (9). Some of these formalisms are described in detail in other chapters of this volume. The computer tool described below is based on the use of *piecewise-linear differential equations*, a formalism originally introduced by Leon Glass and Stuart Kauffman for describing gene regulatory networks (10).

1.3. Piecewise-Linear Models of Gene Regulation

The piecewise-linear (PL) models are differential equation models in which the variables, in our context, denote concentrations of gene products, typically proteins. The rate of change of a concentration at a particular time point is described as the balance of the rates of protein synthesis and degradation. These synthesis and degradation rates are regulated, directly or indirectly, by other proteins, for instance through the activation or inhibition of transcription. The PL models capture the regulatory effects by means of step functions that change their value in a switch-like manner at threshold concentrations of the regulatory proteins. The step functions are approximations of the sigmoidal response functions often occurring in gene regulation. They model the effect of a change in the concentration of a regulatory protein, such as a transcription factor, on the promoter activity of a gene.

The dynamical properties of PL models have been the subject of active research for more than three decades (e.g., (10–18)). The use of step functions allows the qualitative dynamics of the PL models to be analyzed, even in higher-dimensional systems. The thresholds of the concentration variables define a hyperrectangular partition of the state space, such that in every region not located on a threshold, the PL model reduces to an analytically solvable system of differential equations. Moreover, in every such region the derivatives (trends) of the concentration variables have a determinate sign, which is shown to be invariant for rather weak constraints on the parameter values. The definition of these constraints can generally be inferred from available data in the experimental literature or by intuitive reasoning, even in the absence of quantitative information on parameter values.

The properties of the PL models motivate a discrete representation of the continuous dynamics, described by a *state transition graph*. The states in the graph are called *qualitative states* and correspond to the regions in the state space. The transitions between these qualitative states arise from solutions of the PL model that enter one region from another. Each transition thus corresponds to a discrete event, namely the crossing of a threshold by one or more concentration variables, possibly entailing a change in the derivative of these variables.

Genetic Network Analyzer (GNA) is a computer tool specifically developed for the analysis of gene regulatory networks by means of PL models (19–21). It allows the user to define a regulatory network, build a model of this network, determine the steady states of the system, generate a state transition graph starting from an initial state (qualitative simulation of the network dynamics), and analyze the latter graph using model-checking tools. GNA has been used for the analysis of a number of bacterial regulatory networks (see the GNA Web site for examples). In this chapter we specifically focus on the carbon starvation response in *Escherichia coli* (22).

1.4. Piecewise-Linear Models of E. coli Carbon Starvation Response

Glucose is the preferred carbon source of *E. coli*. The adaptation of the bacteria to the depletion of glucose from the growth medium is under the control of a large and complex regulatory network. The role of the metabolic and signaling networks in the adaptation of *E. coli* to carbon source starvation has been extensively studied, but much less has been done at the level of gene expression. In particular, it is not well understood how the backbone of the network, formed by the global regulators of transcription, coordinates the cascades of molecular events driving the adaptation of *E. coli* cells starved for glucose. These transcription factors respond directly or indirectly to glucose depletion, by controlling in a combinatorial fashion the expression of a large number of genes involved in cellular adaptation and survival. In addition, they control each other's expression, thus giving rise to a complex regulatory network.

Figure 1 shows the network of key global regulators involved in the control of the carbon starvation response. It includes well-known pleiotropic transcription regulators, such as the histone-like protein Fis, the catabolic repressor CRP.cAMP (resulting from the expression of genes *crp* and *cya*, and the activation of Cya by carbon depletion), and the general stress response factor RpoS or σ^S (whose stability is regulated by RssB). Changes in DNA topology and its dependence on the relative expression level of the genes *gyrA*, *gyrB*, *gyrI*, and *topA* are also considered, as the three-dimensional structure of DNA modulates the transcription of a large number of genes. Finally, stable RNAs expressed from the *rrn* operons are included as their amount provides a reliable indicator of the growth rate of the cell (high during exponential phase and low during stationary phase).

A PL model of the above network has been developed, as described in refs. 21, 23, building upon an earlier model in ref. 22. The qualitative dynamics of the network has been analyzed by means of GNA, leading to novel insights into the role of some of the feedback loops in the response of the cell to carbon starvation.

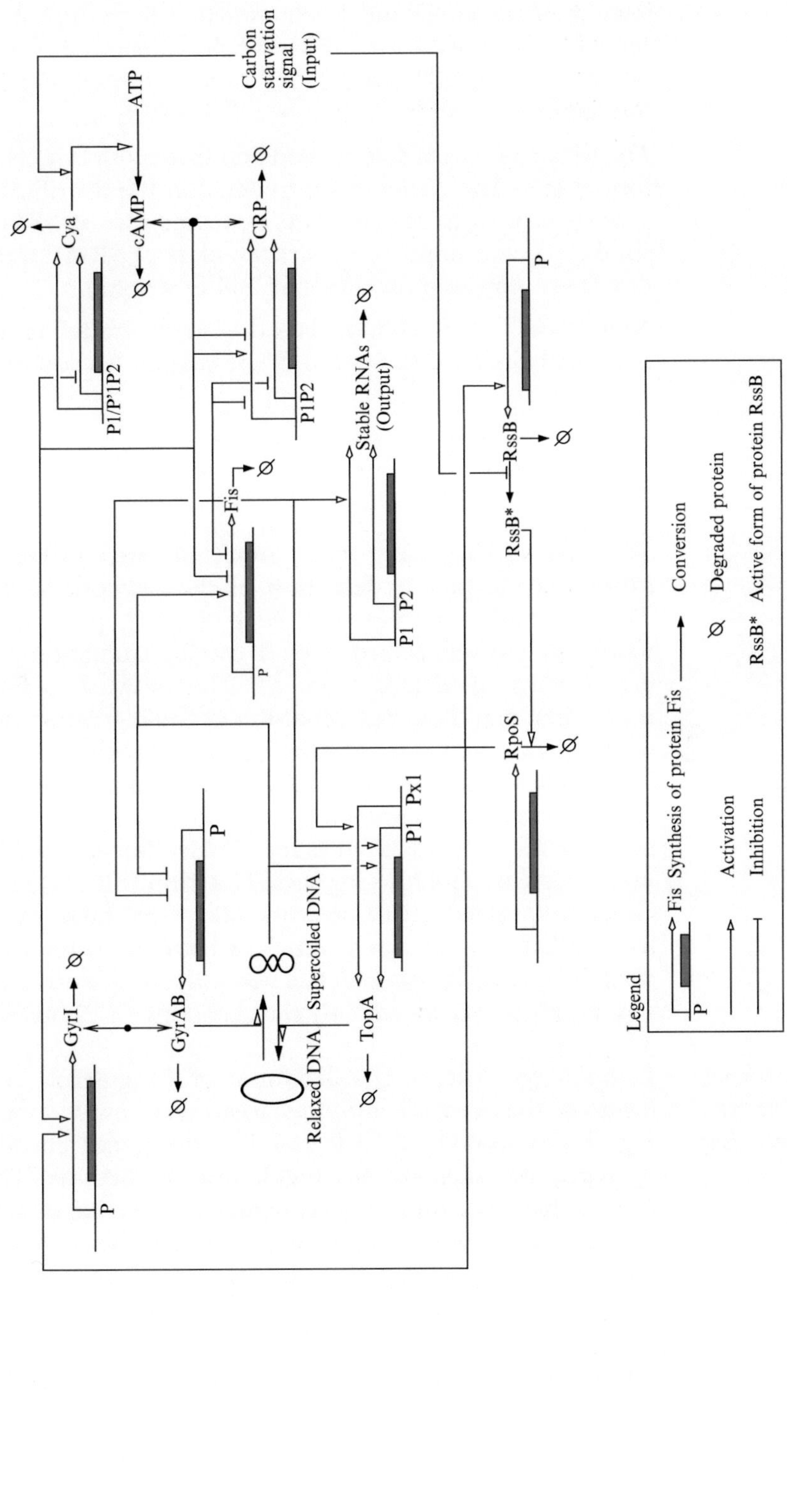

Fig. 1. Network of global regulators involved in the carbon starvation response in *E. coli* [22, 23]. The graphical conventions are explained in the legend.

2. Material

Software GNA is distributed by the company Genostar (http://www.genostar.com) and freely available for nonprofit academic research purposes at http://www-helix.inrialpes.fr/gna. The current version of GNA is 8.1. An online tutorial and installation instructions are available from the GNA Web site.

File formats GNA is able to load and save models in its own XML format (GNAML), described in detail in the distribution. It also exports models in the standard exchange format SBML (http://sbml.org), and imports a restricted class of SBML models (with rate laws using step functions or Hill functions).

Sample files The *E. coli* model used as an illustration in this chapter is available in the distribution of the current version of GNA.

3. Methods

3.1. Defining the Network

The first and arguably most important step in the modeling process is the precise definition of the network to be studied. This demands reflection on the biological questions that one wishes to answer, followed by a careful evaluation of available experimental data (see Note 1). The network definition step results in a graphical representation of the regulatory network, in which genes, proteins, metabolites are specified, as well as the biochemical reactions and regulatory interactions in which they are involved. Several graphical conventions exist for the representation of the network, such as the visualization standard proposed in the Systems Biology Graphical Notation (SBGN) project (24). GNA is equipped with a network editor that takes its inspiration from MIM and adheres as much as possible to the SBGN standard. The modeler can edit the network and reduce it to a representation that is conveniently translated into a PL model.

3.1.1. Defining the Network of Biochemical Reactions and Regulatory Interactions

In order to illustrate the definition of a regulatory network by means of the network editor, we focus on a simple subnetwork of Fig. 1, composed of CRP and Fis, the genes encoding these proteins, and their mutual regulatory interactions. The protein CRP is the major transcription regulator in *E. coli*, controlling the expression of hundreds of genes involved in the adaptive response of the cell to nutrient deprivation as well as in changes of cellular morphology and motility (25). The expression of the gene *crp* is repressed by the protein Fis, although the gene is also expressed at a basal level. Additionally, Fis controls the expression of genes

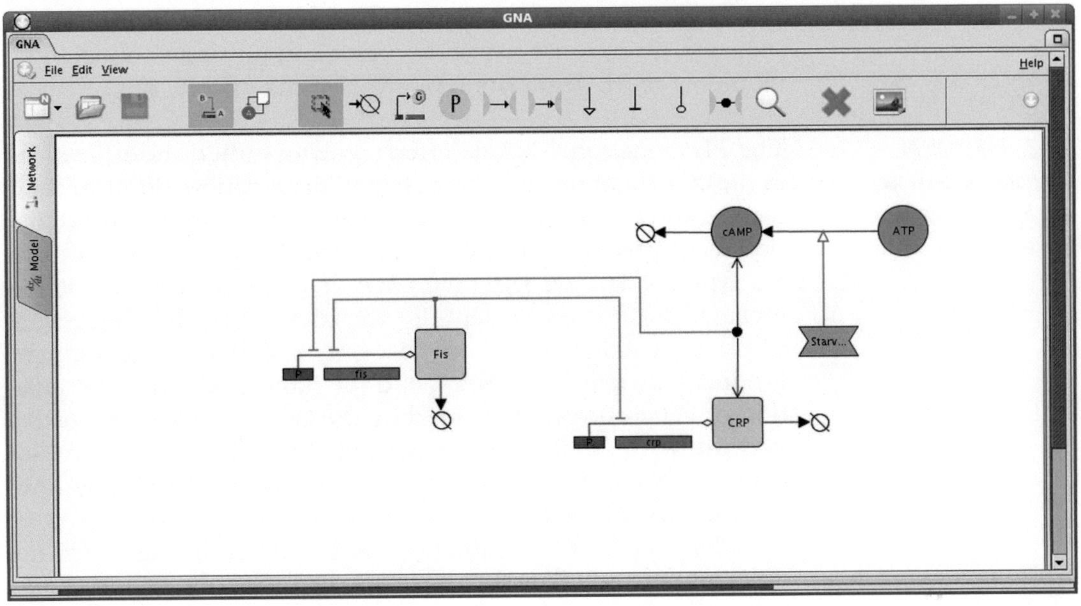

Fig. 2. Subnetwork of the network of global regulators in Fig. 1, defined by means of the network editor of GNA. The subnetwork involves Fis, CRP, and a carbon starvation signal that activates the synthesis of cAMP from ATP. CRP needs to be bound to cAMP to efficiently bind to DNA and thus affect the expression of its targets.

involved in the cellular metabolism (26). For instance, it activates the *rrn* operons, which encode stable RNAs required for protein synthesis. The expression of *fis* is controlled at the transcriptional level, where it is produced from a promoter repressed by CRP and Fis itself. It is known that to efficiently bind DNA, CRP has to be activated by the small metabolite cAMP. The latter is produced from ATP in response to a nutritional stress signal indicating the absence of carbon compounds such as glucose.

The network can be created from the elements shown in the toolbar of GNA, among which we find genes and promoters, proteins, complex formation reactions, degradation reactions, and activating and inhibiting regulatory interactions. The user can select network elements and specify or change their names, as well as define reactions and regulatory interactions involving the elements. GNA automatically positions the network elements, but the user can modify the layout by dragging selected elements to another location, and by adding or removing breakpoints in the lines representing the reactions and interactions. Figure 2 shows the CRP-Fis network mentioned above. It involves biochemical reactions (the conversion of ATP to cAMP), regulation of biochemical reactions (the stimulation of the production of cAMP by a carbon starvation signal), complex formation reactions (the binding of cAMP to CRP), transcriptional regulation (e.g., the inhibition of *fis* by CRP.cAMP), and degradation reactions (e.g., the degradation of CRP and Fis). In the example, only a limited number of biochemical mechanisms are used, but the network editor also supports, for instance, enzymatic reactions, posttranslational modifications, active degradation of proteins and mRNA, etc.

3.1.2. Reduction of the Network to Direct and Indirect Regulatory Interactions

The PL models in GNA describe regulatory interactions involved in the control of protein synthesis and degradation. Some of these may be direct, like the control of gene expression by a transcription factor, while others are indirect, mediated by processes that are typically fast on the time scale of gene expression, such as metabolism. It is convenient, for the later model-building stages, to obtain a reduced network that only includes the genes, gene products (proteins or RNAs), and the direct and indirect regulation of gene expression exerted by the latter. GNA allows such a network to be automatically derived from the network of biochemical reactions and regulatory interactions previously constructed (see Note 2).

The results of the reduction are illustrated in Fig. 3 for the CRP-Fis network. The fast processes mediating the indirect interactions, such as complex formation reactions and metabolic reactions in the CRP-Fis network, are represented as a box. It makes explicit that CRP and the carbon starvation signal together control the expression of *fis*, while hiding the details of the fast processes mediating this effect.

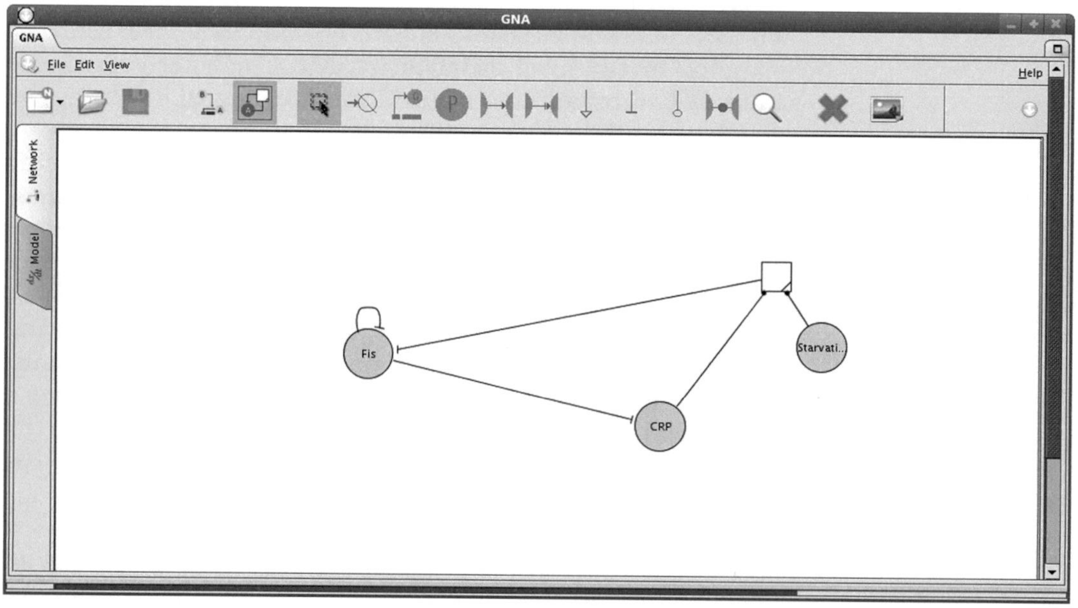

Fig. 3. Reduced network, describing the direct and indirect interactions involved in the control of the expression of *fis* and *crp*. The reduced network is obtained from the CRP-Fis network shown in Fig. 2.

3.2. Building the Model

In the second step of the modeling process, the graphical representation of the network is transformed into a PL model, by adding more information on the regulation of the genes. Several types of information need to be specified in GNA: the variables and parameters of the system, the equations describing the regulatory logic, and parameter inequalities constraining the qualitative dynamics.

3.2.1. Specifying Variables and Parameters

The variables in the model represent the concentrations of the gene products or external signals. Two different types of variables are distinguished in GNA: *state variables* and *input variables.* While the former vary as a function of the other concentration variables in the model, the latter are determined by factors outside the scope of the system. In addition, GNA assumes that they are constant. In the case of the CRP-Fis network, there are two state variables, *CRP* and *Fis*, and one input variable representing the carbon starvation signal, *signal.*

For each variable, a number of parameters need to be defined. A *threshold parameter* indicates the concentration above or below which the regulatory protein associated with the variable has an influence on the expression of a target gene. As a protein may be involved in several regulatory interactions, it may have several different thresholds. For *CRP*, we distinguish a single threshold t_c, corresponding to the repression of *fis*, and for *Fis* two thresholds t_f^1 and t_f^2, corresponding to the repression of *crp* and *fis* itself, respectively. For the input variable *signal* we define a threshold t_s, such that values of the input above this threshold indicate a situation of carbon starvation. GNA additionally defines for each variable a zero and a maximum value.

Finally, if the variable is a state variable, we need to specify so-called *synthesis and degradation parameters.* These define the rate of synthesis and degradation, respectively, of the protein whose concentration is represented by the variable. Because there must always be at least one degradation parameter, GNA creates one by default. In the example, we introduce two synthesis parameters k_c^0 and k_c^1 for CRP, which is produced at a basal rate independently of the regulatory state of the cell, and at a high rate in the absence of Fis. We also add a synthesis parameter k_f for Fis. To complete this step, we define the degradation parameters g_c and g_f for CRP and Fis, respectively.

3.2.2. Defining Model Equations

The actual PL model is defined in the next step, by specifying the regulatory logic of protein synthesis and degradation for each of the state variables. This results in a so-called *state equation*, a differential equation consisting of two parts: a synthesis term and a degradation term.

The synthesis term of the state equation is a sum of step-function expressions, each weighted by a synthesis parameter.

In the case of CRP, we obtain the synthesis term $k_c^0 + k_c^1 s^-(Fis, t_f^1)$, where $s^-(Fis, t_f^1)$ is a step function evaluating to 1 if $Fis < t_f^1$, and to 0 if $Fis > t_f^1$. That is, at a level of Fis below its threshold concentration t_f^1, the gene *crp* is maximally expressed: its product CRP is synthesized at the maximum rate $k_c^0 + k_c^1$. If the concentration of Fis exceeds the threshold t_f^1, *crp* is expressed at its basal rate k_c^0.

The degradation term is defined analogously to the synthesis term, except that we demand that an unregulated, spontaneous degradation term is included, to ensure that the degradation rate of the protein is always positive. If we only have spontaneous degradation or growth dilution, as in the CRP-Fis example, then the degradation term is simply a degradation parameter multiplied by the state variable (e.g., g_c CRP). If the degradation of the protein is also regulated by other proteins, one or more step-function expressions, each weighted by a degradation parameter, are added to the degradation parameter accounting for spontaneous degradation. The resulting sum is multiplied by the state variable.

In general, the step-function expressions in the synthesis and the degradation terms become more complex than for the example of CRP, especially when the number of proteins involved in the regulation of a gene increases. The case of Fis will illustrate some of the reasoning that goes into defining the step-function expressions (see also Note 3). We want to state in the model that the gene *fis* is repressed if (1) CRP is above its threshold concentration under carbon starvation conditions (when cAMP is produced from ATP at high levels and binds to CRP to activate the latter), and (2) the concentration of Fis is below the threshold at which autoregulation sets in. Otherwise, the gene is expressed. The following step-function expression captures the regulatory logic: $k_f(1 - s^+(CRP, t_c)s^+(signal, t_s))s^-(Fis, t_f^2)$. It stipulates that, when Fis is below its threshold ($s^-(Fis, t_f^2) = 1$), if either the concentration of CRP falls below its threshold t_c ($s^+(CRP, t_c) = 0$) or the carbon starvation signal is absent ($s^+(signal, t_s) = 0$), then Fis is synthesized at a rate k_f. Otherwise, the gene is repressed and no protein synthesis occurs. Notice that in the model we state the conditions under which the gene is expressed (and thus not repressed)!

By combining the above synthesis and degradation terms, we obtain the following state equations for the CRP-Fis network:

$$dCRP/dt = k_c^0 + k_c^1 s^-(Fis, t_f^1) - g_c CRP \tag{1}$$

$$dFis/dt = k_f\,(1 - s^+(CRP, t_c)\,s^+(signal, t_s))\,s^-(Fis, t_f^2) - g_f\,Fis \tag{2}$$

3.2.3. Defining Parameter Inequality Constraints

GNA does not require precise numerical values for the threshold, synthesis, and degradation parameters. Instead, inequality constraints on the parameter values need to be specified to predict the qualitative dynamics of the regulatory network. More precisely, GNA demands the user to order the threshold parameters and so-called *focal parameters.* We briefly explain the biological meaning of these parameters and the ordering of the parameters by means of the example.

Recall that we defined two thresholds for Fis: one for the repression of *crp* (t_f^1) and one for the repression of its own gene (t_f^2). In fact, Fis influences *crp* at lower concentrations than its own gene, which gives rise to the inequality $t_f^1 < t_f^2$. The thresholds divide the state space into regions in which the step functions evaluate to either 0 or 1, so that the state equations there reduce to linear and uncoupled differential equations. Inside a region, each protein concentration monotonically converges toward a focal parameter, given by a quotient of synthesis and degradation parameters of the protein (10, 11). The state equation for CRP, for instance, implies that in regions where *Fis* is below t_f^1, *CRP* converges toward the focal parameter $(k_c^0 + k_c^1)/g_c$.

The possible focal parameters in the different regions of the state space can be ordered with respect to the threshold parameters. Intuitively speaking, this defines the strength of gene expression in the regions in a qualitative way, on the scale of ordered threshold concentrations. In the case of variable *CRP*, the possible focal parameters are 0, k_c^0/g_c, and $(k_c^0 + k_c^1)/g_c$. At the highest expression level, corresponding to the focal parameter $(k_c^0 + k_c^1)/g_c$, CRP should be able to repress Fis. This implies that $t_c < (k_c^0 + k_c^1)/g_c$. In addition, we set $t_c < k_c^0/g_c$, to account for the observation that even in situations in which *crp* is weakly expressed, the concentration of CRP is nevertheless high enough to allow repression of *fis* when carbon starvation sets in. From Eq. 2 we infer that this requires CRP to accumulate above t_c, and thus $t_c < k_c^0/g_c$ to hold.

The definition of the parameter inequalities is an important step in the modeling process, as the choices made at this stage strongly influence the analysis results. It may also require some creativity to translate the biological data into mathematical form. Note 4 gives some heuristics that often allow many of the possible orderings to be eliminated. In Fig. 4 a screenshot of GNA shows the results of the modeling process for the variable *Fis*.

3.2.4. Semi-automated Model-Building

The above model-building steps have also been partially automated in a wizard, which transforms the (reduced) network into a PL model by making default choices for variables and parameters, for the regulators of a given state variable, and for the associated regulatory logic (see Note 3). Even if the wizard is

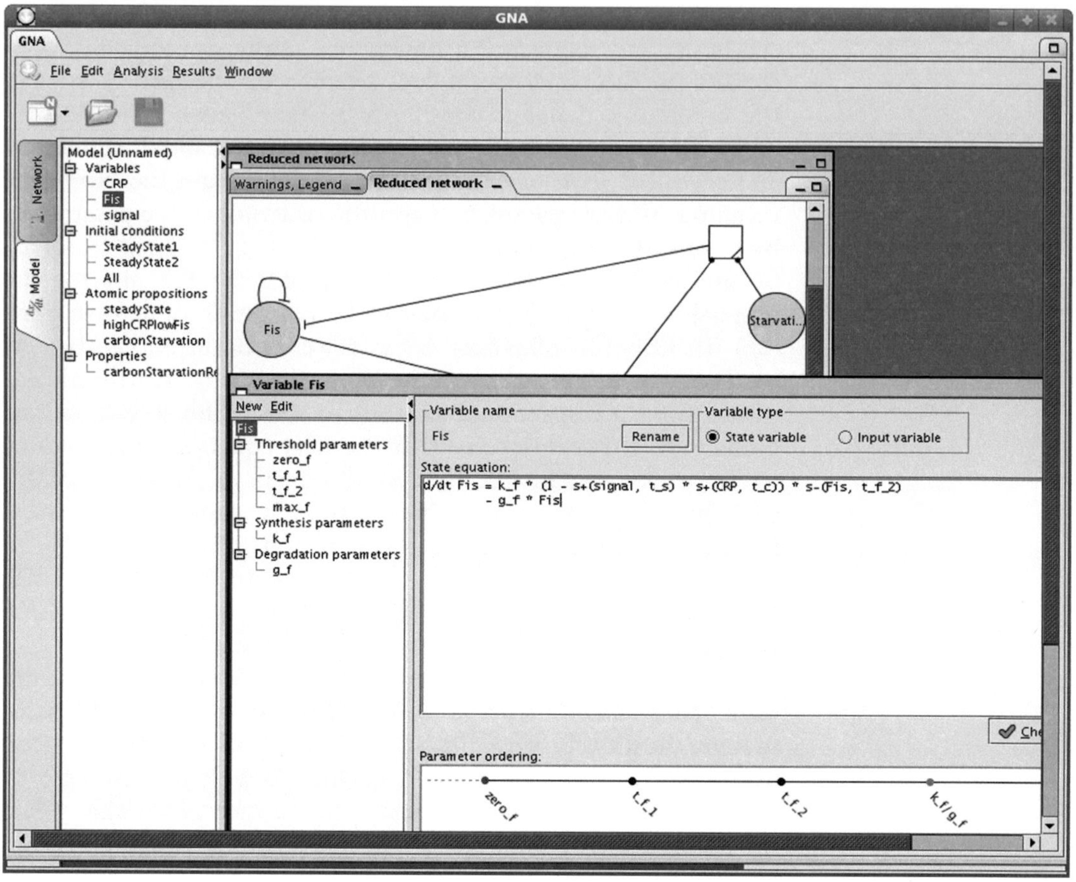

Fig. 4. PL model of CRP-Fis network shown in Fig. 2. The screenshot shows the state equation and parameter inequalities for the variable *Fis*.

capable of dealing with only a restricted range of regulating mechanisms, and may thus not give correct results for complex networks, it is usually worth the effort to run it to prepare a basic – perhaps partial – model that will facilitate and speed up the rest of the modeling process.

3.3. Searching for Steady States

A first insight into the dynamics of the network can be obtained by locating the steady states. These states, in which the derivatives of the concentration variables equal 0, indicate possible functional modes of the system. In general, finding the steady states of large nonlinear ODE models is difficult to achieve other than by exhaustive simulation. However, the special form of the PL models guarantees that in the steady states either all variables equal a focal parameter, or one or more variables are equal to a threshold parameter (see Note 5). Moreover, the inequality constraints are sufficient for efficiently locating all steady states and for determining the stability of some of these (12, 20).

Steady-state analysis by means of GNA identifies two stable steady states for the CRP-Fis network. In the first steady state, the carbon starvation signal is present, and we have $CRP = (k_c^0 + k_c^1)/g_c$ and $Fis = 0$. In the second steady state, the carbon starvation signal is absent, and $CRP = k_c^0/g_c$ and $Fis = t_f^2$. Depending on the growth conditions, the presence or absence of glucose in the medium, the bacteria thus reach one of two opposite gene expression states. The high concentration of CRP under carbon starvation is expected (27), as is the high concentration of Fis under favorable growth conditions (28).

3.4. Simulating the Network

In order to study the dynamics of the regulatory network, GNA allows the user to study the state transition graph, consisting of qualitative states and transitions between qualitative states. The qualitative states correspond to regions in the state space defined by the threshold and focal parameters in the PL model. In each of these regions, the system can be shown to behave in a qualitatively homogeneous way, in the sense that the concentration variables have unique derivative signs (11, 29). A transition between qualitative states occurs when solutions of the PL model reach one region from another.

In practice, we are mostly interested in the subgraph of the state transition graph that is reachable from a set of initial qualitative states. We call the generation of this subgraph, which represents the alternative qualitative behaviors of the system from the given initial qualitative states, a *qualitative simulation* (30). GNA uses the PL model and the inequality constraints to perform a qualitative simulation, and marks the states in the graph that correspond to steady states of the underlying PL model. In addition, it finds the cycles in the graph that may correspond to limit cycles (13, 14, 31). The graphical user interface of GNA has a variety of functionalities to explore the information contained in this graph, such as the selection of paths in the graphs and the visualization of the information associated with each qualitative state, notably the threshold bounds on the variables and the derivative signs.

Figure 5 shows the results of the qualitative simulation of the CRP-Fis network from initial conditions in which CRP and Fis assume the steady-state values $CRP = k_c^0/g_c$ and $Fis = t_f^2$, respectively, but in which the carbon starvation signal is set to $signal > t_s$. These are the initial conditions for studying the response of growing and dividing *E. coli* cells to the onset of carbon starvation. We observe that in this case the graph reduces to a single path, ending up in the other steady state. The system thus switches between the steady states in response to the external signal. The layout of this graph is automatically generated by GNA, but can be changed by the user. Figure 5 also displays how the concentration

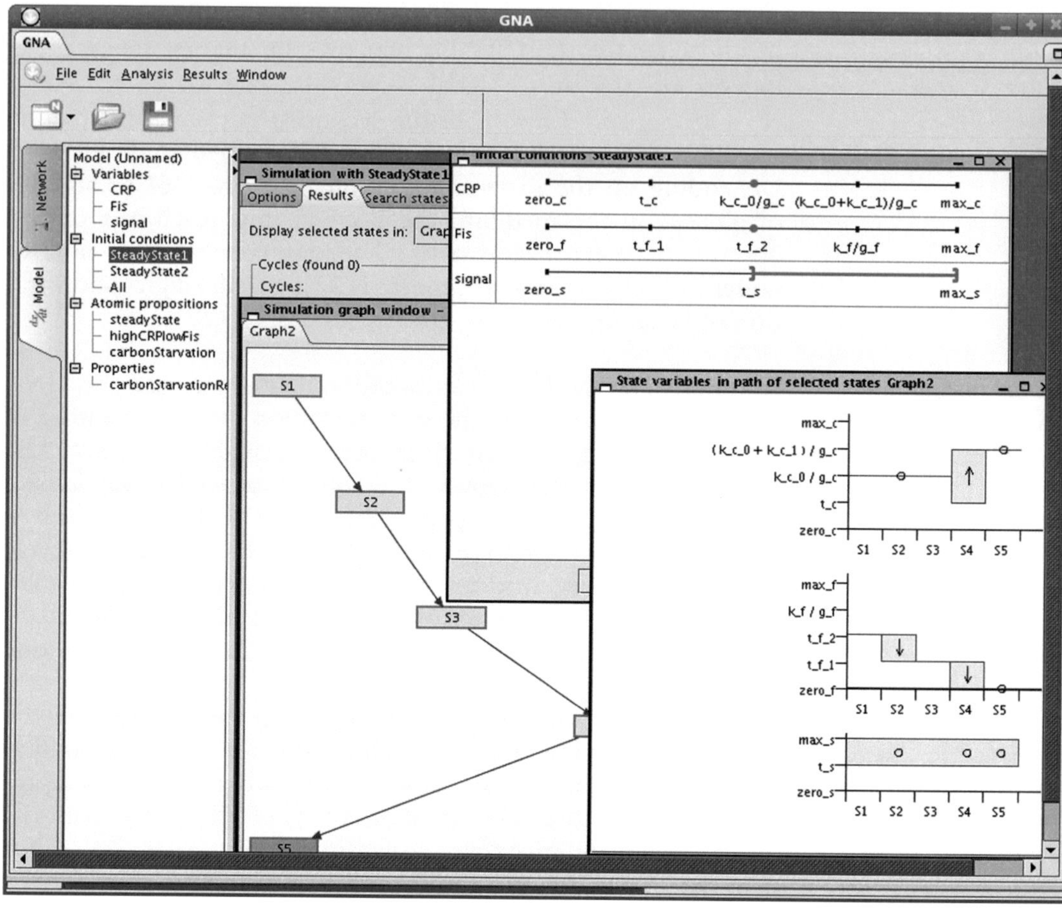

Fig. 5. Qualitative simulation of the CRP-Fis network using the PL model in Fig. 4. The screenshot shows the state transition graph obtained by simulating the response to carbon starvation and the qualitative evolution of the concentration variables along a sample path in the graph.

variables and their derivatives evolve along the path. As can be seen, the concentration of Fis decreases, whereas that of CRP increases.

3.5. Analyzing and Verifying Network Properties

For small networks like the CRP-Fis network it is possible to analyze the state transition graph by hand, using the graphical user interface of GNA. However, the graphs obtained from the qualitative simulation of the carbon starvation network in Fig. 1 consist of hundreds or thousands of states. As a consequence, manual inspection of the graph does not allow one to gain an intuitive understanding of the role of particular feedback mechanisms in the adaptation of the cell to nutritional stress conditions, such as the mutual inhibition of Fis and CRP. The identification of interesting predictions from the model, for instance in the case of a specific mutant or physiological perturbation, is also compromised by the large size of the graphs.

Methods from the field of formal verification provide a promising way to deal with the analysis of large and complex models of genetic regulatory networks. Generally speaking, formal verification proceeds by specifying dynamical properties of interest as statements in temporal logic. Efficient model-checking algorithms exist to determine whether the statements are true or false, and thus whether the properties are satisfied by the model (32). GNA has been extended with a model-checking functionality that allows the user to formulate properties in temporal logic and test these on a state transition graph, associated with a particular model and initial conditions (33, 34). The actual verification of the property takes place through a Web-service based connection of GNA with standard model checkers, like NuSMV (35). However, the user can also export the state transition graph to a text file in a format that is accepted by these model checkers. In order to analyze the state transition graph by means of these tools, the user needs to specify the dynamical properties to be analyzed and launch the model checker.

3.5.1. Specifying Network Properties in Temporal Logic

We first define so-called *atomic propositions*, which are used to describe qualitative states. They indicate, for each variable in the model, its value, the value toward which it converges, and the sign of its derivative. In addition, they indicate, for instance, if the state is a steady state. For the CRP-Fis network, we create atomic propositions corresponding to (1) a steady state, (2) a low concentration of Fis and a high concentration of CRP ($Fis < t_f^1$ and $CRP > t_c$), and (3) carbon starvation ($signal > t_s$). We call these atomic propositions *steadyState*, *highCRPlowFis*, and *carbonStarvation*, respectively.

The atomic propositions are included in *temporal-logic properties*, defined by means of the property editor of GNA. There are basically two different ways to formulate a property. The first is a pattern-based wizard that allows you to formulate a property with limited knowledge of temporal logic. The second is a free-text interface that provides all the flexibility of the temporal logics supported by GNA, notably *Computation Tree Logic (CTL)* and *Computation Tree Regular Logic (CTRL)*. A pattern is a high-level query template that captures recurring questions and can be automatically translated into temporal logic. Four different patterns, each with some variants, are available to the user. Together they cover most of the frequently-asked questions of modelers (21). Once the pattern has been completed, it is automatically translated into temporal logic. This is illustrated in Fig. 6 for the property "If the CRP-Fis network is in the state corresponding to carbon starvation, then it necessarily reaches a state with high CRP and low Fis concentrations". The property is defined as a

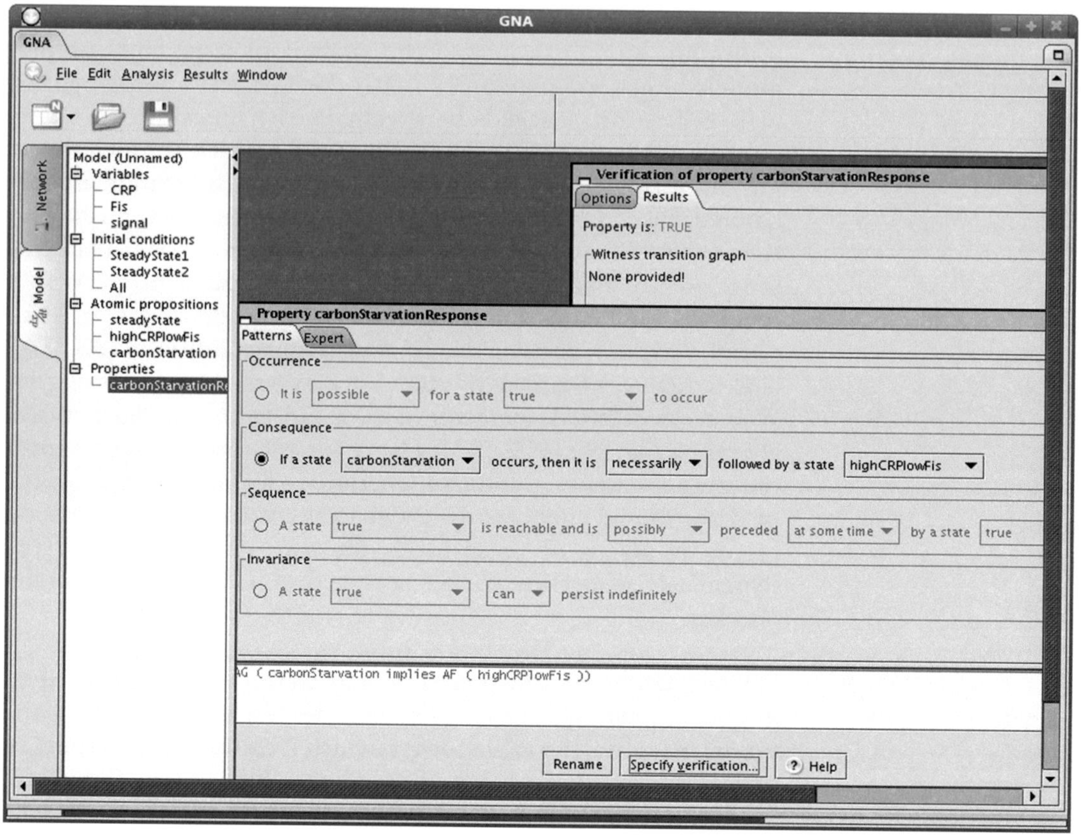

Fig. 6. Verification of the property "If the CRP-Fis network is in the state corresponding to carbon starvation, then it necessarily reaches a state with high CRP and low Fis concentrations", by means of the model checker NuSMV. The property is satisfied by the PL model of the CRP-Fis network.

Consequence pattern (see Note 6), using the previously-defined atomic propositions *highCRPlowFis* and *carbonStarvation*. The corresponding temporal-logic formula reads:

$$AG\,(carbonStarvation \rightarrow AF\,(highCRPlowFis)) \qquad (3)$$

3.5.2. Verifying the Properties by Means of Model Checking

Once the temporal logic formula has been completely specified, we can check its validity by means of the model-checking server. GNA is able to interact with several model checkers. Within the verification module of the program, we have to define which model checker is to be called. GNA uses a priori defined default settings for the model checkers, but the modeler can change some of these (see Note 7). In addition, we have to specify if the verification of the property is to be carried out on an *explicit* or *implicit graph*. An explicit graph is the state transition graph previously obtained by running a qualitative simulation. If the graph is too big, it is not possible to use the model-checker Web server. The implicit graph is a specific model description, containing all the necessary rules for the generation of the successors states by the model checker. This avoids having to run a simulation in GNA and ensures that a smaller file is sent through the Web-server connection.

Once the verification is completed, a result is returned by the model-checker server. The result consists of two elements: a verdict (true/false/error) and a counterexample (if the verdict is false). The counterexample represents a path that contradicts the property. Some model checkers also return a witness (if the verdict is true). Counterexamples and witnesses can be analyzed in the graphical user interface, like any other state transition graph.

The above-mentioned property is true for the CRP-Fis network, as tested on an implicit representation of the state transition graph by means of NuSMV (Fig. 6). The test of the properties in the large networks takes a few seconds to complete, while manual analysis of the graph would not have been possible, as it consists of several thousands of states. Interestingly, the truth of this property is preserved in the PL model of the entire carbon starvation network in Fig. 1.

3.6. Saving and Exchanging the Modeling Work

GNA allows the complete modeling project to be saved in a single file, comprising the regulatory network and its layout, the PL model, various initial conditions, the atomic propositions, and the temporal-logic properties. This file has the GNAML format (Subheading 2). The PL model can also be stored as an SBML file, which makes it possible to load the model in numerical modeling and simulation tools, such as Copasi (36), or in general-purpose mathematical software, such as MATLAB, and thus refine the qualitative analysis started in GNA. All intermediate simulation results in GNA can be stored as pictures in PNG or JPEG format.

4. Notes

1. *Network definition*. There is a temptation to start the modeling process with a network that is as complete as possible, taking into account all regulators known to play a role in the phenomenon under study. In many situations, it is advisable to start exactly the opposite way, focusing on what are perceived, in the literature or by the modeler, as the key regulators. This allows one to start with a small and simple model, the analysis of which may either provide a rough indication of the network dynamics underlying the observed phenomena or single out parts of the network that need to be extended to account for particular observations. It is instructive to bear in mind that some of the largest and most complex models in the literature have grown out of small and simple models, such as the cell-cycle models of Tyson and Novak (compare ref. 37 with ref. 38).

2. *Network reduction*. Two approaches can be followed to identify indirect interactions mediated by metabolism and signal transduction in the reduced network: an intuitive approach and a principled approach.

   The intuitive approach, currently implemented in GNA, simply lists for every regulator of gene expression which gene products or external signals compose it or are involved in the modification of its activity. In the CRP-Fis network we thus find for the effective regulator CRP.cAMP the protein CRP and the carbon starvation signal. Together these are involved in an indirect interaction regulating the expression of *crp*. In the extended network of Fig. 1, we would also include the enzyme Cya in this indirect interaction, as it catalyzes the conversion of ATP into cAMP.

   In the principled approach we place the reduction problem in the context of the kinetic modeling of biochemical reactions (2). We build a kinetic model of the entire system of biochemical reactions in the network, we reformulate the model so as to separate the fast and slow variables, and we apply quasi-steady-state approximations to the fast part of the system. There is a rich literature on this subject, applied to genetic, metabolic, and signal transduction networks (see ref. 2 for an accessible introduction). A complete development of this approach for the *E. coli* network in Fig. 1 can be found in ref. 23.

3. *Definition of step-function expressions*. The step-function expressions used in the PL models account for the regulatory logic of protein synthesis and degradation. They can be thought of as expressing, for example, the conditions under which a gene is expressed or, to formulate it differently, not repressed. The step functions are all-or-none approximations of sigmoidal functions that are often found to govern the response of gene expression to changes in the concentration of regulatory proteins. Generally speaking, step functions are defined as $s^{+}(P, t_P)$ and $s^{-}(P, t_P)$, where P is a protein concentration and t_P a concentration threshold. They are combined into algebraic expressions that are equivalent to the Boolean functions used in logical models. For discussions of the relation between PL and logical models, see ref. 29, 39.

   In practice, it may not be trivial to translate the regulatory logic of gene regulation into step-function expressions, as illustrated by the example of the gene *fis* in Subheading 2. It is often helpful to first formulate the conditions for expression in Boolean logic. For instance, in the case of *fis*, we say that the gene is expressed if

$$NOT\,(s^{+}(CRP, t_c)\;AND\;s^{+}(signal, t_s))\;AND\;NOT\;s^{+}(Fis, t_f^2), \quad (4)$$

that is, there is not enough active CRP and not enough Fis to repress the gene. We then recursively apply the following rules for the translation of Boolean logic expressions b, b_1, b_2 into step-function expressions (40):

$$NOT\, s^+(P, t_P) \equiv s^-(P, t_P) \tag{5}$$

$$NOT\, b \equiv 1 - b \tag{6}$$

$$b_1\, AND\, b_2 \equiv b_1\, b_2 \tag{7}$$

$$b_1\, OR\, b_2 = NOT\,(NOT\, b_1\, AND\, NOT\, b_2) \tag{8}$$

$$\equiv 1 - (1 - b_1)\,(1 - b_2) \tag{9}$$

Application of these rules to the *fis* example in Eq.4 indeed yields the step-function expression used in Eq. 2.

4. *Definition of inequality constraints.* The inequality constraints of a PL model define a total order on the threshold and focal parameters. The PL models are robust to variations in parameter values as long as the latter do not change this order (11, 29). On the contrary, modifications in the total order of the threshold and focal parameters may change the qualitative dynamics of the system. This makes the definition of the parameter inequalities an important step in the analysis of a regulatory network by means of GNA. For each variable, between one and five parameter inequalities are usually needed. For instance, in the CRP-Fis network we have

$$t_c < k_c^0/g_c < (k_c^0 + k_c^1)/g_c \tag{10}$$

$$t_f^1 < t_f^2 < k_f/g_f \tag{11}$$

That is, for both CRP and Fis, two inequalities define a total order on the three threshold and focal parameters. The choice of the inequalities can be based on experimental data and mathematical constraints, but it requires some experience to see which kind of information is helpful for this.

A first heuristic for the ordering of the threshold parameters is that, for a given regulatory protein, the threshold for the negative autoregulation of its own gene is usually higher than the thresholds for the other targets. In fact, the concentration of the regulatory protein cannot rise above the threshold for negative autoregulation, since this would immediately stop the expression of its own gene, and thus the synthesis of the regulatory protein. It would not make sense to assume that other targets of the gene are activated or inhibited at higher levels than the threshold for negative autoregulation, since these levels are not reachable under the assumption. In the example in the paper, we, therefore, set $t_f^1 < t_f^2$. A second heuristic is based on experimental data, in the case that the regulatory protein is a transcription factor. If the transcription factor has a higher affinity for one promoter region than for another, the threshold for the first target will be lower than for the second. That is, the target gene will be activated or inhibited at lower concentrations of the transcription factor.

Similar heuristics can be used for the ordering of the focal parameters with respect to each other and to the threshold parameters. First, there usually exist mathematical constraints between the focal parameters. For instance, since $k_c^0 < k_c^0 + k_c^1$, we necessarily have $k_c^0/g_c < (k_c^0 + k_c^1)/g_c$. Second, the highest concentration that a gene product can theoretically reach is given by the largest focal parameter, which describes the level reached for the maximal synthesis rate and the minimal degradation rate. In the PL model of the example network, these levels are $(k_c^0 + k_c^1)/g_c$ and k_f/g_f, for CRP and Fis, respectively. Obviously, these levels must exceed the highest threshold for the regulatory proteins, otherwise the observed regulation of a target gene at this threshold would not be possible. In the example this motivates the inequality constraints $t_c < (k_c^0 + k_c^1)/g_c$ and $t_f^2 < k_f/g_f$. A third source of information is the analysis of mutant data in the literature, notably mutants that cause one of the synthesis parameters of a regulatory protein to become 0 (e.g., a mutant in the promoter region). If a target of the regulatory protein is still regulated in this mutant background, then the corresponding maximum focal parameter must be larger than the threshold for this target. For instance, if *fis* would still be repressed by CRP in a mutant disabling the activation of *crp* expression by Fis, when $k_c^1 = 0$, we would necessarily have $t_c^1 < k_c^0/g_c$.

The above heuristics can be completed with more context-depending ways of determining parameter inequalities. This may still not be sufficient to fix a total order on the threshold and focal parameters. One way to deal with this problem would be to enumerate the remaining combinations of parameter inequalities (if their number is not too high), and analyze the model for each of these. Model-checking tools may be helpful in systematically exploring the possibilities (41).

5. *Some subtle mathematical properties of PL models.* The step-function approximations result in PL models with favorable mathematical properties, except at the thresholds where discontinuities occur. When crossing a threshold, the derivative of a variable may abruptly switch to another value and even to another sign. While some authors ignore the dynamics at the thresholds, this is potentially dangerous as it may cause steady states and other important dynamical properties of the system

to be missed. In GNA we have employed an approach from control theory, in which PL differential equations are generalized to differential inclusions (11, 15), to deal with these discontinuities. The use of differential inclusions allows for steady states in which one or more variables assume their threshold value. This is, for instance, the case for the second steady state of the CRP-Fis network in Subheading 3, in which *Fis* takes the value t_f^2. In some special situations, the use of differential inclusions leads to steady state sets. These steady state sets can be treated as normal steady states of the system under an appropriate generalization of the notion of stability (12).

6. *Temporal-logic patterns for the specification of network properties.* The problem of posing relevant and interesting questions is critical in modeling in general, but even more so in the context of applying formal verification techniques, due to the fact that it is not easy for nonexperts to formulate queries in temporal logic. The pattern-based property editor of GNA helps the user with this task by allowing the specification of temporal-logic properties in the form of patterns. Patterns are high-level query templates that formulate recurring biological questions using a domain-specific language rather than temporal logic. They were originally introduced in the formal verification field and recently adapted for use in systems biology (21). From a study of the literature on the modeling of regulatory networks, we found that most of the questions asked by experts can be reduced to a set of four patterns concerning the occurrence/exclusion, consequence, sequence, and invariance of events. Notice that these patterns are classes of properties sufficiently generic to be applicable in a variety of systems biology models, and that the aim of these patterns is not to cover all possible questions the modeler can think of, but rather to simplify the formulation of the most frequent or otherwise important questions.

7. *Interpretation of model-checking results.* The state transition graph produced by GNA is a discrete abstraction of the dynamics of the underlying continuous system (11). This means that some of the information contained in the original PL model is lost in the abstraction, which is the price to pay for the possibility to analyze the dynamics of the system in graph form. It requires the modeler to be careful though when interpreting the results returned by a model checker.

   First, the discrete abstraction is guaranteed to preserve all solutions of the underlying PL models as paths in the graphs. In some cases, however, the resulting graph is an over approximation of the network dynamics (11, 29). That is, some paths in the state transition graph may not correspond to any genuine solutions of the PL model. As a consequence, if a query asks for the existence of a path with a certain property, a true verdict cannot guarantee that the property also holds in the underlying continuous system. On the contrary, if the query requires the property to hold for all paths in the graph, a so-called universal property, then a true verdict does carry over to the PL model. The situation is exactly the inverse in the case that a model checker returns a false verdict.

Second, the state transition graph contains so-called self-transitions, from a qualitative state to itself. These transitions occur when the system can stay for some time in the state and is not bound to instantaneously leave it. However, from the mathematical properties of the PL models, we also know that the time the system can stay in such a state is finite, except if it corresponds to a steady state of the PL model. This information is not preserved when making the discrete abstraction from PL models to state transition graphs. As a consequence, when asking if a property possibly indefinitely holds in the future, the model checker may return true if it holds in a qualitative state with a self-transition, even when the latter is not a steady state. In order to avoid mistaken interpretations of such results, we have added so-called *fairness constraints* (32) to the properties verified by means of the NuSMV model checker (33). The latest version of GNA allows the user to specify fairness constraints other than the defaults stated in the user preferences.

Acknowledgments

This work was partially supported by the FCT program (PhD grant SFRH/BD/32965/2006 to PTM). BB and HdJ were supported by the European commission under project COBIOS (FP6-2005-NEST-PATH-COM/043379). HdJ, JG, and DR were supported by the European commission under project EC-MOAN (FP6-2005-NEST-PATH-COM/043235).

References

1. Bolouri H. (2008) *Computational Modelling of Gene Regulatory Networks: A Primer.* Imperial College Press, London.
2. Heinrich R, Schuster S. (1996) *The Regulation of Cellular Systems.* Chapman & Hall, New York.
3. de Jong H. (2002) Modeling and simulation of genetic regulatory systems: a literature review. J Comput Biol, 9:67–103.
4. Fisher J, Henzinger T. (2007) Executable cell biology. Nat Biotechnol, 25(11):1239–1250.

5. Karlebach G, Shamir R. (2008) Modelling and analysis of gene regulatory networks. Nat Rev Mol Cell Biol, 9(10):770–780.
6. Thomas R, d'Ari R. (1990) *Biological Feedback*. CRC Press, Boca Raton, FL.
7. Chaouiya C. (2007) Petri net modelling of biological networks. Brief Bioinform, 8(4): 210–219.
8. Palsson B. (2006) *Systems Biology: Properties of Reconstructed Networks*. Cambridge University Press, New York.
9. Ciocchetta F, Hillston J. (2008) Process Algebras in Systems Biology. In *Proc. 8th International School on Formal Methods for the Design of Computer, Communication, and Software Systems (SFM 2008), Volume 5016 of* LNCS. Edited by Bernardo M, Degano P, Zavattaro G, Springer-Verlag, Berlin, 313–365.
10. Glass L, Kauffman S. (1973) The logical analysis of continuous non-linear biochemical control networks. J Theor Biol, 39:103–129.
11. Batt G, de Jong H, Page M, Geiselmann J. (2008) Symbolic reachability analysis of genetic regulatory networks using discrete abstractions. Automatica, 44(4):982–989.
12. Casey R, de Jong H, Gouzé JL. (2006) Piecewise-linear models of genetic regulatory networks: equilibria and their stability. J Math Biol, 52:27–56.
13. Edwards R. (2006) Analysis of continuous-time switching networks. Physica D, 146 (1–4):165–199.
14. Farcot E, Gouzé JL. (2009) Periodic solutions of piecewise affine gene network models with non uniform decay rates: the case of a negative feedback loop. Acta Biotheor, 57(4):429–455.
15. Gouzé JL, Sari T. (2002) A class of piecewise linear differential equations arising in biological models. Dyn Syst, 17(4):299–316.
16. Mestl T, Plahte E, Omholt S. (1995) A mathematical framework for describing and analysing gene regulatory networks. J Theor Biol, 176(2):291–300.
17. Glass L, Pasternack J. (1978) Stable oscillations in mathematical models of biological control systems. J Math Biol, 6:207–223.
18. Plahte E, Kjøglum S. (2005) Analysis and generic properties of gene regulatory networks with graded response functions. Physica D, 201:150–176.
19. de Jong H, Geiselmann J, Hernandez C, Page M. (2003) Genetic Network Analyzer: qualitative simulation of genetic regulatory networks. Bioinformatics, 19(3):336–344.
20. de Jong H, Page M. (2008) Search for steady states of piecewise-linear differential equation models of genetic regulatory networks. ACM/IEEE Trans Comput Biol Bioinform, 5(2):208–222.
21. Monteiro P, Ropers D, Mateescu R, Freitas A, de Jong H. (2008) Temporal logic patterns for querying dynamic models of cellular interaction networks. Bioinformatics, 24(16): i227–i233.
22. Ropers D, de Jong H, Page M, Schneider D, Geiselmann J. (2006) Qualitative simulation of the carbon starvation response in *Escherichia coli*. Biosystems, 84(2):124–152.
23. Ropers D, Baldazzi V, de Jong H. (2011) Model reduction using piecewise-linear approximations preserves dynamic properties of the carbon starvation response in *Escherichia coli*. ACM/IEEE Trans Comput Biol Bioinform, 8(1):166–181.
24. Novère NL, Hucka M, Mi H, Moodie S, Schreiber F, Sorokin A, Demir E, Wegner K, Aladjem M, Wimalaratne S, Bergman F, Gauges R, Ghazal P, Kawaji H, Li L, Matsuoka Y, Villéger A, Boyd S, Calzone L, Courtot M, Dogrusoz U, Freeman T, Funahashi A, Ghosh S, Jouraku S, Kim S, Kolpakov F, Luna A, Sahle S, Schmidt E, Watterson S, Wu G, Goryanin I, Kell D, Sander C, Sauro H, Snoep J, Kohn K, Kitano H. (2009) The systems biology graphical notation. Nat Biotechnol, 27(8):735–742.
25. MH Saier J, Ramseier T, Reizer J. (1996) Regulation of Carbon Utilization. In *Escherichia coli and Salmonella: Cellular and Molecular Biology*, 2nd ed. Edited by Neidhardt et al., ASM Press, Washington, DC, 1325–1343.
26. Gonzalez-Gil G, Bringmann P, Kahmann R: Fis is a regulator of metabolism in *Escherichia coli. Mol. Microbiol.* 1996, 22:21–29.
27. Ishizuka H, Hanamura A, Inada T, Aiba H. (1994) Mechanism of the down-regulation of cAMP receptor protein by glucose in *Escherichia coli*: role of autoregulation of the *crp* gene. EMBO J, 13(13):3077–3082.
28. Azam TA, Iwata A, Nishimura A, Ueda S, Ishihama A. (1999) Growth phase dependent variation in protein composition of the *Escherichia coli* nucleoid. J Bacteriol, 181 (20):6361–6370.
29. de Jong H, Gouzé JL, Hernandez C, Page M, Sari T, Geiselmann J. (2004) Qualitative simulation of genetic regulatory networks using piecewise-linear models. Bull Math Biol, 66(2):301–340.
30. Kuipers B. (1994) *Qualitative Reasoning: Modeling and Simulation with Incomplete Knowledge*. MIT Press, Cambridge, MA.
31. Farcot E. (2006) Geometric properties of a class of piecewise affine biological network models. J Math Biol, 52(3):373–418.

32. Clarke E, Grumberg O, Peled D. (1999) *Model Checking*. MIT Press, Boston, MA.
33. Batt G, Ropers D, de Jong H, Geiselmann J, Mateescu R, Page M, Schneider D. (2005) Validation of qualitative models of genetic regulatory networks by model checking: analysis of the nutritional stress response in *Escherichia coli*. Bioinformatics, 21(Suppl 1):i19–i28.
34. Monteiro P, Dumas E, Besson B, Mateescu R, Page M, Freitas A, de Jong H. (2009) A service-oriented architecture for integrating the modeling and verification of genetic regulatory networks. BMC Bioinform, 10:450.
35. Cimatti A, Clarke EM, Giunchiglia E, Giunchiglia F, Pistore M, Roveri M, Sebastiani R, Tacchella A. (2002) NuSMV2: An Open-Source Tool for Symbolic Model Checking. In *Proc. 14th Intl. Conf. Computer Aided Verification (CAV 2002), Volume 2404 of* LNCS. Edited by Brinksma D, Larsen K, Springer-Verlag, Berlin, 359–364.
36. Hoops S, Sahle S, Gauges R, Lee C, Pahle J, Simus N, Singhal M, Xu L, Mendes P, Kummer U. (2006) COPASI – a COmplex PAthway SImulator. Bioinformatics, 22(24): 3067–3074.
37. Chen K, Calzone L, Csikasz-Nagy A, Cross F, Novak B, Tyson J. (2004) Integrative analysis of cell cycle control in budding yeast. Mol Biol Cell, 15(8):3841–3862.
38. Tyson J. (1991) Modeling the cell division cycle: cdc2 and cyclin interactions. Proc Natl Acad Sci U S A, 88:7328–7332.
39. Snoussi E. (1989) Qualitative dynamics of piecewise-linear differential equations: a discrete mapping approach. Dynam Stab Syst, 4(3–4):189–207.
40. Plahte E, Mestl T, Omholt S. (1998) A methodological basis for description and analysis of systems with complex switch-like interactions. J Math Biol, 36(4):321–348.
41. Bernot G, Comet JP, Richard A, Guespin J. (2004) Application of formal methods to biological regulatory networks: extending Thomas' asynchronous logical approach with temporal logic. J Theor Biol, 229(3): 339–348.

Chapter 23

Logical Modelling of Gene Regulatory Networks with GINsim

Claudine Chaouiya, Aurélien Naldi, and Denis Thieffry

Abstract

Discrete mathematical formalisms are well adapted to model large biological networks, for which detailed kinetic data are scarce. This chapter introduces the reader to a well-established qualitative (logical) framework for the modelling of regulatory networks.

Relying on GINsim, a software implementing this logical formalism, we guide the reader step by step towards the definition and the analysis of a simple model of the lysis–lysogeny decision in the bacteriophage λ.

Key words: Networks, Logical modelling, Qualitative analysis, Graph, Regulatory circuit, Dynamical simulation

1. Introduction

The logical modelling framework has been applied to the study of a wide range of networks involved in the lysis–lysogeny decision of the bacteriophage λ (1), induction of the type 3 secretion system in *Pseudomonas aeruginosa* (2), tryptophan biosynthesis in *Escherichia coli* (3), drosophila development (4–6), yeast cell cycle control (7, 8), lymphocyte differentiation (9), etc.

Our goal here is to introduce the reader to this modelling framework, relying on GINsim, a dedicated software. The chapter is organised as follows. In the remaining part of the introduction, we briefly describe the biological application that serves as a study case, namely, the lysis–lysogeny decision in the bacteriophage λ. Subheading 2 provides practical information on GINsim, file

Jacques van Helden et al. (eds.), *Bacterial Molecular Networks: Methods and Protocols*, Methods in Molecular Biology, vol. 804, DOI 10.1007/978-1-61779-361-5_23,

formats, and sample files. General guidelines to define and analyse logical models are given in Subheading 3. Next, we proceed with the stepwise construction and analysis of the phage λ switch logical model. The chapter ends with an outlook section.

1.1. Study Case: Lysis–Lysogeny Decision in the Bacteriophage λ

A number of bacterial and viral genes take part in the decision between lysis and lysogeny in temperate bacterial viruses such as the bacteriophage λ. A comprehensive description of this genetic switch is provided in ref. (10). The phage λ regulatory network has been modelled using statistical mechanics (e.g. (11)), stochastic chemical kinetics (e.g. (12)), hybrid Petri nets (e.g. (13)), and the logical formalism (1). Here, we rely on the logical model presented in (1), which focuses on the core of the λ regulatory network involving four main cross-regulating genes: cI coding for λ repressor, the only gene expressed in the lysogenic state; cro, coding for another repressor, expressed in the lytic cycle; cII, an activator of several lysogenic genes, including cI; and N, an anti-terminator factor promoting cII expression (see Fig. 1 for a graphical representation of this core regulatory network).

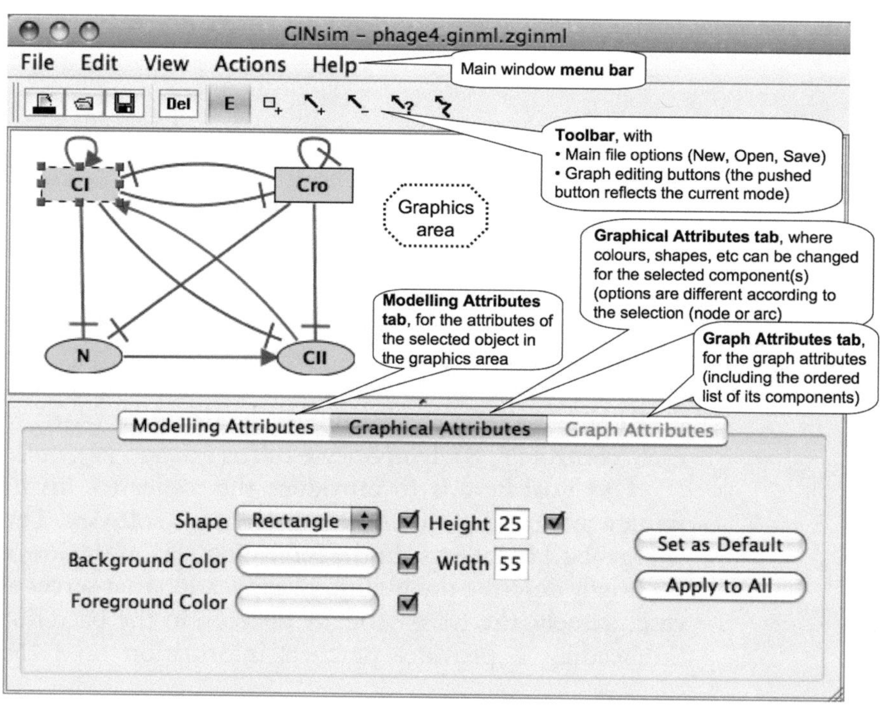

Fig. 1. Phage λ logical regulatory graph displayed in the GINsim main window. The menu bar at the top displays five menu titles. File scrolling menu provides access to classical file management options, to an option for merging the current graph with another one (chosen by selecting the file that stores it), as well as to export facilities towards formats other than GINML. The central area displays the graph, while the lower panel contains three tabs: the Modelling Attributes tab (see also Figs. 2 and 3) and the Graphical Attributes tab (selected here), both related to the selected graph component, as well as the Graph Attributes tab, which relates to the graph as a whole.

2. Material

2.1. GINsim

GINsim supports the definition, the simulation and the analysis of regulatory graphs, based on the (multi-valued) logical formalism. Developed in Java, GINsim is platform-independent and only requires a recent Java Virtual Machine.

GINsim is freely available for academic users. The latest official version is available from the GINsim web site (http://gin.univ-mrs.fr/GINsim) along with documentation and a model repository. The current official version is GINsim-2.3 (14). To get started with GINsim, download `GINsim-# version.jar`, the Java ARchive, from the download section of the web site.

To launch GINsim, double-click on the file icon or use the command: `java -jar GINsim-# version.jar`. You are then ready to define a logical regulatory graph, or to open a pre-existing logical model and analyse its dynamics.

2.2. File Formats

To store logical regulatory graphs and state transition graphs, possibly including graphical attributs, GINsim relies on a specific XML file format called GINML. Furthermore, ZIP archives can be generated (extension `.zginml`), which contain, in addition to the flat file of the model (file `regulatoryGraph.ginml`), additional files for mutants, initial states, and other simulation parameter definitions (files `mutant`, `initialState`, `reg2dyn_parameters`, respectively).

GINsim supplies export facilities in a number of file formats (see Note 1).

2.3. Sample Files

Two sample files are provided for our study case:

1. `phage4.ginml`, the four-variable model as defined in Subheading 1;
2. `phage4.zginml`, the four-variable model together with the simulation scenarios presented in Subheading 4.2.

3. Logical Regulatory Graph Definition and Analysis

In this section, we describe the necessary steps to define a logical model of a regulatory network and to analyse its dynamical properties.

3.1. Defining a Logical Regulatory Graph

A logical model of a regulatory network is defined as a *logical regulatory graph*, where the nodes represent regulatory components (genes, proteins, etc.), whereas the arcs represent regulatory interactions.

1. Define a set of regulatory components $\{g_0,\ldots,g_n\}$. To each component g_i (i belongs to symbol $\{0,\ldots n\}$), associate a *maximal level* max_i, defining a range of functional qualitative levels $\{0,\ldots max_i\}$ for g_i.
2. Define a set of arcs between nodes. An arc represents a regulatory effect of its *source* node onto its *target* node. For each arc, assign a *threshold* θ in $\{1,\ldots max_i\}$ (max_i being the maximal level of g_i, the source of the arc). This threshold indicates the level of g_i at which the regulatory effect on g_j, the target of the interaction, occurs. Further, assign a sign in {+,–,?}, where + denotes an activation, – an inhibition and ? a dual interaction (i.e. an interaction with activatory or inhibitory effect, depending on the presence of cofactors).

   When the maximal level max_i of a node g_i is higher than 1, an outgoing arc pointing from g_i towards a node g_j may be composed of several interactions (i.e. there is a multi-arc from g_i to g_j), denoting situations where the regulatory effect of g_i on g_j is different depending on the level of g_i (see Note 2).

   In addition, one must define the rules governing the evolution of the regulatory component levels. These rules are specified in the form of *logical parameters.*
3. For each node g_i and for each combination of incoming interactions, specify the *target level* of g_i (see Note 3 for the definitions of the basal levels and default values in GINsim). If g_i is subject to p interactions, there are 2^p possible combinations of incoming interactions.

3.2. Dynamical Simulations

The dynamical behaviour of a logical regulatory graph is described as another type of graph called *state transition graph.* In this graph, each node represents a state of the model, defined by a vector s of the component levels (s_i, the i th component of s, is the level of g_i in state s). The arcs represent transitions between states. One core function of GINsim is the automatic construction of this graph. However, to use this function judiciously, it is important to understand the principles underlying this construction, which are outlined hereafter. Afterwards, several options to analyse the dynamics of logical regulatory graphs are presented.

A state s fully determines the interactions that are active (that is to say for which the source levels are above their thresholds). This interaction set determines the logical parameters to be considered in state s. For each component g_i, if its current level s_i is different from the value of the relevant parameter in state s, there is a call to update the level of g_i towards the target level specified by the parameter value. Now, there can be several such components called to change their levels. Two main strategies are commonly used. In the *synchronous updating*, all concerned components change their levels simultaneously in a unique transition towards the next state. By contrast, the *asynchronous updating* generates a successor state for each component updating call. If a state involves *k* updating calls, it will thus have *k* successors, and each successor state will differ from it by the level of a single component (see Note 4 for additional explanations). The introduction of priority classes allows to define more subtle updating schedules (an illustration is provided in Subheading 2; see also Note 5).

Building a state transition graph, one can choose to generate the *full state transition graph*, considering all possible initial conditions, or yet a (sub)graph from specific initial state(s).

In such state transition graphs, it is then possible to determine the stable states (defined as nodes with no outgoing arcs), or more complex attractors (defined as terminal maximal strongly connected components, denoting a stable oscillatory behaviour). Moreover, it may be interesting to search a path between two states, in particular to verify the reachability of a given stable state from an initial state.

3.3. Circuit Analysis

Regulatory circuits are responsible for the emergence of dynamical properties such as multistationarity or stable oscillations (see Note 6). GINsim implements specific algorithms to:

1. Identify all the circuits of a regulatory graph (possibly considering constraints such as length limitation, component exclusion, etc.).
2. Determine the functionality contexts of these circuits, using the method presented in ref. 15.

3.4. Definition of Perturbations

Common perturbations are easily specified within the logical framework:

1. A gene knock-down is specified by driving and constraining the level of the corresponding regulatory node to the value 0.
2. Ectopic expression is specified by driving and constraining the level of the corresponding regulatory component to its highest value (or possibly to a range of values greater than zero, in the case of a multi-valued node).
3. Multiple perturbations can be defined by specifying several of the above constraints.

More subtle perturbations can be defined by changing the parametrisation of the model (i.e. changing the effect of a regulation).

4. Application to the Lysis–Lysogeny Decision of the Bacteriophage λ

Here, we illustrate the use of GINsim through the construction and the analysis of a logical model for the core regulatory network driving the lysis–lysogeny decision of the phage λ.

4.1. Defining the Logical Regulatory Graph

The set of instructions listed below defines a logical model, which can be saved in the GINML format as described in Subheading 2.

1. Add the four nodes corresponding to CI, Cro, CII, and N, specifying their names and maximal levels as defined in Table 1. Figure 1 illustrates this step.
2. Add the arcs between the nodes as defined in Fig. 1, specifying their associated signs and (sets of) intervals (see Table 2 and Note 2). Figure 2 illustrates this step.
3. For each node (select the node to edit it), specify its logical parameters as given in Table 3. This is done by selecting the interactions defining a parameter and clicking on the left arrow to add the corresponding parameter in the parameter list (see Note 3 for default values). Figure 3 illustrates this step. Note that interactions defined in a regulatory graph might have no effect on the target node behaviour (see Note 7 for further discussion).
4. Change the graphical appearance of nodes and arcs of the graph at your convenience. For this, select the object, node or arc, and the Graphical Attributes tab as indicated in Fig. 1.
5. Verify that the order of the variables is: CI, Cro, CII, N (shown by selecting the Graph Attributes tab, accessible when no object is selected in the main window); if this is not the case, modify the component order accordingly, using the arrows close to the component list at the bottom-left of the tab.
6. Save your model, either using the Save option of File menu or the save button in the toolbar (floppy disk icon in Fig. 1).

Table 1
Components and maximum values in the phage λ model

Gene	Maximum
CI	2
Cro	3
CII	1
N	1

Table 2
Interactions and activity ranges in the phage λ model

Source	Target	Range
CI	CI	[2,2]
	Cro	[2,2]
	CII	[2,2]
	N	[1,2]
Cro	CI	[1,3]
	Cro	[3,3]
	CII	[3,3]
	N	[2,3]
CII	CI	[1,1]
N	CII	[1,1]

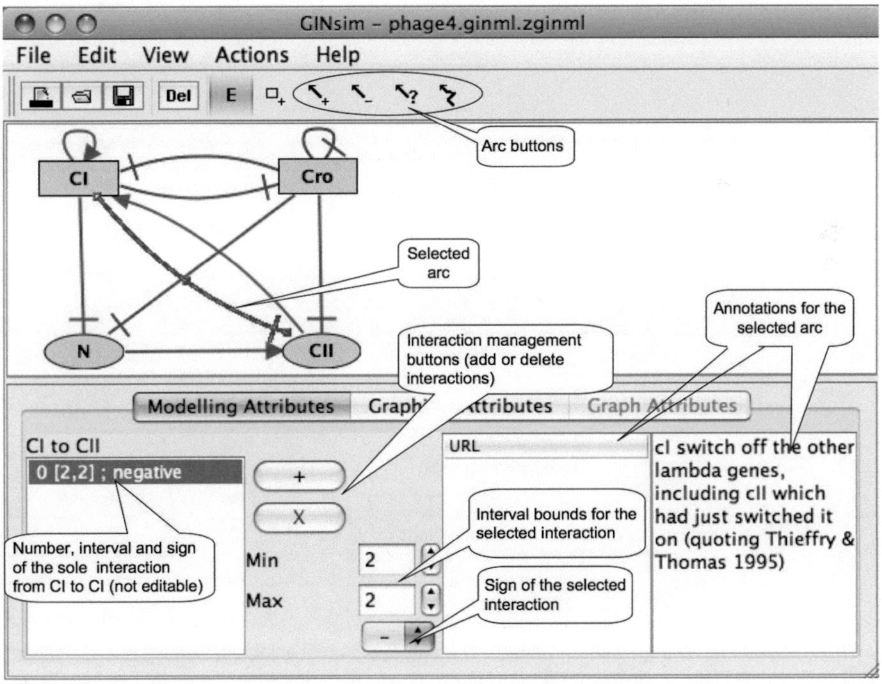

Fig. 2. Regulatory arcs management in GINsim. To add an arc, the corresponding arc button must be pushed (push twice to add multiple arcs in a row), allowing the drawing of an arc between a source component and its target. The interactions encompassed by an arc, with their signs, intervals of a selected arc are defined in the Modelling Attributes tab as shown (see also Note 2).

Table 3
Logical parameters in the phage λ model

Gene	Interaction combination	Target value
CI	∅	2
	CI[2,2]	2
	CII[1,1]	2
	CI[2,2] and CII[1,1]	2
	Cro[1,3] and CII[1,1]	2
	CI[2,2] and Cro[1,3] and CII[1,1]	2
Cro	∅	3
	Cro[3,3]	2
CII	N[1,1]	1
N	∅	1

For each gene, combinations of active interactions (defining the logical parameters) leading to a non-zero target value are specified. Similarly to the notation adopted in GINsim, an interaction targeting a gene is denoted by the name of its source followed by the interval of values for which it is active. The symbol ∅ denotes the basal value (no interaction acting on the component)

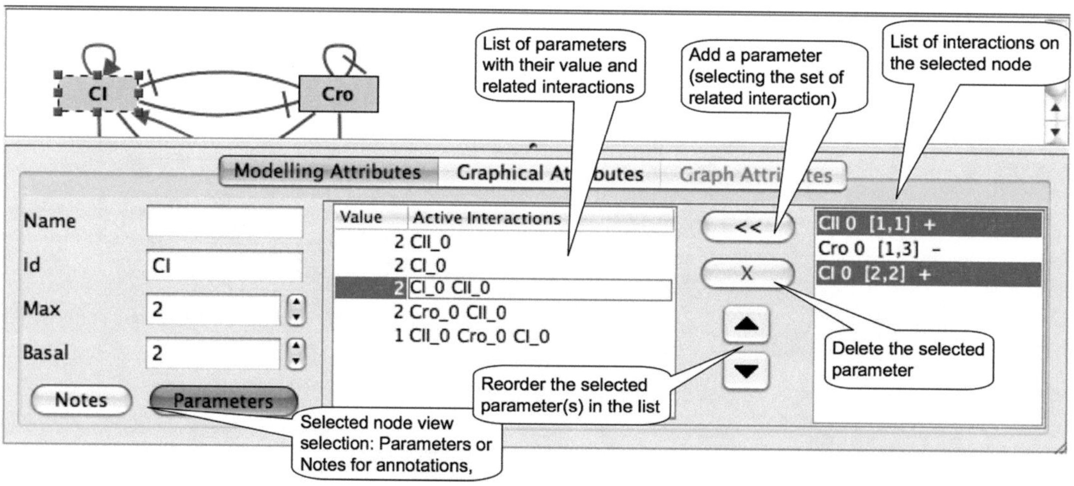

Fig. 3. Regulatory components management in GINsim. The name, maximal level, basal value and logical parameters of a selected node are defined in the Modelling Attributes tab as shown. To define a (non-zero) logical parameter, first select the corresponding interactions combination and add the parameter using the dedicated button (<<). The parameter value is set to 1 by default but can be further edited, in particular for multilevel nodes (see also the Note 3).

4.2. Dynamical Analysis

The simulation of a logical model defined as described above results in a state transition graph (which can be saved as a GINML file). The simulation settings (initial states, updating schemes and perturbations) can also be saved in the archive `zginml` (cf. Subheading 2.3).

Let us first consider the construction of the *full asynchronous state transition graph*:

1. Choose the option "Run Simulation" in the Action menu.
2. Before clicking the Run button, verify that the default settings are as specified in Fig. 4: asynchronous updating, no priority, no mutant selected, no initial state specified.
3. Display the state transition graph obtained. In this new window, you can rearrange the nodes, change the graphical settings, and check outgoing transitions by selecting a state (see Fig. 4). In the default level layout, the nodes with no incoming arc are placed at the top, whereas the nodes with no outgoing arc are placed at the bottom; stable states are further emphasised with different graphical attributes. Here, we obtain a unique stable state denoted (2000) (following the order defined above, this vector states that CI = 2, Cro = CII = N = 0), which corresponds to immunity (lysogenic state with high CI and low levels for the three other components).
4. Launch the computation of the graph of strongly connected components (SCC) for this state transition graph and verify that there are indeed two attractors: the previously identified stable state (2000), and a cyclic attractor encompassing the two states (0200) and (0300), which corresponds to lysis, with an homeostatic expression of Cro around its highest values (see Fig. 4).

   Further analyses can be performed:
5. Using the "Stable States" option of the Action menu of the main window, verify that the unique stable state of this model is indeed (2000) (see Fig. 5); this calculation bypasses the construction of the STG, which is particularly useful for large models.
6. Define a mutant corresponding to an ectopic expression of Cro (such perturbations can be encoded before the computation of stable states or of a state transition graph), and verify that the stable lysogenic state is lost for this perturbation (see Fig. 5).
7. Assign a higher priority to the degradations (decreasing orders, synchronous class) as in the setting illustrated in Fig. 6 (see also Note 5), and enter (0000) as initial state. Verify that the corresponding state transition graph comprises 19 states (to be compared to the 46 generated for an asynchronous updating,

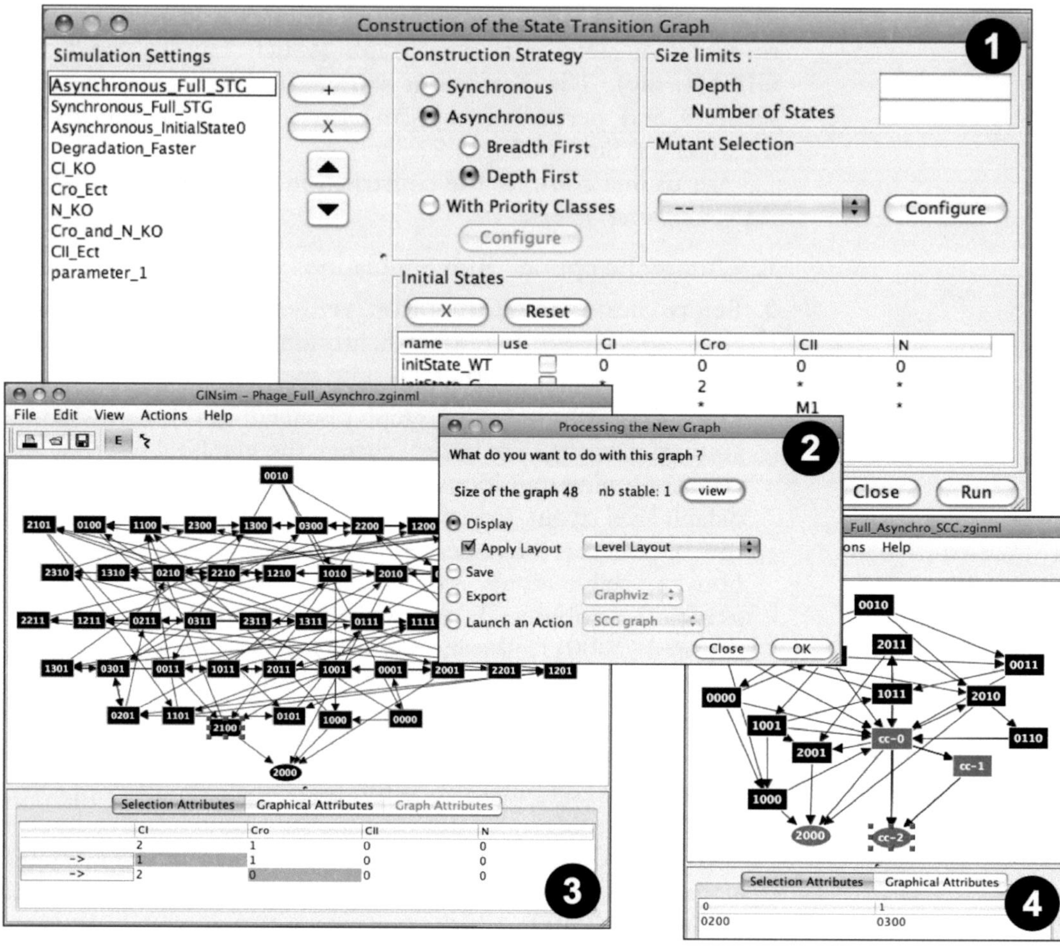

Fig. 4. Asynchronous full state transition graph for the phage λ logical model. The window (1) displays the simulation settings with an asynchronous updating and no initial state specification (meaning that all states are considered). Automatically opened after the construction of the STG, the window (2) indicates the total number of states, the number of stable states (which can be visualised), and a choice of further operations. When the graph is too large, GINsim does not support its visualisation. The window (3) displays the full STG encompassing 48 states, including the stable state (2000) laying at the bottom; the levels of the components for the selected state (2100) and its successors are shown below. The window (4) displays the corresponding graph of strongly connected components (SCC) encompassing 16 nodes, which correspond to single states or strongly connected components (selecting a component displays the states encompassed, the cyclical attractor here); the two attractors are shown at the bottom, with no further successors. To obtain the SCC graph, launch its construction in the Action menu of window (3), or select this action in window (2).

with no priorities and the same initial state). The same attractors are recovered (the lysogenic stable state, and the short cycle denoting lysis, with high levels of Cro). This graph encompasses only single component transitions (multiple updating orders do not occur). It can be easily verified that if syntheses are made synchronous, the lysogenic state cannot be reached from (0000).

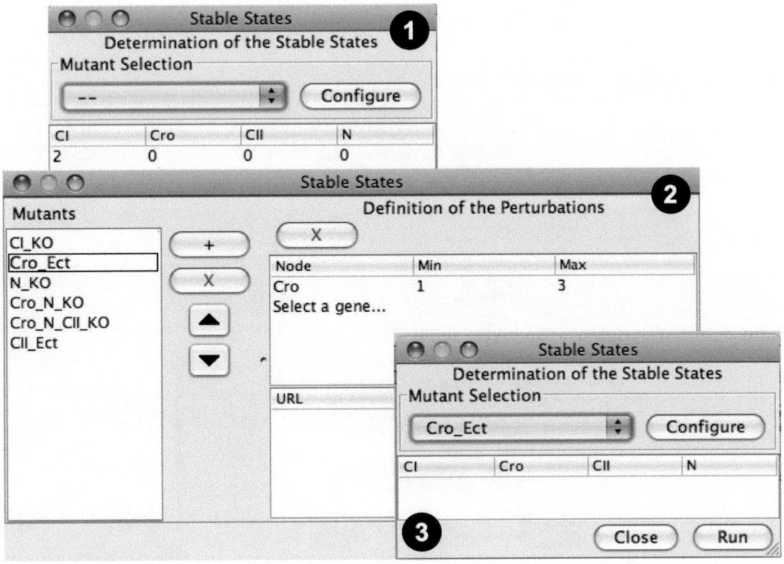

Fig. 5. Stable state determination and perturbation (or mutant) specification. The window (1) of the stable state determination can be open from the main window Action menu. For the wild type case, we obtain a unique stable state with a high level of CI, corresponding to the lysogenic state. The window (2) enables the specification of perturbations. Here the perturbation is called *Cro_Ect* and consists in blocking the level of Cro within the interval [1, 3] (this perturbation specification can be launched from the Stable state window (1) or yet from the STG construction window (1) in Fig. 4). In the presence of ectopic Cro, there is no stable state (window (3)).

8. Using the "Circuit Functionality" option of the Action menu, verify that the regulatory graph contains seven circuits, among which four are functional (i.e. have a non-empty functionality context). Verify that the positive circuit defined by the cross inhibitions between CI and Cro is functional provided CII $= 0$ (indeed, the inhibition of CI by Cro is not functional in the presence of CII). Figure 7 illustrates this analysis. Further details on the analysis of the roles of the regulatory circuits of this model can be found in ref. 1.
9. Define a mutant in which CII is ectopically expressed and verify that the system has the stable state (2010) as unique attractor (by constructing the full asynchronous state transition graph and the corresponding graph of strongly connected components, see Fig. 8). Note that, as the level of CII is blocked to value 1, the CI-Cro circuit is not functional and thus unable to generate multiple attractors.

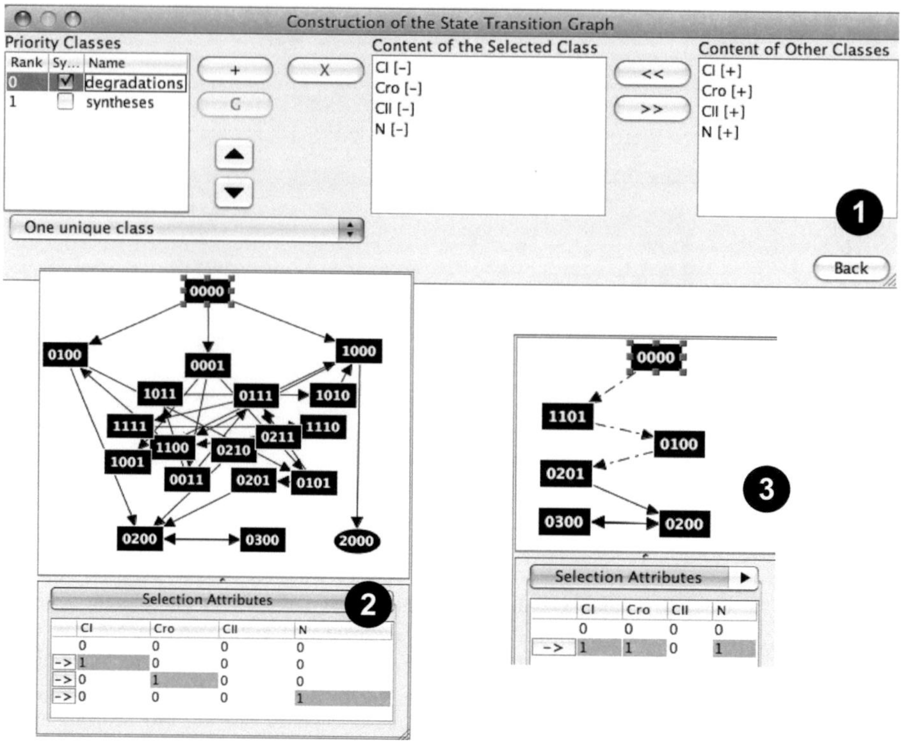

Fig. 6. Priority classes: in window (1), a higher priority is attributed to the degradations (decreasing orders denoted by a sign −), which are considered so fast that they are updated synchronously. Syntheses (increasing orders denoted by a +) are grouped in an asynchronous class with a lower priority. Window (2) displays the STG obtained for this updating scheme, considering the initial state (0000). In the lower panel, the three successors of the initial state are displayed, all corresponding to increasing orders. The STG encompasses 19 states, with the two attractors observed in the full asynchronous STG (Fig. 4). Window (3) shows the STG obtained when both classes are processed synchronously. This STG encompasses six states. Note that the cyclical attractor is lost. In the lower panel, the unique successor of (0000) is displayed, with the three increasing orders simultaneously performed (synchronous transitions are denoted by *dashed lines*).

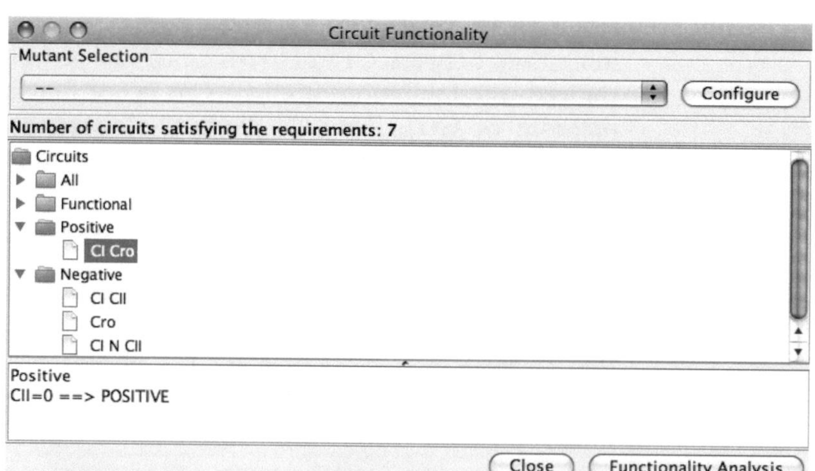

Fig. 7. Circuit analysis for the phage λ logical model: among the seven circuits found in the regulatory graph, four are functional: one is positive, while the other three are negative. The selected circuit (involving CI and Cro) is functional and positive in the absence of CII.

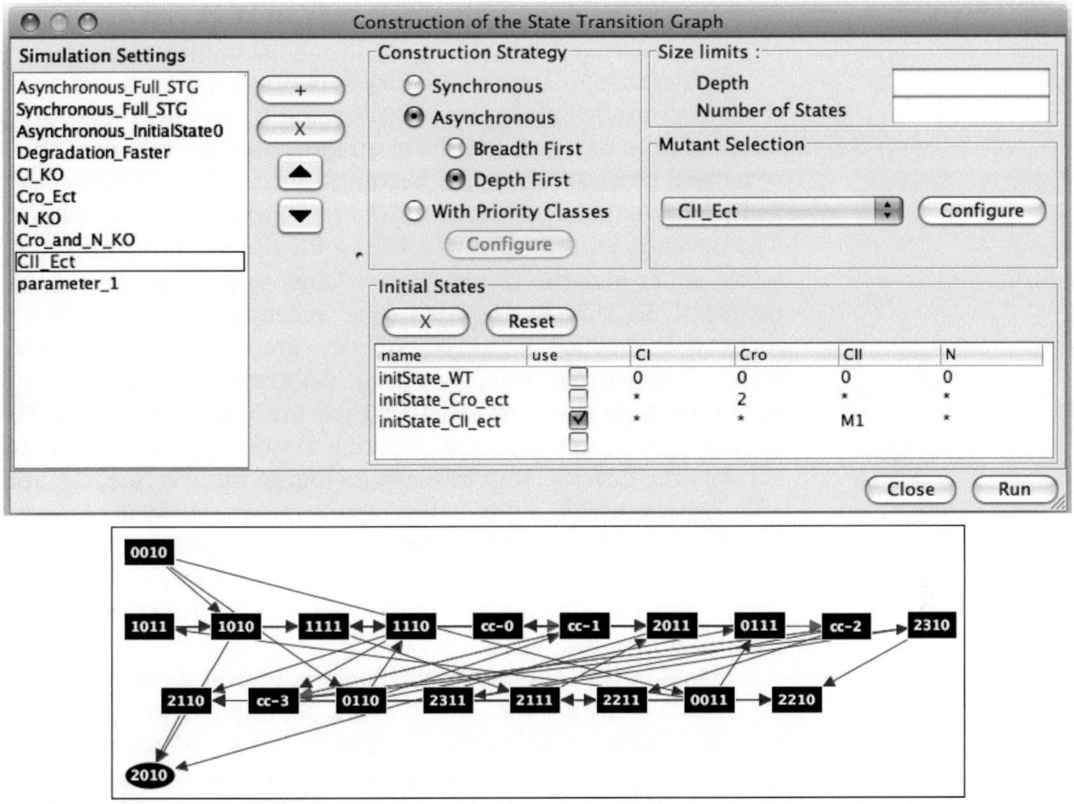

Fig. 8. Simulation in the case of an ectopic expression of CII, considering all initial states with CII expressed. In this condition, there is a unique attractor, which is the stable state (2010). This is easily verified by displaying the graph of strongly connected components (lower panel).

5. Outlook

The logical formalism is particularly useful to model regulatory networks for which precise quantitative information is barely available. In this chapter, we introduce the reader to the definition and analysis of logical models using the dedicated software GINsim, relying upon a concrete application, namely the core regulatory network controlling the lysis–lysogeny decision of the bacteriophage λ. Although very simplified, the resulting model captures important properties of the biological system, such as the existence of two asymptotical behaviours, as well as the roles of the key regulatory components, which can be easily assessed by proper mutant simulations. Furthermore, an original circuit analysis tool enables the delineation of regulatory circuits underlying specific dynamical properties, along with constraints on input components. More specifically, the positive cross-inhibitory circuit between CI and Cro is clearly at the basis of the observed multi-stationarity, whereas the observed homeostatic regulation of Cro in the lytic attractor can be attributed to a negative auto-regulatory loop.

As increasingly large regulatory networks are deciphered, their dynamical modelling leads to challenging problems. Indeed, the construction of state transition graphs, not to mention their structural analysis, quickly becomes intractable as their size grows exponentially with the number of components considered. The development of novel algorithms for the determination of all stable states and the circuit analysis aims at getting around this problem. In this respect, we have recently defined a formal method to reduce logical regulatory graphs, which preserves essential dynamical properties (16). Another promising strategy to handle large networks relies on their intrinsic modularity (17). Several groups are currently working on dedicated methods to support the development of modular logical models (see, e.g. ref. 18). However, this is not sufficient to support a modular view or construction of regulatory networks. What is needed is a compositional framework that would further facilitate the derivation of salient dynamical properties of the whole network from those of its constitutive modules.

Finally, depending on available data, logical models can also be used as a template to design more quantitative models, using dedicated software suites. In this respect, we are regularly adding novel export plugins to facilitate the transposition of logical models into other formal frameworks, including Petri nets, thereby enabling the recourse to other computational methods, such as model checking techniques.

6. Notes

1. GINsim allows the user to export logical regulatory graphs as well as state transition graphs towards various formats, facilitating the use of other softwares:
   (a) Visualisation tools:
      - Graphviz, an open source graph visualisation software offering main graph layout programs (`http://www.graphviz.org/`).
      - BioLayout Express 3D, a tool for the visualisation and analysis of biological networks (`http://www.biolayout.org`).
      - Cytoscape, another open source software platform for visualising molecular interaction networks. Cytoscape further aims at integrating interactions

with gene expression profiles and other state data (http://www.cytoscape.org/).

- Scalable Vector Graphics (SVG) format, first graphical export provided by GINsim, an XML standard for describing two-dimensional graphics (http://www.w3.org/Graphics/SVG/); note that forthcoming GINsim versions will provide export towards PNG image format.

(b) Petri net tools:

- Integrated Net Analyzer (INA) supporting the analysis of Place/Transition Nets (Petri Nets) and Coloured Petri nets (http://www2.informatik.hu-berlin.de/~starke/ina.html) see (Marwan et al., Chapter 21).
- Snoopy, a tool to design and animate hierarchical graphs, among others Petri nets (http://www-dssz.informatik.tu-cottbus.de/DSSZ/Software/Snoopy), see (Marwan et al., Chapter 21).

(c) Other modelling tools (forthcoming GINsim version):

- GNA, a software for the piecewise linear modelling of regulatory networks (http://ibis.inrialpes.fr/article122.html), see (Batt et al., Chapter 22).

2. Logical regulatory graphs may encompass multi-arcs, composed of different interactions having different effects. To each interaction, an interval specifying the range of the source levels for which the interaction occurs must be specified. Intervals assigned to interactions with the same source and target must obviously be disjoint. Moreover, a sign should be specified (positive, negative, or dual) for each regulatory effect.
3. For each node g, each combination of incoming interactions defines a logical parameter. This includes the case where no interaction acts on g (when the value of every regulator of g is below the first threshold of the arc pointing towards g). The logical parameter related to this case is called the *basal level* of g, i.e. the level of g_i in the absence of any specific activation. Considering that many parameters are usually null, zero is the default value in GINsim (i.e. a parameter that is not explicitly assigned takes value zero).
4. Transitions between states of the state transition graphs amount to the update of one (in the asynchronous case)

or several (in the synchronous case) components. In any case, the update (increase or decrease) of a component is unitary (current value +1 or −1). Obviously, this remark applies only for multi-valued components (for which the maximal level is greater than 1).

5. Priority classes allow to refine the updating schemes applied to construct the state transition graphs (19). GINsim users can group components into different classes and assign a priority rank to each of them. In case of concurrent updating transitions (i.e. calls for level changes for several components in the same state), GINsim updates the gene(s) belonging to the class with the highest ranking. For each priority class, the user can further specify the desired updating assumption, which then determines the treatment of concurrent transition calls inside that class. When several classes have the same rank, concurrent transitions are treated under an asynchronous assumption (no priority).

6. A regulatory circuit is defined by a sequence of interactions forming a simple closed directed path. The sign of a circuit is defined as the product of the signs of its interactions. Consequently, a circuit is positive if it has zero or an even number of inhibitions, it is negative otherwise. R. Thomas proposed that positive circuits are necessary to generate multistationarity, whereas negative circuits are necessary to generate stable oscillations (see (20) and references therein). External regulators might prevent the functioning of a circuit imbedded in a more complex network. Naldi et al. (15) presents a method to determine the *functionality context* of a circuit in terms of constraints on the levels of its external regulator. A circuit functionality context can be interpreted as the part of the state space where the circuit is functional, i.e. generates the expected dynamical property (21).

7. In the phage λ model, as defined in Table 3, the CI autoregulation is not *functional*, that is to say, it plays no role in the behaviour of its target node. In other words, combinations of interactions with and without CI auto-regulation lead to the same target value: the interaction from CII alone (CII [1,1]) or the combination of the interactions from CI and CII (CI [2,2] and CII[1,1]) push CI towards the same value 2 for CI (see Table 3). Indeed, although CI has been reported to directly activate its own promoter, this auto-regulation does not appear to be crucial for the lysis–lysogeny decision.

Acknowledgments

A.N. has been supported by a PhD grant from the French Ministry of Research and Technology. This project was further supported by research grants from the EU FP6 (DIAMONDS STREP) and FP7 (APOSYS large scale project) programs, and by the Belgian Science Policy Office (IAP BioMaGNet).

References

1. Thieffry D, Thomas R. (1995) Dynamical behaviour of biological regulatory networks, II. Immunity control in bacteriophage lambda. Bull Math Biol, 57(2):277–297.
2. Filopon D, Mérieau A, Bernot G, Comet JP, LeBerre R, Guery B, Polack B, Guespin-Michel J. (2006) Epigenetic acquisition of inducibility of type III cytotoxicity in aeruginosa. BMC Bioinformatics 2006, 7:272.
3. Simão E, Remy E, Thieffry D, Chaouiya C. (2005) Qualitative modelling of regulated metabolic pathways: application to the tryptophan biosynthesis in *E. coli*. Bioinformatics, 21(Suppl 2):190–196.
4. Chaves M, Albert R, Sontag ED. (2005) Robustness and fragility of Boolean models for genetic regulatory networks. J Theor Biol, 235(3):431–449.
5. Sánchez L, Chaouiya C, Thieffry D. (2008) Segmenting the fly embryo: a logical analysis of the segment polarity cross-regulatory module. Int J Dev Biol, 52(8):1059–1075.
6. González A, Chaouiya C, Thieffry D. (2008) Logical modelling of the role of the Hh pathway in the patterning of the Drosophila wing disc. Bioinformatics, 24:i234–i240.
7. Fauré A, Naldi A, Lopez F, Chaouiya C, Ciliberto A, Thieffry D. (2009) Modular logical modelling of the budding yeast cell cycle. Mol Biosyst, 5:1787–1796.
8. Irons DJ. (2009) Logical analysis of the budding yeast cell cycle. J Theor Biol, 257(4): 543–559.
9. Mendoza L. (2006) A network model for the control of the differentiation process in Th cells. Biosystems, 84(2):101–114.
10. Ptashne M. (2004) *A Genetic Switch: Phage Lambda Revisited*. 3rd ed. Cold Spring Harbor Laboratory Press, Cold Spring Harbor.
11. Ackers GK, Johnson AD, Shea MA. (1982) Quantitative model for gene regulation by lambda phage repressor. Proc Natl Acad Sci U S A, 79(4):1129–1133.
12. Arkin A, Ross J, McAdams HH. (1998) Stochastic kinetic analysis of a developmental pathway bifurcation in phage lambda-infected *Escherichia coli* cells. Genetics, 149:1633–1648.
13. Matsuno H, Doi A, Nagasaki M, Miyano S. (2000) Hybrid Petri net representation of gene regulatory network. In *Pac Symp Biocomput* 2000:341–352.
14. Naldi A, Berenguier D, Fauré A, Lopez F, Thieffry D, Chaouiya C. (2009) Logical modelling of regulatory networks with GINsim 2.3. Biosystems, 97(2):134–139.
15. Naldi A, Thieffry D, Chaouiya C. (2007) Decision diagrams for the representation of logical models of regulatory networks. Lect Notes Comput Sci, 4695:233–247.
16. Naldi A, Remy E, Thieffry D, Chaouiya C. (2009) A reduction of logical regulatory graphs preserving essential dynamical properties. Lect Notes Comput Sci, 5688:266–280.
17. Hartwell L, Hopfield J, Leibler S, Murray A. (1999) From molecular to modular cell biology. Nature, 402:C47–C52.
18. Mirschel S, Steinmetz K, Rempel M, Ginkel M, Gilles ED. (2009) ProMoT: modular modeling for systems biology. Bioinformatics, 25(5):687–689.
19. Fauré A, Naldi A, Chaouiya C, Thieffry D. (2006) Dynamical analysis of a generic Boolean model for the control of the mammalian cell cycle. Bioinformatics, 22(14):124–131.
20. Thieffry D. (2007) Dynamical roles of biological regulatory circuits. Brief Bioinformatics, 8(4):220–225.
21. Remy E, Ruet P. (2008) From minimal signed circuits to the dynamics of Boolean regulatory networks. Bioinformatics, 24(16):i220–i226.

Chapter 24

Modelling the Evolution of Mutualistic Symbioses

Maren L. Friesen and Emily I. Jones

Abstract

Mutualistic microbial symbioses are one of the key innovations in the evolution of biological diversity, enabling the expansion of species' niches and the production of sophisticated structures such as the eukaryotic cell. For some of the best-studied cases, we are beginning to have network models of symbiotic metabolism, but this work is in its infancy and has not been developed with an evolutionary perspective. However, theoreticians have long been interested in how these symbioses arise and persist and have applied modelling approaches from economics, evolution, ecology, and sociobology to a number of fundamental questions. We provide an overview of these questions, followed by specific modelling examples. We cover economic game theory, including the Prisoner's Dilemma, the Snowdrift game, and biological markets. We also describe the eco-evolutionary framework of adaptive dynamics, inclusive fitness, and population genetic models. We aim to provide insight into the strengths and weaknesses of each approach and into how current evolutionary methods can benefit an understanding of the mechanistic basis of host–symbiont interactions elucidated by molecular network models.

Key words: Symbiosis, Mutualism, Game theory, Adaptive dynamics, Inclusive fitness, Biological markets, Population genetics, Cooperative bargaining

1. Introduction

Mutualistic microbial symbioses have played a fundamental role in the evolution and diversification of life. These widespread interactions range from the ancient symbiotic merger of prokaryotic lineages in eukaryotes, to plant colonization of land assisted by fungi, to bacteria that extend the metabolic capabilities of their animal and plant hosts (1–7). The advantage of mutualistic symbiosis is that each species cooperates by contributing commodities that the partner species cannot (cheaply) produce itself, resulting in mutual benefit. We define cooperation broadly as any action that increases the fitness of the recipient (8). Some authors require that cooperation be selected for because of the fitness benefits provided to others (9); we feel that this condition is overly restrictive, not to mention extremely difficult to demonstrate empirically

Jacques van Helden et al. (eds.), *Bacterial Molecular Networks: Methods and Protocols*, Methods in Molecular Biology, vol. 804, DOI 10.1007/978-1-61779-361-5_24, © Springer Science+Business Media, LLC 2012

Table 1
Definitions of terms

Term	Definition
Cooperation	Action by one individual that increases the fitness of a recipient individual (8)
Altruism	Cooperation with a fitness cost to the actor
By-product cooperation	Cooperation without a fitness cost to the actor (8, 71)
Mutualism	Reciprocal cooperation between individuals of different species
Symbiosis	An interaction in which an individual spends a significant portion of its lifecycle within or in close association with an individual of another species
Obligate/facultative symbiosis	Partners must associate with one another/do not have to associate to survive or reproduce
Vertical/horizontal transmission	Symbionts are transmitted from parent to offspring/through the environment between non-relatives
Partner choice	Individuals distinguish between potential partners and interact with those that will provide higher benefits (22)
Partner fidelity feedback	Partners engage in repeated interactions, such that improving the fitness of the partner increases its ability to provide rewards (8)
Sanctions	Inferior partners are not rewarded, or the interaction with inferior partners is terminated (23)

(definitions for this chapter are summarized in Table 1). When cooperation entails a fitness cost to the actor, there is an intrinsic conflict of interest between partners, as each could increase its fitness by giving less or taking more. Indeed, studies of natural symbioses have found that symbiont behaviour varies continuously from mutualistic to parasitic (10) and that hosts can cheat their symbionts through enslavement (11). The conflict of interest can also take the form of a "tragedy of the commons" when multiple symbionts interact with the host, as each symbiont could exploit the cooperation of the others (12).

1.1. Overview of Mutualistic Bacterial Symbioses

Mutualistic symbioses are typically comprised of a multicellular "host" and a microbial "symbiont." We focus here on bacterial symbionts that provide some metabolic capability that their host lacks. A prominent example is nitrogen-fixation by alpha-proteobacteria and cyanobacteria that associate with leguminous plants, lichens, and insects (5, 13). Bacterial symbionts that associate with insects are known to synthesize essential amino acids or B vitamins, to recycle nitrogen, and to expand the host's diet by degrading cellulose or detoxifying plant toxins (14). Photosynthesis by cyanobacteria is used as a nutritional strategy by some fungi, forming holobionts called lichens (15). Symbionts can also help protect their hosts, for example *Vibrio fischeri*, which are housed in a light organ of their host squids and provide chemiluminescence as camouflage (16), and a number of facultative bacteria that protect aphids from attack by fungi and parasitoids as well as heat stress (17). It is common for hosts to harbour multiple symbiont species, such as the case of the sharpshooter and its two symbionts, *Baumannia* and *Sulcia*, that serve complementary metabolic functions (18).

1.2. Network Models of Symbiosis

While metabolic models have been constructed only for the best-studied symbioses, it is already clear that these models can give considerable insight into the interaction between hosts and symbionts (for related modelling approaches, see Chapter 6 by Arita, and Chapter 20 by Behre et al.). Though metabolic network models have not been used to test evolutionary hypotheses, these models can provide estimates of the costs of cooperating with the host, identify potential conflicts of interest between host and symbionts, and suggest previously unidentified benefits to the host. We are aware only of network models constructed for *Buchnera aphidicola*, an aphid symbiont, and two species of rhizobia. A constraint based flux-balance analysis (FBA) model was developed for *B. aphidicola* containing 196 gene products with 240 metabolic compounds and 263 reactions that was parameterized with empirical data on export rates of essential amino acids (19). This model predicts that 94% of *B. aphidicola* genes are essential for growth and shows that amino acid export to the host is costly in terms of biomass production, suggesting that indirect selection for increased host fitness via vertical transmission has shaped this microbe's metabolism. *Rhizobium etli*, a symbiont of pea, has a metabolic network constructed of 363 genes with 371 metabolic compounds and 387 reactions (20). The objective function used to optimize fluxes across this network included amino acid cycling, ammonium export, and the accumulation of poly-beta-hydroxybutyrate (PHB) and glycogen as energy stores. FBA of *in silico* gene deletions gave several predictions consistent with experimental data; one intriguing result that suggests fitness conflict between host and symbiont is that removing the ability to store PHB and/or glycogen increases nitrogen fixation rates. A symbiosis interactome model including 440 proteins with 1,041 interactions was constructed for the *Medicago* symbiont *Sinorhizobium meliloti* (21). This model was used to predict novel symbiotic proteins, of which three out of four were validated experimentally. These models represent an important first step in understanding the mechanistic basis of costs and benefits in mutualistic symbioses.

1.3. Models of Mutualistic Symbiosis

Models attempting to explain the origin and maintenance of mutualistic symbioses typically investigate the question of how hosts and symbionts resist the "temptation to defect," i.e. to exploit a cooperative partner. Cooperation within and between species can be favoured by several mechanisms: (1) partner choice, the selection of one or multiple individuals to associate with on the basis of signals that reflect future cooperative investment (22); (2) partner fidelity, the association of individuals over repeated interactions, such that investment in the partner increases the cooperative benefits received in future iterations of the interaction (8); (3) sanctions, a form of punishment for poorly performing partners in which benefits are withheld or the interaction is terminated (23); and (4) vertical transmission of symbionts or spatially structured interactions, which can align fitness interests through multilevel selection (24). We introduce the modelling approaches that have been used to study mutualistic symbioses, summarized in Table 2. We discuss the context in which each model type is used and illustrate the methods with examples from the literature in Subheading 2; models are presented from the most simple to the most biologically realistic. In Subheading 3 we briefly summarize software used in this research and we conclude in Subheading 4 with future perspectives, including connecting evolutionary models to molecular network models, and guidelines for how to select a modelling approach.

2. Modelling Approaches

2.1. Game Theory

Game theory was originally developed by economists to explain how rational actors should behave in social interactions (25). The central concept is the "unbeatable strategy" that maximizes personal benefit against all potential partner strategies, also known as the "evolutionarily stable strategy" (ESS). In models of mutualistic symbioses, players may either "cooperate" or "cheat," depending on the amount of commodity they contribute to their partners. Game theoretic models are the simplest type of phenotypic model: the genetic basis of traits is not specified and population dynamics are typically assumed to be constant. Nevertheless, these simplifications allow for analysis of otherwise intractably complex games and strategies.

2.1.1. The Prisoner's Dilemma

The classic game used to ask how intra- and interspecific cooperation can evolve is the Prisoner's Dilemma (26, 27). As demonstrated in Fig. 1, the ESS of this game is mutual cheating, but two modifications allow cooperators to persist. First, iterated games between pairs of players can favour cooperation by allowing current investment to depend on the partner's past behaviour.

Table 2
A comparison of modelling approaches

Type	Model	Pros	Cons	References
Economic	Game Theory	Simple models with many extensions (e.g., spatial structure; repeated, flexible interactions)	Typically no ecological dynamics and no genetics	(27, 72, 73)
	Biological Markets	Focus is on partner choice and negotiation	Typically no ecological dynamics and no genetics	(22, 37)
Eco-evolutionary	Adaptive Dynamics	Fitness depends on population dynamics; Analytic characterization of equilibria	No genetic basis; assumes separation of ecological and evolutionary time scales	(74)
Genetic	Inclusive Fitness	Considers intraspecific interactions within the symbiosis	Typically no ecological dynamics	(24, 75)
	Population Genetics	Tracks allele frequencies; explicit genetic architecture	Typically no ecological dynamics	(55, 76)

	Cooperator S (I_S>0)	Cheater S (I_s=0)
Cooperator H (I_H>0)	$b(I_S)$ - $c(I_H)$, $b(I_H)$ - $c(I_S)$	-$c(I_H)$, - $b(I_H)$
Cheater H (I_H=0)	$b(I_S)$, - $c(I_s)$	0,0

Fig. 1. Payoff matrix for the Prisoner's Dilemma game. The payoffs for the host (H) and the symbiont (S) depend on the amounts each player invests in the interaction (I). Investment by each player gives a benefit ($b(I)$) to the partner at a cost ($c(I)$) to itself. In this game, mutual cooperation yields a higher payoff than mutual cheating. If its partner cooperates, a player could gain more by investing nothing, but cooperating with a cheater is costly.

Second, spatial structure can favour cooperation by causing repeated interactions with the same partner, or its close relatives, who are likely to share the same strategy. An example that combines these conceptual mechanisms is the model of Doebeli and Knowlton (28). The benefits ($b(I)$) and costs ($c(I)$) of the amount invested (I) are:

$$b(I) = b_0(1 - e^{-b_1 I})$$
$$c(I) = c_0 I$$

where coefficients b_0, b_1, and c_0 determine the shape of the benefit and cost curves. These benefits and costs yield the initial payoffs for the host ($P_H[1]$) and symbiont ($P_S[1]$) with strategies consisting of the initial investment amounts ($I_H[1]$ and $I_S[1]$):

$$P_H[1] = b_0(1 - e^{-b_1 I_S[1]}) - c_0 I_H[1]$$
$$P_S[1] = b_0(1 - e^{-b_1 I_H[1]}) - c_0 I_S[1]$$

The reward rates (b_H and b_S) determine the effect of the previous payoff on investment in the next iteration, so that investment by one partner is rewarded by investment by the other partner:

$$I_H[2] = I_H[1] + b_H P_H[1]$$
$$I_S[2] = I_S[1] + b_S P_S[1].$$

This type of flexible strategy is further discussed in Note 1. The evolution of cooperation can then be modelled by replacing resident trait values with mutant trait values whenever individuals with a mutated trait perform better than the resident. In non-spatial simulations, investment by both species evolves to zero. However, if hosts and symbionts interact only with neighbours, investment and reward rate increase through coevolution, maintaining the mutualism.

2.1.2. Summary and Extensions

While iterated interactions between the same individuals are unlikely to be a realistic model of bacterial symbionts, which typically associate for multiple bacterial generations within their hosts, these simplified models provide valuable intuition for the processes that enable mutualism to evolve. In this case, the key ingredients are prolonged interactions and spatial structure that act to couple host and symbiont fitness. Another two-player cooperation game is the Hawk–Dove or Snowdrift game, which is the same as the Prisoner's Dilemma except that the payoff from mutual cheating is worse than that from being cheated; thus, even when the partner defects, the best strategy is to cooperate. This game may more accurately describe interactions in which players cooperate by producing a public good (29), making it relevant for cases with a potential "tragedy of the commons." Frequency-dependent selection is generated even in single-round games, such that cooperators and cheaters coexist in the absence of spatial structure (30). Spatial structure has the opposite effect on co-operation in the Snowdrift game as in the Prisoner's Dilemma, with lower frequencies of cooperation than in non-structured populations (29).

2.2. Biological Markets

The games described above assume that individuals are constrained to interact with a given partner. However, individuals often have control over whether they interact and which partners, if any, are chosen. In these cases, commodity exchange resembles a market where many potential partners contribute to the supply of and demand for commodities. In an idealized economic model of resource trade, the divergent capacities of hosts and symbionts for resource acquisition can result in "comparative advantage", i.e. trading the resource that each species can obtain cheaply to enable net benefit for both (31). However, the game theoretic considerations above suggest that additional mechanisms are required to ensure fair trade.

2.2.1. Cooperative Bargaining in the Rhizobia–Legume Symbiosis

Negotiation is a way for partners to reach a mutually satisfactory solution and can be modelled through cooperative bargaining games. In these games, individuals do not exchange commodities until the partner makes an acceptable offer (32, 33). Akçay and Roughgarden (34) take a cooperative bargaining approach to address the evolution of resource exchange in the rhizobia–legume symbiosis. In their model, host-plant growth (g_H) depends on the simultaneous amino acid flux (A_H) and carbon flux (C_H), while rhizobia growth (g_S) depends on the allocation of carbon to the nodules (C_S) minus the amount of energy used to fix nitrogen that is exported to the plant (A_S):

$$g_H = A_H C_H,$$

$$g_S = C_S - A_S.$$

A given set of growth rates, $G = (g_H, g_S)$, is termed a "contract"; altered nutrient fluxes cause different growth rates and are considered "offers", $G' = (g'_H, g'_S)$. Offers that increase both partners' growth are accepted immediately. If an offer decreases one partner's growth, the host and symbiont play a "war of attrition". The length of time each partner stays in the game is relative to (1) the cost of not receiving mutualistic benefits and (2) the benefit received under its preferred contract. This leads to the condition that G' will be accepted if $g'_H g'_S > g_H g_S$. Thus, the host and symbiont reach a "Nash Bargaining Solution" defined as the investment level that maximizes their joint payoff.

2.2.2. Summary and Extensions

Cooperative bargaining is an attractive model for symbiosis as it provides a mechanism for the resolution of conflicts over fair trading that arise in biological markets. Resource exchange fluxes between host and symbiont represent a candidate mechanism for this type of interaction (35). Other models incorporate aspects of negotiation through the formation and/or termination of the interaction. This "partner choice" enables hosts and symbionts to preferentially interact with cooperators (22). Partner choice requires that there be an abundance of partners to choose among and that individuals be able to recognize which partners will be cooperators. The rhizobia–legume symbiosis meets these conditions (36), and while the mechanism is unknown it may involve variation in host–symbiont signalling during the formation of symbiosis. To advertise their cooperativeness, players may "bid" for the commodity offered by the partner species (37). Alternatively, individuals may simply terminate poor interactions and acquire new partners (38, 39). Hosts may have an advantage here, since bacterial symbionts are likely to be killed by the host's defence responses if the interaction goes poorly and the bacteria may not be able to disperse from their hosts if they are being exploited.

2.3. Adaptive Dynamics

Whether a mutant can successfully invade a population is likely to depend on the population densities and traits of both the focal and interacting species. As the population dynamics themselves depend on the phenotypes present, trait evolution can feed-back into the ecological dynamics of a species, producing the coupled ecological and evolutionary dynamics that are the basis of the "Adaptive Dynamics" framework (40, 41). To manage the increased complexity of eco-evolutionary feedbacks, these models simplify evolution by ignoring the genetic basis of traits and assuming that ecological dynamics reach equilibrium in between mutation events. Note 2 further discusses modelling phenotypes rather than genotypes and Note 3 discusses the potential importance of population dynamics. Adaptive dynamics models differ from most other approaches in that they explicitly test for disruptive selection leading populations to become polymorphic through "evolutionary branching." This outcome occurs when fitness is nonlinear and frequency-dependent, such that rare phenotypes have an advantage. In other cases, the adaptive dynamics approach demonstrates that species evolve to ESS trait values identical to those attained in other phenotypic models.

2.3.1. Adaptive Dynamics of Obligate Mutualisms

Ferrière et al. (42) use adaptive dynamics to investigate the evolution of commodity exchange between two obligate mutualist species. The ecological dynamics of two mutualists with population densities H and S are given by per-capita growth rates:

$$\frac{1}{H}\left(\frac{\mathrm{d}H}{\mathrm{d}t}\right) = -g_H(I_H) - d_H H + SI_S(1 - \alpha H),$$
$$\frac{1}{S}\left(\frac{\mathrm{d}S}{\mathrm{d}t}\right) = -g_S(I_S) - d_S S + HI_H(1 - \beta S),$$

where I_H and I_S are the amount of commodities traded between species. The per-capita growth rates are decreased by investment in these commodities ($-g_H(I_H)$ and $-g_S(I_S)$) and by intraspecific competition ($-d_H H$ and $-d_S S$). Meanwhile, the population growth rates are increased by receiving commodities from the partner species ($S\,I_S$ and $H\,I_H$), modified by intraspecific competition ($(1-\alpha H)$ and $(1-\beta S)$). For given trait values, both partners reach ecological equilibrium densities H^* and S^*.

The evolution of commodity investment proceeds by considering the fate of rare mutations in the investment traits of each species. The growth rate of a rare mutant is given by the invasion fitnesses (w_H and w_S) of mutants with phenotypes $I_{H\mathrm{mut}}$ and $I_{S\mathrm{mut}}$ at the ecological equilibrium:

$$w_H(I_{H\mathrm{mut}}, I_H, I_S) = -g_H(I_{H\mathrm{mut}}) - d_H H^* + I_S[1 - \alpha(I_{H\mathrm{mut}} - I_H)H^*]S^*,$$
$$w_S(I_{S\mathrm{mut}}, I_H, I_S) = -g_S(I_{S\mathrm{mut}}) - d_S S^* + I_H[1 - \beta(I_{S\mathrm{mut}} - I_H)S^*]H^*.$$

In addition to affecting each species' intrinsic growth rate, the amount invested into mutualism affects intraspecific competition for partners, described by the functions α and β. Competition is symmetric when α and β are constants and asymmetric when they are functions of the mutation size and direction. The direction of coevolution is found by calculating how invasion fitness changes as a function of the mutant trait; setting these fitness gradients to zero gives the coevolutionary outcome:

$$\left.\frac{\partial w_H}{\partial I_{H\mathrm{mut}}}\right|_{I_{H\mathrm{mut}}=I_H} = -g'_H(I_H) + \alpha' I_S H^* S^* = 0,$$
$$\left.\frac{\partial w_S}{\partial I_{S\mathrm{mut}}}\right|_{I_{S\mathrm{mut}}=I_S} = -g'_S(I_S) + \beta' I_H H^* S^* = 0.$$

The points that satisfy each equality are the trait values that the species evolves towards and can be visualized as evolutionary

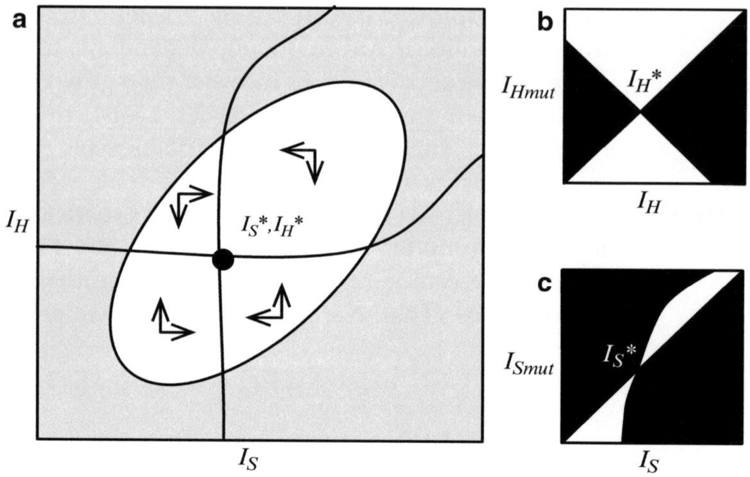

Fig. 2. The adaptive dynamics approach. (**a**) Diagram of the ecological and evolutionary space with respect to host and symbiont levels of investment (I_H and I_S). The species coexist at intermediate investment levels (*white*), but go extinct when investment is too high or too low (*grey*). Evolutionary isoclines (*solid lines*) are the set of trait values that each species evolves towards. The intersection of these isoclines (I_H^*, I_S^*) is the coevolutionary singularity; in this case, it is an attractor, as shown by the vectors of the selection gradient. The evolutionary process can be further visualized using "pairwise invasibility plots" for the host (**b**) and the symbiont (**c**). These plots show the sign of the invasion fitness of a rare mutant in a resident population, given that the trait of the partner species is fixed at the coevolutionary singular point. Combinations of mutant and resident trait values at which the mutant can invade the population are *black*; those at which it cannot invade are *white*. In (**b**), when resident investment (I_H) is above I_H^*, mutants with lower investment can invade (points below the diagonal are *black*); conversely, if I_H is below I_H^*, mutants with higher values can invade (points above the diagonal are *black*). At the intersection of these regions, no strategy can invade (the areas above and below the diagonal are *white*), i.e. this is an ESS. However, in (**c**), the symbiont exhibits different dynamics; at the convergent point I_S^*, mutants with any other investment level can invade (the areas above and below the diagonal are *black*). Consequently, the population experiences disruptive selection that leads to the emergence of polymorphism through "evolutionary branching".

isoclines (Fig. 2a). Where these evolutionary isoclines intersect, there is a coevolutionary singularity (43) (Fig. 2) that the species will evolve to if it is ecologically feasible. If competition is symmetric, α' and β' are zero, therefore investment in mutualism evolves to zero and both obligate mutualists go extinct. When competition is asymmetric and gives better partners a moderate competitive advantage, the trait values in each species evolve to the evolutionary singular point. This point can either be (1) an ESS (Fig. 2b) and evolution will halt or (2) an "evolutionary branching point" (Fig. 2c) and the species will become polymorphic. In this model, the type of singular point depends on whether investment increases or decreases the competitive advantage and whether the costs of investment increase or decrease with investment level. Measuring the average level of investment after evolutionary branching shows that polymorphism is beneficial for the mutualism overall, as highly cooperative partners more than compensate for cheaters.

2.3.2. Summary and Extensions

This model makes the intriguing prediction that diversification of hosts and/or symbionts increases the total amount of resources invested in mutualism, and supports the idea that asymmetric investment into partners based on their cooperativeness (e.g. through partner choice) is important in order for obligate mutualisms to evolve. While most symbiosis models examine how the level of cooperation evolves, other important questions are how vertical transmission and obligate symbioses can evolve from interactions that presumably began as horizontally acquired and facultative. Adaptive dynamics models predict that if avoiding infection is costly to the host, then vertical transmission of symbionts is favoured, and, furthermore, that vertical transmission favours the evolution of dependency (44).

2.4. Inclusive Fitness

Even if cooperation is not reciprocated, cooperative genes can receive indirect benefits by helping relatives. This concept is termed "inclusive fitness" and dates back to Hamilton (45), who showed that for costly helping behaviour to be selected, the costs (c) must be less than the benefits (b) weighted by the average relatedness between actor and recipient (r), i.e. $br\text{-}c > 0$. Inclusive fitness models use this relationship to determine whether selection can favour cooperative traits that do not directly aid the actor. These models typically do not specify the genetic basis for cooperative traits, but rather use Hamilton's relatedness parameter (r), which can be understood as the proportion of shared genes between actor and recipient and is defined more formally as a phenotypic or genotypic regression (46). This general formulation of inclusive fitness theory has been widely applied to study cooperation and conflict in bacteria (47).

2.4.1. Inclusive Fitness of Symbiont–Symbiont Conflict

West et al. (23) take an inclusive fitness approach to determine the role of intraspecific interactions when multiple related symbionts interact with a single host, such as nitrogen-fixing rhizobium bacteria interacting with a legume host-plant. Without allowing the plant strategy to evolve, they consider how host sanctions influence the evolution of symbiont investment. Rhizobium fitness is a function of nitrogen-fixation strategy, I_S: $w_S = I_H(1 - I_S)f(I_S)$, where host investment is a linear function of symbiont strategy ($I_H = a + I_S$) and $f(I_S)$ is the effect of host sanctions on symbiont fitness. The parameter a controls the importance of symbiotically fixed nitrogen to plant fitness.

Differentiating fitness (w_S) with respect to nitrogen-fixation rate (I_S) and using the regression definition of relatedness (r) between rhizobia that infect different nodules on the same plant allows the ESS level of nitrogen-fixation to be calculated as

$$I_S{}^* = 1 - 2a + r + \sqrt{(4a - 4a^2 + (1 + r)^2}/2(2 + r),$$

which is an increasing function of r. Comparing models with sanctions on individual nodules versus the case where plant resources are distributed evenly across the nodules of a plant demonstrates that nodule sanctions dramatically increase the level of N-fixation.

2.4.2. Summary and Extensions

This model demonstrates that sanctions can maintain much higher levels of symbiont cooperation than kin selection alone, providing a mechanism to avert the "tragedy of the commons" when multiple symbionts infect a host. This approach can also be used to model host and symbiont coevolution if relatedness is interpreted as genetic covariance (24). The regression between mutualist genotypes across partner species, r, can be positive due to spatial structure (28) or partner choice (22). In the extended formulation of Frank (24), the condition for host investment I_H to increase is: $b_1 r_{\bar{I}_H,I_H} + b_2 r_{\bar{I}_S,I_H} - c>0$, where $r_{\bar{I}_H,I_H}$ is the regression of sub-population host investment on individual host investment, $r_{\bar{I}_S,I_H}$ is the regression of sub-population symbiont investment on individual symbiont investment, c is the cost to the individual host, and the b_1 and b_2 are the respective benefits. Frank's model (24) demonstrates that spatial structure allows positive correlations between mutualist species to evolve through positive fitness feedbacks acting in local populations. This is similar to multilevel selection models, but explicit models incorporating within- and between-host selection on symbiont traits have only been explored thus far in pathogenic systems (48).

2.5. Population Genetics

Population genetic models differ from the phenotypic modelling approaches described above in that the genetic basis of mutualistic behaviour is specified (see Note 2). The dynamics of genetic models can give insight into the consequences of epistatic selection, which captures how alleles in the two species interact with one another to give non-additive fitness effects. Many population genetic models, however, assume simple single-gene haploid or diploid inheritance, making them similar to models where the genetics are unspecified (49–52). Furthermore, these models typically do not incorporate population dynamics, focusing instead on aspects of the evolutionary process such as rates of adaptation, fixation probabilities of beneficial alleles, and the existence of balanced polymorphisms. However, they do allow evolutionary processes to occur on the same time scale as ecological dynamics and might be better suited to cases of high mutation frequency.

2.5.1. Population Genetics of Coral Symbioses

Day et al. (53) develop a population genetic model to describe how coevolution between corals and zooxanthellae may influence coral bleaching. Although zooxanthellae are photosynthetic dinoflagellates, this model is readily applicable to any facultative symbiosis involving a host and a bacterial partner. Their model demonstrates how ecological and genetic models can be combined. Coral hosts (H) and zooxanthellae symbionts (S) can reproduce both in free-living form and in their associated "holobiont" form (M). One way that bleaching of the holobiont, i.e. failure of the symbiosis, occurs is through dissociation at rate ξ. Hosts and symbionts associate to form new holobionts at rate γ. The population dynamics are thus:

$$\mathrm{d}M/\mathrm{d}t = g_M M - \xi M + \gamma HS,$$
$$\mathrm{d}H/\mathrm{d}t = \xi M + g_H H - \gamma HS,$$
$$\mathrm{d}S/\mathrm{d}t = \xi M + g_S S - \gamma HS.$$

The growth rate of hosts (g_H) is assumed to be negative, while the growth rates of holobionts (g_M) and symbionts (g_S) include logistic density dependence. Both host and symbiont are assumed to be haploid, with a single di-allelic locus that influences the holobiont dissociation threshold through ξ. The holobiont population dynamics can depend on host and symbiont genotypes through the growth parameter (g_M) and the rate of dissociation (ξ). There can be both an additive effect of the genotypes as well as an epistatic interaction term.

The model can be simplified by assuming that the alleles that determine the tolerance of the symbiosis do not affect free-living growth rates. Selection on host and symbiont alleles is described by the parameters w_H and w_S. Under a purely additive model, the host and symbiont allele frequencies (p_{HM} and p_{SM}) in the holobiont are given by:

$$\mathrm{d}p_{HM}/\mathrm{d}t = p_{HM}(1 - p_{HM})w_H + Dw_S,$$
$$\mathrm{d}p_{SM}/\mathrm{d}t = p_{SM}(1 - p_{SM})w_S + Dw_H,$$

where D is the interspecific linkage disequilibrium, whose dynamics are given by

$$\mathrm{d}D/\mathrm{d}t = D[(1 - 2p_{HM})w_H + (1 - 2p_{SM})w_S].$$

Thus, resistance alleles experience not only direct selection, but also indirect selection due to associations with the resistance allele in the other species. The type of genetic effect is important to the evolutionary outcome. With additive genetic effects, interspecific linkage disequilibrium decays and the partners evolve independently. Alternatively, positive epistasis can occur, with the strongest bleaching tolerance when both resistance alleles are present. Increasing linkage between partners' genotypes causes indirect selection to accelerate the evolution of bleaching tolerance. On the other hand, negative epistasis occurs when only a single resistance allele in either host or symbiont is sufficient for bleaching tolerance. In this case, negative genetic correlations between host and symbiont evolve and indirect selection slows the evolution of tolerance.

2.5.2. Summary and Extensions

This example shows that even simplified genetic systems can lead to complex models when multiple species are interacting in an ecological framework. In particular, it demonstrates that epistasis between species can impact the resilience of the symbiosis. More genetically complex models have been developed, such as the class of multi-locus population genetic models (e.g. 54). Analysis of such models based on different types of ecological interactions conclude that only a single polymorphic locus can be maintained with competitive or mutualistic interactions, while host–pathogen interactions yield a variety of outcomes including limit cycles and chaotic allele dynamics (55). Another type of population genetic model considers the combined action of genetic drift and selection; an application of this technique to the aphid-*Buchnera* symbiosis demonstrates that host population size and the number of symbionts transmitted determines the rate of fixation of deleterious, beneficial, and selfish mutations (56).

3. Available Software

While there are no dedicated software packages for evolutionary models of symbiosis, Mathematica (57) is widely used for its ability to perform symbolic mathematics. Simulations can be run in any programming language, but Matlab (58) has user-friendly debugging routines, R (59) has a wide array of built-in statistical distributions, and C (60)/C++ (61) are unmatched for their speed and usability. Meanwhile, an extensive range of programs have been developed for individual- or agent-based simulations, which track the behaviour and status of individuals following programmed rules (reviewed in ref. 62).

4. Perspectives

Thus far, models of the evolution of mutualistic symbioses have focused on broad and relatively general questions: (1) how do species overcome the temptation to defect? (2) how is the tragedy of the commons averted when multiple symbionts interact with a host? and (3) how do symbioses become vertically transmitted and/or obligate? We look forward to the integration of different modelling approaches and a more complete synthesis of how different mechanisms shape these interactions (e.g. 63).

Meanwhile, the applicability of these evolutionary models to biological systems can be dramatically improved by understanding the mechanisms underlying cooperation. For example, our growing understanding of the genetic basis of symbioses makes it possible to develop evolutionary models with parameterizations based on molecular and functional data (64–66). An area ripe for integration is the use of metabolic and protein–protein networks to establish costs and benefits of different strategies and identify the molecular trade-offs that shape symbiosis evolution. Combining network models with existing evolutionary approaches will enable us to make coevolutionary predictions for real-world symbioses, including under global change, and to design novel symbioses. Understanding the structure of these networks may also enable predictions about the speed and feasibility of evolutionary change to the interaction, such as the conditions that would lead mutualistic symbionts to transition into organelles or pathogens. Empirically parameterized models will provide insight into studies of experimental evolution, where the effects of perturbations in the structure of interactions can be tested, molecular coevolution, in order to detect cross-genome epistatic selection, and even comparative phylogenetics, to test hypotheses of diversification and extinction rates with respect to patterns of symbioses (e.g. 67).

5. Notes

When modelling the evolution of a symbiosis, it is important to combine simple analytical equations with more complex simulations as a means of cross-validating the robustness of assumptions and conclusions. In addition, several key decisions must be made, including:

1. Phenotype or Genotype? Phenotypic models that assume quantitative traits are simpler and are likely to give the same results as a single-locus haploid genetic model. Thus, they may be sufficient in many cases. However, genetic models can provide important insight if there is a non-additive interaction between alleles of host and symbiont. They can also model situations where evolution occurs on the same timescale as ecological dynamics.
2. Constant or Dynamic Populations? When symbioses involve non-linear cost and benefit functions, the interactions can be strongly dependent on population densities. Furthermore, interactions that are sometimes or always pathogenic tend to have non-equilibrium population dynamics. In these cases, models that explicitly consider population dynamics will likely give the most insight. Exceptions might be facultative symbioses, wherein population densities are largely determined by other factors.

3. Fixed or Flexible Strategies? Species may have very different options for how they interact. At the most basic level, all models assume that evolving species either invest in the interaction or not. The investment level may be constant or updated over repeated interactions. Additionally, species may have the potential to choose whether to interact at all, what partner(s) to interact with, and how long to interact for. Partners may also communicate to advertise their cooperativeness or to negotiate. Finally, hosts may be able to impose sanctions against cheaters and to control symbiont transmission mode.

In the course of modelling, a number of problems can be encountered. These include the classification of an evolutionary equilibrium (just because it exists does not mean it can be reached, or that it will be stable), double counting of individuals when considering inclusive fitness, and different consequences of synchronous and asynchronous updating in (spatial) simulations. Therefore, we recommend consultation of more in depth resources, including the guides by Otto and Day (68), Edelstein-Keshet (69), and Hastings (70), as well as those listed in Table 2.

Acknowledgments

We thank D. Drown, B. Foley, the volume editors, and an anonymous reviewer for helpful comments on the manuscript. This work was supported by NSF PGRP 0820846 to S. Nuzhdin and NSF DMS 0540524 to R. Gomulkiewicz.

References

1. Cavalier-Smith T. (2006) Origin of mitochondria by intracellular enslavement of a photosynthetic purple bacterium. Proc R Soc B-Biol Sci, 273:1943–1952.
2. Gould SB, Waller RF, McFadden GI. (2008) Plastid evolution. Annu Rev Plant Biol, 59:491–517.
3. Archibald JM. (2009) The puzzle of plastid evolution. Curr Biol, 19:R81–R88.
4. Heckman DS, Geiser DM, Eidell BR, Stauffer RL, Kardos NL, Hedges SB. (2001) Molecular evidence for the early colonization of land by fungi and plants. Science, 293:1129–1133.
5. Kneip C, Lockhart P, Voß C, Maier U. (2007) Nitrogen fixation in eukaryotes: new models for symbiosis. BMC Evol Biol, 7:55.
6. Moya A, Peretó J, Gil R, Latorre A. (2008) Learning how to live together: genomic insights into prokaryote-animal symbioses. Nat Rev Genet, 9:218–229.
7. Moran NA. (2006) Symbiosis. Curr Biol, 16: R866–R871.
8. Sachs JL, Mueller UG, Wilcox TP, Bull JJ. (2004) The evolution of cooperation. Q Rev Biol, 79:135–160.
9. West SA, Griffin AS, Gardner A. (2007) Social semantics: altruism, cooperation, mutualism, strong reciprocity and group selection. J Evol Biol, 20:415–432.
10. van Baalen M, Jansen VAA. (2001) Dangerous liaisons: the ecology of private interest and common good. Oikos, 95:211–224.

11. Frean MR, Abraham ER. (2004) Adaptation and enslavement in endosymbiont-host associations. Phys Rev E, 69:051913.
12. Hardin G. (1968) The tragedy of the commons. Science, 162:1243–1248.
13. Grube M, Cardinale M, de Castro JV, Mueller H, Berg G. (2009) Species-specific structural and functional diversity of bacterial communities in lichen symbioses. ISME J, 3:1105–1115.
14. Douglas AE. (2009) The microbial dimension in insect nutritional ecology. Funct Ecol, 23:38–47.
15. Honegger R. (1991) Functional aspects of the lichen symbiosis. Annu Rev Plant Phys, 42:553–578.
16. Nyholm SV, McFall-Ngai M. (2004) The winnowing: establishing the squid-*vibrio* symbiosis. Nat Rev Microbiol, 2:632–642.
17. Oliver KM, Degnan PH, Burke GR, Moran NA. (2010) Facultative symbionts in aphids and the horizontal transfer of ecologically important traits. Annu Rev Entomol, 55:247–266.
18. McCutcheon JP, Moran NA. (2007) Parallel genomic evolution and metabolic interdependence in an ancient symbiosis. Proc Natl Acad Sci U S A, 104:19392–19397.
19. Thomas GH, Zucker J, Macdonald SJ, Sorokin A, Goryanin I, Douglas AE. (2009) A fragile metabolic network adapted for cooperation in the symbiotic bacterium *Buchnera aphidicola*. BMC Syst Biol, 3:24.
20. Resendis-Antonio O, Reed JL, Encarnación S, Collado-Vides J, Palsson BØ. (2007) Metabolic reconstruction and modeling of nitrogen fixation in *Rhizobium etli*. PLoS Comput Biol, 3:1887–1895.
21. Rodriguez-Llorente I, Caviedes MA, Dary M, Palomares AJ, Cánovas FM, Peregrín-Alvarez JM. (2009) The symbiosis interactome: a computational approach reveals novel components, functional interactions and modules in *Sinorhizobium meliloti*. BMC Syst Biol, 3:63.
22. Noë R, Hammerstein P. (1994) Biological markets: supply and demand determine the effect of partner choice in cooperation, mutualism and mating. Behav Ecol Sociobiol, 35:1–11.
23. West SA, Kiers ET, Simms EL, Denison RF. (2002) Sanctions and mutualism stability: why do rhizobia fix nitrogen? Proc R Soc B-Biol Sci, 269:685–694.
24. Frank SA. (1994) Genetics of mutualism: the evolution of altruism between species. J Theor Biol, 170:393–400.
25. von Neumann J, Morgenstern O. (1944) *Theory of Games and Economic Behavior*. Princeton University Press, Princeton, NJ.
26. Trivers RL. (1971) Evolution of reciprocal altruism. Q Rev Biol, 46:35–57.
27. Axelrod R, Hamilton WD. (1981) The evolution of cooperation. Science, 211:1390–1396.
28. Doebeli M, Knowlton N. (1998) The evolution of interspecific mutualisms. Proc Natl Acad Sci U S A, 95:8676–8680.
29. Doebeli M, Hauert C. (2005) Models of cooperation based on the Prisoner's Dilemma and the Snowdrift game. Ecol Lett, 8:748–766.
30. Doebeli M, Hauert C, Killingback T. (2004) The evolutionary origin of cooperators and defectors. Science, 306:859–862.
31. Schwartz MW, Hoeksema JD. (1998) Specialization and resource trade: biological markets as a model of mutualisms. Ecology, 79:1029–1038.
32. Nash JF. (1950) The bargaining problem. Econometrica, 18:155–162.
33. Nash JF. (1953) Two-person cooperative games. Econometrica, 21:128–140.
34. Akçay E, Roughgarden J. (2007) Extra-pair parentage: a new theory based on transactions in a cooperative game. Evol Ecol Res, 9:1223–1243.
35. Lodwig EM, Hosie AHF, Bordes A, Findlay K, Allaway D, Karunakaran R, Downie JA, Poole PS. (2003) Amino-acid cycling drives nitrogen fixation in the legume-*Rhizobium* symbiosis. Nature, 422:722–726.
36. Heath KD, Tiffin P. (2009) Stabilizing mechanisms in a legume-*rhizobium* mutualism. Evolution, 63:652–662.
37. Noë R, Hammerstein P. (1995) Biological markets. Trends Ecol Evol, 10:336–339.
38. Johnstone RA, Bshary R. (2008) Mutualism, market effects and partner control. J Evol Biol, 21:879–888.
39. Johnstone RA, Bshary R. (2002) From parasitism to mutualism: partner control in asymmetric interactions. Ecol Lett, 5:634–639.
40. Metz JAJ, Nisbet RM, Geritz SAH. (1992) How should we define "fitness" for general ecological scenarios? Trends Ecol Evol 7:198–202.

41. Dieckmann U, Law R. (1996) The dynamical theory of coevolution: a derivation from stochastic ecological processes. J Math Biol, 34:579–612.
42. Ferrière R, Bronstein JL, Rinaldi S, Law R, Gauduchon M. (2002) Cheating and the evolutionary stability of mutualisms. Proc R Soc B-Biol Sci, 269:773–780.
43. Geritz SAH, Metz JAJ, Kisdi E, Meszéna G. (1997) Dynamics of adaptation and evolutionary branching. Phys Rev Lett, 78:2024–2027.
44. Law R, Dieckmann U. (1998) Symbiosis through exploitation and the merger of lineages in evolution. Proc R Soc B-Biol Sci, 265:1245–1253.
45. Hamilton WD (1964) Genetical evolution of social behaviour I. J Theor Biol, 7:1–16.
46. Frank SA (1998) *Foundations of Social Evolution*. Princeton University Press, Princeton, NJ
47. West SA, Diggle SP, Buckling A, Gardner A, Griffins AS. (2007) The social lives of microbes. Annu Rev Ecol Evol System, 38:53–77.
48. Mideo N, Alizon S, Day T. (2008) Linking within- and between-host dynamics in the evolutionary epidemiology of infectious diseases. Trends Ecol Evol, 23:511–517.
49. Nuismer SL, Gomulkiewicz R, Morgan MT. (2003) Coevolution in temporally variable environments. Am Nat, 162:195–204.
50. Gandon S. (2002) Local adaptation and the geometry of host–parasite coevolution. Ecol Lett, 5:246–256.
51. Gandon S, Michalakis Y. (2002) Local adaptation, evolutionary potential and host–parasite coevolution: interactions between migration, mutation, population size and generation time. J Evol Biol, 15:451–462.
52. Nuismer SL, Thompson JN, Gomulkiewicz R. (1999) Gene flow and geographically structured coevolution. Proc R Soc B-Biol Sci, 266:605–609.
53. Day T, Nagel L, Van Oppen MJH, Caley MJ. (2008) Factors affecting the evolution of bleaching resistance in corals. Am Nat, 171: E72–E88.
54. Ridenhour BJ, Nuismer SL (2007) Polygenic traits and parasite local adaptation. Evolution, 61:368–376.
55. Kopp M, Gavrilets S. (2006) Multilocus genetics and the coevolution of quantitative traits. Evolution, 60:1321–1336.
56. Rispe C, Moran NA. (2000) Accumulation of deleterious mutations in endosymbionts: Muller's ratchet with two levels of selection. Am Nat, 156:425–441.
57. Wolfram Research Inc. (2008) *Mathematica Version 7.0*. Wolfram Research Inc., Champaign, IL.
58. MathWorks. (2009) *MATLAB Version 7.9*. MathWorks, Natick, MA.
59. R Development Core Team. (2009) *R: A Language and Environment for Statistical Computing*. R Foundation for Statistical Computing, Vienna, Austria.
60. Ritchie D. (1972) *C Programming Language*. Bell Telephone Laboratories, Murray Hill, NJ.
61. Stroustrup B. (1979) *C++ Programming Language*. Bell Telephone Laboratories, Murray Hill, NJ.
62. Railsback SF, Lytinen SL, Jackson SK. (2006) Agent-based simulation platforms: review and development recommendations. Simul Trans Soc Model Simul Int, 82:609–623.
63. Foster KR, Wenseleers T. (2006) A general model for the evolution of mutualisms. J Evol Biol, 19:1283–1293.
64. Griffin A, West S, Buckling A. (2004) Cooperation and competition in pathogenic bacteria. Nature, 430:1024.
65. Morgan AD, Gandon S, Buckling A. (2005) The effect of migration on local adaptation in a coevolving host–parasite system. Nature, 437:253–256.
66. Buckling A, Brockhurst MA (2008) Kin selection and the evolution of virulence. Heredity, 100:484–488.
67. Sachs JL, Simms EL (2006) Pathways to mutualism breakdown. Trends Ecol Evol, 21:585–592.
68. Otto SP, Day T. (2007) *A Biologist's Guide to Mathematical Modeling in Ecology and Evolution*. Princeton University Press, Princeton, NJ.
69. Edelstein-Keshet L. (1988) *Mathematical Models in Biology*. Random House, New York, NY.
70. Hastings A. (1997) *Population Biology: Concepts and Models*. Springer, New York, NY.
71. Connor RC. (1995) The benefits of mutualism: a conceptual framework. Biol Rev Camb Philos Soc, 70:427–457.
72. Hofbauer J, Sigmund K. (1998) *Evolutionary Games and Population Dynamics*. Cambridge University Press, Cambridge.

73. McElreath R, Boyd R. (2007) *Mathematical Models of Social Evolution: A Guide for the Perplexed*. University of Chicago Press, Chicago, IL.
74. Dercole F, Rinaldi S. (2008) *Analysis of Evolutionary Processes: The Adaptive Dynamics Approach and Its Applications*. Princeton University Press, Princeton, NJ.
75. Taylor PD, Frank SA. (1996) How to make a kin selection model. J Theor Biol, 180:27–37.
76. Roughgarden J. (1979) *Theory of Population Genetics and Evolutionary Ecology: An Introduction*. Macmillan Publishing Company, New York, NY.

Chapter 25

Modelling the Onset of Virulence in Pathogenic Bacteria

Wilfred D. Kepseu, Frédérique Van Gijsegem, and Jacques-Alexandre Sepulchre

Abstract

Bacterial virulence is a multifactorial process. In this chapter, we review some known mechanisms used by bacteria to trigger their production of virulence factors. We develop the idea that although the onset of virulence shows up an abrupt transition, the modelling of this dynamics can be classified in two qualitatively distinct infectious transitions which are respectively called "shift" or "switch." We review methods enabling one to determine the types of behaviour that can be exhibited by a given model and we consider applications in three cases of virulence factor regulation. We conclude that in most cases a "successful" infection would require that the onset of virulence follows an irreversible switch behaviour.

Key words: Pathogenicity, Virulence, Positive feedback, Bistability, Hysteresis, Pectinolytic enzyme, Quorum sensing, Type III secretion system

1. Introduction

Bacterial pathogens provoke diseases that can be very diverse in their symptoms and severities. They can also affect very different hosts including animals, plants or insects, and various tissues or organs within a given host. Nevertheless, many pathogens share common virulence traits. They can, for example possess efficient uptake systems to compete with their host for acquisition of the essential element iron (1, 2) and evolve a battery of enzymes and molecules to cope with the environmental stresses encountered inside the host eukaryotic cell. But perhaps the most effective virulence strategy shared by pathogenic bacteria is the diverse secretory systems used to deliver active proteins such as toxins, degradative enzymes, or effectors that subvert eukaryotic cell functions. In Gram-negative bacteria, these secretion systems are

Jacques van Helden et al. (eds.), *Bacterial Molecular Networks: Methods and Protocols*, Methods in Molecular Biology, vol. 804, DOI 10.1007/978-1-61779-361-5_25, © Springer Science+Business Media, LLC 2012

classified into four major and several minor groups (3) and are devoted to the transport of proteins across the bacterial inner and outer membranes into the environment. In addition, type III and IV systems are also able to directly inject into the eukaryotic cell bacterial effector proteins that can interfere with host cellular functions (4, 5).

The interaction of a pathogen with its host is a dynamic multistep process. In particular, pathogens often can colonize their host and reside latently in it without provoking any symptoms until disease occurs when the environmental conditions are favourable for massive microbial multiplication and induction of symptoms. This transition from latent infection to disease expression must be tightly controlled. Consequently, the expression of behaviours associated with virulence and virulence factors is coordinated at the different stages of infection, in order to meet requirements in energy conservation, appropriate disease development, evasion of host defences and eventual dispersal. The survival of a pathogen, therefore, relies on a controlled global virulence regulation network to express virulence genes at the right time and place. Pathogens use several mechanisms to control the coordinated expression of the virulence genes during the different phases of infection. In this chapter, a short review of three mechanisms of regulation is made from a modelling point of view and the transition principle following a switch-like behaviour to virulence is discussed.

2. Two Different Types of Virulence Onsets: The Shift or The Switch

Bacterial pathogenesis requires the precise spatial and temporal control of gene expression, the dynamics of which is controlled by gene regulatory networks (6, 7). These bacterial molecular networks orchestrate the onset of virulence, which is often characterized by a sharp transition between two states of production of virulence factors. Loosely speaking, we say that a system has a *switch-like* behaviour if, by variation of an infection-inducing parameter such as the concentration of a key molecule, the system changes from one state to another biologically different state. The two potential steady states generally correspond to low and high activity of the system respectively and, in the context of the onset of virulence to the latent and the virulent states. However, two distinct phenomena can lie behind a switch-like behaviour, and, as it is explained below, it may be biologically relevant to distinguish which one occurs in a specific situation.

According to other authors (8, 9) we refer to a *shift* as a transition which characterizes a system with only one but continuously deformable stable state: by increasing some infection parameter, the virulent state is turned on in a sharp manner but

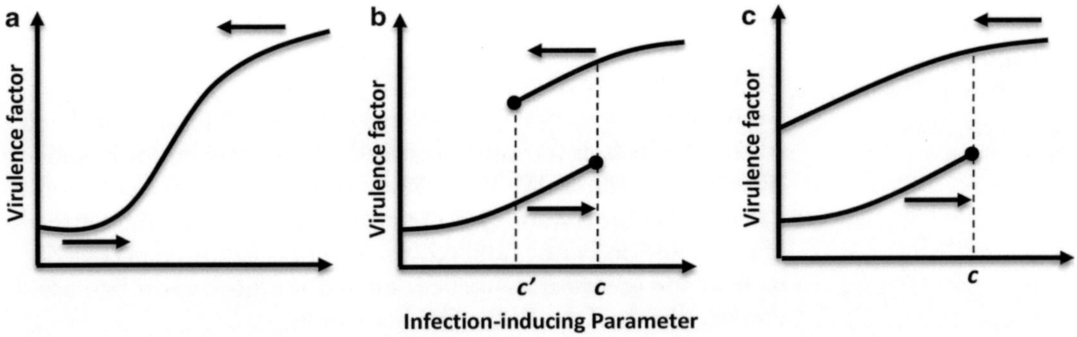

Fig. 1. Schematic representation of the steady state of virulence for monostable and bistable systems. (**a**) Shift transition observed in monostable systems. The signal-response curve is independent of the direction of change in the signal intensity. (**b**) Hysteresis in bistable system. Two stable steady states coexist when the infection parameter is in the (c', c) range. Depending on the direction of change and amplitude in the infection parameter, the system will be characterized by different response-curves. (**c**) Irreversibility in a bistable system.

if the parameter decreases to its initial value, the system returns to the former state (9) (Fig. 1a). In fact, for each value of this parameter, the system follows a *single branch* of stable states but there is a small range of parameters where the change is ultrasensitive. The *switch* is a structurally different type of behaviour (Fig. 1b, c). It is characterized by an abrupt discontinuous transition between *two branches* of stable states when the infection parameter is moved. Thus, a switch is associated with the mathematical concept of "bifurcation in a bistable system." A fundamental characteristic of a switch is the *hysteresis* phenomenon. This means that if the change from stable state A to stable state B occurs at a critical value c when the infection parameter increases, even if this parameter decreases below c the system remains in state B until it possibly reaches another bifurcation point $c' < c$ (Fig. 1b). For any value of the parameter between c and c', the system is bistable. A classical example of this behaviour is the lysis/lysogeny switch of phage λ (10, 11). If the hysteresis is pronounced enough, it can turn into a form of actively maintained irreversibility, with the system remaining "on" even after the stimulus is lowered to zero (12) (Fig. 1c). Within these limits, a bistable system acts like an irreversible switch (12).

Therefore, although at the onset of an infection a sharp transition is usually observed between two biological states (latent/virulent), the actual number of stable states of the system is a fundamental qualitative difference between a shift and a switch. While in the case of a switch two stable states coexist in a region of the parameter space, in the case of a shift the system has a unique stable state for each value of the parameter.

In the next section, we review methodologies to investigate, in a bacterial network, whether the transition between the latent and the virulent states is actually a shift or a switch.

3. Methods to Distinguish Between a Shift and a Switch Transition

In the previous section, we have seen that the difference between a shift and a switch transition is based on the concept of bistability of stationary states. We shall thus start by reviewing a criterion to decide whether bistability is possible in a given bacterial network. If the criterion is not fulfilled, then the transition is necessarily a shift. If the criterion is satisfied, further methods must be considered to decide if the transition is a switch.

In systems biology, the onset of virulence in bacteria is described by gene and metabolic networks. A quantitative knowledge of all the molecular interactions within these networks is generally not available, but experimental data usually allow for describing the biological networks as interaction graphs (6, 13). These graphs are defined according to the following conventions. An interaction is positive if activation or accumulation of a component leads to activation or accumulation of another component. The interaction is negative if activation or accumulation of a component leads to deactivation or depletion of another component. Positive and negative feedbacks may consist of a single component that directly activates or represses its own activity, or they may include several components and involve indirect interactions (8). In a regulatory graph, a circuit is a directed path without repetition of nodes, except that the first node of the circuit is also the last node. A circuit is positive if it contains an even number of inhibitions and is negative otherwise. It is difficult to intuitively understand how the dynamical properties of a network emerge from its interaction graph, especially when several feedback circuits are involved. The Thomas' method offers a simple modelling framework to determine the dynamics of gene networks (14) and states that a necessary condition for the co-existence of several stable states is the presence of a positive circuit in the network (see Note 1).

Let us insist on the fact that the presence of a positive feedback loop is necessary, but not sufficient to get bistability of stationary states (15–17). Once the criterion of a positive circuit is fulfilled, a second task is to determine the actual number of "branches" of stationary states in the system and thereby, whether the hysteresis phenomenon can be observed in the system. Unfortunately, if the system is arbitrary and has many variables, no simple recipe allows to perform this task. Assuming that some key parameters that govern the infectious transition are already identified, some mathematical methods exist to investigate bistability in the system:

1. A general way to investigate the presence of bistability and hysteretic behaviours in biological systems is to use continuation softwares, such as DDE-BIFTOOL, Content, or Matcont (18, 19). These graphical software packages for the

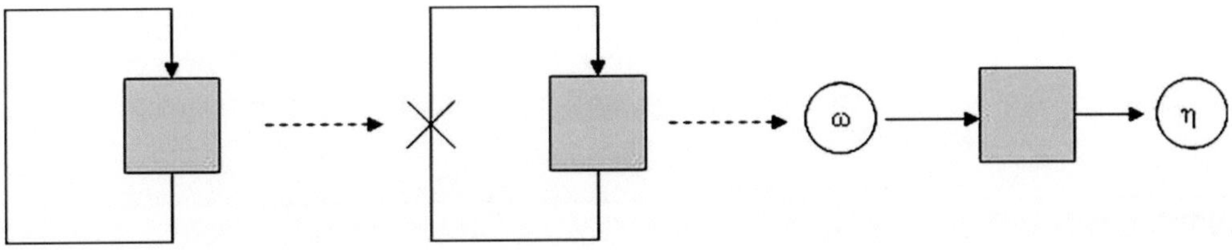

Fig. 2. Schematic view of a feedback system before and after breaking the feedback loop. ω is the input signal of the open-loop system and η the output signal (see ref. 21).

interactive numerical study of dynamical systems use a system of ordinary differential equations for computing equilibrium curves, limit points, Hopf points, limit cycles and period doubling bifurcation points of limit cycles. The main difficulty in this method is to fix the biological parameter values and to establish the appropriate parameters to delineate a possible hysteretic domain. This can require a great number of simulations. However, this method is simple to implement and can be used with quite complicated models.

2. When the system has only two variables, a faster alternative to the continuation method is to apply a standard graphical test, which consists of drawing the phase portrait with the nullclines (see Note 2) of the differential equations describing the network. The intersections of the nullclines correspond to the stationary states of the system. This method can thus be used to visualize under which conditions a particular positive-feedback system will exhibit several steady states. But, this test is valid only if the system contains no more than two variables (20).
3. A recent mathematical method for analysing multistability properties of positive-feedback systems has been proposed by Angeli et al. (21). The method applies to a class of systems called "monotone" (see Note 3) and is based on the behaviour of such systems when the feedback loop is opened. Considering a feedback loop connecting the output to the input nodes of a system, we can analyse the latter by opening the loop of the network (Fig. 2). The variables corresponding to the input and the output nodes are considered respectively as an input parameter ω and an output variable η. Therefore, the corresponding open loop can be mathematically described by a set of differential equations with ω as a parameter. After analysing the behaviour of the output signal as a function of the input parameter, the original system is recovered by replacing ω by η. Here, the number of steady states can be determined by counting the number of zeroes of one-variable function ($\eta = \omega$) constructed from the steady-state curve of the open-loop network in function of a control parameter.

This method is relatively easy to use and can be applied to systems with more than two variables. However, it requires that the system under consideration is monotone.

4. Cases Study

In the three following sections, we discuss the transitions of virulence via three different mechanisms and discuss the possibilities that these transitions appear as a shift or a switch, according to the choices of parameter.

4.1. The Onset of Virulence via the Induction Mechanism

One of the major virulence factors of pectinolytic bacteria such as soft rot enterobacteria, is the tightly controlled synthesis and secretion of a battery of enzymes able to degrade pectin, the main component of the plant cell wall. In *Erwinia chrysanthemi* (*Ech*), the production of these enzymes called pectinases responds to various environmental factors including pH and temperature, but is also controlled by bacterial growth phase and by pectin degradation products. Expression of the genes encoding pectinases is controlled by several regulators. Some of them like the PecS and PecT repressors or the GacA activator are global regulators that control most virulence factors. Others like the global regulator of sugar metabolism CRP or the central iron uptake regulator Fur link pectinolysis to the general metabolic status of the bacterium (22–25). In addition, pectinase production is specifically controlled by pectin degradation products via the KdgR regulator. In the absence of pectin, KdgR represses all genes involved in pectin catabolism. In the presence of pectin, the basal production of the catabolic enzymes allows for the accumulation of the pectin degradation product KDG that binds to and inactivates KdgR (13). In this section, we summarize some of the mechanisms used by *Ech* to produce these pectinolytic virulence factors and present a simple model that describes the production of these virulence factors.

The mechanism leading to the onset of virulence in *Ech* is not yet fully understood but several experimental (26–29) and theoretical (13, 30, 31) studies have been carried out. The basic (sub)network responsible for the production of the main virulence factors, the Pectate Lyases (Pels), has been identified over the past years. This system consists of a positive feedback loop, which involves a metabolic and a genetic part. The circuit is organized such that the *pel* genes are repressed by the main repressor KdgR, that also represses its own gene. In normal conditions, Pels enzymes are synthesized in negligible amount. However, when the bacteria encounter a host plant, a few molecules of Pels enzymes are sufficient to degrade a small amount of pectin which is metabolized into KDG.

KDG then forms a complex with KdgR allowing for the derepression of *pel* genes, thus creating a positive feedback loop. By describing these interactions at the molecular level, a recent work (32) provides a quantitative mathematical model describing the Pels steady states in the presence of a constant amount of pectin. Note that in the plant the quantity of pectin available to the bacteria fluctuates, so that experimental testing of the outcome of our modelling requires the use of a device where the amount of pectin can be fixed to a constant value, such as a bioreactor.

Considering x as the Pel concentration, y as the KdgR concentration, and s as the level of pectin susceptible to be metabolized, which can be considered as an "infection-inducing" parameter, one has the simplified following mathematical model (32):

$$\frac{dx}{dt} = a\frac{1}{1 + y_2/K_6} - \alpha_1\frac{1}{1 + s/K_1}x,$$

$$\frac{dy}{dt} = b\frac{1}{1 + y_2/K_5} - \alpha_2 y,$$

with the following implicit equation defining y_2, the concentration of dimerized KdgR:

$$y = 2y_2\left[1 + \left(c\frac{s}{K_1 + s}x\right)^2 + \frac{1}{K_6 + y_2} + \frac{1}{K_5 + y_2}\right]$$

In these equations, a is the maximum rate of Pel synthesis, b is the maximum rate of KdgR synthesis, c is the non-dimensionalized PGA consumption rate, α_1 is the Pel degradation rate, α_2 is the KdgR degradation rate, K_1 is the Michaelis Menten constant for the Pel.PGA complex formation, K_5 is the dissociation constant of KdgR on his own promoter, and K_6 is the dissociation constant of KdgR on the *pel* promoter.

Mathematical analysis of the model shows that the transition to virulence as a function of the amount of pectin available to degradation can be a switch or a shift depending on the relative affinities of the transcription factor KdgR for the promoter of the *pel* genes (Pr-*pel*) and for its own gene promoter (Pr-*kdgR*) (see Fig. 3). Numerical study of the model shows that, depending on the ratio between the parameters K_6 and K_5 (representing the dissociation constant of the Pr-*pel*·KdgR$_2$ complex and the Pr-*kdgR*·KdgR$_2$ complex, respectively), the system can exhibit a monostable or a bistable behaviour in function of the amount of pectin available. As depicted in Fig. 4, the transition to virulence in function of the pectin parameter can turn from a shift to a switch transition when the affinity of KdgR for its own promoter increases gradually relative to its affinity for the *pel* promoter. Therefore, we clearly see in this example that although the Pel-KdgR circuit is positive in the sense defined above, the bistability

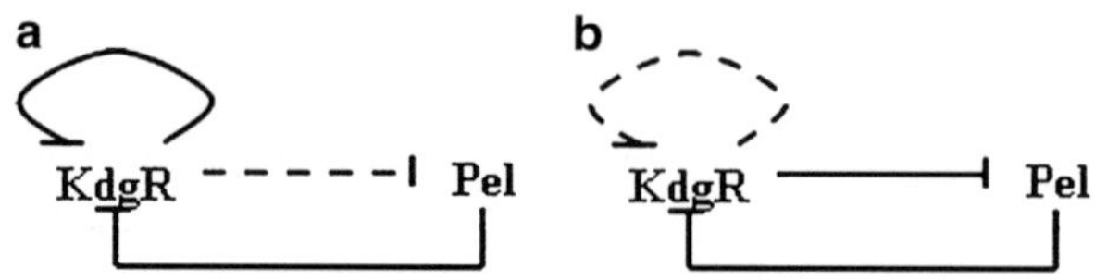

Fig. 3. Simplified description of the occurrence of (**a**) the shift or (**b**) the switch behaviour in the sub-network based on KdgR affinity on the *pel* or *kdgR* promoters. The *dotted line* corresponds to the reduced affinity.

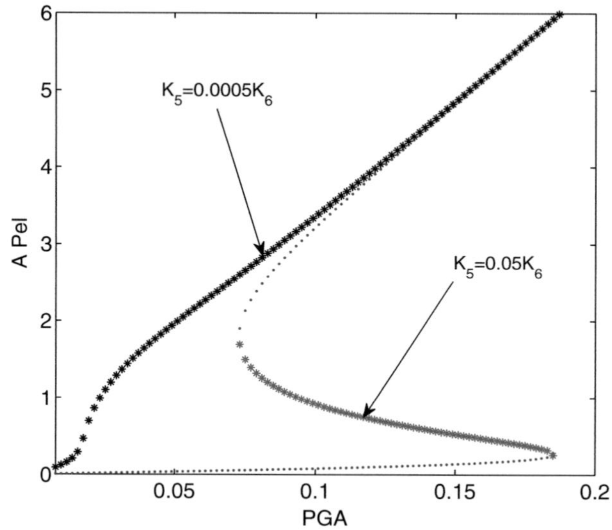

Fig. 4. Effect of varying the ratio of the dissociation constants between KdgR dimer and the promoters of genes *kdgR* and *pel* respectively, on the hysteresis behaviour with $a = 4.04$ μM/min; $b = 1.3 \times 10^{-3}$ μM/min; $c = 2 \times 10^{4}$; $\alpha_1 = 3.21 \times 10^{-4}$/min; $\alpha_2 = 5.78 \times 10^{-4}$/min; $K_1 = 0.2$ g/l; $K_6 = 10^{-3}$ μM. For $K_5 = 0.05K_6$.

property is not automatic, but is rather controlled by the existence of a negative feedback of KdgR on itself. A biological interpretation of the ways the transition turns from a shift to a switch is as follows. (1) If autorepression of KdgR is strong compared to its repression on *pel* genes, then the action of KDG on the Pr-*pel*·$KdgR_2$ complex gradually releases the *pel* genes operator but not much that of the *kdgR* gene, thus allowing for a gradual production of Pel enzymes (shift). (2) If the affinity of KdgR for its own gene promoter is decreased (K_5 increased), then the action of KDG dissociates not only the Pr-*pel*·$KdgR_2$ complex but also the complex Pr-*kdgR*·$KdgR_2$, which tends to increase the production of KdgR. Consequently, Pel production remains low as long as the level of KDG (which depends on the amount of pectin available for degradation) reaches a sufficient threshold to induce an abrupt production of Pel. In this case, the transition to virulence appears to be a switch.

We conjecture that in *Ech* the ratio of K_5/K_6 is in the range enabling a switch transition rather than a shift. A hysteretic transition to high Pel production would be more advantageous for the expression of virulence in *Ech*. Indeed, bacteria hosted in a plant would produce enough KDG by pectin degradation to switch to a massive production of Pels, such that the underlying hysteresis property would guarantee the point of no return of the infection. This means that if for some reason the concentration of available pectin decreases after the transition to virulence, the state of Pel production in the bacteria can remain high. Whereas, if the transition was a shift, a slight fall in the amount of the available pectin would lead to a sharp decrease of the Pel production, possibly leading to a failed infection.

4.2. The Onset of Virulence via the Quorum Sensing Mechanism

In many pathogens, the production of virulence factors increases when the bacterial population reaches a certain density. This regulation typically functions by cell–cell communication mechanisms involving low molecular weight signal molecules thought to freely diffuse across bacterial membranes. Signal concentrations inside and outside the cell will be similar when a population is within a confined space. Thus, intracellular signal concentrations sufficient to trigger a response will occur only when intercellular concentrations have increased after a population increase. This method of gene regulation is now commonly called "quorum sensing" (QS) (32, 33). Several signal molecules have been characterized both in Gram-positive and Gram-negative bacteria, with acyl homoserine lactone derivatives (AHLs) being the most studied. QS is used by a wide variety of terrestrial and marine bacteria (34–36) and has been shown to control a diverse range of microbial behaviours such as production of exoenzymes, antibiotics, cell division, virulence, bioluminescence, swarming motility, plasmid conjugation, and others. In this section, we consider the "generic" model of Dockery–Keener (37) describing how QS controls virulence. We also discuss the type of transitions associated with the QS mechanism.

The infection strategies of pathogenic bacteria often require swift global changes in gene expression and physiology in response to environmental cues. In particular, an efficient strategy relies on cell-to-cell communication to coordinate crucial steps in pathogenesis. For example, the QS phenomenon has been experimentally and theoretically well studied in *Pseudomonas aeruginosa* (37–41). Many mathematical models have been developed to show that QS is endowed with auto-activation in order to act as a switch. The quorum sensing system of *P. aeruginosa* consists of two redundant regulatory systems: the *las (lasI/lasR)* and the *rhl (rhlI/rhlR)* systems. The *las* system regulates the expression of elastase, responsible for elastolytic activity, which destroys elastin-containing human lung tissues and causes pulmonary haemorrhages associated with *P. aeruginosa* infection. The *las* system

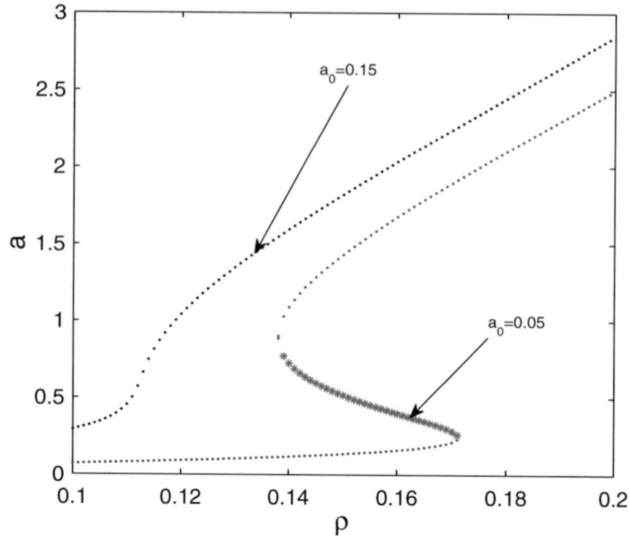

Fig. 5. Switch and shift behaviour described by the QS model with $v_a = 1$; $\theta_a = 1$; $\delta = 0.5$; $k_e = 0.1$; $k_a = 0.02$.

is composed of *lasI*, the autoinducer synthase gene responsible for the synthesis of 3-oxo-C12-HSL autoinducer, and the *lasR* gene that codes for a transcriptional activator protein. *lasR* positively controls *lasI* expression. The *las* system is positively controlled by both GacA and Vfr, which are needed for the transcription of *lasR*.

As a simplified description of the autoregulation mechanism that governs the QS regulator, the authors in ref. 37 consider that the intracellular pheromone produced by bacteria can diffuse outside the bacterium and increase its external concentration. The internal concentration is denoted a, and the external concentration that reflects the bacterial population density is denoted e. The dynamics of the QS system then follow from the equations:

$$\frac{\mathrm{d}a}{\mathrm{d}t} = v_a \frac{a^2}{a^2 + \theta_a} + a_0 - k_a a - \frac{\delta}{\rho}(e - a)$$

$$\frac{\mathrm{d}e}{\mathrm{d}t} = \frac{\delta}{1 - \rho}(a - e) - k_e e$$

where ρ is the bacteria density, δ is the cell membrane conductance, and k_e and k_a are the degradation rates of e and a, respectively. In the absence of trans-membrane diffusion ($\delta = 0$), the synthesis of the autoinducer a, is governed by a constant production term (a_0) and by a Hill function (with v_a and θ_a the production rate and the threshold of production, respectively) which describes the autoactivation process.

By using a continuation method (see above), a simple analysis of the steady states shows that in this system, the switch transition can be readily obtained in function of the cell density ρ (cf. Fig. 5). For small values of ρ, there is a unique stable steady state with a small value of a. As ρ increases, a new branch of stable state appears with larger values of a, whereas as ρ further increases, the branch with small a disappears. Thus, we have a case of bistability, allowing a switch transition to virulence when the bacterial density increases. In this situation, if after the switch transition the bacterial density decreased because of a counter-attack by the host immune system, pheromone synthesis could remain high and thus help the bacteria to succeed in their infection. However, this behaviour depends on the parameter a_0, i.e. the constitutive production of the internal pheromones (with concentration a). For instance, a threefold increase of parameter a_0 as compared with the previous situation leads to a shift transition (Fig. 5). However, such a situation would be biologically precarious for the pathogen because a small variation in the conditions leading to virulence would result in an unsuccessful infection.

Using this simplified mathematical model describing the QS-related induction mechanism in *Ech*, a recent theoretical study (30) shows that the virulence process is due to an abrupt switch of the production of pectates lyases from a low to a high level, with the transition being controlled by bacterial density. Thus, the analysis reported in the previous section shows that in the case of *Ech*, at least two hysteresis phenomena can be involved, namely both an induction mechanism and a QS mechanism, that can be combined for the onset of virulence.

QS switches have been studied in many other bacterial species (41, 42). For example, Williams et al. (41) show that the QS response of *Vibrio fischeri* involves a rapid switch between low and high induction states of the *lux* operon over a narrow range of the autoinducer 3-oxohexanoyl-L-homoserine lactone concentration. In this system, LuxR is an autoinducer-dependent positive regulator of the *lux* operon, which encodes the autoinducer synthase. This creates a positive feedback loop common in many bacterial species that exhibit QS-controlled gene expression. Applying a combination of modelling and experimental analyses, the authors provide evidence for a LuxR autoregulatory feedback loop that allows LuxR to increase its concentration in the cell during the switch to full *lux* activation, thereby allowing for two distinct responses within the same cell population.

4.3. The Onset of Virulence Associated with the Type III Secretion System

The Type III secretion system (T3SS) is a complex molecular machine that consists of about 20 structural proteins. In animal pathogens, this Type III apparatus consists of a needle-like structure that crosses both the inner and the outer bacterial membranes and that extends about 50 nm from the cell surface. The genes encoding the components of the Type III systems are organized in several operons whose expression is co-regulated with the

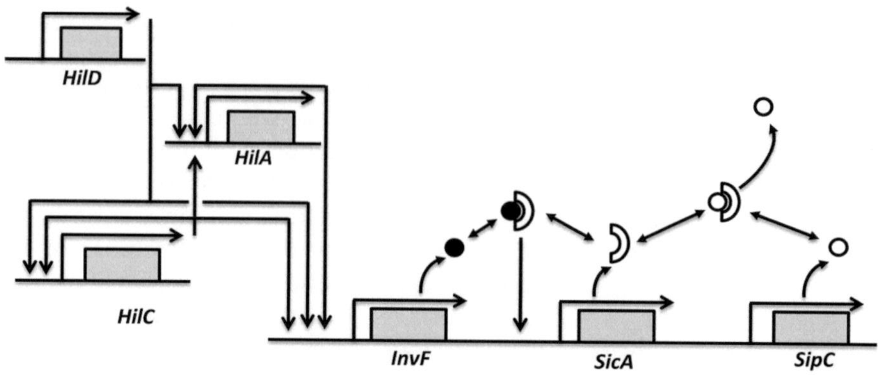

Fig. 6. Regulatory network encoded within SPI-1 (*see* ref. 46). The SPI-1 T3SS is induced by a number of environmental and cell state signals, which differentially affect *hilA, hilC* and *hilD*. These transcription factors co-regulate each other and control different operons in SPI-1. Only HilA regulates the expression of genes forming the inner membrane ring and shaft, but all three proteins converge at two promoters controlling a long operon that begins with the *invF* gene and contains *sicA*, the apoptosis-inducing effector encoding *sipB* gene and *sipC*. Just upstream of *sicA*, there is a promoter internal to this operon, which is activated by InvF when bound to the SicA chaperone. The chaperone is sequestered by the SipC protein until it is exported, after which SicA can bind to InvF, leading to the upregulation of additional effectors genes like *sipB* (see ref. 46).

production of several effectors secreted by the T3SS and their cognate chaperones. Expression of T3SS coding genes is controlled by sophisticated regulatory networks (43, 44). Although several T3SS components are highly conserved in Gram-negative bacteria, a variety of networks control T3SS regulons. Furthermore, several pathogens have evolved an elegant regulation mechanism where the secretion process itself regulates the production of the machinery components (45). In this section, we provide a brief review of a model recently developed to study a network regulating the T3SS in *Salmonella enterica* (46).

T3SS allows pathogens to deliver toxins directly into the host cell cytoplasm. *Salmonella enterica* synthesize two T3SSs that facilitate different stages of an infection. The T3SS-needle encoded in *Salmonella* Pathogenicity Island (SPI)-1 can be induced in culture and is required for bacteria during the early phase of infection to invade and induce rapid apoptosis of mammalian cells (46–49). During the subsequent systemic phase, SPI-1 is strongly repressed and SPI-2 is activated, encoding a different type III apparatus and new effectors. This enables the bacteria to survive and replicate within infected macrophages (50). The network is organized such that environmental and cell state regulators converge on three transcription factors (HilA, HilC, HilD) whose regulation is highly interconnected (Fig. 6) (46, 51–53). These transcription factors form a tangled circuit containing several positive feedback loops (53, 54) that differentially control the transcription of SPI-1 operons. At the core of the circuit is the InvF transcription factor, which is only active when bound to the SicA chaperone. Prior to the completion of functional T3SS-needles, SipC binds to and sequesters SicA. Once the needle is functional, SipC is exported and SicA is free to bind InvF. The InvF:SicA complex then activates the *sicA* promoter, thereby upregulating the expression of additional chaperones and effectors genes, including some that are encoded outside SPI-1. This circuit forms a positive "split" feedback loop, where *sicA* is internal to the loop but *invF* is encoded outside the loop (Fig. 6).

A simple mathematical model of the SicA:InvF positive feedback loop has been recently constructed (46) to quantify the delayed dynamics of the *sicA* expression. The model describes a genetic circuit containing the three proteins InvF, SicA, and SipC where the transcription factor InvF is activated when bound by the chaperone SicA, which also binds to the secreted SipC effector protein. Using the model, and assuming that InvF and sicA form a 1:1 complex, the authors (46) could show that in this case, the transition is a shift. It has been proposed that SicA could interact with InvF as a dimer (55). If this turns out to be the case, then bistability can occur in the model for a very narrow parameter space (46). However, a shift transition is more appropriate in *Salmonella* infection because SPI-1 is only transiently active and during systemic invasion, the production of the apoptosis-inducing SipB effector needs to be downregulated.

5. Perspectives

In molecular networks regulating bacterial infections, the presence of positive feedback loops enhances the activation of virulence factors and thereby facilitates the onset of virulence. Mathematical analysis of positive regulatory circuits enables one to refine this concept and to distinguish between two basic ways in which this enhancement can proceed, through a "shift" or a "switch" transition. In both cases the production of virulence factors can be relatively sharp, but in the first case the transition is gradual and reversible, whereas in the second it is all-or-nothing and hysteretic, and sometimes irreversible (cf. Subheading 2). Therefore, we predict that in the context of pathogenic bacteria, the switch transition is generally more appropriate. This ratchet mechanism enables an efficient infection by setting the bacterial offensive to a point of no return.

The concept of bistability provides a useful framework for thinking about bacterial virulence. Given a bacterial molecular network, it is worthwhile to study in detail the positive circuits and the associated hysteresis phenomena to better determine the virulence potential of the organism. Future experimental work will help us to understand whether bistability is a particular feature of a few biological systems or a recurring theme and unifying principle of cell biology.

6. Notes

1. René Thomas highlighted the predominant dynamical role of positive and negative circuits in a network by stating the following conjectures (14):
   (a) A necessary condition for the co-existence of several stable states is the presence of a positive circuit in the network.
   (b) A necessary condition for the presence of several sustained oscillations is the presence of a negative circuit in the network.

   This conjecture has been mathematically proved by Gouzé (56) and the proof of the first conjecture has been extended to a more general framework by Soulé in ref. 57.
2. Considering a two variables system defined by:

$$\frac{dx}{dt} = f(x, y)$$

$$\frac{dy}{dt} = g(x, y)$$

   In order to investigate bistability or hysteresis, with continuation software, one needs to rewrite these systems of equations in the software language and to compute the hysteresis domain using biologically appropriate parameter values.

   In the case of a phase space analysis, the nullclines can be found by setting $x' = 0$ and $y' = 0$ in turn. The intersections between x and y nullclines are equilibrium points: thus, finding nullclines is a useful way to identify such points, particularly when the system is not amenable to analytical solutions
3. We define a graph to be monotone if every loop in the graph is positive and there is a directed path from each node of the graph to the output node. The property of monotonicity has been well described in ref. 20.

Acknowledgments

This work was supported by the ANR program "Regupath" (N°ANR-07-BLAN-0212) which is financed by the French Ministry of Research. In particular, W.K. has a post-doctoral contract funded by this ANR. Axel Cournac is acknowledged for stimulating discussions about the bistability in Pel regulation.

References

1. Weinberg ED. (2009) Iron availability and infection. BBA, 1790:600–605.
2. Expert D. (1999) Withholding and exchanging iron: interactions between *Erwinia* spp. and their plant hosts. Annu Rev Phytopathol, 37:307–334.
3. Gerlach RG, Hensel M. (2007) Protein secretion systems and adhesins: the molecular armory of gram-negative pathogens. Int J Med Microbiol, 297:401–415.
4. Cornelis GR, Van Gijsegem F. (2000) Assembly and function of type III secretory systems. Ann Rev Microbiol, 54:735–774.
5. Backert S, Meyer TF. (2006) Type IV secretion systems and their effectors in bacterial pathogenesis. Curr Opin Microbiol, 9:207–217.
6. Lazdunski AM, Ventre I, Sturgis JN. (2004) Regulatory circuits and communication in gram-negative bacteria. Nature, 2:581–592.
7. Dayan F, Monticelli M, Pouysségur J, Pécou E. (2009) Gene regulation in response to graded hypoxia: the non-redundant roles of the oxygen sensors PHD and FIH in the HIF pathway. J Theor Biol, 259:304–316.
8. Mitrophanov AY, Groisman EA. (2008) Positive feedback in cellular control systems. Bioessays, 30(6):542–555.
9. Mitrophanov AY, Churchward G, Borodovsky M. (2007) Control of *Streptococcus pyogenes* virulence: modeling of the CovR/S signal transduction system. J Theor Biol, 246: 113–128.
10. Gardner TS, Cantor CR, Collins JJ. (2000) Construction of a genetic toggle switch in *Escherichia coli*. Nature, 403:339–342.
11. Boots M, Hudson PJ, Sasaki A. (2004) Large shifts in pathogen virulence relate to host population structure. Science, 303:842–844.
12. Bagowski CP, Ferrell JE Jr. (2001) Bistability in the JNK cascade. Curr Biol, 11:1176–1182.
13. Sepulchre JA, Reverchon S, Nasser W. (2007) Modeling the onset of virulence in a pectinolytic bacterium. J Theor Biol, 244:239–257.
14. Edited by Thomas R. (1981) *On the Relation Between the Logical Structure of Systems and Their Ability to Generate Multiple Steady States and Sustained Oscillations*, volume 9. Springer, 180–193.
15. Griffith JS. (1968) Mathematics of cellular control processes. II. Positive feedback to one gene. J Theor Biol, 20:209–216.
16. Ferrell JE Jr, Xiong W. (2001) Bistability in cell signaling: how to make continuous processes discontinuous and reversible processes irreversible. Chaos, 11:227–236.
17. Hasty J, McMillen D, Isaacs F, Collins JJ. (2001) Computational studies of gene regulatory networks: in numero molecular biology. Nat Rev Genet, 2:268–279.
18. Engelborghs K, Luzyanina T, Roose D. (2002) Numerical bifurcation analysis of delay differential equations using DDE-BIFTOOL. ACM Trans Math Softw, 28(1):1–21.
19. Dhooge A, Govaerts W, Kuznetsov YA. (2003) MATCONT: a MATLAB package for numerical bifurcation analysis of ODEs. ACM Trans Math Softw, 29(2):141–164.
20. Strogatz S. (2001). *Nonlinear Dynamics and Chaos: With Applications to Physics, Biology, Chemistry, and Engineering (Studies in Nonlinearity)*. Perseus Books Group.
21. Angeli D, Ferrell JE Jr, Sontag ED. (2004) Detection of multistability, bifurcations and hysteresis in a large class of biological positive-feedback systems. PNAS, 101(7):1822–1827.
22. Hugouvieux-Cotte-Pattat N, Condemine G, Nasser W, Reverchon S. (1996) Regulation of pectinolysis in *Erwinia chrysanthemi*. Annu Rev Microbiol, 50:213–257.
23. Lebeau A, Reverchon S, Gaubert S, Kraepiel Y, Simond-Côte E, Nasser W, Van Gijsegem F. (2008) The GacA global regulator is required for the appropriate expression of *Erwinia chrysanthemi* 3937 pathogenicity genes during plant infection. Environ Microbiol, 10: 545–559.

24. Reverchon S, Expert D, Robert-Baudouy J, Nasser W. (1997) The cyclic AMP receptor protein is the main activator of pectinolysis genes in *Erwinia chrysanthemi*. J Bacteriol, 179:3500–3508.
25. Franza T, Michaud-Soret I, Piquerel P, Expert D. (2002) Coupling of iron assimilation and pectinolysis in *Erwinia chrysanthemi* 3937. Mol Plant-Microbe Interact, 15(11): 1181–1191.
26. Nasser W, Reverchon S. (2002) H-NS-dependent activation of pectate lyases synthesis in the phytopathogenic bacterium *Erwinia chrysanthemi* is mediated by the PecT repressor. Mol Microbiol, 43:733–748.
27. Nasser W, Reverchon S, Vedel R, Bocarra M. (2005) PecS and PecT coregulate the synthesis of HrpN and pectate lyases, two virulence determinants in *Erwinia Chrysanthemi*. Mol Plant-Microbe Interact, 18:1205–1214.
28. Lautier T, Nasser W. (2007) The DNA nucleoid-associated protein Fis coordinates the expression of the main virulence genes in the phytopathogenic bacterium *Erwinia chrysanthemi*. Mol Microbiol, 66:1474–1490.
29. Lautier T, Blot N, Muskhelishvili G, Nasser W. (2007) Integration of two essential virulence modulating signals at the *Erwinia chrysanthemi pel* gene promoters: a role for Fis in the growth-phase regulation. Mol Microbiol, 66:1491–1505.
30. Kepseu WD, Woafo P, Sepulchre JA. (2010) Dynamics of the transition to pathogenicity in *Erwinia chrysanthemi*. J Biol Syst, 18(1): 1–31.
31. Kepseu WD, Sepulchre JA, Reverchon S, Nasser W. (2010) Towards a quantitative modeling of the synthesis of the pectates lyases, essential virulence factors in *Dickeya dadantii*. J Biol Chem, 285(37):28565–28676.
32. Waters CM, Bassler BL. (2005) Quorum sensing: cell-to-cell communication in bacteria. Annu Rev Cell Dev, 21:319–346.
33. Mole BM, Baltrus DA, Dangl JL, Grant SR. (2007) Global virulence regulation networks in phytopathogenic bacteria. Trends Microbiol, 15(8):363–371.
34. Nasser W, Reverchon S. (2007) New insights into the regulatory mechanisms of the LuxR family of quorum sensing regulators. Anal Bioanal Chem, 387:381–390.
35. Denny TP. (1995) Involvement of bacterial polysaccharides in plant pathogenesis. Annu Rev Phytopathol, 33:173–197.
36. Denny TP. (1999) Autoregulator-dependent control of extracellular polysaccharide production in phytopathogenic bacteria. Eur J Plant Pathol, 105:417–430.
37. Dockery JD, Keener JP. (2001) A mathematical model for quorum sensing in *Pseudomonas aeruginosa*. Bull Math Biol, 63:95–11.
38. Viretta AU, Fussenegger M. (2004) Modeling the quorum sensing regulatory network of human–pathogenic *Pseudomonas aeruginosa*. Biotechnol Prog, 20:670–678.
39. Von Bodman SB, Bauer WD, Coplin DL. (2003) Quorum sensing in plant pathogenic bacteria. Annu Rev Phytopathol, 41:455–482.
40. Cabrol S, Olliver A, Pier GB, Andremont A, Ruimy R. (2003) Transcription of quorum-sensing system genes in clinical and environmental isolates of *Pseudomonas aeruginosa*. J Bacteriol, 185:7222–7230.
41. Williams JW, Cui X, Levchenko A, Stevens AM. (2008) Robust and sensitive control of a quorum-sensing circuit by two interlocked feedback loops. Mol Syst Biol, 4:234–245.
42. Zhu J, Miller MB, Vance RE, Dziejman M, Bassler BL, Mekalanos JJ. (2002) Quorum-sensing regulators control virulence gene expression in *Vibrio cholerae*. PNAS, 99(5): 3129–3134.
43. Francis MS, Wolf-Watz H, Forsberg A. (2002) Regulation of type III secretion systems. Curr Opin Microbiol, 5:166–172.
44. Tang X, Xiao Y, Zhou JM. (2006) Regulation of the type III secretion system in phytopathogenic bacteria. Mol Plant-Microbe Interact, 19:1159–1166.
45. Brutinel ED, Yahr TL. (2008) Control of gene expression by type III secretory activity. Curr Opin Microbiol, 11:128–133.
46. Temme K, Salis H, Tullman-Ercek D, Levskaya A, Hong SH, Voigt CA. (2008) Induction and relaxation dynamics of the regulatory network controlling the Type III secretion system encoded within *Salmonella* pathogenicity island 1. J Mol Biol, 377:47–61.
47. Clark MA, Jepson MA, Simmons NL, Hirst BM. (1994) Preferential interaction of *Salmonella typhimurium* with mouse Peyer's patch M cells. Res Microbiol, 145:543–552.
48. Jones BD, Ghori N, Falkow S. (1994) *Salmonella typhimurium* initiates murine infection by penetrating and destroying the specialized epithelial M cells of Peyer's patches. J Exp Med, 180:15–23.
49. Eriksson S, Lucchini S, Thompson A, Rhen M, Hinton JCD. (2003) Unravelling the biology of macrophage infection by gene expression profiling of intracellular *Salmonella enterica*. Mol Microbiol, 47:103–118.

50. Bajaj V, Lucas RL, Hwang C, Lee CA. (1996) Coordinate regulation of *Salmonella typhimurium* invasion genes by environmental and regulatory factors is mediated by control of *hilA* expression. Mol Microbiol, 22:703–714.
51. Boddicker JD, Knosp BM, Jones BR. (2003) Transcription of the *Salmonella* invasion gene activator, *hilA*, requires HilD activation in the absence of negative regulators. J Bacteriol, 185:525–533.
52. Lucas RL, Lee CA. (2001) role of HilC and HilD in regulation of *hilA* expression in *Salmonella enterica* sevorar typhimurium. J Bacteriol, 183:2733–2745.
53. Boddicker JD, Jones BR. (2004) Long protease activity causes down regulation of *Salmonella* pathogenicity island I invasion gene expression after infection of epithelial cells. Infect Immun, 72:2002–2013.
54. Olekhnovich IN, Kadner RJ. (2002) DNA-binding activities of the HilC and HilD virulence regulatory proteins of *Salmonella enterica* sevorar typhimurium. J Bacteriol, 184:4148–4160.
55. Darwin KH, Miller VL. (2001) Type III secretion chaperone-dependent regulation: activation of virulence genes by SicA and InvF in *Salmonella typhimurium*. EMBO J, 20:1850–1862.
56. Gouzé JL. (1998) Positive and negative circuits in dynamical systems. J Biol Systems, 6:11–15.
57. Soulé C. (2003) Graphic requirements for multistationnarity. ComplexUs, 1:123–133.

Chapter 26

Spatial and Stochastic Cellular Modeling with the Smoldyn Simulator

Steven S. Andrews

Abstract

This chapter describes how to use Smoldyn, which is a computer program for modeling cellular systems with spatial and stochastic detail. Smoldyn represents each molecule of interest as an individual point-like particle. These simulated molecules diffuse, interact with surfaces (e.g., biological membranes), and undergo chemical reactions much as they would in real biochemical systems. Smoldyn has been used to model signal transduction within bacterial cells, pheromone signaling between yeast cells, bacterial carboxysome function, diffusion in crowded spaces, and many other systems. A new "rule-based modeling" feature automatically generates chemical species and reactions as they arise in simulations due to protein modifications and complexation. Smoldyn is easy to use, quantitatively accurate, and computationally efficient. It is generally best for systems with length scales between nanometers and several microns, time scales from tens of nanoseconds to tens of minutes, and up to about 10^5 individual molecules. Smoldyn runs on Macintosh, Linux, or Windows systems, is open source, and can be downloaded from http://www.smoldyn.org.

Key words: Computational biology, Smoldyn, Particle-based simulation, Spatial modeling, Rule-based modeling

1. Introduction

Computational modeling is becoming an important cell biology research method. Modeling is useful for quantitatively testing hypotheses about systems of interest, helping analyze and interpret experimental data, and developing a deeper understanding of how biochemical systems function. In these and other roles, simulations have helped elucidate cellular systems such as bacterial chemotaxis (1), the eukaryotic cell cycle (2), and actin-based cellular protrusion (3). Active simulation algorithm research and frequent increases in computer power suggest that cellular modeling will

Jacques van Helden et al. (eds.), *Bacterial Molecular Networks: Methods and Protocols*, Methods in Molecular Biology, vol. 804, DOI 10.1007/978-1-61779-361-5_26, © Springer Science+Business Media, LLC 2012

continue gaining importance for many years to come. Recent articles that review simulation methods and software include (4–7).

This chapter describes how to use the Smoldyn simulation program, which simulates biochemical processes with spatial and stochastic detail (8). Smoldyn represents each molecule of interest as an individual point-like particle that has a location in continuous space. Simulated molecules diffuse, undergo chemical reactions, and interact with membranes according to simple biophysical principles and user-specified parameters.

Smoldyn is typically best for models with spatial scales from nanometers to microns, temporal scales from tens of nanoseconds to tens of minutes, and up to about 10^5 individual molecules. The lower ends of these ranges arise from the fact that Smoldyn does not represent excluded volumes, orientations, or momenta of simulated molecules. These limit Smoldyn's spatial resolution to a few molecular radii and temporal resolution to molecules' rotational diffusion time constants (9, 10). The upper ends of the ranges arise from the fact that Smoldyn is too computationally intensive to simulate much larger systems conveniently on current desktop computers (however, recent work on parallelizing Smoldyn for graphics processing units is starting to enable larger simulations).

Smoldyn's level of simulation detail is ideally suited for modeling intracellular organization. More specifically, the so-called particle-based simulation method that Smoldyn uses (along with ChemCell (11) and MCell (12)) has proven useful for modeling single-molecule tracking experiments (13, 14), molecular diffusion in restricted environments (15–17), stochastic signaling noise that arises from spatial organization (6, 8, 18), and the causes and effects of protein localization (19, 20).

As some examples, Lipkow and coworkers developed several Smoldyn simulations of intracellular signal transduction for *Escherichia coli* chemotaxis. Each model included about ten different proteins and about ten chemical reactions. Using these simulations, they found that intracellular crowding accentuates signaling differences to different flagellar motors (17), that the CheZ phosphatase is likely to change its intracellular location upon cellular stimulation (19), and that stable concentration gradients are likely to arise for the CheY signal transducer (20). In another example (Fig. 1), several colleagues and I used Smoldyn to help understand why haploid yeast cells that secrete the pheromone-degrading protease Bar1 are better able to discriminate between potential mating partners than yeast cells that do not secrete Bar1 (21). We found that Bar1 sharpens the local pheromone concentration gradient because pheromone is progressively inactivated as it diffuses through a Bar1 cloud (8). This model included four proteins, six chemical reactions, and about 2×10^5 individual molecules; it also covered about 7,000 μm^3 of volume, simulated about 75 min of time, and took

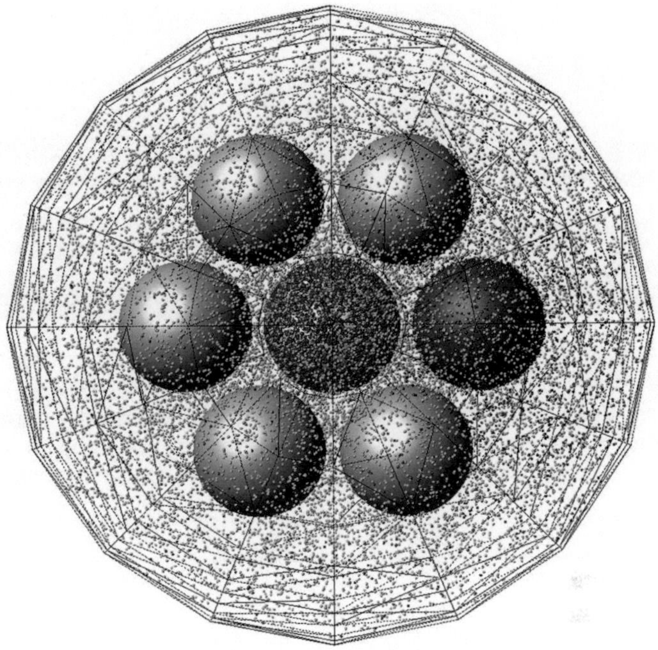

Fig. 1. A model of the effect of the Bar1 protease on yeast signaling. The mesh surface on the outside of the system is a triangulated spherical boundary that absorbs molecules with Smoldyn's "unbounded-emitter" method. The central sphere is a "receiver" cell which is covered with 6,622 receptors (red if bound to pheromone and blue if not, in online version), *light gray* spheres are "challenger" cells which secrete pheromone slowly, and the *dark gray* sphere on the right is a "target" cell which secretes pheromone quickly. *Green* (*light gray* in print version) *dots* are Bar1 proteins and *black dots* are pheromone molecules. This model showed that the pheromone-degrading Bar1 protease improves yeast mating partner discrimination by sharpening the pheromone concentration gradient. Republished with permission from ref. 8. The supplementary information for the original publication presents the model parameter selection process unusually thoroughly.

about 10 h to run. As a final example, Savage is using Smoldyn to investigate carbon transport and fixation in cyanobacteria, which takes place in organelle-like carboxysomes (Fig. 2). His model includes several partially transmitting surfaces.

Within the field of particle-based simulation, Smoldyn is notable for its high accuracy and good computational efficiency (8). Each Smoldyn algorithm is either exact or nearly exact for any length simulation time step, and complete Smoldyn simulations approach exactness as time steps are reduced toward zero (18, 22, 23) (by definition, exact simulation methods produce results that are theoretically indistinguishable from those of the underlying model system). Smoldyn achieves relatively high computational efficiency through judicious approximations for simulating chemical reactions (22) and molecule–surface interactions (23), and through data structures that reduce unnecessary computations.

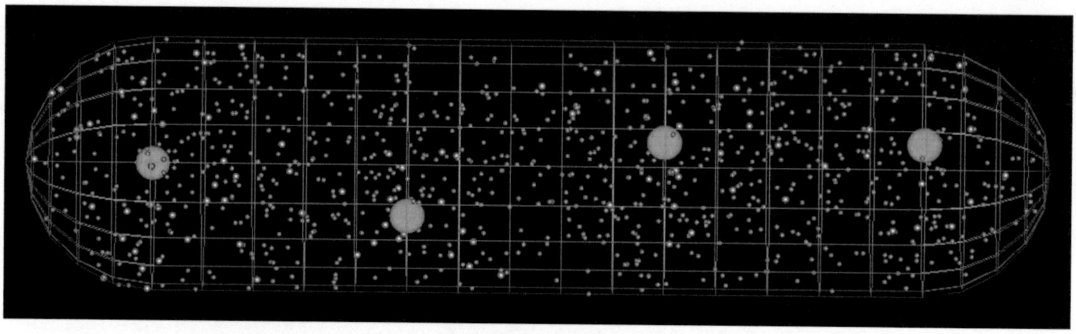

Fig. 2. A model of carboxysome function in a cyanobacterium. The wireframe outside surface depicts the cell membrane (internally, Smoldyn represents it as a smooth continuous surface). The four large green spheres are carboxysomes, which are organelle-like proteinaceous compartments that contain the carbon-fixation apparatus of cyanobacteria. *Smaller red dots* are carbon dioxide molecules and larger *light blue dots* are 3-phosphoglycerate molecules. Savage and coworkers are using this model to investigate the roles of compartmentalization and spatial organization in carbon fixation.

2. Materials

Smoldyn is available at http://www.smoldyn.org. The distribution includes the program source code, complete installation instructions, a user's manual, a programmer's manual, example files, and several utility programs. The Smoldyn User's Manual is much more thorough than this chapter and should be consulted for statement syntax, available options, and all other details. Smoldyn is open source and runs on Macintosh, Linux, and Windows computers (however, the Windows version does not support rule-based modeling, described in Subheading 3.10). Faster computers run simulations faster, of course, although laptops or desktops up to about 5 years old are often adequate.

Smoldyn runs from a shell prompt, which most operating systems supply with a "Terminal" or "Command Prompt" application. Installation on Mac or Linux computers is typically as easy as changing to the Smoldyn download directory and typing `./configure`, `make`, and `sudo make install` at a shell prompt. For Windows, the download includes a prebuilt Smoldyn executable and all necessary "dll" files.

Smoldyn support is available at the on-line forum http://www.smoldyn.org/forum, or via e-mail at support@smoldyn.org.

3. Methods

3.1. Running Smoldyn

Run Smoldyn by typing `smoldyn myfile.txt` at a shell prompt, where `myfile.txt` is your configuration file name. You can also append option flags to this command, such as `-o` to suppress text output, `-p` to just display parameters, or `-t` for text-only operation. Upon starting, Smoldyn reads model parameters from your configuration file, calculates and displays simulation parameters, and runs the simulation. As the simulation runs, Smoldyn displays the simulated system to a graphics window and/or saves quantitative data to one or more output files. Smoldyn stops when the simulation is complete. Box 1 presents an example configuration file and its output.

3.2. The Configuration File

Smoldyn configuration files are plain text files. The first word of each line tells Smoldyn which parameters to set and the rest of the line lists those parameters, separated by spaces or tabs. Usually, the statement sequence does not matter. When it does, it is usually obvious; for example, molecular species need to be defined before their diffusion coefficients. Smoldyn displays error messages and terminates if it cannot parse the configuration file.

Denote comments in your configuration file with a "#" symbol, or bracket multiple comment lines with /* and */. Ideally, your comments should list the model name, its author or authors, its creation and modification dates, citations to relevant references, and the model's distribution terms; if so, your model will obey the "minimum information requested in the annotation of biochemical models" (MIRIAM) (24), which aids model sharing and reuse. Your comments should also list the model units, such as microns and seconds, or nanometers and microseconds. This is because Smoldyn does not assume any particular units, but instead requires that all parameters use the same units. For example, if you choose microns and seconds, then you will need to enter diffusion coefficients in $\mu m^2/s$, first order reaction rate constants in s^{-1}, and second order reaction rate constants in $\mu m^3/s$. See Table 3.2.1 of the Smoldyn User's Manual.

Use `read_file` to separate your model description into multiple configuration files. This can be useful for separating lists of surface or molecule information from the main model file, or for separating fixed model components from ones you wish to vary. End each configuration file with `end_file`.

Several macro statements that substitute text "tokens" with replacement text can simplify model development. These include `define`, `define_global`, `undefine`, `ifdefine`, `ifundefine`, `else`, and `endif`. In a typical use, you would collect the model parameters that you wish to explore to the top of your configuration file with define statements and then refer to them later on by their token names (see Box 1).

3.3. Space and Time

Specify whether your system is 1-, 2-, or 3-dimensional with `dim`. For example, you could model protein motion along cytoskeletal filaments in one dimension, processes on cell membranes

Box 1. A sample configuration file and its output. The sections of the configuration file match those in Methods. The snapshot was saved from Smoldyn's graphical output at 2 seconds. The dynamics were graphed with Microsoft Excel from data that Smoldyn saved to box1out.txt. Below, S is substrate, E is enzyme, ES is enzyme-substrate complex, and P is product

---

Configuration file

```
# Enzymatic reactions on a surface
# by Steve Andrews, October 2009.
# This model is in the public domain.
# Units are microns and seconds.

define K_FWD 0.001
define K_BACK 1
define K_PROD 1
```

configuration file statements

```
dim 2
boundaries 0 -1 1
boundaries 1 -1 1
time_start 0
time_stop 10
time_step 0.01
```

space and time

```
species S E ES P
difc S 3
difc P 3
color S(all) 0 0.8 0
color E(all) 0.5 0 0
color ES(all) 1 0.5 0
color P(all) 0 0 0.5
display_size all(all) 0.02
display_size E(all) 0.03
display_size ES(all) 0.03
max_mol 1000
```

molecules

```
graphics opengl_good
frame_thickness 0
```

graphics

```
max_surface 1
start_surface membrane
  action both all reflect
  color both 0 0 0
  thickness 1
  max_panels sphere 1
  panel sphere 0 0 1 50
end_surface
```

surfaces

reactions

```
reaction fwd E(front) + S(bsoln) -> ES(front) K_FWD
reaction back ES(front) -> E(front) + S(bsoln) K_BACK
product_placement back pgemmax 0.2
reaction prod ES(front) -> E(front) + P(bsoln) K_PROD
```

```
max_compartment 1
start_compartment inside
  surface membrane
  point 0 0
end_compartment
```

compartments

```
compartment_mol 500 S inside
surface_mol 100 E(front) membrane all all
```

more molecules

```
output_files box1out.txt
cmd B molcountheader box1out.txt
cmd N 10 molcount box1out.txt
```

runtime commands

```
end_file
```

configuration file

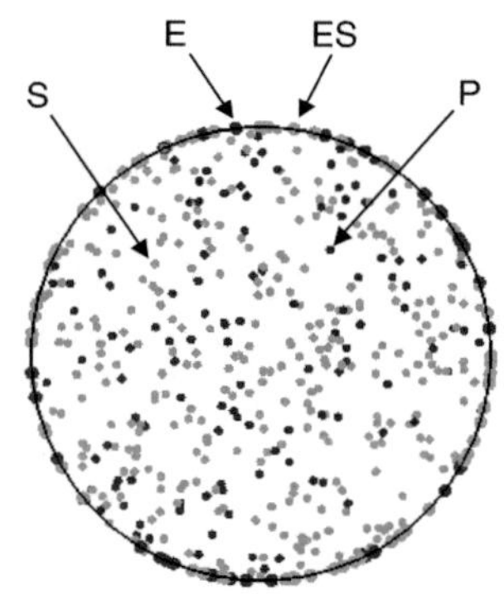

Snapshot of system, *t* = 2 s

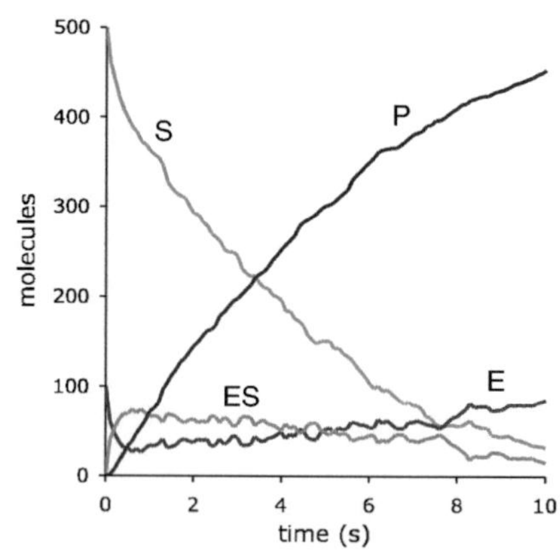

System dynamics

with two dimensions, or cellular systems with three dimensions. Low-dimensional models can also approximate higher dimensional ones.

Define the system's spatial extent with `boundaries`. This does not limit the simulated space but, provided that you keep all molecules within these boundaries, helps Smoldyn run efficiently. Internally, Smoldyn uses the boundaries to spatially partition the system volume so as to reduce the number of molecule–molecule and molecule–surface interactions that it has to investigate (see Note 1 for partition optimization). If your model does not include surfaces (Subheading 3.7), then specify how molecules should interact with the boundaries. The options are reflection, transmission, absorption, or periodic (periodic means that molecules that diffuse out of one side of the system are immediately diffused into the opposite side of the system, which can avoid edge effects in effectively infinite systems). On the other hand, if your model includes any surfaces, then the boundaries defined here will only transmit molecules and you will have to use surfaces to confine molecules.

Specify the simulation starting time, stopping time, and time step with `time_start`, `time_stop`, and `time_step`. Smoldyn only supports fixed-length time steps, in contrast to the adaptive time steps that MCell supports (12) or the event-based time steps that the Green's Function Reaction Dynamics method uses (25). However, you can change time steps midsimulation with the `settimestep` command (Subheading 3.6). See Note 2 for advice on choosing time steps.

3.4. Molecules

Each Smoldyn molecule is a member of a chemical species, and all molecules of a particular species are equivalent. Enter individual species names with `species` statements. Distinct forms of a molecule, such as phosphorylated and unphosphorylated, or monomeric and dimeric, need to be defined as separate species. To simplify the definitions for these multiple molecular forms, Smoldyn can automatically enumerate both them and their chemical reactions using rule-based modeling (Subheading 3.10).

Each molecule, regardless of its species, may be in any of five states. "Solution" state, in which molecules are not bound to surfaces, is the default. Molecules can also bind surfaces in a "front," "back," "up," or "down" state. The former two options represent molecules that bind a single side of a surface, where typical uses include peripheral membrane proteins, GPI-anchored proteins, or adsorbates. The other two options, up and down, represent surface-spanning molecules, such as ion channels or trans-membrane receptors. The separate up and down states allow one to distinguish whether the protein's active side faces the surface's front or back.

Define the diffusion coefficient for each species and state with `difc`. See Note 3 for advice on choosing diffusion coefficients. Smoldyn can also diffuse molecules anisotropically (`difm` statement), such as for confining molecules to a plane or for simulating diffusion in anisotropic environments. In addition, molecular drift in a specified direction (`drift` statement) can represent molecules in flow systems or around motile cells. For example, you could simulate gel electrophoresis with a combination of diffusion and drift.

Internally, Smoldyn keeps track of molecules with a "dead" list of inactive molecules and several "live" lists of simulated molecules. See Note 4 for advice on arranging the live lists.

The following statements add molecules to the starting state of your model: `mol` adds solution phase molecules to either specific or random locations; `surface_mol` (enter after you define surfaces) adds surface-bound molecules to specific surfaces or regions on surfaces; and `compartment_mol` (enter after you define compartments) adds solution phase molecules to specific spatial compartments.

3.5. Graphics

Smoldyn simulation results can only be visualized as they are computed. While this can be inconvenient for generating publication quality figures, the immediate output is very helpful for model development. You can choose from three levels of rendering quality, or no graphical output at all, using the `graphics` statement. Because graphical rendering is time-consuming, you can speed simulations up by using low quality rendering or by using `graphic_iter` to only render graphics periodically. Conversely, you can slow simulations down with `graphic_delay`.

The default background color is white, which you can modify with `background_color`. For everything else, the default color is black. Color the boundary edges with `frame_color`, spatial partitions with `grid_color`, molecules with `color`, and surfaces with `surface` ***name*** `color`. You can also set boundary line weights with `frame_thickness`, spatial partition line weights with `grid_thickness`, molecule sizes with `display_size`, surface edge weights with thickness, and surface drawing styles with `polygon`. With Smoldyn's best rendering quality, called `opengl_better`, you can place light sources in the system, including their ambient, diffuse, and specular colors, using `light`. To make full use of these lights, set surface reflection parameters with `shininess`. As always, see the Smoldyn User's Manual for details. Unfortunately, Smoldyn renders partially transparent surfaces poorly. This means that if you want to see what's inside a surface, such as a cell membrane, it is usually best to render it with a wireframe.

You can manipulate Smoldyn's graphical display during your simulation using key strokes. For example, rotate the system with arrow keys, pan with shift + arrows, zoom in with "=," zoom out with "–," or revert to the default view with "0." Also, the space bar pauses the simulation, "T" saves a TIFF image of the current graphics display, and "Q" quits the simulation.

Smoldyn can save movies of simulations as stacks of TIFF files. Again, each image is a simple copy of the graphics window display. To turn on recording, set the number of simulation time steps per image with `tiff_iter`, name the files with `tiff_name`, and number the files with `tiff_min` and `tiff_max`. ImageJ, QuickTime Pro, and other software programs can assemble TIFF stacks into self-contained movies.

3.6. Runtime Commands

Smoldyn includes a "virtual experimenter" who can observe or manipulate the simulated system. You can instruct this virtual experimenter to perform tasks before, during, or after the simulation, including at periodic intervals (see Table 3.6.1 of the Smoldyn User's Manual). Issue commands using the `cmd` statement, the timing parameters, the name of the specific task, and any task-specific parameters.

The first class of commands are simulation control commands. Examples are `stop` and `pause`, which stop or pause the simulation. Also, `keypress` mimics the behavior of the user pressing a key (see Subheading 3.5), which can be useful for automating the graphical display. These control commands are particularly useful when combined with conditional commands. For example, the statement "`cmd E ifno substrate stop`" tells the virtual experimenter to count the number of substrate molecules at every time step and to stop the simulation if none remain.

The second class, observation commands, save information about the system to text files. For example, `molcount` records the number of molecules for each species and `molcountspace` records histograms of molecule counts along a line in the system, from which concentration profiles can be calculated. A particularly powerful observation command is `savesim`, which saves the entire simulation state to a Smoldyn-readable configuration file. By saving the simulation state regularly, you can abort a simulation mid-run and then restart it later on. Alternatively, you can equilibrate your model in one simulation, save the result with `savesim`, and then run other simulations that start from this equilibrated state. All observation commands require that you explicitly declare the output file names with `output_files`, and also with `output_root` if you do not want the files to be in your working directory. The output file "stdout" sends the output to the terminal window. To save data to a series of files, number the first one with `output_file_number` and progress to subsequent ones with the `incrementfile` command.

Finally, manipulation commands give you wide latitude to modify the system during the simulation, and not necessarily in accordance with biochemical principles. For example, you can instruct the virtual experimenter to add molecules to the system with `pointsource` or `volumesource`, remove molecules from inside spheres with `killmolinsphere`, or replace a specified fraction of one molecular species in some volume with a different species using `replacevolmol`. Also, several commands allow you to create fixed concentration inputs to your model, such as `fixmolcountonsurf` and `fixmolcountincmpt`. The `set` command is extremely powerful because you can follow it with essentially any Smoldyn statement. This allows you to add species, reactions, or surfaces, change the simulation time step, or modify your model in many other ways, all mid-simulation.

3.7. Surfaces

Unlike biological membranes or other real surfaces, Smoldyn surfaces are infinitely thin. This means that molecules in solution phase, or in "front" or "back" surface-bound states, are always on the front or back side of a surface; also, molecules in "up" or "down" states are always exactly at the surface. Each Smoldyn surface is composed of geometric primitives called panels, which can be triangular, spherical, cylindrical, or other shapes. Along with being convenient and computationally efficient, this representation method is also quite versatile because it allows arbitrarily complex surface geometries, disjoint surfaces (e.g., a collection of vesicles), and/or open surfaces (e.g., the fragmented membrane of a lysed cell).

For example, the model shown in Fig. 1 includes four surfaces: (1) a mesh of 480 triangles that surrounds the entire system, (2) a spherical "receiver" cell in the middle, (3) a spherical "target" cell shown toward the right side in dark gray, and (4) five spherical "challenger" cells shown in light gray. Note that this last surface is disjoint. The model shown in Fig. 2 includes two surfaces: (1) the cell membrane, assembled from two hemispheres and a cylinder and (2) four spherical carboxysome protein shells.

Define each surface with a block of statements that starts with `start-surface` and ends with `end_surface` (you can also enter surface parameters with statements that start with `surface` and the surface name, but the block format is usually easier). Within this block, Within this block, list each individual panel with panel. list each individual panel with `panel`. If you want surface-bound molecules to be able to diffuse between neighboring panels, whether they are on the same surface or different surfaces, then list each panel's neighbors with `neighbor`.

Two utility programs included in the distribution help generate panel and neighbor data. The first, wrl2smol, reads Virtual Reality Modeling Language (VRML) files of triangle mesh data and converts them to lists of triangle panels and their neighbors in Smoldyn format. Mathematica, MatLab, ImageJ, and other programs can generate VRML files. The second utility program, SmolCrowd, generates random arrays of nonoverlapping circles or spheres, which can be useful for modeling macromolecular crowding.

Several parameters characterize each surface. These include graphical display parameters, described in Subheading 3.5, and surface–molecule interaction parameters. Many types of interactions are possible. For example, you can make a surface impermeable, transmitting, or irreversibly absorbing for molecules that diffuse into it, and these interactions can differ for different molecular species and the two surface faces. Specify these behaviors with `action`. Alternatively, specify coefficients for adsorption to, desorption from, or partial transmission through surfaces (23) with `rate` (see Note 5 for advice on these coefficients). Two molecular pseudo-states, "fsoln" and "bsoln," allow you to distinguish solution-state molecules on the front side of a surface from solution-state molecules on the back side of a surface. You can also specify rates for transitions between surface-bound states, such as from the front state to the back state. Additionally, you can instruct molecules to change species when they cross a surface; this is particularly useful for models with spatially variable diffusion coefficients, such as between extracellular and intracellular regions, or nonraft and raft regions of membranes. Jump surfaces (defined with `action` and `jump`) magically transport molecules from one panel to another. They are useful for adding holes to otherwise impermeable surfaces, such as pores in a nuclear envelope, and for creating periodic boundaries. Finally, unbounded-emitter surfaces (`unbounded_emitter` statement) irreversibly absorb molecules with coefficients that make internal concentration profiles mimic those that would be seen for an unbounded system (23). The triangulated mesh in Fig. 1 is an unbounded-emitter surface.

3.8. Chemical Reactions

Smoldyn supports zeroth, first, and second order chemical reactions, where the order is just the number of reactants. Use zeroth order reactions to add molecules at random locations within the entire system volume or within a compartment (Subheading 3.9) at a roughly constant rate. These violate mass balance and are thus unphysical; however, they are particularly useful for including protein or mRNA synthesis in models but not the respective synthesis machinery. Use first order, or unimolecular, reactions for spontaneous molecular changes, such as protein conformation, dissociation, and decay processes. These are also useful for pseudo-first order reactions, in which one of two reactants (e.g., ATP) is unmodeled and assumed to be constant and uniformly distributed. Use second order, or bimolecular, reactions for interactions between pairs of molecules, such as association and enzymatic reactions.

Enter reactions with the syntax: "`reaction` *name* `reactants` —> `products` *rate*," where *name* is the reaction name, `reactants` and `products` are lists of species and states that are separated by "+" symbols (use "0" if a list is empty and use the fsoln and bsoln pseudo-states as necessary), and *rate* is the reaction rate constant. Alternatively, use `reaction_surface` or `reaction_cmpt` for reactions that should only occur on specific surfaces or in specific compartments. Smoldyn allows up to 16 products for each reaction. Note 6 offers advice on choosing reaction rate constants and Note 7 describes ways to speed up bimolecular reaction simulation, which is typically Smoldyn's slowest component.

Reversible association reactions, of the form A + B ↔ C, involve an additional complication. When a C molecule dissociates, the new A and B molecules form in close proximity to each other, which makes them especially likely to react with each other. This reaction between dissociation products is called a geminate recombination. Nonspatial treatments of chemical kinetics, including conventional mass-action theory and nonspatial simulations, cannot account for geminate recombinations, so they are often ignored. However, geminate recombinations affect reaction rates both in real systems and in particle-based simulations. For example, the association reaction rate for A + B ↔ C is faster if it is measured at equilibrium than if it is measured with all C removed as it is formed (e.g., with the additional reaction C + D → E, and a high concentration of D), because the former situation includes geminate recombinations and the latter does not. So that Smoldyn can accurately simulate the kinetics of reversible reactions, and other reactions where the products can react with each other, you need to specify how Smoldyn should place dissociation products in the system with `product_placement`. Ideally, you should choose the placement method that represents whether the experimental rate constants were measured at equilibrium or with products removed as they were formed (see Table 3.8.2 of the Smoldyn User's Manual); and you should also give the probability of geminate recombinations. However, these details are almost never known, so I suggest using "`product_placement name pgemmax 0.2`," where name is the reaction name. This states that reaction rates were measured at equilibrium (typically correct) and that reaction products should have up to a 20% probability of recombining with each other. This geminate recombination probability typically leads to efficient simulations and physically sensible simulation parameters (the binding and unbinding radii (18, 22)). Smoldyn uses the same parameters if you do not enter a `product_placement` statement, but issues a warning.

Smoldyn also uses the bimolecular reaction formalism for conformational spread and excluded volume interactions. Use conformational spread interactions to model mechanical coupling between stationary molecules (set the maximum interaction distance with `confspread_radius`). The prototypical conformational spread example occurs in the *E. coli* flagellar motor, in which motor proteins in "active" conformations induce activity in their neighbors and those with "inactive" conformations induce inactivity in their neighbors (26); this positive feedback makes the motor switch-like. Excluded volume interactions (`product_placement` statement) keep molecules spatially separate. They are particularly useful for keeping molecules that are confined to a channel from passing each other. They also permit more general simulations of molecules with excluded volume, although this aspect of Smoldyn is computationally inefficient.

3.9. Spatial Compartments

Smoldyn compartments are regions of volume bounded by surfaces. A cell's cytoplasm, nucleus, or extracellular space are typical examples. Compartments are useful for specifying initial conditions of models, for recording simulation results, for defining reactions that are only relevant in specific regions, and for compatibility with compartment-based simulators (e.g., Virtual Cell (27) and MOOSE (28)). To use compartments, define individual compartments with blocks of statements bracketed by `start_compartment` and `end_compartment`.

You can define a compartment with either of two methods, or with a combination of them. In the first, list the compartment's bounding surfaces (the same surfaces that Subheading 3.7 describes) with `surface` statements and list one or more "inside-defining" points with `point` statements. These points define the compartment's volume as follows: a molecule is defined to be in a compartment if and only if a straight line can be drawn between it and an inside-defining point without crossing one of the compartment's bounding surfaces. For example, we could define a compartment for the model shown in Fig. 2 that included the entire bacterial volume, which we'll call "cell," by making the cell membrane the compartment surface and listing an interior-defining point near the center of the cell. This compartment would behave as one would expect, where any molecule within the bacterium (including molecules in carboxysomes) would be in the "cell" compartment and any molecule outside of the bacterial membrane would not be in the "cell" compartment. We could define another compartment that included all of the carboxysome interiors, called "carboxysome," by setting the compartment surface to the carboxysome surface (which is disjoint) and listing an interior-defining point at the center of each carboxysome.

The other way to define a compartment is with logical combinations of previously defined compartments, using the `compartment` statement. Continuing with the previous examples, we could define the bacterial cytoplasm in Fig. 2 that is not within a carboxysome as a compartment called "cytoplasm" with "`compartment equal cell`" and then "`compartment andnot carboxysome`."

Overall, this compartment definition method is somewhat nonintuitive, but it is easy to use, versatile, and computationally efficient.

3.10. Rule-Based Reaction Network Expansion

Most proteins can adopt any of a very large number of different states, such as by phosphorylation, nucleotide-binding, multiple conformations, or complexation with other proteins. In Smoldyn, as in most simulators, each of these states needs to be modeled as a distinct species, with distinct chemical reactions. Manually listing these species and their reactions is often impractical though, so modelers typically simplify their models to only include the most essential states, while ignoring the rest. The drawbacks of this approach are, of course, that it can overlook important biological interactions and it precludes the study of realistic reaction networks. Thus, several groups have pursued a different approach, in which the simulator generates the species and chemical reactions automatically from "interaction rules" (29, 30). Smoldyn supports this so-called rule-based modeling with a module called libmoleculizer. A notable libmoleculizer feature is that it only generates species and their reactions as they arise; this means that Smoldyn does not need to keep track of multimeric complexes that could theoretically exist, but that never actually form. This speeds simulations, reduces computer memory use, and simplifies models.

In the libmoleculizer formalism, chemical species are assembled from building blocks called mols. Both individual mols and multimeric complexes of mols are chemical species. Each mol can have modification sites, such as for phosphorylation, methylation, or nucleotide-binding. Also, each mol can have binding sites with which it can reversibly bind other mols to form multimeric complexes. These binding sites have "shapes" so that their activity can be modulated through allosteric interactions. For example, a binding site can be put in its "active" shape when a specific modification site is phosphorylated, and in its "inactive" shape when that modification site is unphosphorylated.

Enter network generation rules in your configuration file in a block of statements that starts with `start_rules` and ends with `end_rules`. Within this block, enter rules in the following categories. (1) Under "Modifications," declare all possible post-translational modifications. For example, "`name = PP`" might represent double phosphorylation. (2) Under "Molecules," define the mols, their binding sites, and their modification sites. For example, "`Fus3(ToSte5, ToSte12 {inactive, active}, *PhosSite {None,P,PP})`" states that `Fus3` is the name of a mol, `ToSte5` and `ToSte12` are binding sites, the `ToSte12` binding site can adopt either an inactive or an active shape, and PhosSite is a modification site that can have zero, one, or two phosphate groups. Mols also need to be defined as Smoldyn species (Subheading 3.2). (3) Under "Explicit-Species," assign your own names to specific multimeric complexes, if desired. (4) Under "Explicit-Species-Class," assign names to complexes that share specific characteristics. For example, the class "Fus3 (ToSte12!1).Ste12(ToFus3!1)" includes all molecular species that include a Fus3 bound to a Ste12, using Fus3's ToSte12 site and Ste12's ToFus3 site; this class leaves other modification states and other binding site occupancies unspecified. Use species classes to set the graphical display of newly generated species using the Smoldyn statements (i.e., not entered in the libmoleculizer rule block) species_class_display_size and species_class_color. (5) Under "Association-Reactions," define binding reactions between mols. For example, "Fus3(ToSte12 {active}) + Ste12(ToFus3) —> Fus3 (ToSte12!1).Ste12(ToFus3!1)" defines a possible binding between Fus3 and Ste12, and states that this association can only happen when Fus3's ToSte12 binding site is in its active shape. (6) Under "Transformation-Reactions," list reactions in which mol modification states change, such as between phosphorylated and unphosphorylated states. For example, "Fus3(*PhosSite {None}) —> Fus3(*PhosSite {P})" is a reaction in which Fus3 is spontaneously phosphorylated. (7 and 8) Under "Allosteric-Complexes" and "Allosteric-Omniplexes," specify how binding site shapes should depend on the binding and modification states of a mol or complex. Allosteric-complexes are for specific species and allosteric-omniplexes are for classes of species. An example of the latter is "Fus3(ToSte12 {active <— *}, *PhosSite {PP})" which states that Fus3's ToSte12 binding site should be in its active shape whenever the PhosSite is doubly phosphorylated.

Smoldyn's rule-based modeling support is still very new, so we are continuing to add functionality and improve the configuration file syntax.

4. Notes

1. *Partitioning space.* Smoldyn subdivides the system within its boundaries into a grid of uniformly spaced "virtual boxes." These boxes do not affect the simulated results. Instead, they make simulations more efficient by reducing the number of potential molecule–molecule and molecule–surface interactions that Smoldyn needs to check at each time step. The default partition spacing, which yields an average of about four molecules per box when the simulation starts, is often good but can

usually be improved upon. Improvement is especially important if the model starting state is not representative of its typical state or if molecules are not distributed homogenously.

Optimize the partition spacing by timing simulations (Smoldyn displays the run time when it terminates) that use a range of box sizes, which you can set with `molperbox` or `boxsize`, and choose the fastest size for which Smoldyn does not report any errors. The errors to watch for are those that report that bimolecular reaction binding radii, or analogous distances, are larger than the box widths. When they arise, they indicate that Smoldyn will not detect some bimolecular reactions, which means that those reactions will simulate too slowly, and simulation accuracy will be reduced (see Note 7). Other Smoldyn warnings about box sizes, such as those that comment on unusually many or few molecules per box, are simply suggestions that different partition spacings may improve performance.

2. *Choosing simulation time steps.* When choosing the time step for a simulation, there is an unavoidable trade-off between using shorter time steps to get more accurate results, and using longer time steps for faster simulations. Thus, the succinct answer to the question of what time step to use, is that you should choose the longest time step that yields sufficiently accurate results. You can find it by running trial simulations with a wide range of time steps and graphing representative simulation results (e.g., the maximum amount of a product concentration, time until a substrate concentration is halved, steady-state receptor occupancy, etc.) against the log of the time step. Typically, this plot will show results that are independent of time step lengths for short time steps, that errors increase log-linearly for long time steps, and that these regions are separated by a cross-over region that is about a factor of 10 in width. A judgment call is required at this point to decide how much accuracy, if any, you are willing to sacrifice for faster simulations.

In addition to this heuristic method, it is worth considering how the time step length affects individual simulation algorithms. In particular, (1) Smoldyn displaces each diffusing molecule by about $s = (2D\Delta t)^{1/2}$, where D is the molecule's diffusion coefficient and Δt is the time step, at each time step. The average displacement for the fastest diffusing species, s_{max}, is the simulation's spatial resolution. In general, s_{max} should be significantly smaller than important geometrical features, surface curvature radii, and distances between fixed molecules. (2) Characteristic transition times for unimolecular reactions, molecular desorption, and transitions between surface-bound states are all $\tau = 1/k$, where k is the appropriate rate constant. Also, characteristic times for bimolecular reactions with well-mixed reactants are $\tau = ([A] + [B])/(k[A][B])$, where k is the reaction rate constant and [A] and [B] are the reactant concentrations. And, characteristic times for molecular adsorption and permeability are $\tau = d/\kappa$, where d is the distance between surfaces (e.g., the cell length) and κ is the adsorption or transmission coefficient. To simulate these dynamics accurately, the time step should be smaller than these characteristic times. (3) Smoldyn performs bimolecular reactions between pairs of reactant molecules that are closer than their "binding radius." Binding radii, which Smoldyn computes from reaction rates, diffusion coefficients, and the simulation time step, increase as time steps are made longer and are often larger than physical molecular radii (22). If a molecular species is so concentrated that these binding radii frequently overlap each other, which is especially problematic for clustered surface-bound molecules (e.g., the receptors in Fig. 1 (8)), this decreases accuracy. (4) For fast bimolecular reactions that are far from steady state, Smoldyn's simulated results more closely agree with diffusion-limited kinetics when short time steps are used and with activation-limited kinetics when long time steps are used (22). This is unlikely to affect typical biochemical network models, but may affect reaction biophysics models. For all of these time step considerations, note that Smoldyn displays all internal simulation parameters, several characteristic times, and any warnings for potential problems before it starts simulations. It is prudent to check that this information seems reasonable.

In practice, many researchers use time steps around 0.1 ms (16, 17, 19, 20, 31, 32). Smoldyn has also been used with time steps as short as 0.06 ns, in a test of its reaction algorithms (22), and with time steps up to 10 or 20 ms (8, 33). The latter models, one of which is shown in Fig. 1, were able to use long time steps because they had relatively low molecular densities and large system geometries.

3. *Choosing diffusion coefficients.* In homogeneous solutions, such as water or liquid growth media, diffusion coefficients can often be approximated reasonably well with the Stokes–Einstein equation. It is

$$D = \frac{k_B T}{6\pi\eta r},$$

where D is the diffusion coefficient, k_B is Boltzmann's constant, T is the absolute temperature, η is the solution viscosity, and r is the radius of the diffusing particle, which is assumed to be spherical. This particle radius can be easily calculated from its mass m and density ρ,

$$r = \sqrt[3]{\frac{3m}{4\pi\rho}} \approx 0.0655\sqrt[3]{m}\,\text{nm}.$$

In the latter equality, m is measured in Daltons; also, it assumes an average density of 1.41 g/cm, which is reasonably accurate for proteins above 40 kDa and within 10% for smaller proteins (34). Combining this radius calculation with the Stokes–Einstein equation yields the protein diffusion coefficient estimate

$$D \approx \frac{3270}{\sqrt[3]{m}} \ \mu\text{m}^2/\text{s}.$$

This is for a temperature of 20°C, where water's viscosity is very nearly 1 mPa s. Two examples are: for lysozyme, m is 14.6 kDa, the experimental D is 111 $\mu m^2/s$ (35), and the computed D is 134 $\mu m^2/s$; and for green fluorescent protein (GFP), m is 26.9 kDa, the experimental D is 87 $\mu m^2/s$ (36, 37), and the computed D is 109 $\mu m^2/s$. Of course, theories that account for protein shape yield better predictions (35).

For small molecules in water, experimentally measured diffusion coefficients are readily available. Examples include 2,010 $\mu m^2/s$ for oxygen at 20°C and 380 $\mu m^2/s$ for lactose at 15°C (38). For quick estimates, the Stokes–Einstein equation is usually correct to within a factor of two.

In eukaryotic cytoplasms, nuclei, and mitochondria, experiments by Verkman's group show that macromolecule diffusion coefficients are about 25% of their values in water, for masses up to about 500 kDa (39, 40). Larger molecules and filamentous molecules, such as DNA, diffuse more slowly, likely due to sieving by actin networks. In the *E. coli* bacterial cytoplasm, GFP diffuses about another factor of 5 slower than it does in eukaryotes, now with a diffusion coefficient around 3–8 $\mu m^2/s$ (41, 42). Diffusion is much slower yet in the *E. coli* periplasm, where the 42.5 kDa maltose-binding protein has a diffusion coefficient of about 0.009 $\mu m^2/s$ (43). Finally, a couple of membrane-bound protein diffusion coefficients are 0.09 $\mu m^2/s$ for aquaporin-1 (about 30 kDa) in nonpolarized fibroblast cell membranes (44) and 0.012 $\mu m^2/s$ for histidine kinase PleC (117 kDa, including fluorophore) in *Caulobacter* membranes (14).

From these data, I suggest the rules-of-thumb: use 80% of the Stokes–Einstein value, as calculated above, for proteins in water, and divide the aqueous diffusion coefficient by 4 for eukaryotic cell and organelle cytoplasms, by 15 for bacterial cytoplasms, by 1,000 for bacterial periplasms, by 1,000 for eukaryotic membranes, and by 4,000 for bacterial membranes.

4. *Optimizing molecule lists.* As Subheading 3.4 mentions, Smoldyn stores molecules in several lists. By default, Smoldyn stores all simulated molecules that do not diffuse in a "fixed" list and

those that do diffuse in a "diffusing" list. This scheme is more efficient than a single list because Smoldyn does not need to diffuse or perform surface interactions for molecules in the fixed list; also Smoldyn does not need to check for bimolecular reactions between pairs of molecules in the fixed list.

Adding more lists can speed simulations up further, often by factors of five or more, typically by minimizing the number of potential bimolecular reactions that Smoldyn needs to check. For example, consider the chemical reaction A + B → C. If Smoldyn stored all three species in the same list, then when Smoldyn searched the list to see which AB molecule pairs could react, it would also encounter many nonreactive AA, AC, BB, BC, and CC molecule pairs, each of which would take a small amount of time to check and then ignore. On the other hand, Smoldyn would only encounter AB pairs if A, B, and C were stored in three separate lists. Generalizing this example, more molecule lists reduce the number of unnecessary checks, so typically produce faster simulations. This trend does not continue indefinitely though, because each molecule list also requires some processing time; this becomes important for lists with few molecules. Thus, overall, simulation performance is usually best when each abundant species has its own list and when sparsely populated species share lists. Further list optimization, by considering the potential bimolecular and molecule–surface interactions that Smoldyn has to check at each time step, can produce additional gains.

Create molecule lists with `molecule_lists` and assign species to them with `mol_list`.

5. *Choosing molecule–surface interaction coefficients.* Adsorption coefficients are limited to about the thermal velocity of the adsorbent, which is (45)

$$\kappa_{\max} = \sqrt{\frac{k_B T}{m}} \approx \frac{1.56 \times 10^9}{\sqrt{m}} \quad \mu\text{m/s}.$$

The latter equality assumes that m is measured in Daltons and the temperature is 20°C. For example, the largest possible adsorption coefficient for a 50 kDa protein is about 7×10^6 μm/s. Real adsorption coefficients are likely to be vastly smaller than this maximum. For example, Huang and coworkers chose an adsorption coefficient of 0.025 μm/s for the adsorption of MinD (29.5 kDa) to the inside of the *E. coli* cell membrane in a biochemical model (46). Unfortunately, remarkably few experimental papers present quantitative data on protein adsorption to or desorption from lipid bilayers, despite an extensive literature.

Transmission coefficients for molecules through membranes can be calculated using the equation (47)

$$\kappa = \frac{K_{\mathrm{mem}} D_{\mathrm{mem}}}{d_{\mathrm{mem}}},$$

where K_{mem} is the partition coefficient for the molecule into the membrane from the neighboring solution (e.g., cytoplasm), D_{mem} is the molecule's diffusion coefficient in the membrane, and d_{mem} is the membrane thickness. Qualitatively, this equation shows that hydrophobic molecules are transmitted faster than hydrophilic ones because they partition into the membrane more readily, and that smaller molecules are transmitted faster than larger ones because they diffuse faster. A paper by Paula et al. (47) shows how to compute the necessary parameters. It also lists some experimentally determined coefficients for transmission through lipid bilayers, including: 3.5×10^{-8} μm/s for potassium ions, 0.014 μm/s for urea, 0.027 μm/s for glycerol, 5 μm/s for protons, and 150 μm/s for water molecules; these values are for 2.7 nm thick bilayers, which is about the thickness of typical biological membranes.

Adsorption and transmission coefficients can also be estimated using the respective characteristic times (see Note 2), if these times can be inferred from experiments.

6. *Choosing reaction rate constants.* Experimental bimolecular reaction rates are limited to

$$k_{\mathrm{max}} = 4\pi(D_{\mathrm{A}} + D_{\mathrm{B}})(r_{\mathrm{A}_s} + r_{\mathrm{B}})$$

by the rate at which reactants can diffuse together (48). Here, k_{max} is the diffusion-limited reaction rate constant, D_{A} and D_{B} are the diffusion coefficients of the reactants, and r_{A} and r_{B} are the reactant radii. Highly reactive small molecules sometimes react with rates that are close to this maximum (49), but typical biochemical reaction rate constants are much slower. You can verify that your simulated rates will be well below this maximum by comparing Smoldyn's output parameter labeled "binding radius if dt were 0" to the sum of the physical reactant radii. They will be equal if your simulated reaction is at the maximum rate, and proportionately less for slower reactions.

Other than this one qualitative check, reaction rates usually have to found from the scientific literature that is relevant to your model. Typically, some reaction rates will have been measured, others can be calculated from published data (e.g., association reaction rates can be calculated from dissociation constants and dissociation reaction rates), others can be estimated from characteristic reaction times and published figures, and yet others simply have to be guessed. Previously developed models, whether of your system or similar systems, are often a good source of reaction rates and literature references. For this, the BioModels database (http://www.ebi.ac.uk/biomodels-main) (50) is a particularly good place to start. Finally, curated literature summaries that list quantitative data are rarely available, but can be excellent resources when they are. For example, the Yeast Pheromone Model wiki (http://yeastpheromonemodel.org/wiki/Main_page) lists reaction rates and their references for most reactions in the yeast pheromone response system. A few representative rates from this wiki are pheromone binds Ste2 receptors at 1.8×10^5 M^{-1} s^{-1} (3.1×10^{-4} $\mu m^3/s$) and unbinds at 10^{-3} s^{-1}, the Gpa1 G-protein subunit exchanges GDP for GTP at 6.17×10^{-4} s^{-1}, and the Fus3 MAP kinase binds the Ste5 scaffold protein at 2.3×10^6 M^{-1} s^{-1} (3.8×10^{-3} $\mu m^3/s$) and unbinds at 2.3 s^{-1}. While these are reasonably typical values for these types of interactions, other similar reactions are often faster or slower by several orders of magnitude.

Note that bimolecular reactions that take place in one- and two-dimensional systems, such as along filaments or within membranes, do not have reaction rate constants in the same sense as reactions in three dimensions. Also, current versions of Smoldyn cannot quantitatively simulate low-dimensional reactions.

7. *Lowering bimolecular reaction accuracy for faster simulations.* Smoldyn partitions space into virtual boxes (see Note 1) to reduce the number of potential molecule–molecule and molecule–surface interactions that need checking at each time step. Most molecule–molecule interactions occur within single boxes in typical simulations. However, Smoldyn also has to check for interactions between molecules that are in adjacent boxes to achieve high accuracy. There are typically about 50 times more potential "interbox" than "intrabox" interactions because (1) in three-dimensional systems, each box has 26 neighboring boxes (or eight neighbors in 2-D or two neighbors in 1-D), and (2) there are n^2 possible pairwise interactions for n molecules in each of two neighboring boxes but only about $n^2/2$ possible pairwise interactions for n molecules within a box. The result is that checking for interbox interactions is computationally costly, relative to the few interactions that are actually detected.

   Using the `accuracy` statement, you may be able to speed Smoldyn up by instructing it to ignore some or all potential interbox interactions. Of course, this will cause Smoldyn to overlook some reactions that should occur, so bimolecular reactions will simulate somewhat too slowly. If desired, this problem can be mitigated by determining reaction rate correction factors from trial simulations (alternatively, correction factors can be estimated from the total numbers of intrabox and interbox reactions, which Smoldyn reports upon termination). In practice, the simulation speed improvement is typically much less than the maximum possible factor of 50, and is often just a factor of 2 or less. Thus, the best approach is to try lowering Smoldyn's accuracy parameter and directly assessing how much the simulation speed improved and how much the actual accuracy decreased.

Acknowledgments

I thank Nathan Addy for developing the Smoldyn build system and the libmoleculizer module, David Savage for providing the configuration file for Fig. 2, and Roger Brent for encouragement and helpful discussions. I also appreciate helpful comments and suggestions from many Smoldyn users, including Karen Lipkow and Shahid Khan in particular. This work was supported by MITRE contract number 79729, awarded to Roger Brent.

References

1. Tindall MJ, Porter SL, Maini PK, Gaglia G, Armitage JP. (2008) Overview of mathematical approaches used to model bacterial chemotaxis I: the single cell. Bull Math Biol, 70:1525–1569.
2. Tyson JJ, Novak B. (2008) Temporal organization of the cell cycle. Curr Biol, 18: R759–R768.
3. Mogilner A. (2006) On the edge: modeling protrusion. Curr Opin Cell Biol, 18:32–39.
4. Alves R, Antunes F, Salvador A. (2006) Tools for kinetic modeling of biochemical networks. Nat Biotechnol, 24:667–672.
5. Takahashi K, Arjunan SNV, Tomita M. (2005) Space in systems biology of signaling pathways – towards intracellular molecular crowding in silico. FEBS Lett, 579:1783–1788.
6. Andrews SS, Arkin AP. (2006) Simulating cell biology. Curr Biol, 16:R523–R527.
7. Andrews SS, Dinh T, Arkin AP. (2009) Stochastic Models of Biological Processes. In *Encyclopedia of Complexity and System Science*. Vol. 9. Edited by Meyers RA, Springer, New York, 8730–8749.
8. Andrews SS, Addy NJ, Brent R, Arkin AP. (2010) Detailed simulation of cell biology with Smoldyn 2.1. PLoS Comp Biol, 6: e1000705.
9. Dayel MJ, Hom EFY, Verkman AS. (1999) Diffusion of green fluorescent protein in the aqueous-phase lumen of endoplasmic reticulum. Biophys J, 76:2843–2851.
10. Partikian A, Ölveczky B, Swaminathan R, Li Y, Verkman AS. (1998) Rapid diffusion of green fluorescent protein in the mitochondrial matrix. J Cell Biol, 140:821–829.
11. Plimpton SJ, Slepoy A. (2005) Microbial cell modeling via reacting diffusive particles. J Phys Conf Ser, 16:305–309.
12. Kerr RA, Bartol TM, Kaminsky B, Dittrich M, Chang J-CJ, Baden SB, Sejnowski TJ, Stiles JR. (2008) Fast Monte Carlo simulation methods for biological reaction-diffusion systems in solution and on surfaces. SIAM J Sci Comput, 30:3126–3149.
13. Jin S, Haggie PM, Verkman AS. (2007) Single-particle tracking of membrane protein diffusion in a potential: simulation, detection, and application to confined diffusion of CFTR Cl^- channels. Biophys J, 93:1079–1088.
14. Deich J, Judd EM, McAdams HH, Moerner WE. (2004) Visualization of the movement of single histidine kinase molecules in live *Caulobacter* cells. Proc Natl Acad Sci U S A, 101:15921–15926.
15. Coggan JS, Bartol TM, Esquenazi E, Stiles JR, Lamont S, Martone ME, Berg DK, Ellisman MH, Sejnowski TJ. (2005) Evidence for ectopic neurotransmission at a neuronal synapse. Science, 309:446–451.

16. Grati Mh, Schneider ME, Lipkow K, Strehler EE, Wenthold RJ, Kachar B. (2006) Rapid turnover of stereocilia membrane proteins: evidence from the trafficking and mobility of plasma membrane Ca^{2+}–ATPase 2. J Neurosci, 26:6386–6395.
17. Lipkow K, Andrews SS, Bray D. (2005) Simulated diffusion of CheYp through the cytoplasm of *E. coli*. J Bact, 187:45–53.
18. Andrews SS. (2005) Serial rebinding of ligands to clustered receptors as exemplified by bacterial chemotaxis. Phys Biol, 2:111–122.
19. Lipkow K. (2006) Changing cellular location of CheZ predicted by molecular simulations. PLoS Comp Biol, 2:e39.
20. Lipkow K, Odde DJ. (2008) Model for protein concentration gradients in the cytoplasm. Cell Mol Bioeng, 1:84–92.
21. Jackson CL, Hartwell LH. (1990) Courtship in *S. cerevisiae*: both cell types choose mating partners by responding to the strongest pheromone signal. Cell 63:1039–1051.
22. Andrews SS, Bray D. (2004) Stochastic simulation of chemical reactions with spatial resolution and single molecule detail. Phys Biol, 1:137–151.
23. Andrews SS. (2009) Accurate particle-based simulation of adsorption, desorption, and partial transmission. Phys Biol, 6:46015.
24. Le Novère N, Finney A, Hucka M, Bhalla US, Campagne F, Julio C-V, Crampin EJ, Halstead M, Klipp E, Mendes P, Nielsen P, Sauro H, Shapiro B, Snoep JL, Spence HD, Wanner BL. (2005) Minimum information requested in the annotation of biochemical models (MIRIAM). Nat Biotechnol, 23:1509–1515.
25. van Zon JS, ten Wolde PR. (2005) Green's function reaction dynamics: a particle-based approach for simulating biochemical networks in time and space. J Chem Phys, 123:234910.
26. Duke TAJ, LeNovère N, Bray D. (2001) Conformational spread in a ring of proteins: a stochastic approach to allostery. J Mol Biol, 308:541–553.
27. Slepchenko B, Schaff J, Macara I, Loew LM. (2003) Quantitative cell biology with the Virtual Cell. TRENDS Cell Biol, 13:570–576.
28. Ray S, Deshpande R, Dudani N, Bhalla US. (2008) A general biological simulator: the Multiscale Object Oriented Simulation Environment, MOOSE. BMC Neurosci, 9 (Suppl 1):P93.
29. Blinov ML, Faeder JR, Goldstein B, Hlavacek WS. (2004) BioNetGen: software for rule based modeling of signal transduction based on the interactions of molecular domains. Bioinformatics, 20:3289–3291.
30. Lok L, Brent R. (2005) Automatic generation of cellular reaction networks with Moleculizer 1.0. Nat Biotech, 23:131–136.
31. DePristo MA, Chang L, Vale RD, Khan SM, Lipkow K. (2009) Introducing simulated cellular architecture to the quantitative analysis of fluorescent microscopy. Prog Biophys Mol Biol, 100:25–32.
32. Dobrzynski M, Rodríguez JV, Kaandorp JA, Blom JG. (2007) Computational methods for diffusion-influenced biochemical reactions. Bioinformatics, 23:1967–1977.
33. Palm MM, Steijaert MN, ten Eikelder HMM, Hilbers PAJ. (2009) Modeling molecule exchange at membranes. In *Proceedings of the Third International Conference on the Foundations of Systems Biology in Engineering*, Denver, Colorado.
34. Fischer H, Polikarpov I, Craievich AF. (2004) Average protein density is a molecular-weight-dependent function. Protein Sci, 13:2825–2828.
35. Brune D, Kim S. (1993) Predicting protein diffusion coefficients Proc Natl Acad Sci U S A, 90:3835–3839.
36. Swaminathan R, Hoang CP, Verkman AS. (1997) Photobleaching recovery and anisotropic decay of green fluorescent protein GFP-S65T in solution and cells: cytoplasmic viscosity probed by green fluorescent protein translational and rotational diffusion. Biophys J, 72:1900–1907.
37. Brown EB, Wu ES, Zipfel W, Webb WW. (1999) Measurement of molecular diffusion in solution by multiphoton fluorescence photobleaching recovery. Biophys J, 77:2837–2849.
38. Lide DR. (Ed.) (2004). *CRC Handbook of Chemistry and Physics*. CRC Press, Boca Raton, FL.
39. Verkman AS. (2002) Solute and macromolecule diffusion in cellular aqueous compartments. Trends Biochem Sci, 27:27–33.
40. Dix JA, Verkman AS. (2008) Crowding effects on diffusion in solutions and cells. Annu Rev Biophys, 37:247–263.
41. Elowitz MB, Surette MG, Wolf P-E, Stock JB, Leibler S. (1999) Protein mobility in the cytoplasm of *Escherichia coli*. J Bacteriol, 181:197–203.
42. van den Bogaart G, Hermans N, Karasnikov V, Poolman B. (2007) Protein mobility and diffusive barriers in *Escherichia coli*: consequences of osmotic stress. Mol Microbiol, 64:858–871.
43. Brass JM, Higgins CF, Foley M, Rugman PA, Birmingham J, Garland PB. (1986) Lateral

diffusion of proteins in the periplasm of *Escherichia coli*. J Bacteriol 165:787–794.

44. Crane JM, Verkman AS. (2008) Long-range nonanomalous diffusion of quantum dot-labeled aquaporin-1 water channels in the cell plasma membrane. Biophys J, 94:702–713.
45. Chou T, D'Orsogna MR. (2007) Multistage adsorption of diffusing macromolecules and viruses. J Chem Phys, 127:105101.
46. Huang KC, Meir Y, Wingreen NS. (2003) Dynamic structures in *Escherichia coli*: spontaneous formation of MinE rings and MinD polar zones. Proc Natl Acad Sci U S A, 100:12724–12728.
47. Paula S, Volkov AG, van Hoek AN, Haines TH, Deamer DW. (1996) Permeation of protons, potassium ions, and small polar molecules through phospholipid bilayers as a function of membrane thickness. Biophys J, 70:339–348.
48. von Smoluchowski M. (1917) Versuch einer mathematischen Theorie der Koagulationskinetik kolloider Lösungen. Z Phys Chem, 92:129–168.
49. Cohen B, Huppert D, Agmon N. (2000) Non-exponential Smoluchowski dynamics in fast acid-base reaction. J Am Chem Soc, 122:9838–9839.
50. Le Novère N, Bornstein B, Broicher A, Courtot M, Donizelli M, Dharuri H, Li L, Sauro H, Schilstra M, Shapiro B, Snoep JL, Hucka M. (2006) BioModels Database: a free, centralized database of curated, published, quantitative kinetic models of biochemical and cellular systems. Nucleic Acids Res, 34: D689–D691.

Index

Jacques van Helden et al. (eds.), *Bacterial Molecular Networks: Methods and Protocols*, Methods in Molecular Biology, vol. 804, DOI 10.1007/978-1-61779-361-5, © Springer Science+Business Media, LLC 2012

MIX
Papier aus verantwortungsvollen Quellen
Paper from responsible sources
FSC® C105338

If you have any concerns about our products,
you can contact us on
ProductSafety@springernature.com

In case Publisher is established outside the EU,
the EU authorized representative is:
Springer Nature Customer Service Center GmbH
Europaplatz 3, 69115 Heidelberg, Germany

Printed by Libri Plureos GmbH
in Hamburg, Germany